S0-BDH-326

10.50

INTEGRATED ALGEBRA AND TRIGONOMETRY
WITH
ANALYTIC GEOMETRY

PRENTICE-HALL, INC., ENGLEWOOD CLIFFS, NEW JERSEY

INTEGRATED ALGEBRA AND TRIGONOMETRY
With Analytic Geometry

third edition

ROBERT C. FISHER
Florida International University

ALLEN D. ZIEBUR
State University of New York at Binghamton

INTEGRATED ALGEBRA AND TRIGONOMETRY

With Analytic Geometry   Third Edition

Robert C. Fisher and Allen D. Ziebur

© 1972, 1967, 1958, 1957 by Prentice-Hall, Inc., Englewood Cliffs, N.J.
All rights reserved. No part of this book may be
reproduced in any form or by any means
without permission in writing from the publisher.

10 9 8 7 6 5 4 3 2 1

ISBN: 0-13-468959-3

Library of Congress Catalog Card Number: 70-168619

Printed in the United States of America

QA152
.F553
1972
cl

PRENTICE-HALL INTERNATIONAL, INC., London
PRENTICE-HALL OF AUSTRALIA, PTY. LTD., Sydney
PRENTICE-HALL OF CANADA, LTD., Toronto
PRENTICE-HALL OF INDIA PRIVATE LIMITED, New Delhi
PRENTICE-HALL OF JAPAN, INC., Tokyo

LAKELAND COMMUNITY COLLEGE
LIBRARY

contents

Chapter Eight

SYSTEMS OF EQUATIONS, 255

Chapter Nine

ENUMERATION, THE BINOMIAL THEOREM, AND PROBABILITY, 289

Chapter Ten

SEQUENCES, 333

Chapter Eleven

TOPICS IN ANALYTIC GEOMETRY, 361

preface

In preparing this new edition, we dismantled the chapter on inverse functions and distributed the pieces among the appropriate earlier chapters. The result, we think, is a more logical arrangement of topics. With the same end in view, we split the chapter on trigonometric functions into two chapters, a big one on Circular Functions (analytic trigonometry) and a small one on Trigonometric Functions (geometric trigonometry). The relative size of these chapters reflects our opinion of the relative importance of their contents.

We also did a considerable amount of rewriting of the "unchanged" chapters to tighten and improve the exposition. Some errors and misprints were corrected (and others, no doubt, were introduced). Major changes were made in most of the problem sets, both in the routine and the not-so-routine problems.

None of these changes affects the basic goal of the text; it is still designed to provide a firm foundation for the study of calculus. Although Chapter One contains material, such as sets, inequalities, and absolute values, that may be new to the student, one of its main purposes is to give him an opportunity to review and think about some of the algebra he already knows. In Chapter Two we begin the study of the unifying theme of the book, the concept of a function. Chapters Three and Four deal with specific functions, the exponential, logarithmic, and circular functions. The circular functions are presented analytically, as sets of pairs of numbers. In Chapter Five, we introduce the trigonometric functions from a geometric viewpoint. Chapters Six and Seven bring in the complex numbers and cover the usual topics in the elementary theory of equations. In Chapter Eight we study systems of equations, paying particular attention to linear systems. Chapter Nine is devoted to problems in enumeration and an introduction to the theory of probability. In Chapter Ten, we take up sequences and progressions. Many of the ideas of analytic geometry—coordinate systems, graphs, lines, and so on—have come up naturally in the first ten chapters. These are summarized at the beginning of Chapter Eleven, and then we turn to the standard topics of conics, polar coordinates, and transformations of axes. Our discussion of analytic geometry stops just short of the introduction of vectors. The student will be ready for calculus after Chapter Eleven.

Most classes contain students of varying mathematical backgrounds and ability. We have attempted to take this variation into account by arranging our sections so that the most basic and elementary material comes first. Occasionally, the last few paragraphs of a section (or the last few problems at the end of the section) may treat more difficult ideas—ones which the more able student might like to develop but which are not essential to the continuity of the book. Every student should attempt to work all the Review Problems, and superior students will profit by tackling some of the Miscellaneous Problems at the end of each chapter.

R. C. F.
A. D. Z.

INTEGRATED ALGEBRA AND TRIGONOMETRY
WITH
ANALYTIC GEOMETRY

The System
of
Real Numbers

one

Much of the world of mathematics is a world of numbers, and in order to work with numbers effectively, we must know the rules that govern their use. You already have a good deal of experience with the basic rules of algebra; in fact, you probably are so familiar with them that you apply them mechanically, without thinking about them. Therefore, in this first chapter we have placed our discussion of the important properties of the real numbers, and our review of the rules of algebra, in an unfamiliar setting. We use inequalities instead of equations, and we bring in the concept of absolute value. Introducing these topics not only acquaints you with some ideas and techniques that are fundamental to later work in mathematics, but, equally important, enables you to perform familiar algebraic operations under new circumstances, forcing you to think about what you are doing, and thereby increasing your understanding of algebra. We also introduce, and use, the language of sets.

It is not our purpose in this chapter to go into a precise, detailed explanation of the development of the real number system. Your familiarity with real numbers and your ability to use them will carry you over any gaps in our presentation.

1

The system of real numbers can be considered as an extension of certain other number systems. As we take up these various systems of numbers, we shall emphasize three points. First, we must be able to identify the numbers of each of the systems. Second, since the systems with which we are concerned form a chain leading up to the system of real numbers (and beyond, in Chapter Six), we shall want to be able to tell how a given system is related to a preceding one. Third, we must consider how the various mathematical operations, such as addition and multiplication, are performed.

The simplest system of numbers consists of the **positive integers**—namely, the numbers 1, 2, 3, 4, and so on. You already know how to add and multiply positive integers, of course. There are, however, certain fundamental rules that are well worth noting explicitly. If the letters a, b, and c represent any positive integers, then the following laws of arithmetic are true.

(1-1) $\quad (a + b) + c = a + (b + c)$. *The associative law of addition.*

EXAMPLE $\quad (3 + 4) + 5 = 3 + (4 + 5)$.

(1-2) $\quad (ab)c = a(bc)$. *The associative law of multiplication.*

EXAMPLE $\quad (3 \cdot 4) \cdot 5 = 3 \cdot (4 \cdot 5)$.

(1-3) $\quad a + b = b + a$. *The commutative law of addition.*

EXAMPLE $\quad 3 + 4 = 4 + 3$.

(1-4) $\quad ab = ba$. *The commutative law of multiplication.*

EXAMPLE $\quad 3 \cdot 4 = 4 \cdot 3$.

(1-5) $\quad a(b + c) = ab + ac$. *The distributive law.*

EXAMPLE $\quad 3(4 + 5) = 3 \cdot 4 + 3 \cdot 5$.

Notice that Rule 1-4 may be used to rewrite Rule 1-5 in the form

(1-6) $\qquad\qquad (b + c)a = ba + ca$.

The operations of subtraction and division are defined in terms of the operations of addition and multiplication. Thus the number d that satisfies the equation $a = b + d$ is called the **difference** $a - b$. The number q is the **quotient** a/b if $a = bq$. Observe that for any two positive integers a and b there is always a third, c, such that $a + b = c$; but there is not necessarily a positive integer d such that $a - b = d$. The expression $3 - 5$, for example, cannot represent a positive integer, since the equation $3 = 5 + d$ is satisfied for no positive integer d. We can overcome this difficulty by introducing the **negative integers** and the number **zero** to form, together with the positive integers, the system of all **integers,** ... $-3, -2, -1, 0, 1, 2, 3, 4, \ldots$ In this system, subtraction is always possible, and every number has a negative. The **negative** or **additive inverse** of a number a (which we write as $-a$) is the number that satisfies the equation $a + x = 0$.

Extending the system of positive integers to the system of integers still leaves us with a number system in which division is not always possible. This situation is remedied by a further extension that we will discuss in the next section; here we want to confine our attention to the integers.

Clearly, the positive integers are part of the system of all integers; now the problem is to define the arithmetic operations on the latter system. Let us emphasize the word "define." The arithmetic operations on a set of numbers are "man-made," not "God-given." There is no law of nature that says, for example, that $(-1)a$ *is* $-a$ for each number a; that is one of our definitions. These man-made rules of arithmetic for the integers (which we assume you know) are chosen so that mathematical operations with integers obey the same fundamental laws (Rules 1-1 to 1-6) that are true for the positive integers. It can be shown that the definitions of the arithmetic operations on the integers are the only possible ones that preserve the operations on the positive integers and also the associative, commutative, and distributive laws.

The distributive laws, Equations 1-5 and 1-6, are used extensively in the familiar process called *factoring*. An algebraic expression is in **factored form** if it is written as a product. To **expand** a product means to write it as an algebraic sum. The following examples illustrate how the various laws are used in factoring.

EXAMPLE 1-1   Factor $x(y + z) + u(y + z)$.

Solution   If we use Formula 1-6 with $a = y + z$, $b = x$, and $c = u$, we obtain

$$x(y + z) + u(y + z) = (x + u)(y + z).$$

EXAMPLE 1-2   Factor $x^2 - y^2$.

Solution
$$x^2 - y^2 = x^2 - xy + xy - y^2 = x^2 - xy + yx - y^2$$
$$= x(x - y) + y(x - y) = (x + y)(x - y).$$

The following examples illustrate how the distributive law and other laws can be used to expand products.

EXAMPLE 1-3   Expand $(x + 5)(x - 4)$.

Solution
$$(x + 5)(x - 4) = (x + 5)x + (x + 5)(-4)$$
$$= x^2 + 5x - 4x - 20 = x^2 + (5 - 4)x - 20$$
$$= x^2 + x - 20.$$

3

EXAMPLE 1-4 Expand $(x - y)(x^2 + xy + y^2)$.

Solution

$$(x - y)(x^2 + xy + y^2) = x(x^2 + xy + y^2) - y(x^2 + xy + y^2)$$
$$= x^3 + x^2y + xy^2 - yx^2 - xy^2 - y^3$$
$$= x^3 - y^3.$$

Notice that this result tells us how to write $x^3 - y^3$ in factored form.

It is sometimes easier to expand an expression if we recognize it as the factored form of a known expression.

EXAMPLE 1-5 Expand $(x + 2y)(x - 2y)(x^2 + 4y^2)$.

Solution Here we use the factoring formula $a^2 - b^2 = (a - b)(a + b)$ twice:

$$(x + 2y)(x - 2y)(x^2 + 4y^2) = (x^2 - 4y^2)(x^2 + 4y^2)$$
$$= x^4 - 16y^4.$$

**PROBLEMS
1**

1. Calculate the following numbers.
   (a) $-(-5) - (4 - 7)(2 - 4)$
   (c) $16^2 - 15^2$
   (b) $[2 - (1 - 6)][-(2 - 3) + 1]$
   (d) $202^3 - 200^3$

2. Find the negatives of the following numbers.
   (a) $-(-4)$
   (b) $-a$
   (c) $-x^3$
   (d) $(-x)^3$
   (e) $x - y$
   (f) $-[u - (v - w)]$

3. What law allows us to write a sum of three numbers without using parentheses?

4. Simplify the following expressions.
   (a) $-(x - 2) - [x - (2x - 2)]$
   (b) $2x + [x - (3x + y)]$
   (c) $-[x - 2(x - 1)]$
   (d) $4(3 - 2x) - 3(4x - 2)$
   (e) $(x + 5)^2 - (x - 5)^2$
   (f) $x - [x - (x - 1)]$

5. Expand the following products.
   (a) $(x - y)(x^2 + xy + y^2)$
   (b) $(x - 1)(x + 1)(x - 2)$
   (c) $(2x - y)^3$
   (d) $(.1x - .2y)^2$
   (e) $(2x + 3y)(3y - 2x)$
   (f) $(x - y)[(x + y) - (x + y)(x - y)]$

6. Factor the following expressions.
   (a) $x^3 - y^3$
   (b) $4x^3 - 16xy^2$
   (c) $27y^3 + \dfrac{8}{x^3}$
   (d) $12a^2 + 7ab - 10b^2$
   (e) $9x^2 - .01$
   (f) $x^4 - y^4$
   (g) $128 - 2x^6$
   (h) $8x^2 - 24xy + 18y^2$
   (i) $9z^2 - 4x^2 + 4xy - y^2$
   (j) $16 + 8x - 2x^3 - x^4$
   (k) $18xyz + 6x^2z - 6z^2y - 2z^2x$
   (l) $z - zx - 6 + 7x - x^2$

4

7. Let x denote an integer. For what choices of x do the following expressions represent even integers? Negative integers?

   (a) $1 + x/3$ (b) $(4x + 2) + (4x + 1)$

   (c) $x^2 - 1$ (d) $(x^2 - 1)/4$

8. Prove that the equation $(a - b) - c = a - (b - c)$ is not a law of arithmetic.

9. Suppose that A, B, and C represent things that can be added and multiplied according to the following rules: $A + B = A$ and $AB = B$. Which of the laws of arithmetic (Equations 1-1 through 1-6) hold?

2

THE RATIONAL NUMBERS

Given two integers a and b, we can subtract b from a; that is, we can find an integer d such that $a = b + d$. But we cannot always divide a by b if the only numbers we use are integers; that is, we cannot always find an integer q such that $a = bq$. For example, there is no integer q that satisfies the equation $5 = 3q$. In this section we will consider an extension of the system of integers, the system of **rational numbers,** in which the operation of division fails in only one case. The presence of the integer 0 creates this special difficulty, and we will dispose of it first.

"Division by zero" illustrates the fact that the arithmetic operations are *defined*, and that the definitions are chosen so as to be consistent with previously developed rules of arithmetic. Let a be any number except 0. What number q should we choose so that $a/0 = q$? From our definition of a quotient in Section 1, the number q should be such that $a = q \cdot 0$. But this last equation cannot be valid, because $q \cdot 0 = 0$ no matter what number we choose for q, and we have assumed specifically that $a \neq 0$ (a is not equal to zero). Since we cannot assign any number to the symbol $a/0$ that will be consistent with the method of assigning numbers to other quotients, we simply do not define the symbol $a/0$. A somewhat different problem confronts us in the case of the quotient $0/0$. A number q that would be suitable for this quotient would be one for which $0 = 0 \cdot q$. But this last equation is true for any number q, and there is no particular reason to perfer one number over any other. Again we avoid the difficulty by leaving the symbol $0/0$ undefined. We say that *the symbols $a/0$ and $0/0$ are meaningless*, by which we merely mean that *we have assigned no meaning to them.*

We will now consider the rational number system in which division by numbers other than zero is always possible. A **rational number** can be written in the form a/b, where a and b are integers and b is not 0. Thus $\frac{2}{3}$ and $\frac{-11}{7}$ are rational numbers. The word "rational" refers to the fact that these numbers are written as *ratios* of integers. If we agree that the rational number $a/1$ is the "same" as the integer a, then we can consider the system of integers as part of the system of rational numbers.

The definitions of the arithmetic operations that involve the rational numbers are, as usual, motivated by the desire to preserve the fundamental laws we mentioned in Section 1. In order to refresh your memory of the arithmetic of fractions we will list some of the rules that govern the operations with rational numbers. For the purposes of this section, you may assume that the letters represent integers, but the rules are equally valid for other numbers. In no case, of course, may a denominator be zero.

(2-1)
$$\frac{a}{1} = a.$$

(2-2)
$$\frac{a}{b} = \frac{c}{d} \text{ if, and only if, } ad = cb.$$

(2-3)
$$\frac{a}{c} + \frac{b}{c} = \frac{a+b}{c}.$$

(2-4)
$$\frac{a}{b} \cdot \frac{c}{d} = \frac{ac}{bd}.$$

The first rule is merely the convention that establishes the integers as a part of the system of rational numbers. Equations 2-2, 2-3, and 2-4 are the definitions of equality, addition, and multiplication of rational numbers. The rule for division is

(2-5)
$$\frac{a}{b} \div \frac{c}{d} = \frac{ad}{bc}.$$

This rule is established by showing that $a/b = (ad/bc)(c/d)$.

The "cancellation" rule is

(2-6)
$$\frac{ca}{cb} = \frac{a}{b}.$$

This rule stems directly from the definition of equality. In accordance with our agreement that 0 cannot appear in a denominator, we are here assuming that $c \neq 0$. (We will not specifically mention this agreement again; you are simply to assume in Example 2-1 below that x and y are not numbers whose sum is 0, and so on.)

The use of the rules governing mathematical operations with rational numbers is illustrated by the following examples.

EXAMPLE 2-1   Simplify $\dfrac{x^2 - y^2}{x+y}$.

Solution   Notice how Rules 2-6 and 2-1 are used in this solution:

$$\frac{x^2 - y^2}{x+y} = \frac{(x+y)(x-y)}{x+y} = x - y.$$

EXAMPLE 2-2 Simplify $\dfrac{x+\dfrac{1}{y}}{x-\dfrac{1}{y}}$.

Solution

$$\frac{x+\dfrac{1}{y}}{x-\dfrac{1}{y}} = \frac{y\left(x+\dfrac{1}{y}\right)}{y\left(x-\dfrac{1}{y}\right)} = \frac{yx+1}{yx-1} .$$

EXAMPLE 2-3 Perform the following addition: $\dfrac{2x}{x^2-4} + \dfrac{3}{x^2-5x+6}$.

Solution

$$\frac{2x}{x^2-4} + \frac{3}{x^2-5x+6} = \frac{2x}{(x-2)(x+2)} + \frac{3}{(x-2)(x-3)}$$

$$= \frac{2x(x-3)+3(x+2)}{(x-2)(x+2)(x-3)}$$

$$= \frac{2x^2-3x+6}{(x-2)(x+2)(x-3)} .$$

EXAMPLE 2-4 Perform the following subtraction: $\dfrac{3x}{x^2-4} - \dfrac{4}{2-x}$.

Solution

$$\frac{3x}{x^2-4} - \frac{4}{2-x} = \frac{3x}{(x-2)(x+2)} + \frac{4}{x-2}$$

$$= \frac{3x+4(x+2)}{(x-2)(x+2)}$$

$$= \frac{7x+8}{(x-2)(x+2)} .$$

EXAMPLE 2-5 Perform the following division: $1 \left/ \dfrac{x+y}{x^2} \right.$.

Solution

$$1 \left/ \frac{x+y}{x^2} \right. = \left(\frac{1}{1}\right) \cdot \frac{x^2}{x+y} = \frac{x^2}{x+y} .$$

You should regard any legitimate "canceling" to simplify fractions as an application of Rule 2-6. You will avoid a very common error if you will note that *there is no rule which states that* $(a+c)/(b+c)$ *is the same as* $(a+1)/(b+1)$.

If $a \neq 0$, the number $1/a$ is called the **reciprocal** (or **multiplicative inverse**) of a. For example, the reciprocal of $(x+y)/x^2$ is $x^2/(x+y)$. Analogous to the negative of a, the reciprocal of a is the number that satisfies the equation $ax = 1$.

**PROBLEMS
2**

1. Explain why each of the following numbers is rational.

   (a) 17 (b) .17 (c) $4 + \frac{2}{3}$ (d) $\frac{2}{7} - \frac{1}{4}$

   (e) .3333 . . . (f) .7777 . . . (g) 17.76 (h) 149.2

2. Reduce the fraction $\frac{126}{66}$ to "lowest terms." Which of the rules given in this section did you use?

3. Use our rules to show that

   $$\frac{a}{b} = \frac{-a}{-b} \quad \text{and that} \quad \frac{a}{-b} = \frac{-a}{b} = -\frac{a}{b}.$$

4. Simplify the following expressions.

   (a) $\dfrac{uv - uw + uz}{v^2 - vw + vz}$ (b) $\dfrac{x - \dfrac{1}{x}}{x^2 - \dfrac{1}{x^2}}$ (c) $\dfrac{\dfrac{b}{a} + \dfrac{c}{a}}{ab + ac}$

   (d) $\dfrac{x^2 + 2 + 1/x^2}{x^2 + 1}$ (e) $\dfrac{x + y}{x^2 - y^2 + x + y}$ (f) $\dfrac{1}{2 - \frac{1}{3}}$

   (g) $1 - \dfrac{1}{1 + \dfrac{1}{x}}$ (h) $x - \dfrac{x}{x - \dfrac{1}{x}}$

5. Find the reciprocal of each of the following expressions and simplify it.

   (a) $x + 1$ (b) $\dfrac{1}{x} + 1$ (c) $\dfrac{1}{\dfrac{1}{x} + 1}$

   (d) $\dfrac{x - 1}{1 - \dfrac{1}{x}}$ (e) $\dfrac{1 - \dfrac{1}{x}}{x - 1}$ (f) $x + \dfrac{1}{x}$

6. Perform the indicated operations and simplify the result.

   (a) $\left(\dfrac{x^2 - y^2}{x - y}\right) / \left(\dfrac{x + y}{x}\right)$ (b) $x + y + \dfrac{x^2}{y - x}$

   (c) $\dfrac{x + 15}{x^2 - 9} - \dfrac{x + 12}{x^2 - x - 6}$ (d) $\dfrac{3x - 4}{3x^2 + x - 2} - \dfrac{2x + 3}{2x^2 - 3x - 5}$

   (e) $\dfrac{4x + 2}{x(1 - x)} + \dfrac{2}{x - 1} + \dfrac{2}{x}$ (f) $\dfrac{2z}{x + y} - \left(\dfrac{z}{x} + \dfrac{z}{y}\right)$

   (g) $\left(9 - \dfrac{1}{x^2}\right) / (3x + 1) + 1/x^2$ (h) $\left(x^2 - \dfrac{1}{x}\right) / \dfrac{x - 1}{x} + x$

7. Is $(x^3 - 1)/(x - 1)$ equal to $x^2 + x + 1$ for each real number x?

8. If $a \neq 0$, then $a/0$ is not defined. Does the symbol $0/a$ represent a number?

9. Suppose that $a \neq 0$. Show that the reciprocal of the reciprocal of a is a. What is the negative of the negative of a?

10. Which of the following equations (if any) are identities?
    (a) $x/y = y/x$ (b) $x/(y/z) = (x/y)/z$
    (c) $x/(y+z) = (x/y) + (x/z)$ (d) $(y+z)/x = (y/x) + (z/x)$

11. Let $y = x/(2x+1)$. What can you say about x if
    (a) x is an integer and y is a rational number?
    (b) x is an integer and y is an integer?
    (c) x is a rational number and y is an integer?
    (d) x is a rational number and y is a rational number?

3
THE
REAL
NUMBERS

The rational numbers, together with the operations of addition and multiplication, form a system in which the inverse operations of subtraction and of division by non-zero numbers are always possible. From the point of view of mathematical applications, however, this system still lacks certain desirable properties, and we must extend it further. It is much more difficult to describe the extension of the rational numbers to the system of real numbers than it is to describe the extension of the positive integers to the integers, or the integers to the rational numbers. Therefore, our discussion in this section will be mostly intuitive and geometric.

The easiest way to describe the **real numbers** is to say that they are the numbers that can be written as unending decimal expressions. For example, the real numbers π, 2, and $\frac{1}{3}$ can be written as $\pi = 3.14159\ldots$, $2 = 2.000\ldots$, and $\frac{1}{3} = .333\ldots$. To obtain a decimal expression for a given rational number, we simply divide its numerator by its denominator. Thus we see that the system of rational numbers is part of the system of real numbers.

The **number scale** provides us with a convenient geometric representation of our real number system. In constructing the scale, two points of a horizontal line are chosen; one point is labeled 0 and the other is labeled

1. The point labeled 0 is termed the **origin,** and the point labeled 1 is called the **unit point.** The unit point is usually placed to the right of the origin. If we consider the distance between the origin and the unit point as the unit of distance, it is easy to locate the point that corresponds to a given integer. For example, the point two units to the right of the origin corresponds to the number 2, while the point two units to the left of the origin represents the number -2 (Fig. 3-1). We can find points that correspond

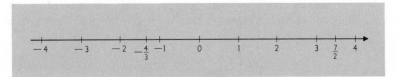

FIGURE 3-1

to other rational numbers just as easily. Thus, to find the point that corresponds to $-\frac{4}{3}$, we divide the segment from -1 to -2 into thirds. The first division point to the left of -1 then corresponds to the rational number $-\frac{4}{3}$. Logically, a number and its corresponding point of the number scale should be distinguished as two different things. The number associated with a point is called the **coordinate** of the point. In practice, however, it is customary simply to refer to "the point 5", instead of using the more cumbersome phrase "the point whose coordinate is 5."

It is not hard to see that each point of the number scale determines an unending decimal and hence represents a real number; a single example will show us how. Let us consider the point P in Fig. 3-2. This point lies between the points 2 and 3. Figure 3-3 is an enlarged view of part of the number scale and shows the interval between the points 2 and 3 divided into ten equal parts by the points 2.1, 2.2, . . . , 2.9. The point P lies between the points 2.4 and 2.5. Figure 3-4 is a magnified picture of the part of the

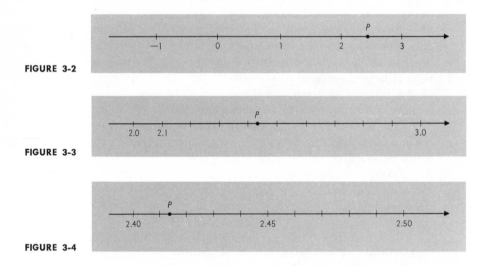

FIGURE 3-2

FIGURE 3-3

FIGURE 3-4

number scale between the points 2.4 and 2.5 and shows that the point P is between the points 2.41 and 2.42. The decimal expression determined by P is, to two places, 2.41. Theoretically, we could continue this process indefinitely, if we agree that any subinterval of the number scale, no matter how small, can be subdivided into ten equal parts. This process generates the unending decimal expression that corresponds to the point P. If at some stage we see that P is a division point, then all further digits in the expression are chosen to be zero. For example, the unending decimal expression that we associate with the point $\frac{3}{4}$ would be .75000 . . . . The example we have just described shows how a given point determines an unending decimal. The converse is also true; each unending decimal determines a unique point

10

of the number scale, and so we think of the number scale as a "picture" of the set of real numbers.

You have worked with decimals for years, and therefore you can do arithmetic with real numbers. The rules of arithmetic for real numbers are such that all the laws we have listed earlier remain valid.

We have seen that the rational numbers form part of the system of real numbers. There are some real numbers, however, that are not rational numbers. These numbers are the **irrational numbers.** As an example of such a number, consider the number r, the length of the hypotenuse of the right triangle with unit legs that is shown in Fig. 3-5. According to the Pythagorean Theorem, we have $r^2 = 1^2 + 1^2 = 2$; that is, r is the real number $\sqrt{2}$.

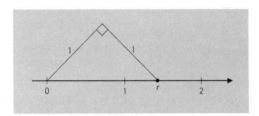

FIGURE 3-5

We will now show that this number cannot be a rational number. For suppose it were; that is, suppose that there were positive integers a and b such that $\sqrt{2} = \dfrac{a}{b}$. In other words, suppose that there were positive integers a and b such that

(3-1) $$a^2 = 2b^2.$$

The integers a and b can each be written as a product of a unique set of prime numbers (a prime number is a positive integer that is not equal to 1 and has no factors other than 1 and itself). Thus $a = p_1 \cdot p_2 \cdots p_k$ and $b = q_1 \cdot q_2 \cdots q_l$, where the p's and q's are prime numbers (not necessarily all different). Therefore, Equation 3-1 can be written as

(3-2) $$p_1 \cdot p_1 \cdot p_2 \cdot p_2 \cdots p_k \cdot p_k = 2q_1 \cdot q_1 \cdot q_2 \cdot q_2 \cdots q_l \cdot q_l.$$

Now we count the number of times that the prime number 2 can appear on each side of this equation. It is clear that it appears either an even number of times or not at all on the left-hand side, whereas there are an odd number of 2's on the right-hand side. This situation is impossible, and hence an equation such as Equation 3-1 cannot hold. Therefore, there is no rational number whose square is 2; the number $\sqrt{2}$ is irrational.

11

The following diagram will, perhaps, make it easier for you to picture the hierarchy of number systems we have developed thus far.

Real numbers

$1, \frac{1}{2}, -2, \sqrt{2}, \ldots$

Rational numbers   *Irrational numbers*

$1, \frac{1}{2}, -2, \ldots$ $\sqrt{2}, \ldots$

Integers

$0, 1, -2, \ldots$

Positive integers

$1, 2, \ldots$

PROBLEMS
3

1. On the number scale, what is the number that is
   (a) one third of the way from 3 to 7?   (b) one third of the way from 3 to -7?
   (c) one third of the way from 3 to x?   (d) one third of the way from c to x?

2. Horizontal and vertical number scales with equal units of length are drawn so that their point of intersection is the zero point of each number scale. The point 7 of the horizontal scale and the point 2 of the vertical scale determine a line L.

   (a) A line parallel to L and containing the point 5 of the horizontal scale intersects the vertical scale in what point?

   (b) A line perpendicular to L and containing the point 2 of the vertical scale intersects the horizontal scale in what point?

3. Explain how you could use a 45°–45°–90° drawing triangle to locate the point $\sqrt{18}$ of a number scale marked in units.

4. Use long division to find an unending repeating decimal expression for the rational number $\frac{1}{11}$. By a "repeating decimal expression" we mean one in which, after a certain point, the digits keep recurring in successive "blocks." An example is the number $23.46731731731\ldots$, where we assume that the "block" 731 is repeated indefinitely.

5. If $a = .141414\ldots$ (repeating), then $100a = 14.141414\ldots$. Therefore,

   $$100a - a = 14.141414\ldots - .141414\ldots$$

   or $99a = 14.$

   Thus, $a = \frac{14}{99}$. Use a similar argument to write $b = .142142\ldots$ and $c = 2.9999\ldots$ as fractions. (It is a fact that all rational numbers have repeating decimal expressions, and all repeating decimals represent rational numbers.)

6. Is the sum of two rational numbers necessarily rational? Is the sum of two irrational numbers necessarily irrational? What about the sum of a rational number and an irrational number?

7. Replace the word "sum" with the word "product" in the preceding questions, and answer them.

12

8. Is there a smallest positive integer? Is there a smallest positive rational number?

9. A vertical line containing the right-angle vertex of the triangle in Fig. 3-5 intersects the number scale in a point s. Prove that s is not a rational number.

10. Are the following numbers rational or irrational?

(a) $3\sqrt{2}$ (b) $\sqrt{3}$ (c) $\sqrt{3}+\sqrt{2}$ (d) $\sqrt[3]{97}$

11. Is it possible to tell by physical measurement whether the length of a given stick is a rational or an irrational number?

4
THE ORDER RELATION

When the number scale is displayed with the usual left-to-right orientation, as in Fig. 3-1, it graphically illustrates the *order relation* among the real numbers. Smaller numbers lie to the left of larger ones. A formal definition of the order relation which does not rely on a picture reads as follows.

Definition 4-1   *We say that **a is less than b (a < b)** and **b is greater than a (b > a)** if $b - a$ is a positive number. The symbols < and > are called **inequality signs.***

According to this definition, the statements "a is positive" and "$a > 0$" say exactly the same thing. Notice how this formal definition is related to our geometric idea that smaller numbers lie to the left of larger ones. The number scale shows the point -5 to be to the left of -3, and $-3 - (-5)$ is the positive number 2, which means, according to Definition 4-1, that -5 is less than -3. The notation $a < b < c$ means that $a < b$ and $b < c$. For obvious geometric reasons, we say that b is **between** a and c. Thus the number 4 is between 2 and 7, since $2 < 4 < 7$.

The following fundamental properties of the order relation can be proved by using Definition 4-1 and some basic facts about positive and negative numbers.

(4-1) For any two numbers a and b, one, and only one, of the following relations is true:

$$a < b, \quad \text{or} \quad a = b, \quad \text{or} \quad a > b \quad (\textit{Trichotomy}).$$

(4-2) If $a < b$ and $b < c$, then $a < c$ (*Transitivity*).

(4-3) If $a < b$, then $a + c < b + c$ for any number c, positive, negative, or zero.

(4-4)   If $c > 0$ and $a < b$, then $ac < bc$.

(4-5)   If $c < 0$ and $a < b$, then $ac > bc$.

Property 4-1 follows from the fact that each real number is either positive, or zero, or its negative is positive. Thus the number $b - a$ is either positive ($a < b$), or 0 ($a = b$), or $-(b - a)$ is positive ($a > b$). The other properties stem from the fact that the sum and product of two positive numbers are positive. Thus, to prove Property 4-2 we think of the inequalities $a < b$ and $b < c$ as the equivalent statements that $b - a$ and $c - b$ are positive numbers. Hence, their sum $(c - b) + (b - a) = c - a$ is positive. But this statement means that $a < c$, and so Property 4-2 is verified. We leave the verification of the other properties to you in the problems.

Rules 4-4 and 4-5 state that when we "multiply both sides" of an inequality, it makes a difference whether the multiplier is positive or negative. Thus the rules tell us that when we multiply both sides of the inequality $3 < 4$ by 2 we get $2 \cdot 3 < 2 \cdot 4$, whereas if we multiply by -2 we obtain $(-2) \cdot 3 > (-2) \cdot 4$. These last two inequalities ($6 < 8$ and $-6 > -8$) are, of course, correct. The most common mistakes that students make with inequalities are violations of Rules 4-4 and 4-5.

The following examples show how the properties of the order relation may be used to derive new relations.

EXAMPLE 4-1 Show that if $0 < a < 1$, then $a^2 < a$.

Solution The relation $0 < a < 1$ means that $a > 0$ *and* $a < 1$. Since a is positive, Property 4-4 states that we may multiply both sides of the inequality $a < 1$ by a to get $a^2 < a$.

EXAMPLE 4-2 If $1 < a$, show that $a < a^2$.

Solution Since $a > 1$ and $1 > 0$, a is positive (Property 4-2). Hence we get the desired result by multiplying both sides of the inequality $1 < a$ by a in accordance with Property 4-4.

The notation $a \leq b$ or $b \geq a$ means that a is a number that is either less than b or equal to b; that is, a *is not greater than* b. For example, $4 \leq 6$ and $6 \leq 6$, since neither 4 nor 6 is greater than 6. The properties of $<$ may be extended to the $\leq$ relation.

(4-6) If $a \leq b$ and $b \leq c$, then $a \leq c$.

(4-7) If $a \leq b$, then $a + c \leq b + c$ for any number c.

(4-8) If $c \geq 0$ and $a \leq b$, then $ac \leq bc$.

(4-9) If $c \leq 0$ and $a \leq b$, then $ac \geq bc$.

In addition to these rules, we list some examples of others that involve various possible combinations of the $<$ and $\leq$ relations.

(4-10) If $a \leq b$ and $b < c$, then $a < c$.

(4-11)   If $c > 0$ and $a \leq b$, then $ca \leq cb$.

(4-12)   If $c < 0$ and $a \leq b$, then $ca \geq cb$.

EXAMPLE 4-3   Show that the sum of the squares of two real numbers is not less than twice the product of the two numbers.

Solution   Let a and b be any two real numbers. We are to show that $a^2 + b^2 \geq 2ab$. Property 4-7 says that this inequality is equivalent to the inequality $a^2 - 2ab + b^2 \geq 0$, or, when we factor the left-hand side, $(a - b)^2 \geq 0$. When we say that this last inequality is *equivalent* to our desired inequality, we mean that each is true when the other is. But the inequality $(a - b)^2 \geq 0$ is true for *every* pair of real numbers a and b, since the square of a real number cannot be negative. In fact, our argument shows that $a^2 + b^2 > 2ab$ if, and only if, $a \neq b$.

EXAMPLE 4-4   Under what conditions does the inequality $1/a \leq 1/b$ imply that $b \leq a$?

Solution   If $ab > 0$, we may multiply both sides of the first inequality by ab to get the second (Rule 4-11), whereas if $ab < 0$, such multiplication produces the reverse inequality $a \leq b$. Therefore, the inequality $1/a \leq 1/b$ implies that $b \leq a$ *provided that* $ab > 0$. (Why didn't we have to consider the case $ab = 0$?)

Frequently we must deal with rational numbers that are approximations of other numbers. The accuracy with which we can measure any physical quantity is always limited to some extent by inadequacies in the instruments we use, or by human factors. Consequently, scientific measurements are usually expressed in some such form as, "The ratio of the mass of a proton to the mass of an electron is 1838 ± 1." This sentence means that the stated ratio is a number r that satisfies the inequalities $1837 \leq r \leq 1839$. Another instance of this type of approximation is the statement that $\sqrt{2} = 1.414$ or $\pi = 3.14$. We already know that $\sqrt{2}$ is not a rational number (and neither is π), so these equalities are not strictly correct. We write them, of course, but only with the understanding that the equation $\sqrt{2} = 1.414$ means that

$$1.4135 \leq \sqrt{2} \leq 1.4145.$$

Now consider the following calculation:

$$\pi\sqrt{2} = (3.14)(1.414) = 4.43996.$$

You should recognize immediately that the given answer to this problem is not as accurate as it seems. To find out just how accurate it is, we use the rules dealing with inequalities. We have

$$1.4135 \leq \sqrt{2} \leq 1.4145 \quad \text{and} \quad 3.135 \leq \pi \leq 3.145.$$

15

Therefore,

$$(1.4135)(3.135) \le (1.4135)\pi$$
$$\le \sqrt{2}\,\pi \le \sqrt{2}\,(3.145) \le (1.4145)(3.145).$$

If we multiply out the products on the extreme ends of these inequalities, the result, correct to two decimal places, is

$$4.43 \le \sqrt{2}\,\pi \le 4.45.$$

**PROBLEMS
4**

1. Which of the following statements are true?
   (a) $-4 < -2$ (b) $-(x-2) < x - 2$ for every x
   (c) If $5 < x$, then $-2 > -x$.   (d) $\frac{7}{9} < .8$
   (e) $.23 < \sqrt{5} - 2 < .24$ (f) If $x \le -y$, then $y \le -x$.
   (g) If $x^2 < y^2$, then $x < y$.   (h) $.333\ldots \le \frac{1}{3}$

2. (a) If $a < b$ and $c < d$, is $a + c < b + d$?
   (b) If $a < b < 0$ and $c < d < 0$, is $ac > bd$?
   (c) If $a < b < 0$ and $0 < c < d$, is $ac < bd$?
   (d) If $a < b < 0$ and $0 < c < d$, is $bd < ac$?

3. If $b > c$ and $ab > ac$, does it necessarily follow that $a > 0$?

4. Find rational numbers x and y that satisfy the given inequalities.
   (a) $x < 1 < y$ and $y - x < \frac{1}{100}$ (b) $1 < x < y$ and $y - 1 < \frac{1}{100}$

5. Can you convince yourself that the sum of a positive number and its reciprocal is never less than 2?

6. The dimensions of a certain table are approximately 3.817 feet by 7.06 feet. What can you say about its perimeter? Its area?

7. (a) Show that if x and y are two numbers such that $x + y = 10$, then $xy \le 25$.
   (b) Show that if $xy = 16$, and x and y are positive, then $x + y \ge 8$.

8. The total resistance R of an electrical circuit that contains two resistances R_1 and R_2 in parallel is given by the equation

$$\frac{1}{R} = \frac{1}{R_1} + \frac{1}{R_2}.$$

   If one resistance, say R_1, is 50 ohms, what are the allowable values for R_2 if the total resistance R is to be between 5 and 10 ohms?

9. If a/b and c/d are rational numbers with b and d positive, show that $a/b < c/d$, if and only if, $ad < bc$.

10. Find the largest integer n such that $3n - 5 < 10$. Prove that there is no smallest integer n such that $3n - 5 < 10$.

11. Write out proofs of Rules 4-3, 4-4, and 4-5.

We will find it convenient to use the notation and terminology of set theory in our discussion of inequalities (and many of the other topics that we take up later), so let us introduce it here. The language of sets is the language of much of mathematics today, and now is a good time for you to learn it if you don't know it already.

There is nothing very complicated about the notion of a set; for example, we talk about the set of real numbers or a set of points in the plane. The most straightforward way to specify a set is to list all its **elements** or **members.** Thus, $\{2, 4, 6, 8\}$ is the set whose members are the four numbers 2, 4, 6, and 8. We enclosed the members of this set in braces $\{\ \}$ to indicate that we think of the set as an entity in itself. Just as we frequently use letters to denote such mathematical entities as numbers and angles, we often name sets by letters. For example, we might say "let Q be the set of rational numbers" or "let P by the set of positive integers." We will consistently use the symbol R^1 to denote the set of real numbers. (The superscript 1 indicates that, graphically speaking, the real numbers have "dimension 1"; that is, we visualize them as a line.)

To indicate that an element a **is a member of the set A,** we write $a \in A.$ We also say that a **belongs to** A or that a **is contained in A.** For example, $\frac{1}{2} \in Q$, where Q is the set of rational numbers. To denote the fact that $\sqrt{2}$ is *not* a member of Q (is not a rational number), we write $\sqrt{2} \notin Q$.

If every member of a set A is also a member of a set B, we say that A **is a subset of B** or A **is contained in B,** and we write $A \subseteq B.$ Thus the fact that positive integers are rational numbers is expressed by the inclusion $P \subseteq Q$. Notice that it is possible to have $A \subseteq B$ *and* $B \subseteq A$; these inclusions are both true if, and only if, the sets A and B contain the same elements, in which case we say that $A = B.$

The **union** of two sets A and B is the set $A \cup B$ that we obtain by lumping together the elements of the sets A and B into a single set. Thus, if O is the set of odd positive integers and E is the set of even positive integers, then the set P of positive integers is given by the equation $P = O \cup E$. More precisely, $A \cup B$ is the set of elements that belong to *at least one* of the sets A and B. An element belongs to the union of A and B if it is a member of A *or* of B (or of both A and B). For example, $\{2, 4, 6, 8\} \cup \{6, 8, 10, 12\} = \{2, 4, 6, 8, 10, 12\}$. Notice that for any set A we have $A \cup A = A$.

The **intersection** of two sets A and B is the set $A \cap B$ whose members are the elements that the sets A and B have in common. For example, $\{2, 4, 6, 8\} \cap \{6, 8, 10, 12\} = \{6, 8\}$. If P is the set of positive integers and Q is the set of rational numbers, then $P \cap Q = P$, and so on. An element belongs to the intersection of A and B if it is a member of A *and* a member of B. Therefore, you should think of the symbol $\cup$ (often read as "cup") as being associated with the word "or" and the symbol $\cap$ (read as "cap") as being associated with the word "and."

Many times we specify a set without listing its members, perhaps because there are too many or because we don't know them. For example, we can't list the set of all the integers and we don't want to list the set of all the even integers between 1 and 1 million because there are too many. We may denote the latter set by the symbols $\{n \mid n$ is an even integer between 1 and 1,000,000$\}$. In this notation, the braces indicate that we are dealing with a set. To the left of the vertical line appears a "typical" element of the set, and to the right of the line is the condition that this element be a member of the set. Thus, the symbols $x \mid$ stand for the phrase "x is a member of the set if, and only if." In our example, we have expressed this condition in words; usually we employ mathematical symbols. For example, $\{x \mid x < -3\}$ is the set of numbers that are less than -3. We see that

$$-5 \in \{x \mid x < -3\}, \quad \text{whereas} \quad 2 \notin \{x \mid x < -3\}.$$

EXAMPLE 5-1 Let $A = \{x \mid x > -2\}$ and $B = \{x \mid x < 3\}$. Describe these sets as collections of points of the number scale. What is $A \cap B$? $A \cup B$?

Solution The set A consists of all numbers that are greater than -2, so a point belongs to A if, and only if, it lies to the right of the point -2 of the number scale. Similarly, we think of B graphically as the set of points to the left of 3. The intersection $A \cap B$, that is, the set of points in *both* A and B, is illustrated in color in Fig. 5-1. It consists of the points between

FIGURE 5-1

-2 and 3, so we have the set equation

$$\{x \mid x > -2\} \cap \{x \mid x < 3\} = \{x \mid -2 < x < 3\}.$$

Every real number belongs to at least one of the sets A or B, since every real number is either greater than -2 or less than 3 (and some numbers are both). In other words, the union $A \cup B = R^1$, the set of all real numbers.

When we specify a set by describing some property of its members, we have to be alert to the possibility that our set may be the **empty set** $\emptyset$, the set that has no members. Thus $\{x \mid x^2 = 2$ and x is a rational number$\} = \emptyset$. We could also describe $\emptyset$ as the set of all circles of radius -3. The following statements are true for any set A: $\emptyset \subseteq A$, $\emptyset \cup A = A$, and $\emptyset \cap A = \emptyset$.

Sometimes we abbreviate our set notation if no misinterpretation is possible. Thus we might write $\{x^2 \mid x \in P\}$, instead of $\{y \mid y = x^2$, where $x \in P\}$, to denote the set of squares of positive integers: $\{1, 4, 9, 16, \ldots\}$. Often we leave out the symbols $x \mid$ and simply write, for example, $\{x < 7\}$ instead of $\{x \mid x < 7\}$. You will get used to these variations as you see them in action.

EXAMPLE 5-2   What is the set $\{x < -2\} \cap \{x > 3\}$?

Solution   An element belongs to the intersection of two sets if, and only if, it belongs to both of them. Thus, in order for a number to belong to our intersection, it would have to be both less than -2 and greater than 3. There is no such number, so the intersection is the empty set; that is,

$$\{x < -2\} \cap \{x > 3\} = \varnothing.$$

A **solution** of an equation is, of course, a number that satisfies the equation. For example, 2 is a solution of the equation $x^3 - 4x^2 + x + 6 = 0$, because the left-hand side reduces to 0 when we substitute 2 for x. The collection of all the solutions of an equation is the **solution set** of the equation. Thus, in our example, the solution set is $\{x \mid x^3 - 4x^2 + x + 6 = 0\}$. The inclusion

$$2 \in \{x \mid x^3 - 4x^2 + x + 6 = 0\}$$

is just a symbolic way of saying that 2 is a solution of the equation. The inclusion

$$\{-1, 2, 3\} \subseteq \{x \mid x^3 - 4x^2 + x + 6 = 0\},$$

which tells us that $\{-1, 2, 3\}$ is a subset of the solution set, is a symbolic way of saying that each of the numbers -1, 2, and 3 is a solution of our equation. Actually, the symbol $\subseteq$ in our last inclusion could be replaced by an equals sign; that is, $\{-1, 2, 3\}$ is the solution set of our equation. In other words, the numbers -1, 2, and 3 are the only solutions of the equation. You can verify this statement by studying the given equation when the left-hand side is written in factored form: $(x + 1)(x - 2)(x - 3) = 0$.

Similarly, a solution of an inequality is a number that satisfies the inequality, and the solution set is the collection of all numbers that satisfy the inequality. Thus, $\{x \mid 2x + 3 < 9\}$ (or simply $\{2x + 3 < 9\}$) is the solution set of the inequality $2x + 3 < 9$. To find this set, we proceed as follows. We first observe (Rule 4-3) that $2x + 3 < 9$ if, and only if, $2x < 6$. This last inequality is equivalent to the inequality $x < 3$ (Rule 4-4). Therefore,

(5-1) $$\{2x + 3 < 9\} = \{2x < 6\} = \{x < 3\},$$

which says that the set B of Example 5-1 is the solution set of the inequality $2x + 3 < 9$. As this example illustrates, solving an inequality is a matter of reducing the description of the solution set to as simple a form as possible. Thus, our final expression in Equations 5-1 makes it clear that the solution set is the set B illustrated in Fig. 5-1, but this fact is not immediately apparent from the first expression. We will solve more complicated inequalities in the next section.

EXAMPLE 5-3   Solve the inequality $3x - 4 < 5x + 7$.

Solution   By subtracting $3x$ from both sides of our given inequality, we obtain the equivalent inequality $-4 < 2x + 7$. Now we subtract 7 from both sides, and we get the equivalent inequality $-11 < 2x$. Finally, we divide by the positive number 2, and we have $-\frac{11}{2} < x$. Thus,

$$\{3x - 4 < 5x + 7\} = \{-\tfrac{11}{2} < x\};$$

our solution set consists of all points to the right of the point $-\frac{11}{2}$ of the number scale.

PROBLEMS 5

1. Illustrate the following sets as collections of points of a number scale.
   (a) $\{x > 3\} \cap \{x < 5\}$ (b) $\{x < -2\} \cup \{x > 7\}$
   (c) $\{-1 < x \le 3\} \cup \{2 < x \le 4\}$ (d) $\{-1 < x \le 3\} \cap \{2 < x \le 4\}$

2. Let $A = \{2, 4, 6, 8, 10\}$, $B = \{8, 10, 12, 14\}$, and $C = \{7, 8, 9, 10, 11\}$.
   (a) Show that $A \cap (B \cup C) = (A \cap B) \cup (A \cap C)$.
   (b) Show that $A \cup (B \cap C) = (A \cup B) \cap (A \cup C)$.
   (c) Show that (the number of elements of A) + (the number of elements of B) = (the number of elements of $A \cup B$) + (the number of elements of $A \cap B$).

3. Solve the following inequalities.
   (a) $x + 6 < 4 - 3x$ (b) $1 - x > 3 - 2x$
   (c) $2 + 3x \le x + 14$ (d) $5x + 1 \ge 6x - 3$
   (e) $(5x + 3) + x \le (7x + 1) - x$ (f) $(5x + 3) + x > (7x + 1) - x$
   (g) $2 < 4x < x + 9$ (h) $x + 1 < 2x - 1 < 3 + x$

4. Verify the following inclusions.
   (a) $\{-1, 2\} \subseteq \{x^5 - x^4 - x^3 - 4x^2 + x + 6 = 0\}$
   (b) $10 \in \{2^x > 1000\}$
   (c) $\{x^2 = 3\} \subseteq \{x^4 = 9\}$
   (d) $\{x^2 < 3\} \subseteq \{x^4 < 9\}$

5. Find the solution sets of the following relations.
   (a) $\dfrac{2x - 1}{x - 3} = 3 + \dfrac{5}{x - 3}$ (b) $\dfrac{3x - 7}{x - 6} = 5 + \dfrac{11}{x - 6}$
   (c) $\dfrac{2x - 1}{x - 3} < 3 + \dfrac{5}{x - 3}$ (d) $\dfrac{3x - 7}{x - 6} < 5 + \dfrac{11}{x - 6}$
   (e) $\dfrac{1}{x} + 5 = \dfrac{5}{x} + 1$ (f) $\dfrac{1}{x} = \dfrac{1}{2 - x}$
   (g) $\dfrac{1}{x} + 5 < \dfrac{5}{x} + 1$ (h) $\dfrac{1}{x} > \dfrac{1}{2 - x}$

6. Suppose that $a \in A$, $b \in B$, $c \in C$, $A \subseteq B$, and $C \subseteq B$. Which of the points a, b, and c surely belong to the following sets?
   (a) B (b) $A \cup C$ (c) $C \cap (A \cup B)$
   (d) $C \cup (A \cap B)$ (e) $C \cap (A \cap B)$ (f) $(A \cup C) \cap B$

7. Which of the following statements are true?

(a) If $A \subseteq B$, then $A \cap B = A$.
(b) If $A \subseteq B$, then $A \cup B = B$.
(c) If $A \cap B = A$, then $A \subseteq B$.
(d) If $A \cup B = B$, then $A \subseteq B$.
(e) If $A \cup B = \emptyset$, then $A = \emptyset$.
(f) If $A \cap B = \emptyset$, then $A = \emptyset$.

8. Let M denote the larger and m denote the smaller of two numbers a and b. Discuss the relationships between the sets $\{x < a\}$, $\{x < b\}$, $\{x < M\}$, and $\{x < m\}$.

9. Let P be the set of positive integers and $A_k = \{n \mid n \in P$ and n divides $k\}$. Thus, A_{30} consists of the divisors of 30: $A_{30} = \{1, 2, 3, 5, 6, 10, 15, 30\}$. List the numbers in the following sets.

(a) $A_6 \cap A_{12}$ (b) $A_6 \cup A_{12}$
(c) $A_6 \cap (A_{12} \cup A_{18})$ (d) $(A_6 \cap A_{12}) \cup (A_6 \cap A_{18})$
(e) $A_6 \cup (A_{12} \cap A_{18})$ (f) $(A_6 \cup A_{12}) \cap (A_6 \cup A_{18})$

10. Suppose that the symbols in Equations 1-1 through 1-5 are reinterpreted so that a, b, and c are sets, $a + b$ is the union of a and b, and ab is the intersection of a and b. Do these "laws of arithmetic" hold with this interpretation?

6
SOLVING
INEQUALITIES

As we pointed out in the last section, solving an equation or an inequality amounts to describing the solution set of the equation or inequality as simply and as explicitly as possible. Thus, the symbols $\{x^3 - 4x^2 + x + 6 = 0\}$ and $\{-1, 2, 3\}$ both describe the solution set of the equation $x^3 - 4x^2 + x + 6 = 0$, and the second description is obviously more explicit than the first. Now let us try our hand at some inequalities that are more complicated than those we tackled in the last section.

EXAMPLE 6-1   Solve the inequality $x^2 - x - 2 > 0$.

Solution   First, we factor the left-hand side of this inequality and write the solution set as

$$\{x^2 - x - 2 > 0\} = \{(x + 1)(x - 2) > 0\}.$$

Now we turn to Fig. 6-1, which illustrates our problem. The number

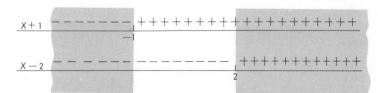

FIGURE 6-1

scales in the figure show the signs of the factors of the given expression.

Since the product will be positive if, and only if, both factors have the same sign, it is clear from the figure that $x^2 - x - 2 > 0$ if x is less than -1 or if x is greater than 2, and conversely. In symbols,

(6-1)
$$\{x^2 - x - 2 > 0\} = \{x < -1\} \cup \{x > 2\}.$$

This set equation gives the solution to our problem, so it is proper to stop at this point. However, it may be instructive to look a little more closely at Equation 6-1. The union symbol $\cup$ on the right-hand side stands for the word "or" in the statement "x is less than -1 *or* x is greater than 2." Now notice that for numbers in the set $\{x < -1\}$ the first *and* second factor of our given product are negative, and so

$$\{x < -1\} = \{x + 1 < 0\} \cap \{x - 2 < 0\}.$$

Here, the intersection symbol $\cap$ takes the place of the word "and." Similarly, the first *and* second factors of the given product are positive for numbers in the set $\{x > 2\} = \{x + 1 > 0\} \cap \{x - 2 > 0\}$. Thus, the right-hand side of Equation 6-1 is a simplified description of the set

$$(\{x + 1 < 0\} \cap \{x - 2 < 0\}) \cup (\{x + 1 > 0\} \cap \{x - 2 > 0\}).$$

You should be able to read from this notation that a number x belongs to the set if both factors of the product $(x + 1)(x - 2)$ are negative or if both factors are positive.

EXAMPLE 6-2   Solve the inequality $x^2 - x - 2 \leq 0$.

Solution   This inequality is obtained from our previous one by replacing the symbol $>$ with the "complementary" symbol $\leq$. Therefore, it is true precisely when the other is false, and so its solution set consists of those real numbers that *do not* belong to the solution set of the preceding inequality. Hence, a glance at Fig. 6-1 shows us that

$$\{x^2 - x - 2 \leq 0\} = \{-1 \leq x \leq 2\}.$$

EXAMPLE 6-3   Solve the inequality $\dfrac{x + 1}{2x - 3} \geq 1$.

Solution   The first step is to subtract 1 from both sides of the given inequality and hence write it in the equivalent form

$$\frac{x + 1}{2x - 3} - 1 = \frac{x + 1}{2x - 3} - \frac{2x - 3}{2x - 3} = \frac{-x + 4}{2x - 3} \geq 0.$$

Again we draw a figure. The number scales in Fig. 6-2 show the signs of

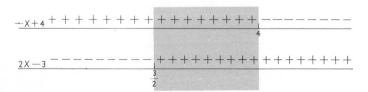

FIGURE 6-2

22

the numerator and denominator of our last quotient. We see that these expressions have the same sign (and hence the quotient is not negative) for numbers between $\frac{3}{2}$ and 4. Thus, we conclude that

$$\left\{\frac{x+1}{2x-3} \geq 1\right\} = \{\tfrac{3}{2} < x \leq 4\}.$$

(Why do we exclude the possibility $x = \frac{3}{2}$ from our considerations?)

By an analysis similar to the one we applied to the right-hand side of Equation 6-1, we can see that the notation $\{\frac{3}{2} < x \leq 4\}$ is a simple way of describing the set

$$(\{-x+4 \geq 0\} \cap \{2x-3 > 0\}) \cup (\{-x+4 \leq 0\} \cap \{2x-3 < 0\}).$$

You should be able to translate these symbols into a verbal description of the condition that a point belongs to the set. (Notice that the second set in the union is the empty set, a fact which is obvious from a glance at Fig. 6-2.)

Let us explicitly mention the fundamental principles that we use when we solve equations or inequalities. If $A_1, A_2, \ldots, A_n$ are n numbers, then

(i) $A_1 \cdot A_2 \cdots A_n = 0$ *if, and only if, at least one of the factors is 0;*

(ii) $A_1 \cdot A_2 \cdots A_n < 0$ *if, and only if, the number of negative factors is odd and no factor is 0;*

(iii) $A_1 \cdot A_2 \cdots A_n > 0$ *if, and only if, the number of negative factors is even and no factor is 0.*

For example, statement (i) tells us that $(x+1)(x-2)(x-3) = 0$ if, and only if, at least one of the equations $x+1 = 0$, $x-2 = 0$, or $x-3 = 0$ holds. Thus, the solution set of the original equation is the union of the solution sets of these simple equations; that is, it is the set $\{-1, 2, 3\}$.

EXAMPLE 6-4 Solve the inequality $(x+1)(x-2)(x-3) < 0$.

Solution The number scales in Fig. 6-3 show the signs of the factors of our given

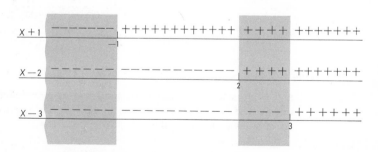

FIGURE 6-3

expression. Now we simply read from this figure (the number of negative factors must be odd) that

$$\{(x+1)(x-2)(x-3) < 0\} = \{x < -1\} \cup \{2 < x < 3\}.$$

We close this section with an obvious remark. In the symbols $\{x \mid x$ has a certain property$\}$ that we use to describe a set, we can replace x with any other letter. For example, $\{x \mid x > 2\} = \{z \mid z > 2\}$, since both sets of symbols describe the set of numbers that are greater than 2. In our shorthand notation, this equation becomes $\{x > 2\} = \{z > 2\}$.

1. Solve the following inequalities.

(a) $x^2 + 4x + 1 \le x^2 + 3x - 2$ (b) $x^2 + 3 \ge x(x + 1)$

(c) $(x - 1)(x + 1) < 0$ (d) $x^2 - 3x \ge 0$

(e) $\dfrac{x - 2}{x^2} \le 0$ (f) $(x + 4)(x - 1) \le (x + 5)(x - 2)$

2. Find the solution set of each of the following inequalities.

(a) $x^2 - x < 2$ (b) $(x + 1)(x - 2) < (x + 1)(2x + 1)$

(c) $(x - 1)(x + 1) < -1$ (d) $(x - 1)(x + 1) \ge -1$

(e) $\dfrac{x - 1}{x + 1} \le 2$ (f) $\dfrac{x^2 + 4}{4x} \le 1$

3. The **complement** of a set A of real numbers is the set

$$cA = \{x \mid x \in R^1 \text{ and } x \notin A\}.$$

For example, $c\{x < 2\} = \{x \ge 2\}$. Find the complements of the following sets.

(a) R^1 (b) $\varnothing$ (c) $\{x > 3\}$

(d) $\{-1 < x < 5\}$ (e) $\{x^2 - 2x + 7 \le 0\}$ (f) $\{x^2 \ne 1\}$

4. Use the definition of the complement of a set given in the preceding problem and decide which of the following statements are true.

(a) $c(cA) = A$ (b) If $A \subseteq B$, then $cA \subseteq cB$.

(c) $c(A \cap B) = cA \cap cB$ (d) $c(A \cap B) = cA \cup cB$

5. Use sketches like those in Example 6-4 to help you solve the following inequalities.

(a) $\dfrac{x}{(x - 1)(x + 2)} < 0$ (b) $\dfrac{x + 1}{(x + 2)(x - 1)} < 0$

(c) $\dfrac{(x + 2)(x + 3)(x + 4)}{x + 5} \le 0$ (d) $\dfrac{(x + 2)^2(x - 3)(x - 4)}{x^3} < 0$

6. Suppose that $a < b$. Show that $\{(x - a)(x - b) < 0\} = \{a < x < b\}$.

7. If A and B are sets of real numbers, the set $A - B$ is defined to be the set of numbers that belong to A but not to B. Use the definition of the complement of a set given in Problem 6-3 to express the following sets without using a minus sign.

(a) $A - B$ (b) $R^1 - A$

(c) $R^1 - (A - B)$ (d) $(A - B) - A$

8. Discuss the truth of the statement that $A - (A - B) = B$.

Another result of the representation of real numbers as points of the number scale is the introduction of the geometric concept of *distance* into our number system. Let x be a point of the scale. Then the number of units between this point and the origin is called the **absolute value of x** and is denoted by the symbol $|x|$. Thus, $|2| = 2$, $|-3| = 3$, $|\pi| = \pi$, and so on. The absolute value of a number gives only its distance from the origin, not its direction. If $|x| = 5$, for example, we know that x is either the number 5 or the number -5, since both these numbers, and no others, are 5 units from the origin. Thus, $\{|x| = 5\} = \{-5, 5\}$.

We can also formulate a definition of absolute value that does not rely on geometric notions. In every case, this definition gives the same result as the geometric definition.

Definition 7-1 *For any real number a,*

$$|a| = a \text{ if } a \geq 0, \quad and \quad |a| = -a \text{ if } a < 0.$$

From this definition, we see that $|2| = 2$, since $2 > 0$, while $|-3| = -(-3) = 3$, since $-3 < 0$. These examples illustrate the fact that *the absolute value of a number is never negative*.

The concept of absolute value is already familiar to you, even if the name is not. For example, the rule for multiplying two real numbers reads, "Multiply the absolute values of the numbers and then multiply this result by 1 or -1, depending on whether the numbers have like or opposite signs." Thus,

$$(2)(-3) = (-1)(|2| \cdot |-3|) = (-1)(2 \cdot 3) = -6.$$

The following facts about absolute value are important for you to know.

(7-1) $$|-a| = |a|.$$

(7-2) $$|a| > 0 \text{ if } a \neq 0 \text{ and } |0| = 0.$$

(7-3) $$|ab| = |a|\,|b|.$$

(7-4) $$\left|\frac{a}{b}\right| = \frac{|a|}{|b|}.$$

(7-5) $$|a|^2 = a^2.$$

Property 7-1 simply states that the points a and $-a$ are equally far from the origin. Property 7-2 states that the number 0 is the only number whose distance from the origin is 0; all others are some positive distance away. Properties 7-3 and 7-4 stem from the definitions of multiplication and division. For example, the product ab is either the number $|a|\,|b|$ or the number $-|a|\,|b|$. The absolute value of both these numbers is $|a|\,|b|$, and so in either case $|ab| = |a|\,|b|$. If a is a positive number, then $a = |a|$; if a is negative, $a = -|a|$. In either event, $a^2 = |a|^2$, as Property 7-5 states.

If you think of a and b as points of a number scale, you can readily convince yourself that *the number* $|b - a|$ *is the distance between* a *and* b. Thus, the equation $|(-2) - 3| = 5$ expresses the fact that the points -2 and 3 are 5 units apart. Observe from Property 7-1 that $|a - b| = |b - a|$. Geometrically, we interpret this equation as saying that the distance between a and b is the same as the distance between b and a. Since $a + b = a - (-b)$, *the number* $|a + b|$ *is the distance between* a *and* $(-b)$.

If x is a number such that $|x - 5| = 2$, then x is 2 units from the point 5 of the number scale. Therefore, either $x = 7$ or $x = 3$. We can get the same result by applying the more formal Definition 7-1. Using set language, we have

$$\{|x - 5| = 2\} = \{x - 5 = -2\} \cup \{x - 5 = 2\} = \{3\} \cup \{7\} = \{3, 7\}.$$

If p is a positive number, then the statement $|x| < p$ means that the point x is less than p units from the origin; that is, x is between $-p$ and p. Thus the inequalities $|x| < p$ and $-p < x < p$ are equivalent, a fact which we express by means of the set equation

$$(7\text{-}6) \qquad \{|x| < p\} = \{-p < x < p\}.$$

Now let a be a real number, and replace x in Equation 7-6 with $(x - a)$. We see then that the inequality $|x - a| < p$ is equivalent to the inequalities $-p < x - a < p$. If we add a to each term, these last relations become $a - p < x < a + p$. Hence, we have the set equation

$$(7\text{-}7) \qquad \{|x - a| < p\} = \{a - p < x < a + p\}.$$

Now let us look at the inequality $|x| > p$. Geometrically, this statement says that the distance between the point x and the origin is *more than* p units. Therefore, the number x is either less than $-p$ *or* greater than p. The appearance of the word "or" suggests that we write the solution set of the inequality $|x| > p$ as a union. Thus, we have

$$(7\text{-}8) \qquad \{|x| > p\} = \{x < -p\} \cup \{x > p\}.$$

Notice that we *cannot* describe this set wholly in terms of inequality symbols as we did in Equation (7-6). In particular, it is absurd to write $-p > x > p$, for these inequalities would imply that $-p > p$.

Since the inequalities $-p < x < p$ say that $-p < x$ *and* $x < p$, we can express the set that appears in Equation 7-6 as an intersection:

$$\{|x| < p\} = \{-p < x\} \cap \{x < p\}.$$

The notation in Equation 7-6 is somewhat simpler, however. It should be clear that the above equations are valid if each symbol $<$ is replaced by $\leq$ and each $>$ is replaced by $\geq$.

EXAMPLE 7-1 Express the inequality $|x| < 3$ without using absolute value signs.

Solution According to Equation 7-6, the given inequality is equivalent to the inequalities $-3 < x < 3$.

EXAMPLE 7-2 Express the inequality $|2x - 1| < 5$ without using absolute value signs.

Solution According to Equation 7-7, the given inequality is equivalent to the inequalities $-4 < 2x < 6$. We divide each term of these inequalities by 2 to obtain the equivalent inequalities $-2 < x < 3$.

PROBLEMS 7

1. Solve the following equations.
   (a) $|x| = \frac{3}{2}$ (b) $|x| = -1$ (c) $|x + 2| = 0$
   (d) $|x| = |-1|$ (e) $|3x| = 9$ (f) $3|x| = 9$
   (g) $|x - 1| = 1$ (h) $|x + 1| = 2$ (i) $|3 - x| = 1$
   (j) $|2 - 3x| = 4$

2. Find the solution set of the equation.
   (a) $|3x| = 3x$ (b) $|3x| = 6x$ (c) $|x - 3| = x - 3$
   (d) $x - 1 = |1 - x|$ (e) $|x + 1| = |x| + 1$ (f) $|x - 1| = |x| - 1$
   (g) $|x^2 - 1| = x^2 - 1$ (h) $|4x^2 - 1| = 1 - 4x^2$

3. If the point x is 3 units away from 5, how far is it from 7?

4. If you know that $|x + 1| = 4$, what can you say about the following numbers?
   (a) $|x|$ (b) $|x - 1|$ (c) $|x - 4|$ (d) $|x + 4|$

5. Show how Equation 7-5 follows from Equation 7-3 and the definition of absolute value.

6. Under what conditions are the following statements true?
   (a) $x \leq |x|$ (b) $|x| \leq -x$ (c) $-|x| \leq x$ (d) $x \leq -|x|$
   (e) $\frac{1}{2}(a + b + |a - b|) = a$ (f) $\frac{1}{2}(a + b - |a - b|) = a$
   (g) $|a + b| = |a| + |b|$ (h) $|a - b| = |a| - |b|$

7. Express the following inequalities without using absolute value signs.
   (a) $|3x| < 2$ (b) $\left|\frac{x}{3}\right| < 2$ (c) $|x - 1| < 3$ (d) $|x + 3| < 7$
   (e) $|2 - x| < 1$ (f) $|2x - 7| \leq 5$ (g) $|x - \frac{1}{2}| \leq \frac{2}{3}$ (h) $\left|\frac{2x^2 - x}{x}\right| \leq 4$
   (i) $|x^3| \leq x^2$ (j) $x^3 \leq |x^2|$

8. Express the following sets without using absolute value signs.
   (a) $\{|2x| > 3\}$ (b) $\{|\frac{1}{2}x| > 3\}$ (c) $\{|x - 1| > 8\}$
   (d) $\{|4x - 5| \geq 2\}$ (e) $\{|x - \frac{1}{2}| \geq \frac{3}{2}\}$ (f) $\{|2x - 1| \geq 3\}$

9. Show that $\{|x - a| > p\} = \{x < a - p\} \cup \{x > a + p\}$.

10. Sketch the following sets as sets of points of the number scale.
    (a) $\{|x - 3| < 2\}$ (b) $\{|3x - 4| < 5\}$ (c) $\{|x + 1| \leq 2\}$
    (d) $\{|x - 2| > \frac{3}{2}\}$ (e) $\{|2x - 7| \geq 1\}$ (f) $\{|7x - 2| \leq 1\}$

11. Solve the following inequalities.
    (a) $2|x| + 3 < 3|x| - 2$ (b) $|x^2 - 5x - 3| < 3$
    (c) $|x^2 - 5| < 4$ (d) $||x| - 3| < 2$

12. Solve the following inequalities.

    (a) $\left|\dfrac{1}{x} - 2\right| < 2$ (b) $\left|\dfrac{1}{x} - 2\right| < 3$

    (c) $\left|\dfrac{1}{x} - 2\right| < 1$ (d) $\left|\dfrac{1}{x} + 1\right| < 2$

8

**INEQUALITIES
INVOLVING
ABSOLUTE
VALUES**

We often use inequalities to locate a point between two extreme limits. We might, for example, say that x is a number such that $3 < x < 5$, to indicate that x lies somewhere in the interval from 3 to 5. Another way of approximately locating a point is to say that it is within a certain distance of a specified "base point." For example, if x is a number such that $|x - m| < p$, then x must lie in the interval between the points $m - p$ and $m + p$, and conversely. Observe that the point m is the *midpoint* of this interval and that the length of the interval is $2p$. Thus the equivalence of the inequalities $m - p < x < m + p$ and $|x - m| < p$ is simply an expression of the geometrically obvious fact that a point lies in a certain interval of the number scale if, and only if, it is less than half the length of the interval from the midpoint.

Now let us consider two numbers a and b, where $a < b$, to be the endpoints of an interval of the number scale. The length of this interval is clearly $b - a$, and suppose that m is its midpoint. Since the distance from a to m is the same as the distance from m to b, we have $m - a = b - m$. When we solve this equation for m, we find that $m = \frac{1}{2}(a + b)$. Thus the numbers $\frac{1}{2}(b - a)$ and $\frac{1}{2}(b + a)$ play the roles of the numbers p and m of the preceding paragraph, and accordingly we see that *the inequalities $a < x < b$ and $|x - \frac{1}{2}(a + b)| < \frac{1}{2}(b - a)$ are equivalent.* In set notation, we have

(8-1) $$\{a < x < b\} = \{|x - \tfrac{1}{2}(a + b)| < \tfrac{1}{2}(b - a)\}.$$

EXAMPLE 8-1   Replace the inequality $1 < x < 3$ by a single inequality involving an absolute value.

Solution   We may use Equation 8-1 with $a = 1$ and $b = 3$ to see that the inequalities $1 < x < 3$ and $|x - 2| < 1$ are equivalent.

EXAMPLE 8-2   Show that $a^2 \leq b^2$ if, and only if, $|a| \leq |b|$.

Solution   Before you read the formal proof that follows, try some numerical examples to convince yourself that the statement is true. We wish to show that the number $b^2 - a^2$ is not less than 0 if, and only if, $|b| - |a|$ is not less than 0, so let us find an equation that relates these differences. Since $b^2 - a^2 = |b|^2 - |a|^2 = (|b| - |a|)(|b| + |a|)$, we have

$$b^2 - a^2 = (|b| - |a|)(|b| + |a|).$$

The factor $(|b| + |a|)$ is positive (unless $a = b = 0$), so it is clear that $b^2 - a^2$ and $|b| - |a|$ have the same sign, and our assertion is proved.

We can use the result of the last example to derive one of the basic inequalities of mathematical analysis, the *triangle inequality*. When we expand $(a + b)^2$ and write $a^2 = |a|^2$ and $b^2 = |b|^2$, we obtain the equation

(8-2)
$$(a + b)^2 = |a|^2 + 2ab + |b|^2.$$

Now observe that $ab \leq |a|\,|b|$, so the right-hand side of Equation 8-2 is not greater than $|a|^2 + 2|a|\,|b| + |b|^2 = (|a| + |b|)^2$. In symbols,

$$(a + b)^2 \leq (|a| + |b|)^2.$$

But according to our result in Example 8-2, this inequality is equivalent to the relation

$$|a + b| \leq \big||a| + |b|\big|.$$

Of course, the outer absolute value signs on the right-hand side of this inequality are unnecessary, and when we remove them, we obtain the **Triangle Inequality**

(8-3)
$$|\boldsymbol{a} + \boldsymbol{b}| \leq |\boldsymbol{a}| + |\boldsymbol{b}|.$$

This inequality leads to a number of other useful relations. For example, suppose that we replace b with $-b$ in Inequality 8-3. Since $|-b| = |b|$, we thereby derive the inequality

(8-4)
$$|\boldsymbol{a} - \boldsymbol{b}| \leq |\boldsymbol{a}| + |\boldsymbol{b}|.$$

This inequality has a simple geometric interpretation. Thus, $|a - b|$ is the distance between a and b, the point a is $|a|$ units from the origin, and b is $|b|$ units from the origin. Hence, Inequality 8-4 says that a and b cannot be more than $|a| + |b|$ units apart. Indeed, it is clear that two numbers, each a given distance from 0, will be farthest apart if they are on opposite sides of the origin, so that $|a - b| = |a| + |b|$ only when a and b have opposite signs.

Many of the problems at the end of this section ask you to determine the correctness of a given inequality. Your first step in solving such a problem should be to test a few specific choices of a and b. Such testing may turn up a "counter-example," or it may suggest that the inequality is true for all numbers. You will often find that some form of the triangle inequality is just the tool you need to verify a particular inequality.

EXAMPLE 8-3   What can you say about the number $|x+1|$ if you are given that

$$|x - 1| < 2?$$

Solution   We have

$$
\begin{aligned}
|x+1| = |x-1+2| &\quad \text{(algebra)} \\
\leq |x-1|+2 &\quad \text{(triangle inequality)} \\
< 2+2 = 4 &\quad (|x-1| < 2 \text{ was given}).
\end{aligned}
$$

EXAMPLE 8-4   What can you say about $|x^2 - 1|$ if you are given that $|x - 1| < 2$?

Solution   Here

$$
\begin{aligned}
|x^2 - 1| = |(x-1)(x+1)| = |x-1|\,|x+1| &\quad \text{(factor)} \\
< 2 \cdot 4 = 8 &\quad \text{(the given inequality and the result of Example 8-3).}
\end{aligned}
$$

**PROBLEMS
8**

1. Replace each of the following inequalities with a single inequality involving an absolute value.

    (a) $1 < x < 3$ (b) $-2 \leq x \leq 4$ (c) $1 < 2x+1 < 6$

    (d) $3 < 1 - 2x < 5$ (e) $-3 < 2x - 1 < \frac{1}{2}$ (f) $-3 < 1 - 2x < \frac{1}{2}$

2. If $|x - 3| < 1$, what can you say about $|x - 2|$?

3. If $|x - 2| < 3$, what can you say about $|x^2 + x - 6|$?

4. Show that if $|x - 1| < \frac{2}{3}$, then $|3 - x| \geq \frac{4}{3}$.

5. Prove that $x - 2 \leq |x + 2|$.

6. Prove that the following inequalities are true for every two numbers a and b.

    (a) $|a| - |b| \leq |a - b|$ (b) $-(|a| - |b|) \leq |a - b|$

    (c) $||a| - |b|| \leq |a - b|$ (d) $|a| - |b| \leq |a + b|$

7. Solve the following inequalities.

    (a) $|x - 1| \leq |x| + 1$ (b) $|x - 1| < |x| - 1$

    (c) $|x| - 2 > |x - 2|$ (d) $||x| - 2| < |x + 2|$

8. Some of the following inequalities are true for every choice of x, and some are not. Prove the true ones, and find counter-examples to the false ones.

    (a) $|x - 2| \leq |x - 3| + 1$ (b) $|x - 4| \leq |x - 3| - 1$

    (c) $|x - 2| \leq |x - 1| - 1$ (d) $|x - 4| \leq |x - 5| + 1$

    (e) $|2x - 1| \leq 2|x - 1| + 1$ (f) $|3x + 2| \leq 3|x - 1| + 5$

9. Prove or disprove.

    (a) $|x - y| \leq |x + y|$ (b) $|x - y| \leq |x| - |y|$

10. Find a positive number d such that $|x - 2| < d$ implies that $|x^2 - 4| < .1$.

9

EXPONENTS You are accustomed to using positive integral exponents to denote products of a number of identical factors. A brief review of the rules that govern the operations with these exponents will set the stage for an extension of the exponent notation to the other integers and fractions.

Definition 9-1 *If a is any number, and n is a positive integer, then*

$$a^n = a \cdot a \cdots a \quad (n \text{ factors}).$$

The equation $5^3 = 5 \cdot 5 \cdot 5$ is a numerical illustration of Definition 9-1; here, $a = 5$ and $n = 3$.

The following laws of exponents are immediate consequences of this definition. We assume that m and n are positive integers, and a and b can be any numbers (with the obvious restriction that no denominator may be 0).

(9-1)
$$a^m a^n = a^{m+n}.$$

(9-2)
$$(a^n)^m = (a^m)^n = a^{mn}.$$

(9-3)
$$(ab)^n = a^n b^n.$$

(9-4)
$$\left(\frac{a}{b}\right)^n = \frac{a^n}{b^n}.$$

(9-5)
$$\frac{a^n}{a^m} = \begin{cases} a^{n-m} \text{ if } n > m. \\ \dfrac{1}{a^{m-n}} \text{ if } n < m. \\ 1 \text{ if } n = m. \end{cases}$$

The example $a^3 \cdot a^4 = (a \cdot a \cdot a)(a \cdot a \cdot a \cdot a) = a^7 = a^{3+4}$ illustrates Rule 9-1, and the equation $(a^3)^4 = a^3 \cdot a^3 \cdot a^3 \cdot a^3 = a^{12} = a^4 \cdot a^4 \cdot a^4 = (a^4)^3$ illustrates Rule 9-2. It will be easy for you to convince yourself that all these rules are true.

Definition 9-1 applies only if n is a positive integer. The symbols 7^0, 3^{-2}, and $5^{1/2}$ are, in terms of this definition, meaningless. They will acquire meaning only as the result of new definitions, which we will now formulate. These definitions are chosen to preserve Rules 9-1 to 9-5. It can be shown that they are the only ones that allow us to continue to use these rules when dealing with exponents that are not necessarily positive integers.

Definition 9-2 *If $a \neq 0$, then $a^0 = 1$.*

No meaning is assigned to the symbol 0^0. Thus $\pi^0 = 1$, $(-1)^0 = 1$, and so on.

At first glance, this definition may seem somewhat arbitrary. Let us show first that Rule 9-1 is still valid. Let a be a non-zero number and n be a positive integer or 0, and let $m = 0$. The rule then reads $a^{0+n} = a^0 \cdot a^n$.

Since $0 + n = n$, and $a^0 = 1$ (Definition 9-2), this last equation is correct. Hence Rule 9-1 is valid if either (or both) of the integers m and n is 0. Now we will show that Definition 9-2 is the *only* definition of the symbol a^0 that preserves Rule 9-1. Suppose that a is a number different from 0 and that a value has been assigned to the symbol a^0. Let $m = 0$ and $n = 1$ in Rule 9-1. If this rule still applies, then $a^{0+1} = a^0 \cdot a^1$; that is, $a = a^0 \cdot a$. Since $a \neq 0$, we can divide by it to get $a^0 = 1$. Therefore, if we insist on preserving Rule 9-1, the only value that we can assign to the symbol a^0 is 1. As a matter of fact, we can also show that Definition 9-2 does not violate Rules 9-2 to 9-5.

The way we define negative exponents is also motivated by Rule 9-1. Suppose $a \neq 0$ and n is a positive integer. We wish to assign a value to the symbol a^{-n} so that we may still use Rule 9-1. According to that rule, we should have $a^n \cdot a^{-n} = a^{n+(-n)} = a^0 = 1$. Hence $a^n \cdot a^{-n} = 1$, which suggests the following definition.

Definition 9-3   *If $a \neq 0$, then $a^{-n} = 1/a^n$.*

For example,

$$5^{-2} = \frac{1}{5^2} = \frac{1}{25} \quad \text{and} \quad \left(\frac{1}{2}\right)^{-3} = \frac{1}{(\frac{1}{2})^3} = 8.$$

Once we have decided on the definition, we must still check to see whether or not the rules of exponents are true. We find that they are. Indeed, we can simplify Rule 9-5, by removing the restrictions on m and n, to read

(9-6) $$\frac{a^n}{a^m} = a^{n-m} \text{ for any integers } m \text{ and } n.$$

For example,

$$\frac{a^5}{a^7} = a^{5-7} = a^{-2} = \frac{1}{a^2} \quad \text{and} \quad \frac{a^{-5}}{a^{-7}} = a^{-5-(-7)} = a^2.$$

The statement $(a/b)^{-n} = (b/a)^n$ is an **identity** in the sense that it is true whenever the letters are replaced by numbers for which the expressions involved have meaning. We must verify the statement by using arguments that do not depend on a particular choice of a, b, and n:

$$\left(\frac{a}{b}\right)^{-n} = \frac{1}{(a/b)^n} \quad \text{(Definition 9-3)}$$

$$= \frac{1}{(a^n/b^n)} \quad \text{(Rule 9-4)}$$

$$= \frac{b^n}{a^n} \quad \text{(division of fractions)}$$

$$= \left(\frac{b}{a}\right)^n \quad \text{(Rule 9-4)}.$$

We assume that neither a nor b is 0, for then the expressions are meaningless. But otherwise, the result is true.

Suppose now that we are asked to show that two expressions, for example $(x^{-1} + y^{-1})$ and $(x + y)^{-1}$, are *not* identically equal. This problem is intrinsically easier than the preceding one, in which we were asked to demonstrate equality for all possible substitutions (with certain obvious exceptions). Here we must deny such equality. Equality cannot *always* hold if there is *at least one* situation in which it does not. Hence it is sufficient to find numbers for which the given expressions have meaning but are not equal. If we let $x = 1$ and $y = 2$ in the expressions $(x^{-1} + y^{-1})$ and $(x + y)^{-1}$, then the first expression becomes $1 + \frac{1}{2} = \frac{3}{2}$, and the second becomes $\frac{1}{3}$. Since these two numbers are not equal, we can conclude that the expressions $(x^{-1} + y^{-1})$ and $(x + y)^{-1}$ do not always represent the same number.

The following examples illustrate the use of exponents.

EXAMPLE 9-1 Simplify the quotient $\dfrac{(x^{-2}y^4)^3}{(xy)^{-3}}$.

Solution

$$\frac{(x^{-2}y^4)^3}{(xy)^{-3}} = \frac{x^{-6}y^{12}}{x^{-3}y^{-3}} = x^{-3}y^{15}.$$

EXAMPLE 9-2 Simplify the quotient $\dfrac{2x^0}{(2x)^0}$.

Solution Notice that the exponent in the numerator applies only to the letter x and so we have

$$\frac{2x^0}{(2x)^0} = \frac{2 \cdot 1}{1} = 2.$$

EXAMPLE 9-3 Simplify the expression $xy(x^{-1} + y^{-1})$.

Solution

$$xy(x^{-1} + y^{-1}) = xyx^{-1} + xyy^{-1} = y + x.$$

EXAMPLE 9-4 Simplify the expression $(3^{-1} + 2^{-1})^{-2}$.

Solution

$$(3^{-1} + 2^{-1})^{-2} = (\tfrac{1}{3} + \tfrac{1}{2})^{-2} = (\tfrac{5}{6})^{-2} = (\tfrac{6}{5})^2 = \tfrac{36}{25}.$$

EXAMPLE 9-5 Write the expression $(x + y^{-1})^{-1}$ without using negative exponents.

Solution

$$(x + y^{-1})^{-1} = \left(x + \frac{1}{y}\right)^{-1} = \left(\frac{xy+1}{y}\right)^{-1} = \frac{y}{xy+1}.$$

33

1. Simplify the following expressions.
   (a) -2^{-3}
   (b) $(-2)^{-3}$
   (c) $3^{-2}+2^{-1}$
   (d) $(3^{-2}+2^{-1})^{-1}$
   (e) $(2^{-1}/3^{-2})^0$
   (f) $2^{-1}/(3^{-1}+2^{-1})$

   (g) $(x^{-1}y^{-3})(x^2y^3)$
   (h) $(a^{-2}b^3c)^2(ab^2c^3)^{-3}$
   (i) $\dfrac{2xy^{-4}}{3x^{-1}y}$

   (j) $\dfrac{(2x^{-1})^3}{x^3}$
   (k) $[(x^{-2})^3]^{-1}$
   (l) $\left[\dfrac{u^{-2}v^3}{u^{-4}v^{-1}}\right]^{-2}$

2. Reduce each of the following expressions to the simplest form not involving zero or negative exponents.
   (a) $(x/2)^0$
   (b) $3a^0/(3a)^0$
   (c) $(a^2 \cdot a^{-3})^3/a^4$
   (d) $(x+x^{-1})^2$
   (e) $(x^2-y^2)(x-y)^{-1}$
   (f) x^2+y^{-2}

   (g) $\dfrac{4^{-1}+3^{-2}}{2}$
   (h) $x^2y^3(x^{-1}+y^{-2})$
   (i) $\dfrac{x^{-1}+x^{-2}}{x}$

   (j) $\dfrac{x}{x^{-1}+x^{-2}}$
   (k) $\dfrac{b^{-1}c^{-1}}{b^{-1}-c^{-1}}$
   (l) $\left(\dfrac{a^{x+2y}}{a^y}\right)^{(x+y)-1}$

3. Write an expression that involves no positive exponents and is equivalent to each of the following.

   (a) x^2-y^2
   (b) $\dfrac{x^2}{y^3}$
   (c) $x^2+\dfrac{1}{y^3}$
   (d) $(x+1)^2$

   (e) $\left(\dfrac{3}{x}\right)^2$
   (f) $\left(\dfrac{a^2}{3}\right)^{-2}$
   (g) $\dfrac{1}{x^2+y^2}$
   (h) $\dfrac{x^2-y^2}{x^2+y^2}$

4. Which of the following are expressions for the same number?

   (a) xy^{-3} and $(xy)^{-3}$
   (b) $1+x^{-1}$ and $\dfrac{1}{1+x}$
   (c) $\dfrac{1}{1+x^{-1}}$ and $1+x$

   (d) $\dfrac{1}{x^{-1}+y^{-1}}$ and $x+y$
   (e) $(2x)^0$ and $(2/x)^0$
   (f) $-x^{-3}$ and $1/(-x)^3$

5. Solve the following inequalities.
   (a) $|x-1|<3^{-1}$
   (b) $2x-4^{-2}<2^{-1}x+1$
   (c) $|x|^{-1}<2^{-1}$
   (d) $|x|^{-1}>3^{-2}$
   (e) $(x-1)^0 \le x-2$
   (f) $x+x^{-1}\le 2$

6. Show that Definition 9-2 does not violate Rules 9-2 to 9-5.

7. Which of our rules of exponents would be preserved if we had defined a^0 to be 0 for each $a \ne 0$? Which rules would be violated?

8. Show that Definition 9-3 does not violate our rules of exponents.

9. Show that if n is an integer and a is a number, then $|a^n| = |a|^n$. Under what circumstances is $(-a)^n = -a^n$?

10. Suggest a value for 0^0 and defend your choice.

Following the pattern of the previous section, we now seek to assign a value to the symbol $a^{1/n}$, where a is a given number and n is a positive integer, in such a way that the laws of exponents are preserved. If such a choice can be made, then according to Rule 9-2 we should have

$$(a^{1/n})^n = a^{(1/n)n} = a^1 = a.$$

We should therefore like $a^{1/n}$ to represent a number r that satisfies the equation $r^n = a$. We need to discuss this last equation a little before we can proceed with the consideration of rational exponents.

Definition 10-1   *If n is a positive integer and $r^n = a$, then r is an **nth root of a**.*

For example, 3 is a 4th root of 81, since $3^4 = 81$, and -2 is a 5th root of -32, since $(-2)^5 = -32$. There may be no (real) nth roots of a given number, or there may be more than one. Both 2 and -2, for example, are square roots of 4, since $2^2 = 4$ and $(-2)^2 = 4$. We won't discuss exactly how many nth roots a number has and how to describe them until Chapter 6, but we will take on faith the following fundamental property of the real number system. *If a is a positive number and n is a positive integer, then there exists exactly one positive nth root of a. If a is a negative number and n is an odd positive integer, then there exists exactly one negative nth root of a.* Each of these nth roots, whether a is positive or negative, is designated by the symbol $\sqrt[n]{a}$, and is called the **principal nth root of a.**

The symbol $\sqrt[n]{a}$ is called a **radical,** and the number a is called the **radicand.** Note one fact. If n is even and a is positive, there are always two real numbers (one positive and one negative) that satisfy the equation $r^n = a$. Only the *positive* number is called the *principal nth root* and is designated by the symbol $\sqrt[n]{a}$. Thus, $\sqrt{4} = 2$, not -2 and not ± 2. In particular, the radical $\sqrt{x^2}$ is not the number x if x is negative. Hence you must write $\sqrt{x^2} = |x|$ unless you are sure that x is not negative. For example, $\sqrt{(-3)^2} = |-3| = 3$, not -3. If a is positive and n is even, the negative nth root of a is designated by $-\sqrt[n]{a}$.

No even power of a real number can be negative. Thus if a is a negative real number and n is an even integer, there is no real nth root of a. For example, there is no real number whose square is -1. We simply do not define the symbol $\sqrt[n]{a}$ as a real number if a is negative and n is even, so the symbol $\sqrt{-1}$ has no meaning here. It is possible to enlarge the number system beyond the real numbers to include numbers whose even powers are negative (we do so in Chapter 6), but for the moment we will not consider such numbers.

We are now ready to define fractional exponents. Recalling from our earlier discussion that $a^{1/n}$ must satisfy the equation $r^n = a$ if Rule 9-2 is to remain true, we make the following definition.

Definition 10-2 *If n is a positive integer, and if $\sqrt[n]{a}$ exists, then*

$$a^{1/n} = \sqrt[n]{a}.$$

Thus, $32^{1/5} = \sqrt[5]{32} = 2$, $(-8)^{1/3} = \sqrt[3]{-8} = -2$, but $(-4)^{1/2}$ is undefined.

The definition of the symbol $a^{m/n}$ also stems from our wish to preserve Rule 9-2. It is clearly desirable to have $a^{m/n} = (a^{1/n})^m = (\sqrt[n]{a})^m$. Also we can demonstrate without too much trouble that $(\sqrt[n]{a})^m = \sqrt[n]{a^m}$ whenever $\sqrt[n]{a}$ exists. We therefore formulate the following definition.

Definition 10-3 *If m and n are both integers, with $n > 0$, and if $\sqrt[n]{a}$ exists (as a real number), then*

$$a^{m/n} = (\sqrt[n]{a})^m = \sqrt[n]{a^m}.$$

This definition preserves Rules 9-1 to 9-5 in the sense that they apply when all terms involved represent real numbers. The next two examples illustrate what can happen when the rules are applied improperly.

EXAMPLE 10-1 According to Rule 9-1,

$$(-2)^{1/2}(-2)^{1/2} = (-2)^{1/2+1/2} = (-2)^1 = -2.$$

According to Rule 9-3 and Definition 10-2,

$$(-2)^{1/2}(-2)^{1/2} = [(-2)(-2)]^{1/2} = 4^{1/2} = 2.$$

But $2 \neq -2$; what is wrong?

Solution We only claimed that our rules of exponents apply when *all the terms involved represent real numbers*. Since $(-2)^{1/2}$ does not represent a real number, there is no reason to expect the rules to hold, and they don't.

EXAMPLE 10-2 What is wrong with the following reasoning? If we replace a with -1, m with 2, and n with $\frac{1}{2}$ in the equation $(a^m)^n = a^{mn}$, we obtain the equation

$$[(-1)^2]^{1/2} = (-1)^{2(1/2)} = (-1)^1 = -1.$$

On the other hand, $[(-1)^2]^{1/2} = 1^{1/2} = 1$, so $-1 = 1$.

Solution We didn't use *all* of Rule 9-2. If we make our substitution in the entire Equation 9-2, we obtain the equation

$$[(-1)^{1/2}]^2 = [(-1)^2]^{1/2} = (-1)^{2(1/2)}.$$

Now we see that we have the same difficulty that cropped up in the last example. Not all the symbols represent real numbers.

The difficulties that arose in Examples 10-1 and 10-2 were the result of attempting to take even roots of negative numbers. When a and b are *positive* numbers, we don't run into undefined symbols, and our rules of exponents can be used freely.

The following examples illustrate the use of the rules of exponents with fractional exponents.

EXAMPLE 10-3   Simplify the quotient $\sqrt{x}/\sqrt[4]{x}$. Write the result in exponential notation.

Solution   Here, since we are dealing with even roots, we tacitly assume that x is positive. If we use the exponent form and the rules of exponents, we have

$$\frac{\sqrt{x}}{\sqrt[4]{x}} = \frac{x^{1/2}}{x^{1/4}} = x^{1/2-1/4} = x^{1/4}.$$

EXAMPLE 10-4   Simplify the radical $\sqrt{\sqrt[3]{x^4}}$. Write the result in radical form.

Solution

$$\sqrt{\sqrt[3]{x^4}} = \left(x^{4/3}\right)^{1/2} = x^{2/3} = \sqrt[3]{x^2}.$$

It is sometimes convenient to write quotients containing radicals in such a way that no radical appears in the denominator; for example,

$$\frac{1}{\sqrt{2}} = \frac{1}{\sqrt{2}} \cdot \frac{\sqrt{2}}{\sqrt{2}} = \frac{\sqrt{2}}{2}.$$

The following examples illustrate methods used to **rationalize denominators.**

EXAMPLE 10-5   Rationalize the denominator in the quotient $1/\sqrt[5]{x^2}$.

Solution

$$\frac{1}{\sqrt[5]{x^2}} = \frac{1}{x^{2/5}} = \frac{1}{x^{2/5}} \cdot \frac{x^{3/5}}{x^{3/5}} = \frac{x^{3/5}}{x} = \frac{\sqrt[5]{x^3}}{x}.$$

EXAMPLE 10-6   Express the product $1/\sqrt[3]{x^2} \cdot 1/\sqrt[4]{x}$ in simplest radical form, rationalizing the denominator.

Solution

$$\frac{1}{\sqrt[3]{x^2}} \cdot \frac{1}{\sqrt[4]{x}} = \frac{1}{x^{2/3}} \cdot \frac{1}{x^{1/4}} = \frac{1}{x^{11/12}} = \frac{1 \cdot x^{1/12}}{x^{11/12} \cdot x^{1/12}} = \frac{x^{1/12}}{x} = \frac{\sqrt[12]{x}}{x}.$$

EXAMPLE 10-7   Rationalize the denominator in the quotient $1/(\sqrt{x} - \sqrt{y})$.

Solution   The trick here is to multiply both the numerator and the denominator by $\sqrt{x} + \sqrt{y}$, for then

$$\frac{1}{\sqrt{x} - \sqrt{y}} \cdot \frac{\sqrt{x} + \sqrt{y}}{\sqrt{x} + \sqrt{y}} = \frac{\sqrt{x} + \sqrt{y}}{x - y}.$$

1. Find the numerical value of each of the following.

   (a) $27^{2/3}$ (b) $49^{3/2}$ (c) $8^{4/3}$ (d) $-81^{3/4}$

   (e) $64^{-5/6}$ (f) $32^{-3/5}$ (g) $(.04)^{-1/2}$ (h) $(.125)^{-1/3}$

   (i) $(-64)^{-5/3}$ (j) $(32)^{.4}$ (k) $(1024)^{-.3}$ (l) $(1,048,576)^{.15}$

2. Use the laws of exponents to prove the following laws of radicals.

   (a) $\sqrt[n]{ab} = \sqrt[n]{a}\sqrt[n]{b}$ (b) $\sqrt[n]{\sqrt[m]{a}} = \sqrt[nm]{a}$ (c) $\sqrt[km]{a^{kn}} = \sqrt[m]{a^n}$

3. Write in radical form without negative exponents, rationalizing denominators.

   (a) $(x^{-1/3})^{3/4}$ (b) $x^{-1/6}/x^{2/3}$ (c) $(a^{-2/3}x^{1/2})^2$

   (d) $(x^2 + 2xy + y^2)^{-1/2}$ (e) $\dfrac{1}{x + y^{1/2}}$ (f) $\dfrac{2^{-1/2} + 3^{1/2}}{2^{1/2} - 3^{-1/2}}$

4. Compute the numbers $(5^{-2} + 12^{-2})^{1/2}$ and $(5^{-2})^{1/2} + (12^{-2})^{1/2}$. Are they equal?

5. Write in fractional exponent form with no denominators.

   (a) $\sqrt[3]{\dfrac{x}{y^2}}$ (b) $\dfrac{2}{\sqrt[4]{8}}$ (c) $\sqrt[4]{x}\sqrt[3]{x^2/y}$

   (d) $\sqrt[3]{-x}\sqrt[5]{x^3}$ (e) $\dfrac{1}{\sqrt{3} - \sqrt{2}}$ (f) $\sqrt{x^3}/(\sqrt[4]{x} + x)$

   (g) $(\sqrt{x} - \sqrt{y})(\sqrt{x} + \sqrt{y})^2$ (h) $\sqrt[3]{a\sqrt{a}}$

6. Write as a fraction with a rational denominator.

   (a) $\dfrac{\sqrt{6} + 2}{\sqrt{6} - 2}$ (b) $\dfrac{\sqrt{5} + 3}{\sqrt{5} - 3}$ (c) $\dfrac{3 + \sqrt{2}}{\sqrt{2} - 3}$

   (d) $\dfrac{x\sqrt{y} + y\sqrt{x}}{\sqrt{x} + \sqrt{y}}$

7. Compute 2^{3^2}.

8. Is the equation $|a^r| = |a|^r$ an identity (assuming that a is real and r is rational)?

9. Show that $(\sqrt[n]{a})^m = \sqrt[n]{a^m}$ if $\sqrt[n]{a}$ exists.

10. Show that $\sqrt{a} \leq \sqrt{b}$ if, and only if, $0 \leq a \leq b$.

11. Can you conclude that $a < b$ if you know that $\sqrt[3]{a} < \sqrt[3]{b}$? Discuss the situation when 3 is replaced by any integer greater than 3.

REVIEW PROBLEMS, CHAPTER ONE

You should be able to answer the following questions without referring back to the text.

1. Sketch the following sets as points of a number line.

   (a) $(\{-1 < x\} \cap \{x \leq 2\}) \cup (\{1 < x\} \cap \{x \leq 2\})$

   (b) $\{x \mid x^2 < x\}$ (c) $\{x \mid \sqrt{x^2} = x\}$

QA152
,F553
1972
cl

2. Write the following expressions in factored form.
   (a) $3x^2 - 12$ (b) $2x^2 - 10x + 12$ (c) $2x^6 - 128$
   (d) $x^3 + x^2 - 2x - 2$ (e) $a^9 + 8b^6$

3. Simplify.
   (a) $8x^{1/2}y^{5/6}/(16x)^{1/4}y^{1/3}$ (b) $(4x^2y^{3/4}z^{1/6}/32x^{-1}y^0z^{-5/6})^{-1/3}$

4. Suppose that n is a positive integer. Find a rational number r such that

$$\frac{1}{n+1} < r < \frac{1}{n}.$$

5. What can you say about the relative position of the points x and y of a number scale if $|x| > y > 0$?

6. Solve the following inequalities.
   (a) $|3x - 2| < 2$ (b) $(x^3 + x)/(x - 2) > 0$

7. Find m and p such that $\{3 < x < 11\} = \{|x - m| < p\}$.

8. For which pairs of numbers is the inequality true?
   (a) $x + y \geq 2\sqrt{xy}$ (b) $x + y < 2\sqrt{xy}$

9. Show that if $|x| \geq |y|$, then $x^2 \geq xy$.

10. Suppose that a, b, and c are positive numbers. Which is larger, $\dfrac{a}{b}$ or $\dfrac{a+c}{b+c}$?

MISCELLANEOUS PROBLEMS, CHAPTER ONE

These exercises are designed to test your ability to apply your knowledge of numbers to somewhat more difficult problems.

1. Suppose that a, b, c, and d are numbers such that $a < b$ and $c < d$. Which of the following statements are true, which false, and which may be either true or false depending on the values of a, b, c, and d?
   (a) $a - b < d - c$ (b) $b - c < a - d$ (c) $a - c < b - d$
   (d) $a - b < c - d$ (e) $b - c < d - a$ (f) $a - c < d - b$

2. Show that $x^2 - xy + y^2 \geq 0$ for each pair of real numbers x and y.

3. The integers $x = 0$ and $y = 0$ satisfy the equation $xy = x + y$. Are there any other pairs of integers that satisfy this equation?

4. Sketch $\left\{\dfrac{x^2 + x + 2}{x + 1} > x + 2\right\}$ as a set of points of a number scale.

5. Show that $\left\{\left|\dfrac{1}{x} - 2\right| < 1\right\} = \{|\tfrac{2}{3} - x| < \tfrac{1}{3}\}$.

6. Factor $x^3 - 7x + 6$.

7. To install certain piping would cost \$1000 for labor plus the amount spent for materials. Suppose that copper pipe would cost \$1200, while iron pipe

39

would cost $400. If the copper pipe would last more than twice as long as the iron, which installation would be cheaper on a cost-per-year basis? Assuming that other factors are equal, what can you say about the price of the iron pipe if the cost-per-year is the same for the two kinds of pipe?

8. Show that $|a+b|+|a-b| = |a|+|b|+||a|-|b||$.

9. Show that $\frac{1}{2}(|a+b|+|a-b|)$ is the larger of $|a|$ and $|b|$.

10. A man is $\frac{1}{4}$ of the way across a railroad trestle when he sees a train that is 1 trestle length behind him approaching the trestle. Which way should he run? At what point of the trestle is it immaterial which way he runs?

11. Prove that the following inequalities are true for non-negative numbers a and b.

(a) $\frac{1}{2}(a+b) \geq (ab)^{1/2}$ (b) $\frac{1}{3}(a+2b) \geq (ab^2)^{1/3}$ (c) $\frac{1}{4}(a+3b) \geq (ab^3)^{1/4}$

12. Prove that if $x > 0$, then $x^3 + 2/x\sqrt{x} \geq 3$.

Functions

two

Scientists in every field try to establish quantitative relationships. Engineers develop formulas which tell them *how many* inches certain beams bend when subjected to a certain *number* of pounds of load; geneticists ask for the *number* of mutations associated with a *measured* dose of radiation; chemists are interested in the relation between the *rate* of a chemical reaction and the *quantity* of each reacting substance.

The law of a falling body is a quantitative relationship. If a body is released from rest in a vacuum, and near the surface of the earth, then one second after its release it will be 16 feet below its starting point, two seconds after its release, 64 feet, and so on. If we know *how many* seconds the body has fallen, the formula $s = 16t^2$ tells us *how many* feet it has fallen. You are surely familiar with many other examples of quantitative relationships.

In this chapter we concentrate on the essential features of such quantitative relationships. Basically, the idea involves pairing the members of two sets of numbers. We start with two sets, and with each number in one of them we pair one of the numbers in the other. In the example of the falling body, the numbers in one set would represent durations of fall, and the numbers in the other set would represent distances. Then with the number 1 (second) is paired the number 16 (feet), and so forth. In mathematics, we

strip away any physical meaning that may be attached to the numbers involved; we simply discuss sets of numbers and sets of pairs of numbers.

One of the main purposes of this chapter is to introduce you to the
definitions and terminology of the mathematical relationships known as
functions. Although some of the concepts may seem rather abstract, they
are motivated by distinctly practical problems, such as the expression of the
law of a falling body. Since the topics covered in this chapter will be cropping
up again and again throughout your entire study of mathematics, your later
work will be easier if you master these topics as early as possible.

11 FUNCTIONS IN GENERAL

Let us begin our discussion by looking at a simple physical example that
illustrates the practical origins of the abstract mathematical idea of a function. In elementary physics we learn that if we apply E volts across a 5-ohm
resistance, then a current of $\frac{1}{5}E$ amperes flows through the resistor (Ohm's
Law). Suppose we have a voltage generator that can produce any voltage
from 0 to 20 volts. Now if we apply 10 volts across the terminals of the
resistor, we get a current of 2 amperes. Thus with the number 10 we pair the
number 2. Similarly, with each number E between 0 and 20 the equation
$I = \frac{1}{5}E$ pairs a number I (between 0 and 4). The collection of these number
pairs is a *function*.

In general, the notion of a function involves three things: (1) a set D
called the *domain* of the function, (2) a set R called the *range* of the function,
and (3) a rule of correspondence that assigns to each element $x \in D$ an
element $y \in R$. The set of pairs $\{(x, y) \mid x \in D, y$ is the corresponding element of $R\}$ is the function. We usually use letters such as f, g, F, G, and ϕ
to name functions. Thus, in a particular context, f might be the name of the
function whose domain is the set $\{1, 2, 3, 4\}$, whose range is the set $\{4, 5, 6\}$,
and whose rule of correspondence that pairs elements of the range with
elements of the domain is given by the following table:

ELEMENT OF THE DOMAIN OF f	CORRESPONDING ELEMENT OF THE RANGE OF f
1	5
2	4
3	4
4	6

If x is an element of the domain of a function f, then the corresponding
element of the range is denoted by $f(x)$. Thus, in the present example,
$f(1) = 5$, $f(2) = 4$, $f(3) = 4$, and $f(4) = 6$. The symbol $f(x)$ is read "f of x",
and it is called the **value** of f at x. You can think of the equation $f(1) = 5$
as a symbolic abbreviation for the sentence, "In our function f, the element

that corresponds to 1 is 5." Notice that f and $f(x)$ are two quite different things; f is the name of the function, whereas $f(x)$ is an element of its range. The function f is a set of pairs. For instance, in the example of this paragraph, f is the set $\{(1, 5), (2, 4), (3, 4), (4, 6)\}$.

As far as the general definition of a function is concerned, we can use any kinds of objects we like to make up the sets D and R (points, angles, numbers, and so on, are perfectly acceptable). However, it is a fact that in most of the functions we meet in our early study of mathematics, the sets D and R are sets of real numbers. It will make our discussion of functions more direct and more concrete if for the present we restrict our attention to such functions. You will find it easy to make any adjustments in terminology that may be required when we deal with other types of functions later. For the moment, then, we will consider that a function is a collection of pairs of real numbers. Not every such collection is a function, though. To each number x in the domain of a function there corresponds just one number y; that is, the first member of each number pair determines the second. Thus a function cannot contain two number pairs with the same first number, and we have the following formal definition.

Definition 11-1   *A set f of pairs of real numbers is called a* **function** *if it does not contain two pairs with the same first number. In other words, if $(x, y_1) \in f$ and $(x, y_2) \in f$, then $y_1 = y_2$. The set of all the first numbers of the pairs of f is the* **domain** *of f, and the set of second numbers is the* **range** *of f.*

EXAMPLE 11-1   Let $f = \{(x, y) \mid x \in R^1, y = |x|\}$ and $g = \{(x, y) \mid x \in R^1, |y| = x\}$. Are f and g functions?

Solution   The set f *is* a function because if $(x, y_1) \in f$ and $(x, y_2) \in f$, then $y_1 = |x|$ and $y_2 = |x|$, so that $y_1 = y_2$. The set g *is not* a function because there are different pairs with the same first number in g; for example, $(1, 1)$ and $(1, -1)$. Notice that a function may contain two pairs with the same second number; for example, the function f contains the pairs $(1, 1)$ and $(-1, 1)$.

To specify a function, we must give its domain, its range, and the rule that pairs with an element x of its domain an element y of its range. Usually this rule is written as a mathematical equation, as in the next example.

EXAMPLE 11-2   Let f be the function whose domain is the set of all real numbers, whose range is the set of all numbers greater than or equal to 2, and whose rule of correspondence is given by the equation $f(x) = x^2 + 2$. Find $3f(0) + f(-1)f(2)$.

Solution   The symbols $f(0)$, $f(-1)$, and $f(2)$ represent numbers. We must find these numbers, multiply the first by 3, and add the result to the product of the other two. By definition, $f(0)$ is the number that corresponds to 0, so we have $f(0) = 0^2 + 2 = 2$. Similarly, $f(-1) = (-1)^2 + 2 = 3$ and $f(2) = 2^2 + 2 = 6$. Therefore, $3f(0) + f(-1)f(2) = 3 \cdot 2 + 3 \cdot 6 = 24$.

The rule of correspondence in the example above is expressed by the equation $f(x) = x^2 + 2$. To find the number in the range that is associated with any particular number in the domain, we merely replace the letter x wherever it appears in the equation $f(x) = x^2 + 2$ with the given number. We followed this procedure in the solution of the example. Suppose, for instance, that a, $|a|$, and $(2 + h)$ represent numbers in the domain of f. Then the corresponding numbers in the range are

$$f(a) = a^2 + 2, \qquad f(|a|) = |a|^2 + 2 = a^2 + 2,$$

and

$$f(2 + h) = (2 + h)^2 + 2 = 6 + 4h + h^2.$$

It is sometimes helpful to think of the correspondence in this case as being defined by the equation $f(\ ) = (\ )^2 + 2$, where any symbol representing a number in the domain of f may be inserted in both parentheses.

Although the domain and range are essential parts of a function, they are frequently not mentioned explicitly, especially when the correspondence between these sets is given in terms of a mathematical formula. In the example above, for instance, there is no need to specify the range explicitly. If in the formula $x^2 + 2$ we substitute the numbers of the domain of f (the set of all real numbers), we obtain its range (the set of numbers greater than or equal to 2). Actually, when the domain of a function consists of all the real numbers for which the rule of correspondence yields real numbers, we usually don't specify the domain, either. We just say that the function is defined by the rule of correspondence. Thus our function f in Example 11-2 is defined by the equation $f(x) = x^2 + 2$. When we define a function by means of an equation, we read the domain from the equation. For example, the equation $g(x) = \sqrt{25 - x^2}$ defines a function g whose domain is understood to be the set of real numbers for which the expression $\sqrt{25 - x^2}$ yields real numbers, that is, the set of numbers between -5 and 5, inclusive. The range of this function turns out to be the set of numbers between 0 and 5, inclusive (use inequalities to verify this statement).

We use similar notational shortcuts when we write the definition of a function as a set of number pairs. For example, we might define the function g of the last paragraph by the equation $g = \{(x, \sqrt{25 - x^2})\}$, rather than the more formal equation $g = \{(x, y) \mid -5 \le x \le 5, \ y = \sqrt{25 - x^2}\}$. The equation $g = \{(x, \sqrt{25 - x^2})\}$ tells us that $g(x) = \sqrt{25 - x^2}$, and, since we are not told otherwise, we are to assume that the domain of g is $\{x \mid \sqrt{25 - x^2}$ is a real number$\}$. Similarly, we could define the function f of Example 11-1 by the equation $f = \{(x, |x|)\}$. This equation tells us that $f(x) = |x|$ for each real number x.

EXAMPLE 11-3   Find the domain D and the range R of the function $h = \left\{ \left(x, \dfrac{x}{|x|} \right) \right\}$.

Solution   We can replace x in the equation $h(x) = \dfrac{x}{|x|}$ with any number except 0, so D is the set of non-zero real numbers. If x is negative, $h(x) = \dfrac{x}{|x|} = \dfrac{x}{-x} = -1$, and if x is positive, $h(x) = \dfrac{x}{|x|} = \dfrac{x}{x} = 1$. Thus, there are only the two numbers -1 and 1 in the range of h; $R = \{-1, 1\}$.

PROBLEMS
11

1. Each of the following statements defines a function. In each case, give the function a letter as a name, determine its domain and range, and find a formula that pairs a number in the range with a number in the domain.

   (a) With each integer is paired its negative.
   (b) With each positive integer is paired its reciprocal.
   (c) With each positive number is paired the sum of its negative and its reciprocal.
   (d) With each number in the set $\{\sqrt{.05} < x < \sqrt{.07}\}$ is paired the first digit of its decimal expression.

2. What are the domains of the following functions, if we assume that the domain is the set of real numbers for which the defining formula yields real numbers?

   (a) $f(x) = x - x^{-1}$ \qquad\qquad (b) $f(x) = (x - x^{-1})^{-1}$
   (c) $f(x) = (2x + 5)^{-2}$ \qquad\quad (d) $F(x) = (x^2 + x - 2)^{2/3}$
   (e) $G(x) = (x^2 + x - 2)^{3/2}$ \quad (f) $H(x) = \sqrt{5/x - 1}$

3. Determine the range of f if its domain is the set $\{1, 2, 3\}$.

   (a) $f = \{(x, 3)\}$ \qquad (b) $f = \{(x, |3 - x|)\}$ \qquad (c) $f = \{(x, 2^{x-2})\}$
   (d) $f = \{(x, x + x^{-1})\}$ \quad (e) $f = \{(x, (-1)^x x)\}$ \quad (f) $f = \{(x, 8^{1/x})\}$

4. If $f(x) = 2x - 3$, evaluate each of the following.

   (a) $f(\sqrt{y})$ \qquad\qquad (b) $\sqrt{f(y)}$ \qquad\qquad (c) $f(z^{-1})$
   (d) $(f(z))^{-1}$ \qquad\quad (e) $f(f(x))$ \qquad\quad (f) $f(x + 1) - f(x)$

5. If $f(x) = x^2 + 3$, find the following numbers.

   (a) $f(3 - 2)$ \quad (b) $f(3) - f(2)$ \quad (c) $f(3 \cdot 2)$ \quad (d) $f(3) \cdot f(2)$
   (e) $f(\sqrt{3})$ \quad (f) $\sqrt{f(3)}$ \quad\quad (g) $f(f(1))$ \quad\quad (h) $f(1/f(1/2))$

6. Suppose that $h = \{(x, 1/x)\}$. Show that $h(a) = b$ if, and only if, $h(b) = a$.

7. Let $f(x) = \sqrt{x}$ and $g(x) = x^2$. Then let $F(x) = f(g(x))$ and $G(x) = g(f(x))$. Are F and G equal?

8. What is the relation between the function p that is defined by the equation $p(q) = 3q^2 + 2q^3$ and the function q that is defined by the equation $q(p) = 3p^2 + 2p^3$?

9. Which of the following sets of pairs of numbers are functions?

   (a) $\{(x, y) \mid |y| = |x|\}$ \quad (b) $\{(x, y) \mid y = \sqrt{x - 1}\} \cup \{(x, y) \mid y = \sqrt{1 - x}\}$
   (c) $\{(u, v) \mid u^2 = v^3\}$ \quad (d) $\{(r, s) \mid r^3 = s^2\}$

10. Let f be the function that assigns the area of an isosceles triangle inscribed in a circle of radius 2 to its altitude h. Find a formula for $f(h)$.

45

12
EXAMPLES OF FUNCTIONS

To help you become familiar with the concept and terminology of functions, we devote this section to a number of illustrations.

Illustration 12-1   The **identity function** assigns to each real number the number itself. In other words, it is the function $\{(x, x) \mid x \in R^1\}$, or in our abbreviated notation, the function $\{(x, x)\}$. If we name this function I, we have $I(x) = x$ for each real number x.

Illustration 12-2   The domain of the **absolute value function** $\{(x, |x|)\}$ is the set of all real numbers, and its range is the set of all non-negative real numbers. The following table gives a few examples of the correspondence between numbers in the domain and the range of the absolute value function.

x	$|x|$
1	1
$-\frac{1}{2}$	$\frac{1}{2}$
π	π
0	0

Illustration 12-3   The symbol $[\![x]\!]$ denotes the **greatest integer** n such that $n \leq x$. To find $[\![x]\!]$, therefore, we "round off" the decimal expression for x to the next lowest whole number. Thus, $[\![\pi]\!] = 3$, $[\![\sqrt{2}]\!] = 1$, $[\![3]\!] = 3$, $[\![-3]\!] = -3$, $[\![-\frac{1}{2}]\!] = -1$, and $[\![-\pi]\!] = -4$. The domain of the greatest integer function $\{(x, [\![x]\!])\}$ is the set R^1 of real numbers, and its range is the set of all integers. Almost every time you state your age you are using the greatest integer function. You say you are 18, for example, when you are actually $18\frac{1}{4}$, and $18 = [\![18\frac{1}{4}]\!]$. When a dealer prices a certain item at \$4.95, he hopes you will read the tag as $[\![4.95]\!] = 4$ dollars. If x is a given real number, then $[\![x]\!]$ can be determined graphically by locating x on the number scale and choosing the first integer at or to the left of this point.

Illustration 12-4   When a certain iron rod 1 inch in cross section and 10 inches long is subjected to a tension of x pounds it will stretch, say, s inches. Hooke's Law of Elasticity states that the relation between the numbers x and s is given by the equation $s = \dfrac{x}{3 \cdot 10^6}$. Thus a force of 30 pounds will stretch the rod 10^{-5} inches. Notice that the mathematical function determined by this last equation makes sense for any real value of x; that is, its domain is the set of all real numbers. The engineer knows, though, that this equation represents the behavior of the rod only for certain values of x. It certainly will not apply, for instance, if the force is so great that the rod breaks. Thus, for practical purposes, specifying the domain of a function may be important.

Illustration 12-5   Let us assume that the materials used in making a cylindrical tin can cost .012¢ per square inch for the sides and .021¢ per square inch for the top or bottom. Suppose we want to make a can with a capacity of 54π cubic inches. But the volume alone won't completely determine the dimensions of the can; it may be tall and thin, or short and squat. We may select any positive number r and make the radius of the base of the can r inches. Since the volume is already fixed, the choice of r determines the height of the can, and hence the dimensions of the can are completely determined. The cost of the can depends on its dimensions, so we see that the cost of the can depends on our choice of the base radius r. Let us denote the cost, in cents, of a can of base radius r inches by c and find a formula that expresses c in terms of r.

Clearly, $c =$ (area of sides) $\times$ .012 $+$ (area of ends) $\times$ .021. If the can is h inches high and has a base radius of r inches, then the area of the sides is $2\pi r h$ square inches and the area of the top (or bottom) is πr^2 square inches. Therefore,

$$(12\text{-}1) \qquad c = (2\pi r h)(.012) + (2\pi r^2)(.021).$$

We have now expressed the cost of the can in terms of the dimensions r and h, but we wanted to express c in terms of r alone. We therefore write a formula for h in terms of r. Since the volume of the can is 54π cubic inches, $\pi r^2 h = 54\pi$ and hence $h = 54/r^2$. Now replace h in Equation 12-1 by $54/r^2$ to obtain the equation

$$(12\text{-}2) \qquad c = \left(\frac{108\pi}{r}\right)(.012) + (2\pi r^2)(.021).$$

From this equation, we can calculate the cost of the can if the base radius is 1 inch, 2 inches, 3 inches, or 4 inches, as 4.21 cents, 2.57 cents, 2.54 cents, and 3.13 cents. A question often asked in a course in calculus is, "For what value of r is c least?" That is, "What are the dimensions of the cheapest can?" In the present case, the answer turns out to be a radius of 2.49 inches. The cost of such a can is 2.45 cents.

Illustration 12-6   Consider the function f whose domain is the set of positive integers and whose rule of correspondence is, "With each positive integer n, pair the digit in the nth decimal place of the unending decimal representation of the number π." Since $\pi = 3.14159\ldots$, we have

$f(1) = 1$, $f(2) = 4$, $f(3) = 1$, $f(4) = 5$, $f(5) = 9$, and so on. The rule of correspondence is easy to state, but there is no simple algebraic formula that defines the function for all positive integers (see Problem 12-11).

1. Find the following numbers if $f(x) = [\![x - |x|]\!]$.
   (a) $f(2)$ (b) $f(-2)$ (c) $f(\frac{3}{2})$ (d) $f(-\frac{3}{2})$
   (e) $f(-\sqrt{2})$ (f) $\sqrt{-f(-2)}$ (g) $f([\![-\frac{1}{2}]\!])$ (h) $|f(-\frac{1}{2})|$

2. Let f be the absolute value function ($f(x) = |x|$). Which of the following statements are true for all numbers in the domain of f?
   (a) $f(x^2) = (f(x))^2$ (b) $f(x + 2) = f(x) + f(2)$
   (c) $f(|x|) = |f(x)|$ (d) $f(2x) = f(2)f(x)$

3. If f denotes the identity function, which of the statements in Problem 12-2 are true?

4. If f denotes the greatest integer function, which of the statements in Problem 12-2 are true?

5. A rectangular region of 4000 square feet is to be fenced on three sides with fencing costing 50¢ per foot, and on the fourth side with fencing costing \$2 per foot. If x denotes the length of the fourth side and y denotes the corresponding cost of the fence in dollars, express y in terms of x. What is the domain of the resulting function? Calculate y when x belongs to the set $\{40, 50, 55, 60\}$.

6. A ship is steaming due north at 12 miles per hour. At midnight, a lighthouse is sighted 3 miles directly west of the ship. If the distance between the ship and the lighthouse t hours later is d miles, find a formula that expresses d in terms of t. Does the formula also give the distance between the ship and the lighthouse t hours *before* midnight?

7. An open box with a square base is to be made of wood costing 4¢ per square foot for the sides and 5¢ per square foot for the bottom. The volume of the box is to be 10 cubic feet. If the bottom of the box is to be x feet by x feet, express the cost, C cents, in terms of x. Find C when $x = 5$ and when $x = 6$.

8. An open box is to be made from a rectangular piece of tin 10 inches long and 8 inches wide by cutting pieces x inches square from each corner and bending up the sides. Express the volume, V cubic inches, of the box in terms of x. What is the domain of the associated function?

9. A spherical snowball with a 2-foot radius starts to melt at a rate that decreases its radius 1 inch per hour.
   (a) Find a formula for its volume, V cubic inches, t minutes later. (The volume of a spherical ball is $\frac{4}{3}\pi r^3$.)
   (b) Find a formula for the area of its surface, S square inches, t minutes later. (The area of a sphere is $4\pi r^2$.)
   (c) Find a formula for the volume if you are given that the rate of melting decreases the surface 20π square inches per hour.

10. What are the domains and ranges of the functions that are defined by the following equations?

(a) $f(x) = [[x]] + [[-x]]$ (b) $g(x) = [[x]][[-x]]$

11. Convince yourself that if f is the function of Illustration 12-6, then $f(n) = [[10^n \pi]] - 10[[10^{n-1} \pi]]$.

12. An airplane leaves home at noon, flying at 200 miles per hour to a city 400 miles away. Can you convince yourself that t hours after noon the plane is $200 + 100t - 100|t - 2|$ miles from home? What would the formula be if the plane flew at 100 miles per hour?

13

CARTESIAN COORDINATES AND THE DISTANCE FORMULA

The number scale lets us think of real numbers in terms of points of a line. Now we will introduce a coordinate system into the plane, which will, among other things, let us think of functions in terms of sets of points of the plane.

We begin by drawing two number scales with a common origin and meeting at right angles. We take the positive direction to be upward on one scale and to the right on the other (Fig. 13-1). These number scales are called

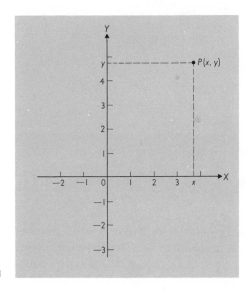

FIGURE 13-1

coordinate axes. The horizontal scale is the **X-axis,** and the vertical scale is the **Y-axis.** Let P be any point of the plane and construct lines through P and perpendicular to the axes. If x is the foot of the perpendicular to the X-axis and y is the corresponding point of the Y-axis, then the pair of numbers (x, y) is associated with P. Conversely, let (x, y) be any pair of numbers. Then construct a line perpendicular to the X-axis at the point x, and construct a line perpendicular to the Y-axis at the point y. The intersection of these two lines will determine exactly one point P to be associated with the pair of numbers (x, y). In summary, with each point of the plane we associate a pair of numbers (x, y) and with each pair of numbers we associate a point P of the plane. The numbers x and y are the **coordinates** of P. We usually ignore the logical distinction between a point and its coordinates and speak of the "point (x, y)," rather than "the point whose coordinates are (x, y)." Our coordinate system is known as a **cartesian coordinate system,** after the seventeenth-century French philosopher and mathematician René Descartes.

From now on, we will use the symbol R^2 to denote the set of all real number pairs. Thus, $R^2 = \{(x, y) \mid x \in R^1 \text{ and } y \in R^1\}$. The superscript 2 indicates that the set of number pairs has a two-dimensional geometric representation, the cartesian plane.

The points $(3, 1)$, $(-2, 3)$, $(-2, -1)$, and $(4, -2)$ are shown in Fig. 13-2. Note carefully that the *first* number of the number pair (a, b) is the

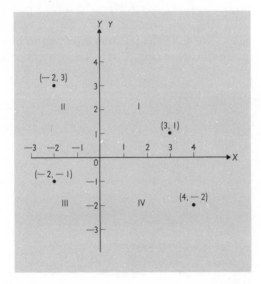

FIGURE 13-2

X-coordinate, and the *second* is the Y-coordinate. The two axes divide the plane into four regions or **quadrants.** These quadrants are numbered I, II, III, and IV, as shown in Fig. 13-2. For example, the point $(-2, 3)$ belongs to the second quadrant.

We have seen that the distance between two points a and b of the number scale is given by the expression $|a - b|$. We can also calculate the distance between two points of the plane.

EXAMPLE 13-1   The point P_1 has coordinates $(-2, -1)$ and the point P_2 has coordinates $(2, 2)$. Find the distance $\overline{P_1P_2}$ between these points.

Solution   The points P_1 and P_2 are plotted in Fig. 13-3. Let P_3 be the point $(2, -1)$. It is apparent that the points P_1, P_2, and P_3 are the vertices of a right triangle, P_3 being the vertex of the right angle. Since the points P_2 and P_3 lie in the same vertical line, you can easily see that the distance be-

50

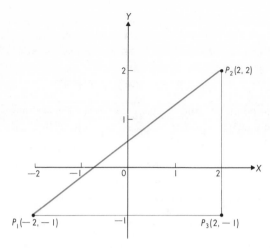

FIGURE 13-3

tween them is 3 units. Similarly, the distance $\overline{P_1P_3} = 4$. Now according to the Pythagorean Theorem,

$$\overline{P_1P_2}^2 = \overline{P_1P_3}^2 + \overline{P_3P_2}^2 = 16 + 9 = 25.$$

It follows that $\overline{P_1P_2} = 5$.

The concept of the distance between two points is so important that we shall develop a formula for it. The arguments we use are the same as the ones we used in Example 13-1.

Theorem 13-1 *Let P_1 and P_2 with coordinates (x_1, y_1) and (x_2, y_2) be any two points of the plane. Then the distance $\overline{P_1P_2}$ is given by the formula*

(13-1)
$$\overline{P_1P_2} = \sqrt{(x_2 - x_1)^2 + (y_2 - y_1)^2}.$$

Proof As in Example 13-1, the auxiliary point P_3 with coordinates (x_2, y_1) is introduced (see Fig. 13-4) in such a way that the points P_1, P_2, and P_3

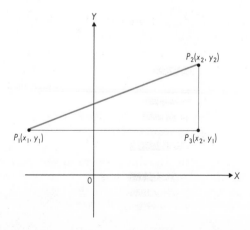

FIGURE 13-4

51

form a right triangle, with P_3 the vertex of the right angle. The legs of this triangle are $\overline{P_1P_3}$ and $\overline{P_2P_3}$ units long, while the length of the hypotenuse, $\overline{P_1P_2}$, is the distance we wish to find. Again, according to the Pythagorean Theorem,

(13-2)
$$\overline{P_1P_2}^2 = \overline{P_1P_3}^2 + \overline{P_2P_3}^2.$$

Since the points P_1 and P_3 have the same Y-coordinate, you can easily convince yourself that $\overline{P_1P_3} = |x_2 - x_1|$, and hence

$$\overline{P_1P_3}^2 = |x_2 - x_1|^2 = (x_2 - x_1)^2.$$

Similarly,

$$\overline{P_2P_3}^2 = |y_2 - y_1|^2 = (y_2 - y_1)^2.$$

We can therefore write Equation 13-2 as

$$\overline{P_1P_2}^2 = (x_2 - x_1)^2 + (y_2 - y_1)^2,$$

which is equivalent to Equation 13-1.

You will use the distance formula so often that you should memorize it.

EXAMPLE 13-2   Find the distance between the points $(-3, 2)$ and $(2, -3)$.

Solution   You should realize that it makes no difference which point is designated P_1. If the first point is labeled P_2 and the second P_1, the distance formula yields

$$\overline{P_1P_2} = \sqrt{(-3-2)^2 + (2+3)^2} = 5\sqrt{2}.$$

EXAMPLE 13-3   Find the distance between the point (x, y) and the origin.

Solution   Let P_1 be the point $(0, 0)$ and P_2 be the point (x, y), and apply the distance formula. The distance turns out to be

$$\sqrt{(x-0)^2 + (y-0)^2} = \sqrt{x^2 + y^2}.$$

If P_1, P_2, and P_3 are any three points, then

(13-3)
$$\overline{P_1P_3} \leq \overline{P_1P_2} + \overline{P_2P_3},$$

and the equality sign holds if, and only if, the three points are collinear; that is, if they lie in a line. This property of distance is called the **triangle inequality** for reasons that will be obvious to you from a glance at Fig. 13-5. The truth of this inequality is apparent from the geometry of the situation, but we can prove it by using Formula 13-1 and considerable algebra.

EXAMPLE 13-4   Find the point Q that is $\frac{3}{4}$ of the way from the point $P(-4, -1)$ to the point $R(12, 11)$ along the segment PR.

Solution   Figure 13-6 illustrates the situation; we are to find the numbers x and y, the coordinates of Q. To find these two numbers, we might write the two

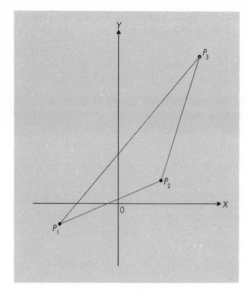

FIGURE 13-5

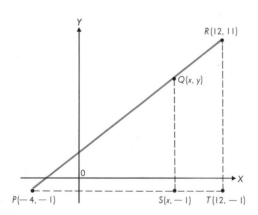

FIGURE 13-6

equations $\overline{PQ} = \frac{3}{4}\overline{PR}$ and $\overline{QR} = \frac{1}{4}\overline{PR}$ in terms of x and y and solve. Although this method will work, it is easier to use a little geometry. If we introduce the auxiliary points $S(x, -1)$ and $T(12, -1)$ shown in Fig. 13-6, we obtain the similar triangles PSQ and PTR. Therefore,

$$\frac{\overline{PS}}{\overline{PT}} = \frac{\overline{PQ}}{\overline{PR}} \quad \text{and} \quad \frac{\overline{QS}}{\overline{RT}} = \frac{\overline{PQ}}{\overline{PR}}.$$

From our figure, we see that $\overline{PS} = x + 4$, $\overline{PT} = 16$, $\overline{QS} = y + 1$, and $\overline{RT} = 12$, and it is a condition of the problem that $\dfrac{\overline{PQ}}{\overline{PR}} = \dfrac{3}{4}$. Hence,

$\dfrac{x+4}{16} = \dfrac{3}{4}$ and $\dfrac{y+1}{12} = \dfrac{3}{4}$, from which it follows that $x = 8$ and $y = 8$.

53

1. Sketch the following pairs of points in a cartesian coordinate system, and find the distance between them.

   (a) $(-3, 2)$ and $(1, 5)$ (b) $(-1, -3)$ and $(5, 5)$
   (c) $(1, 1)$ and $(-4, -11)$ (d) $(-\frac{3}{2}, 2)$ and $(0, 0)$
   (e) $(1, 2)$ and $(-3, -7)$ (f) $(1, \pi)$ and $(-1, 1/\pi)$

2. Find the distance between the given pairs of points.

   (a) $(a+b, a-b)$ and $(b-a, b+a)$ (b) $(0, 0)$ and $(\sqrt{a+b}, \sqrt{a-b})$
   (c) $(a^{1/2}, b^{1/2})$ and $(-a^{1/2}, -b^{1/2})$ (d) $(t, |t|)$ and $(-|t|, t)$
   (e) $(1+t^2, t^3)$ and $(1-t^2, t)$ (f) $(a^{3/2}, \frac{3}{2})$ and $(a^{-3/2}, -\frac{1}{2})$

3. Let O and P denote the points $(0, 0)$ and $(4, 2)$ of a cartesian coordinate system.

   (a) Use the distance formula to find the point Q of the X-axis such that the angle OPQ is a right angle.
   (b) Find the point R of the Y-axis such that the segment OP is the base of an isosceles triangle ORP.

4. Show that the point midway between the points (a, b) and (c, d) is
$$\left(\frac{a+c}{2}, \frac{b+d}{2}\right).$$

5. Two vertices of a square are $(2, 1)$ and $(2, 5)$. Find all other possible sets of vertices

6. Answer the preceding question if the points $(2, 1)$ and $(2, 5)$ are replaced by the points (a, b) and (a, c)

7. Find the area of the triangular region whose vertices are the points $(-1, 2)$, $(3, 2)$, and $(-4, 6)$.

8. The point $(x, 2)$ is 5 units from the point $(2, 6)$. Find the number x.

9. Find the point of the X-axis that is equidistant from the points $(0, -2)$ and $(6, 4)$.

10. Show that $(\sqrt{2}, 2)$ is a point of the circle whose center is the origin and which also contains the point $(-1, \sqrt{5})$.

11. Use the distance formula to determine whether or not the listed points are collinear. Use a figure to check your answer.

    (a) $(4, -3)$, $(-5, 4)$, and $(0, 0)$
    (b) $(2, -3)$, $(-4, 2)$, and $(-1, \frac{1}{3})$
    (c) $(3, 2)$, $(-\frac{4}{3}, \frac{5}{9})$, and $(6, 3)$

12. The set $A = \{(x, y) \mid xy < 0\}$ is the set of pairs of numbers with different signs. Thus $(-2, 3) \in A$, but $(-2, -3) \notin A$. The coordinates of a point of the cartesian plane belong to A if, and only if, the point belongs to Quadrant II or to Quadrant IV. Geometrically, therefore, $A = $ Quadrant II $\cup$ Quadrant IV. Shade the regions of a coordinate plane that represent the following sets of number pairs.

    (a) $\{(x, y) \mid x > 0 \text{ and } y < 0\}$
    (b) $\{(x, y) \mid x > 0 \text{ or } y < 0\}$
    (c) $\{(x, y) \mid 0 < x < 2 \text{ and } 2 < y < 3\}$

(d) $\{(x, y) \mid 0 < x < 2 \text{ or } 2 < y < 3\}$
(e) $\{(x, y) \mid -1 \le x < 0 \text{ and } y \ge 0\}$
(f) $\{(x, y) \mid -1 \le x < 0 \text{ or } y \ge 0\}$
(g) $\{(x, y) \mid |x| \ge 2 \text{ and } |y| \ge 1\}$
(h) $\{(x, y) \mid |x| \ge 2 \text{ or } |y| \ge 1\}$
(i) $\{(x, y) \mid |y| < 1\}$
(j) $\{(x, y) \mid y < 1\}$

14
GRAPHS
OF
FUNCTIONS

The functions with which we are dealing are sets of pairs of real numbers, that is, subsets of R^2. Graphically, we view subsets of R^2 as sets of points in a coordinate plane, so we can picture our functions as subsets of the plane. The point set that represents a function f is called the **graph** of f. Thus in Section 11 we considered the function $f = \{(1, 5), (2, 4), (3, 4), (4, 6)\}$, and its graph consists of the four points shown in Fig. 14-1.

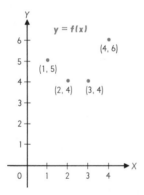

FIGURE 14-1

Most often, the domain of a function does not consist of merely a finite number of points, as in the example above. The function $\{(x, x^2)\}$, for example, has the set of all real numbers as its domain. For each number in the domain there is a number in the range, and hence a point of the graph. The graph of this function therefore consists of an infinite number of points, and it is clearly impossible to plot them all. In such a case, we plot what appears to be a representative sample of the points of the graph and sketch in other points "by inspection." The more we know about the function with which we are dealing, of course, the more nearly this graph we get "by inspection" is likely to approximate the actual graph.

EXAMPLE 14-1   Sketch the graph of the function $\{(x, x^2)\}$.

Solution   In the table for Fig. 14-2 we have listed some of the pairs of numbers that belong to our function. We plotted these points and then joined them to obtain the part of the graph shown in the figure.

EXAMPLE 14-2 Sketch the graph of the greatest integer function.

Solution The table for Fig. 14-3 lists the points we used to sketch the accompanying graph. Convince yourself that our graph is correct.

A point (x, y) belongs to the graph of a function f if, and only if, $y = f(x)$. Thus we can test whether or not a given point belongs to the graph of f by substituting its coordinates into the equation $y = f(x)$. If the coordinates satsify this equation, the point belongs to the graph; otherwise, not. For

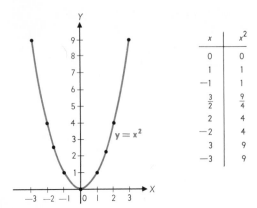

x	x^2
0	0
1	1
-1	1
$\frac{3}{2}$	$\frac{9}{4}$
2	4
-2	4
3	9
-3	9

FIGURE 14-2

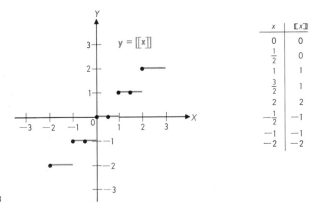

x	$[\![x]\!]$
0	0
$\frac{1}{2}$	0
1	1
$\frac{3}{2}$	1
2	2
$-\frac{1}{2}$	-1
-1	-1
-2	-2

FIGURE 14-3

example, the point $(\sqrt{2}, 1)$ belongs to the graph of the greatest integer function, since the equation $y = [\![x]\!]$ is satisfied when x is replaced by $\sqrt{2}$ and y by 1. The coordinates of the point $(-\sqrt{2}, -1)$ *do not* satisfy the equation, so the point does not belong to the graph of the greatest integer function. These remarks illustrate the concept of the **graph of an equation** or, more generally, the **graph of a relation.** For example, the graph of the equation $x^2 + y^2 = 4$ is the set of points—such as $(0, 2)$, $(\sqrt{2}, \sqrt{2})$, $(-1, \sqrt{3})$, and so on—whose coordinates satisfy the equation. The graph of the inequality

$x < y$ is the set of points—such as $(0, 2)$, $(-1, 5)$, $(-5, -3)$, and so on—whose coordinates satisfy the inequality. In this terminology, the graph of a *function f* is the same as the graph of the *equation* $y = f(x)$.

EXAMPLE 14-3   Sketch the graph of the equation $x^2 + y^2 = 4$.

Solution   We have already found several points of this graph, and we have plotted these points in Fig. 14-4. We could plot more points and then "fill in" the rest of the graph, but instead we will use some of our knowledge of the coordinate plane. Our given equation is equivalent to the equation $\sqrt{x^2 + y^2} = 2$. In Example 13-3, we found that the number $\sqrt{x^2 + y^2}$ is the distance between the point (x, y) and the origin. Thus, in words, our

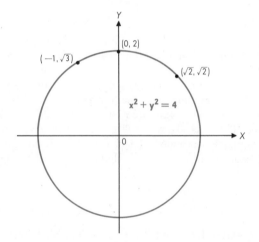

FIGURE 14-4

equation says, "The distance between the point (x, y) and the origin is 2." Clearly, the set of points that are two units from the origin is the circle whose center is the origin and whose radius is 2. This circle is therefore the graph of our given equation, and we have drawn it in Fig. 14-4.

Most of the graphs of equations that we deal with in mathematics are "one-dimensional" figures, such as the circle we just drew. In these cases, we say that the graph is a *curve*. Not all simple looking equations in x and y, however, have graphs that are simple curves.

EXAMPLE 14-4   Sketch the graph of the equation $y + |y| = x + |x|$.

Solution   An easy way to find the points of this graph is to check each quadrant separately. If $x \geq 0$ and $y \geq 0$, we have $|x| = x$ and $|y| = y$, and our equation becomes $2y = 2x$; that is, $y = x$. As you can see, those points with equal coordinates (such as $(1, 1)$, $(7, 7)$, and so on) lie in a line that bisects the first quadrant, as shown in Fig. 14-5. Each point of Quadrant II

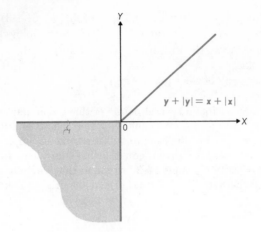

$$y + |y| = x + |x|$$

FIGURE 14-5

has the form $(-a, b)$, where a and b are positive, and for such a point the left-hand side of our equation becomes $b + b = 2b > 0$, while the right-hand side becomes $-a + a = 0$. Thus there are no points of our graph in Quadrant II. Similarly, there are no points of the graph in Quadrant IV. Every point of the form $(-a, -b)$, where $a > 0$ and $b > 0$, however, satisfies the equation; for in this case we have $-b + b = -a + a$. Thus, as shown in Fig. 14-5, the graph of the equation $y + |y| = x + |x|$ is the union of Quadrant III, the negative X-axis, the negative Y-axis, and the half-line that contains the origin and bisects Quadrant I.

PROBLEMS 14

1. Let $D = \{-2, -1, 0, 1, 2\}$ and $D_1 = D - \{0\}$. Sketch the graphs of the following functions.
   (a) $\{(x, y) \mid y = |x|, x \in D\}$ (b) $\{(x, y) \mid y = -2x + 1, x \in D\}$
   (c) $\{(x, y) \mid y = x^2 - 1, x \in D\}$ (d) $\{(x, y) \mid y = x^3, x \in D\}$
   (e) $\{(x, y) \mid y = 1/x, x \in D_1\}$ (f) $\{(x, y) \mid y = 1/x^2, x \in D_1\}$

2. Use your answers to the preceding problem to help you sketch the graphs of the functions defined by the following equations.
   (a) $f(x) = |x|$ (b) $g(x) = -2x + 1$ (c) $h(x) = x^2 - 1$
   (d) $F(x) = x^3$ (e) $G(x) = 1/x$ (f) $H(x) = 1/x^2$

3. Suppose that $f(x) = -\frac{1}{2}(x - 1)$ and $g(x) = -x - 1$. Sketch the graphs of the functions f and g in the same coordinate plane, and use your result to determine graphically a point that belongs to both functions. Check your answer algebraically.

4. Suppose (a, b) is a point of the graph of the function $\{(x, 3/x)\}$. Which of the following points also belong to the graph: (b, a), $(-a, b)$, $(-a, -b)$, $(a, -b)$, and $(-b, -a)$?

5. Sketch the graph of the function $\{(x, \sqrt{x}) \mid 0 \le x \le 4\}$ as well as you can, and use the resulting curve to determine the following numbers.
   (a) $\sqrt{1.5}$ (b) $\sqrt{\pi}$ (c) $2^{1/4}$ (d) $\frac{1}{2}\sqrt{5}$

6. Suppose that $g(x) = 2x^2 + 1$, and P and Q are points of the graph of g. How far apart are P and Q if the X-coordinate of P is 1 and the X-coordinate of Q is 2?

7. How are the graphs of the functions f, g, and h related if $f(x) = x^2$, $g(x) = x^2 + 3$, and $h(x) = x^2 - 2$?

8. (a) Which of the points $(1, 0)$, $(-1, 0)$, $(4, 4)$, and $(9, 17)$ belong to the graph of the equation $y = x^{3/2} - x$?

   (b) Which of the points $(4, 3)$, $(-4, 2\sqrt{2})$, $\left(2, 2 + \dfrac{1}{\sqrt[4]{2}}\right)$, $\left(\dfrac{1}{3}, \sqrt{\dfrac{1 + 3\sqrt{3}}{6}}\right)$

   belong to the graph of the equation $y = \left[\dfrac{2x^0}{x^2 + x^{1/2}}\right]^{-1/2}$?

9. Sketch the graphs of the following equations.

   (a) $|x| + |y| = 0$ (b) $|y| = |x|$ (c) $2y = |x| + x$

   (d) $y = -[\![-x]\!]$ (e) $y = x|x|$ (f) $y = [\![x]\!](1 - 2x + [\![x]\!])$

10. Sketch the graphs of the following relations.

   (a) $y > x$ (b) $|y| > |x|$ (c) $(|x| + x)y = 0$

   (d) $|x| + x < 2y$ (e) $[\![y]\!] = [\![x]\!]$ (f) $[\![x]\!][\![y]\!] = 0$

   (g) $|y| + y \geq |x| + x$ (h) $|y| + y < |x| + x$

15

VARIATION   Among the simplest types of functions are the ones that are determined by equations of the form $y = mx$, where m is a given number. Such functions have many scientific applications: distance = rate × time (constant rate), work = force × distance (constant force), force = mass × acceleration (constant mass), energy = mass × c^2 (c^2 is a constant), and many others.

Definition 15-1   *To say that **y is directly proportional to x** or that **y varies directly as x** means that there is a number m such that $y = mx$ for every related pair (x, y). The number m is called the **constant of proportionality**.*

EXAMPLE 15-1   Express y in terms of x if y is directly proportional to x, and $y = 2$ when $x = 3$.

Solution   We are told that y is directly proportional to x and therefore, from Definition 15-1, we know that $y = mx$, so the problem is to find the number m. If we substitute 3 for x and 2 for y in the equation $y = mx$, we find that $2 = 3m$, and hence $m = \frac{2}{3}$. It follows that y and x satisfy the equation $y = \frac{2}{3}x$.

If f is a given function, we cannot assume that $f(2x)$ and $2f(x)$ are the same number for every number x. Nor is it generally true that $f(2 + 3)$ and $f(2) + f(3)$ are the same. We will now prove that both these statements are true for a function f in which the equation that relates $f(x)$ and x has the form $f(x) = mx$.

Theorem 15-1   *If m is a given number, and $f = \{(x, mx)\}$, then $f(ab) = af(b)$ for any two numbers a and b.*

Proof   Since $f(x) = mx$, we have

$$f(ab) = m \cdot (ab) = a \cdot (mb) = af(b).$$

If we replace a in the equation $f(ab) = af(b)$ with 2, we obtain the equation $f(2b) = 2f(b)$. Thus if we double a number in the domain of f, we double the corresponding number in its range. Theorem 15-1 says that this statement applies to any multiple of a number in the domain of f. It is not hard to see (try it) that the converse of Theorem 15-1 is true, too. That is, *if $f(ab) = af(b)$, then there is a number m such that the numbers $f(x)$ and x are related by the equation $f(x) = mx$.*

Theorem 15-2   *If m is a given number, and $f = \{(x, mx)\}$, then $f(a + b) = f(a) + f(b)$ for any two numbers a and b.*

Proof   Since $f(x) = mx$, we have

$$f(a + b) = m \cdot (a + b) = ma + mb = f(a) + f(b).$$

Theorem 15-2 says that if we add two numbers in the domain of f and then find the corresponding number in the range, or if we find the corresponding numbers first and then add, we end up with the same number. It is a common mistake for students to act as if *all* functions have this property. Under stress, many students will equate $(a + b)^2$ and $a^2 + b^2$, or $\sqrt{a + b}$ and $\sqrt{a} + \sqrt{b}$. But you know that these numbers are not in general equal; that is, the square function and the square root function do not possess the property expressed by Theorem 15-2. Functions that are defined by an equation of the form $f(x) = mx$ are the only elementary functions that have this property; you will study mathematics for a long time before you meet another example of such a function.

Just because y increases when x increases does not mean that y is necessarily directly proportional to x. You may easily convince yourself that if $y = x^3 + x$, then y increases when x does; but y is not directly proportional to x. Nevertheless when a scientist knows only that two quantities are so related that one increases when the other does, he is likely to guess that one is directly proportional to the other, because direct variation is such a simple relationship. He then tests the accuracy of his guess by performing experiments. For example, we can stretch a steel rod by subjecting it to a pulling force. By increasing the force, we can increase the stretch. It is then natural to guess that the amount of stretch is proportional to the pulling force, and experiments show that this guess is, for practical purposes, correct (Hooke's Law).

EXAMPLE 15-2   The velocity of a body falling from rest is directly proportional to the time it falls. If a body attains a speed of 48 feet per second after falling

$1\frac{1}{2}$ seconds, how fast will it be falling 2 seconds later? Is the distance it falls directly proportional to the time of fall?

Solution If the body reaches a velocity of v feet per second at the end of t seconds, then, by our assumption of direct proportionality, $v = mt$. Since $v = 48$ when $t = \frac{3}{2}$, we have $48 = m \cdot \frac{3}{2}$ and hence $m = 32$. Therefore, $v = 32t$, and from this equation we find $v = 112$ feet per second when $t = \frac{7}{2}$.

Suppose the body falls s feet in t seconds. If distance were directly proportional to time, there would be a number r such that $s = rt$. From this equation it would follow that the body falls the same number, r, of feet during its second second of fall as it does during its first. But since it is going faster during its second second, it obviously falls farther, so an equation of the form $s = rt$ cannot be valid.

The terminology of direct variation is also associated with the function determined by the equation $y = mx^2$. In this case, y *is directly proportional to x^2*, or y *varies directly as x^2*. A common example of this type of variation is found in area formulas. The area A of a circle, for instance, is directly proportional to the square of its radius r. Thus, $A = mr^2$. If we measure the radius in certain units (inches, yards, and so forth) and the area in square units of the same type (square inches, square yards, and the like), then the constant of proportionality is, of course, π. We also use the language of variation if y is a multiple of powers of x other than 1 or 2. For example, y *is directly proportional to $x^{1/2}$ (or to the square root of x) if $y = mx^{1/2}$*.

There are a number of important practical situations in which an increase in one quantity leads to a *decrease* in the other. Thus, the centrifugal force on an automobile rounding a curve at a given speed can be expressed in terms of the radius of the curve. An increase in the radius of the curve produces a decrease in the centrifugal force. If air in a cylinder is kept at a constant temperature and is compressed by a piston, the air pressure decreases as the volume increases. Both of these examples, and many others, can be described in terms of *inverse variation*, which we now consider in its mathematical context.

Definition 15-2 *To say that **y is inversely proportional to x** or that **y varies inversely as x** means that there is a number k such that $y = k/x$ for every related pair (x, y).*

EXAMPLE 15-3 Express y in terms of x if y is inversely proportional to x and $y = 2$ when $x = 3$.

Solution Since we are told that y is inversely proportional to x, we know from Definition 15-2 that $y = k/x$. Our problem will be solved when we determine the number k. If we substitute 3 for x and 2 for y in the equation $y = k/x$, we find $2 = k/3$ and hence $k = 6$. It follows that the equation relating x and y is $y = 6/x$.

EXAMPLE 15-4 The electrical resistance of a wire of given length and material is inversely proportional to its cross-sectional area. The resistance in a circuit composed of wire that has a cross section of 82 square millimeters is 210 ohms. What would the resistance have been if the wire had a cross section of 70 square millimeters?

Solution   Let R denote the resistance in ohms corresponding to wire with a cross-sectional area of A square millimeters. Then $R = k/A$. Now $R = 210$ when $A = 82$, so $210 = k/82$, and hence $k = 82 \cdot 210$. The equation determining R is therefore $R = 82 \cdot 210/A$. It is now clear that $R = 82 \cdot 210/70 = 246$ when $A = 70$.

EXAMPLE 15-5   Sketch the graph of the equation $y = 2/x$.

Solution   The graph is shown in Fig. 15-1 and is an example of an *equilateral hyper-*

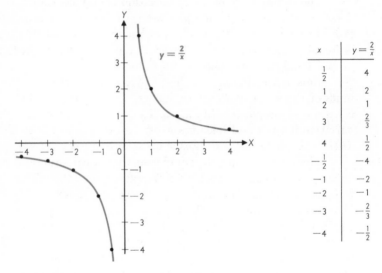

x	$y = \frac{2}{x}$
$\frac{1}{2}$	4
1	2
2	1
3	$\frac{2}{3}$
4	$\frac{1}{2}$
$-\frac{1}{2}$	-4
-1	-2
-2	-1
-3	$-\frac{2}{3}$
-4	$-\frac{1}{2}$

FIGURE 15-1

bola. Notice the appearance of the graph when x is near 0 (x cannot *be* 0, since $\frac{2}{0}$ is not defined). What can you say about $|y|$ when $|x|$ is very large?

The terminology of inverse variation is also used for relations of the type $y = k/x^p$. Thus if $y = k/x^2$, we say that y is inversely proportional to the square of x; if $y = k/\sqrt[3]{x}$, we say that y is inversely proportional to the cube root of x, and so on.

EXAMPLE 15-6   Sketch the graph of the function $\{(x, 1/x^2)\}$.

Solution   The graph is shown in Fig. 15-2.

62

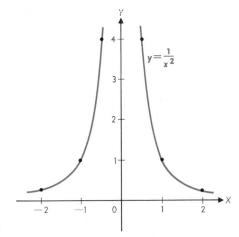

FIGURE 15-2

1. The point $(4, 9)$ belongs to the graph of the equation $y = f(x)$. Find the formula for $f(x)$ if

   (a) y is directly proportional to x.
   (b) y is inversely proportional to x.
   (c) y is directly proportional to x^2.
   (d) y is inversely proportional to $\sqrt{x}$.

2. Suppose that y is directly proportional to x, and let $y = f(x)$.
   (a) Show that $f(x_1)/f(x_2) = x_1/x_2$ for any two numbers x_1 and x_2 ($f(x_2) \neq 0$).
   (b) Does $f(1/a) = 1/f(a)$?   (c) Does $f(ab) = f(a)f(b)$?
   (d) Does $f(a+1) = f(a)+1$?   (e) Does $f(x^2) = (f(x))^2$?
   (f) Show that $\dfrac{f(a+h) - f(a)}{h} = f(1)$ (if $h \neq 0$).

3. Suppose that y is inversely proportional to x and that $y = f(x)$.
   (a) What is the domain of f?   (b) What is the range of f?
   (c) Show that $f(a)/f(b) = b/a$ (assume that $f(b) \neq 0$).
   (d) Does $f(1/a) = 1/f(a)$?   (e) Does $f(a+b) = f(a)+f(b)$?
   (f) Does $f(ab) = af(b)$?   (g) Show that $f(f(x)) = x$ if $f(x) \neq 0$.
   (h) Show that y varies inversely as $3x$.

4. If an automobile travels 1 mile at 30 miles per hour and 1 mile at 40 miles per hour, what is its average velocity?

5. If y varies inversely as x, show that $1/y$ varies directly as x.

6. If y varies directly as x^2, and $y = f(x)$, does $f(ab) = af(b)$?

7. The area of a sphere is directly proportional to the square of the radius. If a sphere of radius 3 inches has an area of 36π square inches, deduce the formula for the area of a sphere.

63

8. Suppose that u and v are inversely proportional to x and that $u = f(x)$ and $v = g(x)$. Further, suppose that $f(1) > g(1)$. Which of the following numbers is larger?

    (a) $f(-2)$ or $g(-2)$ (b) $f(-1)$ or $g(-1)$ (c) $f(\frac{1}{2})$ or $g(\frac{1}{2})$
    (d) $f(2)$ or $g(2)$

9. A body on the surface of the earth is acted on by a gravitational force that is directly proportional to its mass. If a mass of 1 gram encounters a force of 980 dynes, what is the formula relating gravitational force and mass?

10. The current in a certain circuit varies inversely as the resistance in the circuit. If the current is 10 amperes when the resistance is 24 ohms, what will the current be when the resistance is increased to 30 ohms? What values of the resistance will insure that the current is less than 1 ampere?

11. If u is directly proportional to x and v is directly proportional to x, what can you say about:

    (a) uv? (b) u/v? (c) $u + v$? (d) $u - v$?

12. The force of attraction of two oppositely charged bodies is inversely proportional to the square of the distance between them. If the two bodies are 10 centimeters apart, the attractive force is 40 dynes. How far apart should the bodies be moved to make the force of attraction 16 dynes? What is the attractive force when they are 1 meter apart?

13. Kepler's third law states that the time it takes a planet to revolve about the sun varies directly as the $\frac{3}{2}$ power of the maximum radius of its orbit. Using 93 million miles as the maximum radius of the Earth's orbit and 142 million miles as the maximum radius of Mars' orbit, how may days does it take Mars to make one revolution about the sun?

14. The number of oscillations per second of a pendulum varies inversely as the square root of its length. If a pendulum 8 feet long makes 1 oscillation every 3 seconds (that is, $\frac{1}{3}$ oscillation per second), what length pendulum will make 1 complete oscillation per second?

16
LINEAR FUNCTIONS

A function of the form $\{(x, mx + b)\}$, where m and b are given numbers, is called a **linear function.** Thus, the functions of the form $\{(x, mx)\}$ that we studied in the last section are simply linear functions for which $b = 0$. Linear functions are so named because the graph of a linear function is a line. The proof of this assertion may be easier to follow if we look at an example first.

EXAMPLE 16-1 Suppose $f = \{(x, 2x - 3)\}$. Choose any three points of the graph of f and show that they lie in a line.

Solution We are asked to choose *any* three points, so let us arbitrarily take $x = 0$, $x = 1$, and $x = 2$ to find the three points $P_1(0, -3)$, $P_2(1, -1)$, and $P_3(2, 1)$ of the graph of f (see Fig. 16-1). The points will lie in a line if the

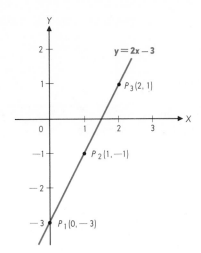

FIGURE 16-1

distance P_1P_3 is equal to the sum of the distances $\overline{P_1P_2}$ and $\overline{P_2P_3}$. (The shortest path between 2 points is a line; see Inequality 13-3.) Now

$$\overline{P_1P_3} = \sqrt{2^2 + 4^2} = \sqrt{20} = 2\sqrt{5},$$

$$\overline{P_1P_2} = \sqrt{1^2 + 2^2} = \sqrt{5},$$

$$\overline{P_2P_3} = \sqrt{1^2 + 2^2} = \sqrt{5},$$

and therefore $\overline{P_1P_3} = \overline{P_1P_2} + \overline{P_2P_3}$.

We shall now apply the same argument to the general linear function.

Theorem 16-1 *The graph of a linear function f is a line that is not parallel to the Y-axis.*

Proof The words "f is a linear function" mean that there are numbers m and b such that $f(x) = mx + b$. We have to show that any three points of the graph of f lie in a line, so suppose that x_1, x_2, and x_3 are three numbers such that $x_1 < x_2 < x_3$, and consider the corresponding three points of the graph of f. These three points are $P_1(x_1, mx_1 + b)$, $P_2(x_2, mx_2 + b)$, and $P_3(x_3, mx_3 + b)$ (Fig. 16-2), and they lie in a line if the distance $\overline{P_1P_3}$ is equal to the sum of the distances $\overline{P_1P_2}$ and $\overline{P_2P_3}$. Now

$$\overline{P_1P_3} = \sqrt{(x_3 - x_1)^2 + [(mx_3 + b) - (mx_1 + b)]^2}$$

$$= \sqrt{(x_3 - x_1)^2 + m^2(x_3 - x_1)^2}$$

$$= \sqrt{(x_3 - x_1)^2(1 + m^2)}$$

$$= \sqrt{(x_3 - x_1)^2}\sqrt{1 + m^2}.$$

Since $x_3 > x_1$, $x_3 - x_1$ is positive, and hence $\sqrt{(x_3 - x_1)^2} = x_3 - x_1$. We therefore have

$$\overline{P_1P_3} = (x_3 - x_1)\sqrt{1 + m^2}.$$

65

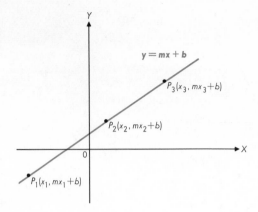

FIGURE 16-2

In exactly the same way we could calculate

$$\overline{P_1P_2} = (x_2 - x_1)\sqrt{1+m^2}$$

and

$$\overline{P_2P_3} = (x_3 - x_2)\sqrt{1+m^2}.$$

We can therefore write

$$\overline{P_1P_2} + \overline{P_2P_3} = (x_2 - x_1)\sqrt{1+m^2} + (x_3 - x_2)\sqrt{1+m^2}$$
$$= [(x_2 - x_1) + (x_3 - x_2)]\sqrt{1+m^2}$$
$$= (x_3 - x_1)\sqrt{1+m^2} = \overline{P_1P_3},$$

so our three points do lie in a line. This line is not vertical (parallel to the Y-axis), since the graph of a function cannot contain two points with the same X-coordinate.

EXAMPLE 16-2   Sketch the graph of the equation $y = 3x - 5$.

Solution   The graph in question is a line, and therefore it is only necessary to find two points of the graph in order to draw it. Two points are $(2, 1)$ and $(1, -2)$, and the graph is shown in Fig. 16-3.

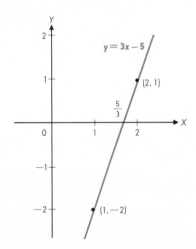

FIGURE 16-3

EXAMPLE 16-3   What can be said about a linear function f if $f(3) = f(1) + f(2)$?

Solution   Since f is a *linear* function, we know that there are numbers m and b such that $f(x) = mx + b$. Therefore, $f(3) = 3m + b$, $f(1) = m + b$, and $f(2) = 2m + b$. The equation $f(3) = f(1) + f(2)$ is valid only if $3m + b = (m + b) + (2m + b)$; that is, $3m + b = 3m + 2b$. But this equation says that $b = 2b$, and hence $b = 0$. Therefore, the function f is defined by an equation of the form $f(x) = mx$. We don't have enough information to determine the value of m. If m is any number, then $f(3) = f(1) + f(2)$. For example, the functions $\{(x, 5x)\}$, $\{(x, -12x)\}$, and $\{(x, 0)\}$ all satisfy this equation. So does the function $\{(x, [\![\frac{1}{5}x]\!])\}$, but it is not linear. Find another non-linear function that satisfies this equation.

Theorem 16-1 tells us that if m and b are given numbers, then the graph of the equation $y = mx + b$ is a line that is not parallel to the Y-axis. The two numbers m and b determine this line, so let us see what geometric significance these numbers have. Since the point $(0, b)$ is one of the points of our line, we see that the number b is the Y-coordinate of the point of intersection of the line and the Y-axis. The number b is called the **Y-intercept** of the line.

To interpret the number m geometrically, let us choose two points (x_1, y_1) and (x_2, y_2) of our line. The coordinates of these points must satisfy the equation of our line, and so

$$y_1 = mx_1 + b \quad \text{and} \quad y_2 = mx_2 + b.$$

When we solve these equations for m, we find

(16-1)
$$m = \frac{y_2 - y_1}{x_2 - x_1}.$$

Equation 16-1 tells us that the number m in the equation $y = mx + b$ is the ratio of the difference of the Y-coordinates to the difference of the X-coordinates of any two points of the line that is the graph of the equation. The number m is called the **slope** of the line. We also say that m is the slope of any line *segment* that contains the points (x_1, y_1) and (x_2, y_2). Equation 16-1 is meaningless if these points lie in a line (or line segment) that is parallel to the Y-axis. We shall simply say that such lines, or segments, *have no slope*. (This statement does not say that such lines have slope 0. What lines have slope 0?)

Figure 16-4 shows that if we move along our line from the point (x_1, y_1) to the point (x_2, y_2), then we move $y_2 - y_1$ units in the Y-direction and $x_2 - x_1$ units in the X-direction. Therefore, the quotient $m = \dfrac{y_2 - y_1}{x_2 - x_1}$ is the number of units moved in the Y-direction for each unit moved in the X-direction, or as we say, the *rate* at which the line rises (or falls).

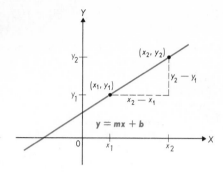

FIGURE 16-4

EXAMPLE 16-4   The equation $F = \frac{9}{5}C + 32$ relates the Fahrenheit and Celsius temperature scales. What do the numbers $\frac{9}{5}$ and 32 represent?

Solution   The number 32 tells us that when the Celsius thermometer reads 0, the Fahrenheit thermometer reads 32. The number $\frac{9}{5}$ is the slope of the line we would draw if we graphed our equation in an axis system in which Celsius temperatures are measured on the horizontal axis and Fahrenheit temperatures are measured on the vertical axis. Thus the number $\frac{9}{5}$ is the number of units of Fahrenheit temperature rise per unit of Celsius temperature rise. If a body's temperature increases 1°C., then it increases $\frac{9}{5}$°F. If a body's temperature increases $-10°$ (decreases 10°)C., then it increases $\frac{9}{5}(-10)° = -18°$F.

Now let us consider a simple, but important, theorem about lines.

Theorem 16-2   *Two lines with equations $y = m_1x + b_1$ and $y = m_2x + b_2$ are parallel if, and only if, $m_1 = m_2$.*

Proof   Non-parallel lines intersect in just one point; that is, there is exactly one pair (x, y) of numbers that satisfy both of the equations $y = m_1x + b_1$ and $y = m_2x + b_2$. By equating these two expressions for y, we see that our lines are *not* parallel provided that the equation $m_1x + b_1 = m_2x + b_2$, or equivalently,

$$(m_1 - m_2)x = b_2 - b_1,$$

has just one solution. The lines are parallel and distinct if this equation has no solutions, and the lines are coincident if it has more than one solution. Clearly, if $m_1 - m_2 \neq 0$, this equation has the single solution $x = \dfrac{b_2 - b_1}{m_1 - m_2}$. If $m_1 - m_2 = 0$, the equation has no solution if $b_2 \neq b_1$ (the lines are parallel and distinct), and every number satisfies it if $b_2 = b_1$ (the lines coincide).

The numbers (if any) in the domain of a function f that satisfy the equation $f(x) = 0$ are called the **zeros** of the function. To find the zeros of the linear function $\{(x, mx + b)\}$ we must solve the equation $mx + b = 0$, so such equations are called **linear equations.** If $m \neq 0$, our linear function has only one zero, $x = -b/m$. From the geometric viewpoint, a zero of a function is the X-coordinate of a point in which the graph of the function meets the X-axis.

68

EXAMPLE 16-5 Find the zeros of the function f if $f(x) = 3x - 5$.

Solution The solution of the equation $f(x) = 3x - 5 = 0$ is $x = \frac{5}{3}$, and this number is the only zero of f (see the graph of f in Fig. 16-3).

PROBLEMS 16

1. Find the formula for $f(x)$ if f is a linear function whose graph contains the following points.
   (a) $(0, 3)$ and $(2, 7)$
   (b) $(-2, 0)$ and $(2, 8)$
   (c) $(k^2, 2k)$ and $(2k^2, 3k)$
   (d) $(c+1, 0)$ and $(0, c-1)$

2. Sketch the graphs of the linear functions that are defined by the following equations. Find the slope of the graph, its Y-intercept, and the zero of f in each case.
   (a) $f(x) = -2x + 3$
   (b) $f(x) = 1 - 5x$
   (c) $f(x) = 2(x-1) + 5$
   (d) $f(x) = \frac{2}{3}x + \frac{3}{2}$

3. What can you say about the formula for $f(x)$ if f is linear and
   (a) $2f(x) = f(2x)$ for every number x?
   (b) $f(x+1) = f(x) + 1$ for every number x?
   (c) $f(2x+1) = f(2x) + 1$ for every number x?
   (d) $f(3) = 2$?
   (e) $f(3) = 2$ and $f(-1) = 1$?
   (f) $f(2) = -4$ and $f(-1) = -f(2)$?
   (g) $|f(x)| = f(|x|)$ for every number x?
   (h) $[\![f(x)]\!] = f([\![x]\!])$ for every number x?

4. Find $\{(x, 2x-3)\} \cap \{(x, \frac{1}{2}(x+3))\}$ both algebraically and graphically.

5. Sketch the set $\{(x, y) \mid y = 2 - x$ and $x < 2\} \cup \{(x, y) \mid y = 2x - 7$ and $x > 2\}$. Is your picture the graph of a function? a linear function?

6. Write an equation of a line that is parallel to the line $y = 2x + 3$ and contains the point $(1, -1)$.

7. Choose k so that the graphs of the following linear equations are parallel to the line whose equation is $y = 3x + 2$.
   (a) $y = (k/3)x + 5$
   (b) $y = (3/k)x - 12$
   (c) $y = k^{1/3}x - 2$
   (d) $y = -k^3x$

8. Sketch the set $\{(x, y) \mid y = 3x - 4\}$ in the coordinate plane and use your figure to solve the inequality $3x - 4 < 0$.

9. Sketch the sets $\{(x, y) \mid y = 2x - 7\}$ and $\{(x, y) \mid y = 4 - 3x\}$ in the coordinate plane, and use your sketch to solve the inequality $4 - 3x < 2x - 7$.

10. A projectile fired straight up attains a velocity of v meters per second after t seconds of flight, and the relation between the numbers v and t is a linear one. If the projectile is fired at a velocity of 100 meters per second and reaches a velocity of 80.4 meters per second after 2 seconds of flight, find the formula for v in terms of t. When does the projectile reach its highest point?

11. If the temperature h meters above the surface of the earth is $T°$ Celsius, then for practical purposes the associated function can be assumed to be linear. Suppose that the temperature on the surface is 20° Celsius and the temperature at 1000 meters is 15° Celsius. What is the temperature at 2000 meters?

69

12. If the total accumulation (principal + interest) of an investment of $50 at $5\frac{1}{2}\%$ simple interest at the end of t years is A dollars, then the associated function is linear. Find a formula for A and determine how long it will take to double the investment.

13. If the quantity of heat in calories required to change 1 gram of solid ice at $0°$ Celsius to water at $T°$ Celsius is denoted by Q, then the associated function is linear when $0 \le T \le 100$. If $Q = 90$ when $T = 10$, and $Q = 150$ when $T = 70$, what is the quantity of heat required to transform the ice into $0°$ water?

14. Suppose f and g are linear functions and that for some number a, $f(g(a)) = g(f(a))$. Show that $f(g(x)) = g(f(x))$ for every real number x. What can you conclude if the slope of the graph of f is 1?

17

INVERSE
FUNCTIONS

The graph of the equation $y = 3x - 5$ shown in Fig. 16-3 is the graph of a linear function f. As with any function, if we pick a number in the domain of f, a corresponding number of the range is thereby determined. For example, the number 2 of the domain determines the number 1 in the range. But this function f has a special feature; its rule of correspondence may also be read backward. Each number in the range of f determines a number in the domain. For example, if we pick the number -2 in the range, it determines the number 1 in the domain, because 1 is the only choice of x such that $f(x)$ is equal to -2. There are many functions for which the rule of correspondence can be reversed, but there are also many functions for which it cannot. For example, suppose that g is the square function, whose graph is shown in Fig. 14-2, and let us choose the number 4 in the range of g. This choice does not determine exactly one number x of the domain such that $g(x) = 4$, since both $g(2)$ and $g(-2)$ are equal to 4.

These examples illustrate the situation in general. We know that the equation $y = f(x)$ determines the number y (in the range of f) that corresponds to a given number x of the domain. Now we want to know if this equation also determines x when y is given. If for each given y in the range of f there is just one number x in the domain such that $y = f(x)$, then we have a rule that pairs with the numbers of the range of f the numbers of its domain. We therefore have all the ingredients of a function. Since we obtained this new function by interchanging the domain and range of f and reading its rule of correspondence backward, we naturally call this new function the *inverse* of f, in accordance with the following definition.

Definition 17-1   *If for each number y in the range of a function f there is exactly one number x in the domain of f such that $y = f(x)$, then f has an **inverse function** f^{-1}. The domain of f^{-1} is the range of f, and the range of f^{-1} is the domain of f. The number that corresponds to a given number y in the range of f is the number x that satisfies the equation $y = f(x)$. Thus,*

(17-1) $$x = f^{-1}(y) \quad \textit{if, and only if,} \quad y = f(x).$$

70

If we view our given function f as a collection of pairs of numbers, it is easy to construct the set of pairs that make up the inverse function f^{-1}. We simply interchange the members of each pair of f:

$$(17\text{-}2) \qquad f^{-1} = \{(y, x) \mid (x, y) \in f\}.$$

To say that for each number y in the range of f the equation $y = f(x)$ has *just one* solution x in the domain means that two different pairs in the set f^{-1} do not have the same first number, which is our criterion that a set of pairs of numbers should constitute a function.

Obviously, not every function has an inverse. For example, the equation $f(x) = 3$ defines a function whose domain is the entire set of real numbers and whose range is the set $\{3\}$. If we pick the number 5 in the domain, we obtain the number 3 in the range, but of course we can't go backward. The question of whether or not a given function has an inverse has a simple graphical answer. Figure 17-1 shows the graph of a function f whose domain

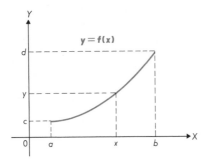

FIGURE 17-1

is the interval $\{a < x < b\}$ and whose range is the interval $\{c < y < d\}$. As with the graph of any function, a line that is parallel to the Y-axis and intersects the X-axis in a point of the domain of f intersects the graph of f in exactly one point. This statement is simply a graphical way of saying that a point x of its domain determines the pair (x, y) of f. For a function with an inverse, the number y also determines this pair. Thus, lines parallel to the X-axis and intersecting the Y-axis in a point of the range also intersect the graph of f in just one point. For a function without an inverse, there is at least one line that is parallel to the X-axis and that intersects the graph in more than one point. Horizontal lines do not intersect the graph in Fig. 17-1 more than once, so that function f has an inverse.

EXAMPLE 17-1   Let f be the linear function that is defined by the equation $f(x) = 3x - 5$. Find the equation that defines the inverse function f^{-1}.

Solution   It is clear from its graph (see Fig. 16-3) that f has an inverse. To find the value of f^{-1} at a given number y, we must solve the equation $y = f(x)$; that is, $y = 3x - 5$ for x. Thus, we have $x = \frac{1}{3}y + \frac{5}{3}$, or $f^{-1}(y) = \frac{1}{3}y + \frac{5}{3}$. Of course, the letter that we use to denote a number in the domain of the inverse function is of no importance whatsoever, so this last equation can be rewritten as $f^{-1}(u) = \frac{1}{3}u + \frac{5}{3}$, or $f^{-1}(s) = \frac{1}{3}s + \frac{5}{3}$, or even $f^{-1}(x) = \frac{1}{3}x + \frac{5}{3}$, and it will still define the same function f^{-1}.

EXAMPLE 17-2   Let $f(x) = x^n$, where n is a positive integer. Does f have an inverse?

Solution   The function f has an inverse if, and only if, the equation $y = x^n$ has exactly one solution x for each real number y of the range. And that depends on n. If n is an even integer, then (see Section 10) the equation $y = x^n$ has two solutions, $\sqrt[n]{y}$ and $-\sqrt[n]{y}$, for each positive number y. So the power function does not have an inverse in this case. But if n is odd, the equation $y = x^n$ *does* have exactly one solution for each real number y, namely, the number $\sqrt[n]{y}$. Thus, if n is odd, our power function f has an inverse, and $f^{-1}(x) = \sqrt[n]{x}$.

Suppose now that f is a function with an inverse. From Equation 17-2 we see that (u, v) is a point of the graph of f^{-1} if, and only if, (v, u) is a point of the graph of f. You can easily see that the line $y = x$ is the perpendicular bisector of the segment whose endpoints are (u, v) and (v, u). We say that these points are *symmetric with respect to this line*. (See Section 82 for a fuller discussion of symmetry.) If we should fold our graph along the line, the points would coincide. Therefore, *the graph of f^{-1} is obtained by reflecting the graph of f about the line that bisects the first and third quadrants*. Figure 17-2

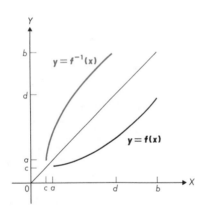

FIGURE 17-2

illustrates what we mean for the function whose graph appears in Fig. 17-1. It is clear that if we reflect the graph of f^{-1} about the line $y = x$, we will come back to the graph of f, which is a geometric way of saying that *the inverse of the function f^{-1} is the function f*.

EXAMPLE 17-3   The domain of a function G is the set of numbers $\{-1 \le x \le 0\}$, and $G(x) = \sqrt{1 - x^2}$. Find the graph of G^{-1}.

Solution   The graph of the equation $y = G(x)$ is the black curve in Fig. 17-3. We obtain the graph of G^{-1} (the colored curve) by reflecting the graph of G about the line $y = x$.

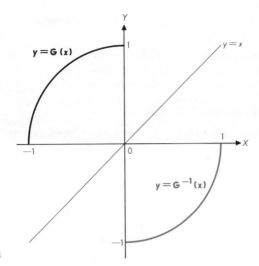

$y = G(x)$

$y = x$

$y = G^{-1}(x)$

FIGURE 17-3

PROBLEMS
17

1. Determine whether or not the function f has an inverse f^{-1}, and if it does, find the formula for $f^{-1}(x)$.

   (a) $f(x) = 2x + 3$ (b) $f(x) = 3 - 2x$ (c) $f(x) = x^6$
   (d) $f(x) = 32x^5$ (e) $f(x) = |x|$ (f) $f(x) = 3/x$
   (g) $f(x) = [\![x]\!]$ (h) $f(x) = x - [\![x]\!]$ (i) $f(x) = x|x|$

2. What is the domain of the function defined by the equation $f(x) = \dfrac{3 - x}{2 + x}$?
   What is its range? Find the formula for $f^{-1}(x)$. What is the domain and what is the range of f^{-1}?

3. Suppose that the function f has an inverse f^{-1}. If u is a number in the domain of f, and v is a number in its range, explain why $f^{-1}(f(u)) = u$ and $f(f^{-1}(v)) = v$.

4. Verify the equations in the preceding problem for the function defined by the given equation.

   (a) $f(x) = 3x - 6$ (b) $f(x) = 1/x^3$
   (c) $f(x) = 1/(x + 1)$ (d) $f(x) = x/(1 + x)$

5. Sketch the graph of f and f^{-1} if f is determined by the given equation.

   (a) $f(x) = x - 2$ (b) $f(x) = x^3$
   (c) $f(x) = 2 - x$ (d) $f(x) = \sqrt[3]{x}$

6. Find f^{-1} if $f(x) = 4(x - 2)|x - 2| + 1$. Sketch the graphs of f and f^{-1}.

7. In each of the following examples, show that $f = f^{-1}$.

   (a) $f(x) = x$ (b) $f(x) = b - x$
   (c) $f(x) = \dfrac{b}{x}$ (d) $f(x) = \dfrac{x + b}{cx - 1}$

8. Suppose f and g are functions that have inverses and are such that the range of f is a subset of the domain of g, and let $p(x) = g(f(x))$. Show that p has

73

an inverse, and find the equation that expresses values of p^{-1} in terms of values of f^{-1} and g^{-1}.

9. A function f is **increasing** in a set S if for each pair of distinct numbers a and b of S, we have $(f(b) - f(a))(b - a) > 0$. What does this inequality tell you about the graph of f? Convince yourself that if f is increasing in its domain, then f has an inverse. A function g is **decreasing** if $-g$ is increasing. Do decreasing functions have inverses?

10. Find an example of a function that is neither increasing nor decreasing and yet has an inverse.

REVIEW PROBLEMS, CHAPTER TWO

You should be able to answer the following questions without referring back to the text.

1. Let $f(x) = x + 1$, and define g by the equation $g(x) = f(x - 3)/f(x)$. Find $g(2)$. What is the domain of g?

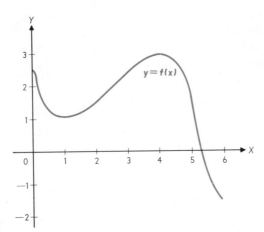

FIGURE II-1

2. The graph of a function f is shown in Fig. II-1.
   (a) What is the domain of f?   (b) What is the range of f?
   (c) Find $f(f(4))$.   (d) Find $f(2)$ and $f(\pi)$.
   (e) If $f(x) = 2$, what is x?   (f) If $f(f(x)) = 2$, what is x?

3. Describe the set $\{(x, y) \mid |x| < 1\} \cup \{(x, y) \mid |y| < 1\}$ as a subset of the cartesian plane.

4. The graph of a linear function contains the point $(0, 1)$ and is parallel to the line with equation $y = 4x - 1$. Find the formula for $f(x)$.

5. The perimeter of a Norman window (rectangle surmounted by a semi-circle) is 240 inches. The vertical side of the window is h inches. Find a formula that expresses the area A (in square inches) in terms of h.

6. Let $F = \{(x, y) \mid y = x^2 - 1 \text{ and } x \geq 0\}$. Find a formula for $F^{-1}(x)$. Sketch the graphs of F and F^{-1}. What is the domain of F^{-1}?

7. Describe the set $\{(x, y) \mid y = 3x + 1\} \cap \{(x, y) \mid |y| < 2\}$ as a subset of the cartesian plane.

8. If u is directly proportional to x and v is inversely proportional to x, what can you say about the product uv?

9. If y is directly proportional to x, is x directly proportional to y? If z is inversely proportional to x, is x inversely proportional to z?

10. What are the zeros of the function $\{(x, |x - 1| - 2)\}$?

11. Define the function F as follows: With the page number of each of the first 10 pages of this book, associate the number of letters in the first complete word in the text on the page. What is the domain of this function? What is its range? Plot its graph.

12. If x is a real number and if P denotes the corresponding point of the graph of the equation $y = x^2$, find an expression for the distance d between the origin and the point P.

13. Denote by $f(i)$ the length in feet of an object that is i inches long. What is the relation between the numbers i and $f(i)$?

14. Three vertices of a parallelogram are $(-1, -1)$, $(1, 4)$, and $(2, 1)$. Find all possibilities for the fourth vertex.

15. Suppose that f is a linear function with an inverse. Use the relationship between the graphs of f and f^{-1} to give a geometric argument that f^{-1} is a linear function.

MISCELLANEOUS PROBLEMS, CHAPTER TWO

These exercises are designed to test your ability to apply your understanding of functions and graphs to somewhat more difficult problems.

1. Sketch the graphs of the following relations.
   (a) $|3x - 4y| = 5$ (b) $[\![3x - 4y]\!] = 5$
   (c) $|3x - 4y| \le 5$ (d) $[\![3x - 4y]\!] \le 5$

2. If a number of electrical resistances are connected in parallel, then the reciprocal of the resistance of the entire circuit is the sum of the reciprocals of the individual resistances. If resistances of 5, 10, and x ohms are connected in parallel, find a formula for the resulting resistance R. Sketch a graph of the equation for $x > 0$. What can you say about R if x is a very small number? A very large number?

3. At a temperature of $0°$, the length of an iron bar is L_0 inches, and at a temperature of $t°$, its length is L inches. If the ratio of the change in length of the bar to its original length is directly proportional to the temperature t, what is the formula for L?

4. Let F be the function that is defined by the equation $F(x) = [\![x]\!] + x$. Does F have an inverse? If so, define F^{-1} by giving its domain and its rule of correspondence.

5. Show that for any two real numbers a and b, $[\![a]\!] + [\![b]\!] \le [\![a+b]\!]$.

6. Let $f(x)$ denote the distance between x and the nearest even integer. Thus, $f(17) = 1$, $f(31\frac{1}{4}) = \frac{3}{4}$, and so on. Sketch the graph of f. See if you can find the formula for $f(x)$.

7. Let f and g be linear functions, and define new functions p and q by means of the equations $p(x) = f(g(x))$ and $q(x) = g(f(x))$. Show that the graphs of p and q are parallel lines.

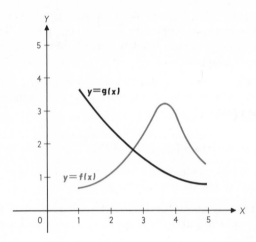

FIGURE II-2

8. The graphs of two functions f and g are shown in Fig. II-2.
   (a) Solve the equations $g(x) = f(2)$ and $f(x) = g(2)$ graphically.
   (b) What are the domain and range of the function p that is defined by the equation $p(x) = g(f(x))$?
   (c) Sketch the graph of the equation $y = [\![g(x)]\!]$.

9. Sketch the graph of the function $\{(x, y) \mid y = 3[\![x]\!] - 5\}$ and find its zeros.

10. Suppose the graphs of the equations $y = m_1x + b_1$ and $y = m_2x + b_2$ are lines that intersect in the point (x_0, y_0). Call this point P_0, and let P_1 be the point of the first line and P_2 be the point of the second line whose X-coordinate is $x_0 + 1$. Show that the triangle $P_1P_0P_2$ is a right triangle whose vertex is P_0 if, and only if, $m_1m_2 = -1$. Thus, the two lines are perpendicular if, and only if, the product of their slopes is -1.

11. If f and g are functions with the same domain D, we define their sum and product by means of the equations $f + g = \{(x, y) \mid x \in D, y = f(x) + g(x)\}$ and $fg = \{(x, y) \mid x \in D, y = f(x)g(x)\}$. Which of Equations 1-1 to 1-6 are valid for these operations?

12. If the functions f and g are related by the inclusion $f \subseteq g$, we say that g is an **extension** of f (or that f is a **restriction** of g). Show that $f \subseteq g$ for the following function pairs.
    (a) $f(x) = x/x$, $g(x) = 1$ (b) $f(x) = \sqrt{x}$, $g(x) = \sqrt[4]{x^2}$
    (c) $f(x) = 2(\sqrt{x})^2$, $g(x) = x + |x|$ (d) $f(x) = 2/(x + |x|)$, $g(x) = 1/x$

76

Exponential
and
Logarithmic Functions

three

Under favorable conditions a single cell of the bacterium *Escherichia coli* will divide into two about once every 20 minutes. If a plate contains 3000 of the organisms at a certain time, we can expect to find $2 \cdot 3000$ organisms if we inspect it 20 minutes later, $2 \cdot (2 \cdot 3000) = 3000 \cdot 2^2$ organisms 40 minutes later, $2 \cdot (3000 \cdot 2^2) = 3000 \cdot 2^3$ organisms 60 minutes later, and so on. Thus, if there are t 20-minute periods after our initial observation, then there are $3000 \cdot 2^t$ bacteria on our plate. The number of bacteria is given by the number originally present times a factor, 2^t, in which t appears as an exponent. Equations of the form $N = kb^t$ describe many kinds of growth—dollars invested at compound interest, charges on an electrical condenser, grams of a decaying radioactive substance, and so forth.

The function f defined by the equation $f(t) = 2^t$ is an example of an *exponential function*. Such functions, and the closely related *logarithmic functions*, form the subject of this chapter.

77

When we say that the equation $f(t) = 2^t$ defines a function f, we take it for granted that the domain of f consists of those numbers that can be used as exponents of 2. Thus, the domain of f includes at least the rational numbers, for we saw in Chapter One that $f(4) = 2^4 = 16$, $f(-3) = 2^{-3} = \frac{1}{8}$, $f(\frac{1}{2}) = 2^{1/2} = \sqrt{2}$, $f(0) = 2^0 = 1$, $f(1.41) = 2^{1.41} = (\sqrt[100]{2})^{141}$, and so on. But irrational numbers, such as π and $\sqrt{2}$, are not to be considered in the domain of f until we can reach an agreement about the meaning of such symbols as 2^π and $2^{\sqrt{2}}$.

A detailed discussion of irrational exponents belongs to a course in calculus. There you will learn a rule that assigns to each positive number b and real number x (rational or irrational) a positive number b^x. This rule is an extension of our definition of rational powers of b. In other words, if p/q is a rational number and you compute $b^{p/q}$ by the method you learn in calculus, you end up with the number $\sqrt[q]{b^p}$. Furthermore, the laws of exponents remain valid; that is, the equations $(b^x)^y = b^{xy}$ and $b^{x+y} = b^x b^y$ are true for every pair x, y of real numbers.

The definition of exponents is such that the graph of an exponential function is a "continuous curve." Thus, to draw the graph of the equation $y = 2^x$ in Fig. 18-1, we plotted the listed points, each of which is obtained by replacing x with a rational number. There is just one "natural way" to join these plotted points and so obtain our graph. After we have drawn the graph, we can use it to approximate irrational powers of 2. For example, Fig. 18-1 shows us that $2^{\sqrt{3}}$ is approximately equal to 3.3.

Let us, then, accept the fact that to each positive number b and real number x, a positive number b^x has been assigned. This assignment defines the **exponential function with base b.** Its domain is the set of all real numbers, R^1, its range is the set of positive real numbers, and it is denoted by **exp$_b$.** Thus,

(18-1) $$\mathbf{exp}_b\ (\boldsymbol{x}) = \boldsymbol{b}^{\boldsymbol{x}} \qquad (b > 0).$$

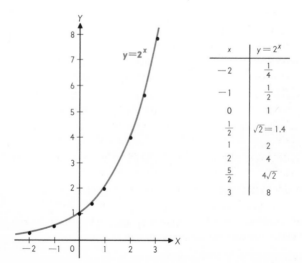

FIGURE 18-1

If $b > 1$, the graph of $\exp_b$ looks very much like the graph in Fig. 18-1. Notice that for $x < 0$, $b^x < 1$, and for $x > 0$, $b^x > 1$. If $r < s$, then $b^r < b^s$.

If $b < 1$, the graph of $\exp_b$ will have a different appearance. A typical example is the case in which $b = \frac{1}{3}$. The graph of the equation $y = (\frac{1}{3})^x$ is shown in Fig. 18-2. For $x < 0$, $\exp_{1/3}(x) > 1$; $\exp_{1/3}(0) = 1$; and for $x > 0$, $\exp_{1/3}(x) < 1$. Furthermore, whenever $r < s$, then $\exp_{1/3}(r) > \exp_{1/3}(s)$. Since $(\frac{1}{3})^x = 3^{-x}$, we can also say that the graph in Fig. 18-2 is typical of the graph of an equation of the form $y = b^{-x}$, where $b > 1$.

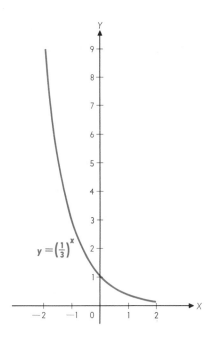

$$y = \left(\tfrac{1}{3}\right)^x$$

FIGURE 18-2

In function notation, the law of exponents $b^{r+s} = b^r \cdot b^s$ is written as $\exp_b(r + s) = \exp_b(r) \cdot \exp_b(s)$. Thus, if we *add* two numbers r and s and then compute the corresponding function value $\exp_b(r + s)$, we get the same number we obtain by first computing the values $\exp_b(r)$ and $\exp_b(s)$ and then *multiplying*. We state this result as a theorem.

Theorem 18-1   *For any $b > 0$,*

$$\mathbf{exp}_b(r + s) = \mathbf{exp}_b(r) \cdot \mathbf{exp}_b(s)$$

for any two numbers r and s.

This theorem is closely related to Theorem 15-2. The only elementary functions that have the property stated in Theorem 18-1 are the exponential functions; you won't encounter another such function unless you study mathematics for many more years.

79

EXAMPLE 18-1   The graph of $\exp_b$ contains the point $(2, 9)$. What is b?

Solution   We know that $\exp_b (x) = b^x$, where b is a positive number that we are to determine. We are given that $\exp_b (2) = 9$, and hence $b^2 = 9$. It follows that $b = 3$.

EXAMPLE 18-2   A certain radioactive salt decays at such a rate that at the end of a year there is only $\frac{500}{501}$ times as much as there was at the beginning of the year. If there are 50 milligrams of the salt at a certain time, how much will be left t years later?

Solution   Let $m(t)$ be the number of milligrams at the end of t years. Then $m(0) = 50$, $m(1) = (\frac{500}{501})50$, $m(2) = (\frac{500}{501})(\frac{500}{501})50 = (\frac{500}{501})^2 50$, and so on. It is then clear that $m(t) = (\frac{500}{501})^t 50$. Thus, $m(t) = 50b^t$, with $b = \frac{500}{501}$.

PROBLEMS 18

1. Find each of the following numbers.
   (a) $\exp_2 (3)$ (b) $\exp_3 (2)$ (c) $\exp_b (0)$ (d) $\exp_b (1)$

2. Discuss the exponential function $\exp_1$. Why is b restricted to positive numbers in our definition of $\exp_b$?

3. Find b if the graph of $\exp_b$ contains the given point.
   (a) $(1, \pi)$ (b) $(-2, 1/4)$ (c) $(2, 8)$ (d) $(2, 1/4)$
   (e) $(0, 1)$ (f) $(10, .00001)$ (g) $(.5, .4\sqrt{5})$ (h) $(-.5, .6)$

4. Use the graph of the function $\exp_2$ in Fig. 18-1 to help you answer the following questions.
   (a) Find $2^{\sqrt{2}}$.   (b) Find $\sqrt[4]{2}$.
   (c) Solve the equation $2^x = 5$.   (d) Solve the equation $2^x = \pi$.
   (e) Sketch the graph of the equation $y = -2^x$.
   (f) Sketch the graph of the equation $y = 2^{-x}$.
   (g) Sketch the graph of the equation $y = 3 \cdot 2^x$.
   (h) Sketch the graph of the equation $y = (\sqrt{2})^x$.
   (i) Sketch the graph of the equation $y = [\![2^x]\!]$.

5. (a) Let p be a real number. Show that $(\exp_b (x))^p = \exp_b (px)$ for each number x.
   (b) Show that $\exp_b (x - y) = \exp_b (x)/\exp_b (y)$ for any two numbers x and y.

6. How many real solutions do each of the following equations have?
   (a) $3^x = 4$ (b) $3^x = -4$ (c) $3^{-x} = -4$ (d) $3^{-x} = 4$

7. Suppose that $b > 1$, and define the functions s_b and c_b by the equations

$$s_b(x) = \tfrac{1}{2}(\exp_b (x) - \exp_b (-x))$$

and

$$c_b(x) = \tfrac{1}{2}(\exp_b (x) + \exp_b (x)).$$

   (a) Prove that $s_b(-x) = -s_b(x)$ and $c_b(-x) = c_b(x)$.

(b) Use the preceding equations and Fig. 18-1 to help you sketch the graphs of the functions s_b and c_b.

(c) Show that

$$s_b(x + y) = s_b(x)c_b(y) + c_b(x)s_b(y)$$

and

$$c_b(x + y) = c_b(x)c_b(y) + s_b(x)s_b(y)$$

for any two numbers x and y.

(d) Define the function t_b by the equation $t_b(x) = s_b(x)/c_b(x)$. What is the domain of t_b? the range?

8. The gas in a balloon is escaping at such a rate that at the end of any minute the amount is .6 of what it was at the beginning of the minute. If the balloon contains 3000 cubic feet of gas at 1 P.M., find a formula for the amount of gas in the balloon t minutes after 1 P.M. How much gas is in the balloon at 1:05 P.M.?

9. Radium A, with an atomic weight of 214, undergoes radioactive decay at such a rate that at the end of any minute there is only .8 as much as there was at the beginning of the minute. Write a formula giving the amount that is left from N_0 grams after t minutes of disintegration. Illustrate the relation between the amount of Radium A present and the time graphically. The **half-life period** of a radioactive substance is the time required for one-half of the active material present at any time to decay. Use your graph to approximate the half-life period of Radium A.

10. Let f be the function whose graph is the graph of $\exp_2$ shifted one unit to the right. How is the Y-coordinate of a point P of the graph of f related to the Y-coordinate of the point of the graph of $\exp_2$ with the same X-coordinate as P?

19

**DEFINITION
OF THE
LOGARITHM**

If, for a given positive number y, the equation $2^x = y$ had two different solutions, r and s, one of them would be the larger, say $s > r$. But then, as we saw in the preceding section, we would have $2^s > 2^r$, contradicting the fact that both of the numbers 2^s and 2^r are equal to y. In short, the equation $2^x = y$ cannot have two solutions. But this equation can also be written as $y = \exp_2(x)$, so the statement that it has a unique solution is simply a way of saying that the function $\exp_2$ has an inverse. This inverse function is called the *logarithmic function with base* 2, and we denote it by log_2. The domain of $\exp_2$ is R^1, and its range is the set of positive numbers. For the inverse log_2, the roles of these sets are reversed. The set of positive numbers is the domain of log_2, and R^1 is its range.

To find a value of log_2, we must solve the equation $2^x = y$ for x, something that is usually easier said than done. Sometimes we can do it by inspection. For example, if $2^x = 8$, then $x = 3$, and it follows that $log_2(8) = 3$. But we cannot solve the equation $2^x = 10$ so easily. To find that $x = 3.322$ is an approximate solution, we must turn to tables that we will study in Section 22.

Even though it is not easy to solve an exponential equation of the form $N = b^x$ for x when N and b are given, it is not hard to convince ourselves that the equation does have a solution. Figures 18-1 and 18-2 show the form of the graph of the equation $y = b^x$ when b is greater than 1 and when b is between 0 and 1. To solve the equation $N = b^x$, we construct a line perpendicular to the Y-axis at the point N, and the solution of the equation is the X-coordinate of the point in which this line intersects our graph. From the figures it is clear that if N is positive, there is just one point of intersection. Thus, if $b \neq 1$, the function $\exp_b$ has an inverse. This inverse function is called the **logarithmic function with base b,** and it is denoted by **$\log_b$.** Its domain is the set of positive numbers, and its range is R^1. The function value $\log_b (N)$ is usually written without parentheses as **$\log_b N$** and is called the **logarithm of N to the base b.** In symbols, we have

$$\log_b N = x \qquad \text{if, and only if, } N = \exp_b (x).$$

So that we can refer to it later, let us give this definition a number.

Definition 19-1   *The equations*

$$\boldsymbol{\log_b N = x} \qquad and \qquad \boldsymbol{b^x = N}$$

are equivalent.

Many problems in logarithms involve nothing more than shifting from one of these equations to the other, so you should always keep both of them in mind. Also remember that we always assume that *a logarithmic base b is a positive number different from* 1. Finally, since $b^x > 0$ for every real number x, the equation $b^x = N$ does not have a solution when $N \leq 0$. *Only logarithms of positive numbers are defined.*

EXAMPLE 19-1   Find $\log_2 8$ and $\log_2 10$.

Solution   According to Definition 19-1, these numbers are the solutions of the equations $2^x = 8$ and $2^x = 10$. Earlier, we stated that these solutions are 3 and approximately 3.322; hence $\log_2 8 = 3$ and $\log_2 10 = 3.322$ (to three decimal places).

EXAMPLE 19-2   If $\log_3 N = 2$, find N.

Solution   According to Definition 19-1, the equation $\log_3 N = 2$ is equivalent to the equation $3^2 = N$. Hence $N = 9$.

EXAMPLE 19-3   Find $\log_9 (\frac{1}{27})$.

Solution   The desired number satisfies the equation $9^x = \frac{1}{27}$. Since $9 = 3^2$ and $\frac{1}{27} = 3^{-3}$, this last equation may be written $3^{2x} = 3^{-3}$. Therefore, $2x = -3$, so $x = -\frac{3}{2}$.

EXAMPLE 19-4   Find the base b for which $\log_b 16 = \log_6 36$.

Solution   According to Definition 19-1, the number $\log_6 36$ is the solution of the equation $6^x = 36$, and hence $\log_6 36 = 2$. The given equation is therefore $\log_b 16 = 2$. But this equation is equivalent to the equation $b^2 = 16$, and hence $b = 4$ or $b = -4$. Since only a positive number can be a base, $b = 4$.

Since we have defined the number $\log_b N$ as the solution of the equation $b^x = N$, we can replace x by $\log_b N$ in this equation to obtain

(19-1) $$b^{\log_b N} = N.$$

This equation can be written in functional notation as

$$\exp_b (\log_b (N)) = N,$$

which is a special case of the equation $f(f^{-1}(x)) = x$ that we mentioned in Problem 17-3.

EXAMPLE 19-5   If a is a positive number and x is any number, express a and a^x as powers of 10.

Solution   According to Equation 19-1, we have $a = 10^{\log_{10} a}$, and therefore $a^x = 10^{(\log_{10} a)x}$.

It should not be difficult for you to use the definition of a logarithm to show that the following equation is true for any base b:

(19-2) $$\log_b b = 1.$$

As we pointed out earlier, the number 3.322 is only an approximation of the solution of the equation $2^x = 10$, and hence the equation $\log_2 10 = 3.322$ is not strictly correct. When we write it we mean, of course, that 3.322 is the best three-place approximation to $\log_2 10$; that is,

$$3.3215 \leq \log_2 10 \leq 3.3225.$$

We will use many similar approximations in this chapter.

PROBLEMS 19

1. Write the following equations in logarithmic form.

    (a) $3^4 = 81$ (b) $10^0 = 1$ (c) $M^k = 5$ (d) $5^k = M$

    (e) $5^M = k$ (f) $M^5 = k$ (g) $k^M = 5$ (h) $k^5 = M$

2. Solve the following equations.

    (a) $\log_5 x = 3$ (b) $\log_8 x = \frac{2}{3}$ (c) $\log_5 25 = x$

    (d) $\log_9 3 = x$ (e) $\log_x 16 = 4$ (f) $\log_x 10 = 3$

(g) $\log_2 \frac{1}{8} = x$ (h) $\log_5 x = -3$ (i) $\log_{25} |x| = .5$

(j) $\log_5 |x| = -2$ (k) $|\log_5 x| = 2$ (l) $\log_x 4 = \log_{2x} 2$

3. Solve the following equations.

(a) $\log_b x = 0$ (b) $\log_x 1 = 0$ (c) $\log_x x = 1$

(d) $x^{\log_x 5} = 5$ (e) $\exp_x (\log_x x) = 5$ (f) $\exp_5 (\log_x 5) = 5$

(g) $\exp_4 (\log_4 (\exp_2 (x))) = 4$

4. Use the graph of the function $\exp_2$ shown in Fig. 18-1 to estimate the following numbers.

(a) $\log_2 3$ (b) $\log_2 6$ (c) $\log_2 .75$ (d) $\log_4 6$

5. Why isn't 1 a suitable number for a logarithmic base?

6. If $\log_b x = 2$, what is $\log_{1/b} x$?

7. If $\log_b x = 2$, what is $\log_b (1/x)$?

8. Prove that $\log_{1/b} x = \log_b (1/x)$.

9. Solve for x.

(a) $2^{\log_2 x} = 5$ (b) $3^{\log_3 5} = x$ (c) $5^{\log_x 7} = 7$

(d) $x^{\log_7 9} = 9$ (e) $\exp_9 (\log_x 7) = 9$ (f) $\exp_7 (\log_x 9) = 9$

(g) $\exp_b (2) = \exp_2 (x)$ (h) $\log_2 x = \log_b 2$

10. Solve for x.

(a) $\log_5 5^3 = x$ (b) $\log_5 5^x = 3$ (c) $\log_5 x^7 = 7$ (d) $\log_x 5^7 = 7$

11. Show that for every real number x, $5^x = \exp_{10} (x \log_{10} 5)$.

12. Show that for any base b and for any number p, $\log_b b^p = p$. (Notice that this equation is a generalization of Equation 19-2.) Write this equation in terms of f and f^{-1} if $f = \exp_b$.

13. Show that $\log_{10} 2$ is not equal to .301 exactly, but only approximately.

20

**FUNDAMENTAL
PROPERTIES OF
LOGARITHMS**

The two basic properties of logarithms are nothing more than rewordings of two basic laws of exponents. We state them here in the following two theorems.

Theorem 20-1   *If M and N are positive numbers and b is any base, then*

(20-1) $$\log_b M \cdot N = \log_b M + \log_b N.$$

Proof   According to Equation 19-1, we have $M = b^{\log_b M}$ and $N = b^{\log_b N}$. Therefore,

$$M \cdot N = b^{\log_b M} b^{\log_b N} = b^{(\log_b M + \log_b N)}.$$

This last equation says that the number $x = \log_b M + \log_b N$ satisfies the equation $b^x = M \cdot N$, and hence (Definition 19-1)

$$\log_b M \cdot N = \log_b M + \log_b N.$$

Theorem 20-2 *If N is a positive number, p is any real number, and b is any base, then*

(20-2)
$$\log_b N^p = p \log_b N.$$

Proof Since $N = b^{\log_b N}$, we have

$$N^p = (b^{\log_b N})^p = b^{p \log_b N}.$$

The number $x = p \log_b N$ therefore satisfies the equation $b^x = N^p$, and so (Definition 19-1) $\log_b N^p = p \log_b N$.

EXAMPLE 20-1 If $\log_b 2 = .69$ and $\log_b 3 = 1.10$, find $\log_b 6$ and $\log_b 8$.

Solution According to Theorem 20-1,

$$\log_b 6 = \log_b 2 + \log_b 3 = .69 + 1.10 = 1.79.$$

According to Theorem 20-2,

$$\log_b 8 = \log_b 2^3 = 3 \log_b 2 = 3(.69) = 2.07.$$

EXAMPLE 20-2 Using the information given in Example 20-1, find $\log_b (1/\sqrt{3})$ and $\log_b (\sqrt[3]{16})$.

Solution According to Theorem 20-2,

$$\log_b \left(\frac{1}{\sqrt{3}}\right) = \log_b 3^{-1/2} = -\frac{1}{2} \log_b 3 = \left(-\frac{1}{2}\right)(1.10) = -.55.$$

According to Theorem 20-2,

$$\log_b (\sqrt[3]{16}) = \log_b 2^{4/3} = \tfrac{4}{3} \log_b 2 = (\tfrac{4}{3})(.69) = .92.$$

EXAMPLE 20-3 Derive the identity $\log_b \left(\frac{1}{N}\right) = -\log_b N$.

Solution According to Theorem 20-2,

$$\log_b \left(\frac{1}{N}\right) = \log_b N^{-1} = (-1) \log_b N = -\log_b N.$$

The following theorem, which stems directly from Theorems 20-1 and 20-2, is frequently useful.

Theorem 20-3 *If M and N are positive numbers and b is any base, then*

(20-3)
$$\log_b \frac{M}{N} = \log_b M - \log_b N.$$

Proof From Theorems 20-1 and 20-2 we have

$$\log_b \frac{M}{N} = \log_b MN^{-1} = \log_b M + \log_b N^{-1} = \log_b M - \log_b N.$$

The logarithms of numbers to a few particular bases have been tabulated. A table of logarithms to the base 10 is provided at the end of this book. Using just one such table, we can find the logarithm of any positive number to any other base by means of a formula that we shall now derive.

Theorem 20-4 *If a and b are two bases and if N is any positive number, then*

(20-4)
$$\log_a N = \frac{\log_b N}{\log_b a}.$$

Proof We may write $N = a^{\log_a N}$. Hence, $\log_b N = \log_b (a^{\log_a N})$. According to Theorem 20-2, this equation may be written $\log_b N = (\log_a N)(\log_b a)$, and this equation is equivalent to Equation 20-4.

EXAMPLE 20-4 Show that $\log_a b = \dfrac{1}{\log_b a}$.

Solution If we let $N = b$, Equation 20-4 becomes

$$\log_a b = \frac{\log_b b}{\log_b a}.$$

But $\log_b b$ is 1, so we have the result we wanted.

EXAMPLE 20-5 If $\log_{10} 2 = .3010$ and $\log_{10} 3 = .4771$, find $\log_3 2$.

Solution Here we let $N = 2$, $a = 3$, and $b = 10$ in Equation 20-4 to obtain

$$\log_3 2 = \frac{\log_{10} 2}{\log_{10} 3} = \frac{.3010}{.4771} = .6309.$$

You can avoid some common misunderstandings if you keep it clearly in mind that *a logarithm is a number*. In particular, Equation 20-4 says that the number $\log_a N$ is obtained by *dividing* the number $\log_b N$ by the number $\log_b a$.

The laws of logarithms are frequently used to change the form of an equation that involves logarithms.

EXAMPLE 20-6 If $\frac{1}{2} \log_3 M + 3 \log_3 N = 1$, express M in terms of N.

Solution When we multiply both sides by 2, the given equation becomes

$$\log_3 M + 6 \log_3 N = 2.$$

Thus, $\log_3 M + \log_3 N^6 = 2$, or $\log_3 MN^6 = 2$. Hence, from Definition 19-1, $MN^6 = 3^2 = 9$, and therefore $M = 9N^{-6}$.

1. Equation 20-1 may be generalized to apply to a product of any number of factors. Prove that

$$\log_b L \cdot M \cdot N = \log_b L + \log_b M + \log_b N.$$

2. Given that $\log_b 2 = .69$, $\log_b 3 = 1.10$, $\log_b 5 = 1.61$, and $\log_b 7 = 1.95$, find the following numbers.
   (a) $\log_b \frac{2}{3}$ (b) $(\log_b 2)/(\log_b 3)$ (c) $\log_b 2^2$
   (d) $(\log_b 2)^2$ (e) $\log_b 9$ (f) $\log_b 15$
   (g) $\log_b 24$ (h) $\log_b 30$ (i) $\log_b 90$
   (j) $\log_b 350$ (k) $\log_b \frac{1}{3}$ (l) $\log_b \sqrt{\frac{2}{3}}$
   (m) $\log_b \frac{27}{25}$ (n) $\log_b 70/b$

3. Simplify the following expressions.
   (a) $\log_b x^3 - \log_b \sqrt{x}$ (b) $\log_b (x^2 - 1) - \log_b (x - 1)$
   (c) $\log_b x - .75 \log_b x + \log_b 3x$ (d) $\log_b (b/\sqrt{x}) - \log_b \sqrt{x/b}$
   (e) $\log_b |x^3 + y^3| - \log_b |x + y|$ (f) $\frac{1}{2} \log_b (x^2 + y^2) - \log_b |x + y|$

4. If $\log_b 2 = .69$ and $\log_b x = 1.22$, find the following numbers.
   (a) $\log_2 b$ (b) $\log_2 x$ (c) $\log_x 2$ (d) $\log_2 \sqrt{b}\, x^3$

5. Given that $\log_{10} 5 = .699$, find the following numbers.
   (a) $\log_{10} 2$ (b) $\log_{10} 80$ (c) $\log_{10} .025$ (d) $10^{.233}$

6. Solve the following equations for x.
   (a) $\log_e \frac{18}{5} + \log_e \frac{10}{3} - \log_e \frac{6}{7} = \log_e x$
   (b) $2 \log_b x = 2 \log_b (1 - a) + 2 \log_b (1 + a) - \log_b \left(\frac{1}{a} - a \right)^2$
   (c) $\log_b x = 2 - a + \log_b \dfrac{a^2 b^a}{b^2}$
   (d) $\log_b x = \log_x b$
   (e) $|\log_5 x| = 2$
   (f) $|\log_5 x| + 2 \log_5 |x| = 3$

7. If $\log_e I = -(R/L)t + \log_e I_0$, show that $I = I_0 e^{-(R/L)t}$.

8. If $\log_b y = \frac{1}{2} \log_b x + c$, show that $y = b^c \sqrt{x}$.

9. Given that $\log_{10} 2 = .3010$ and $\log_{10} 3 = .4771$, find the logarithms of the numbers 4, 5, 6, 8, and 9. (*Hint:* $\log_{10} 10 = 1$)

10. If y is directly proportional to x^p, show that the relation between the numbers $\log_b y$ and $\log_b x$ is linear.

11. Verify the following identities.
    (a) $a^{(\log_b N)/(\log_b a)} = N$ (b) $a^{\log_b N - \log_b a} = (N/a)^{\log a}$
    (c) $a^{(\log_b N)(\log_b a)} = N^{(\log_b a)^2}$ (d) $a^{\log_b N + \log_b a} = (aN)^{\log_b a}$

12. The equations $f(x) = \log_{10} x^2$ and $g(x) = 2 \log_{10} x$ define functions f and g. Are these functions the same or different?

According to Definition 19-1, the exponential equation $N = b^x$ and the logarithmic equation $x = \log_b N$ are equivalent. Hence there might seem to be no reason to study both exponents and logarithms, since whenever we are dealing with one we are at the same time concerned with the other. But there are some cases in which the exponential form of an equation seems more "natural" and other instances in which the logarithmic form seems more appropriate; here is an example. If the plates of a charged electrical condenser are connected by a wire with a certain resistance, the condenser will discharge at a rate depending on the resistance of the wire and certain characteristics of the condenser. When calculus is used to compute the charge q on one plate of the condenser t seconds after the circuit is made, the equation that arises "naturally" is $\log_e q - \log_e q_0 = -kt$, where q_0 represents the original charge on the plate, k is a positive number associated with the circuit, and e is a logarithmic base. The exponential form of this equation is $q = q_0 e^{-kt}$, and in this form it is easier to see how q depends on t. To solve a problem such as this one, we must understand both exponential and logarithmic functions and the relation between them.

We saw in Section 18 that some of the properties of an exponential function depend on whether the base is a number greater than 1 or less than 1. The same is true for logarithmic functions. To avoid discussing two cases, we shall assume in the remainder of this chapter that *every logarithmic base is a number greater than* 1. The logarithmic bases that are used most often are greater than 1.

As we saw in our discussion of inverse functions in Section 17, the graph of f^{-1} can be obtained by reflecting the graph of f about the line that bisects the first and third quadrants. Since the function $\log_b$ is the inverse of the function $\exp_b$, we can obtain the graph of a logarithmic function by reflecting the graph of an exponential function. For example, the graph of the equation $y = \log_2 x$ that appears in Fig. 21-1 is the reflection of the curve in Fig. 18-1 about the line $y = x$.

Although Fig. 21-1 shows the graph of the logarithmic function with the particular base 2, it is representative of the graph of a logarithmic function with any base greater than 1. The graph illustrates a number of the characteristics of such a logarithmic function:

    (i) If $x < 1$, then $\log_b x < 0$.

    (ii) If $x > 1$, then $\log_b x > 0$.

    (iii) If $r < s$, then $\log_b r < \log_b s$, and conversely.

    (iv) Any line parallel to the X-axis cuts the curve in exactly one point. Thus for a given number y there is just one number x such that $\log_b x = y$ (namely, $x = b^y$). If, therefore, *we know that* $\log_b r = \log_b s$, *then we can conclude that* $r = s$.

    (v) The curve is very "steep" when x is small, but when x is large the curve, although rising, does so much more slowly.

For computational purposes, the number 10 is used as a logarithmic base, but a certain irrational number is a more suitable base to use in problems

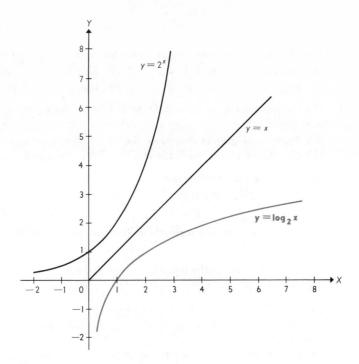

FIGURE 21-1

that involve calculus. This number is denoted by the letter e (just as we use the letter π to denote an irrational number that arises naturally in certain geometry problems). Figure 21-2 shows a graph of the equation $y = \log_e x$.

EXAMPLE 21-1   From the graph in Fig. 21-2, find as well as you can (a) $\log_e 1.5$, (b) $\log_e .5$, (c) the number x for which $\log_e x = 1.5$, and (d) the value of e.

Solution   We see from Fig. 21-2 that the answers to (a), (b), and (c) are .4, $-.7$, and 4.5. The number e satisfies the equation $\log_e x = 1$, and from the figure, it appears that $e = 2.7$ (actually, to 5 decimal places, $e = \mathbf{2.71828}$).

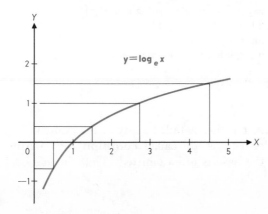

FIGURE 21-2

1. Use the graph shown in Fig. 21-1 to estimate the following numbers.
   (a) $\log_2 3$ (b) $\log_2 4.5$ (c) $\log_2 \frac{2}{3}$ (d) $\log_2 30$
   (e) $\log_{1/2} 5$ (f) $\log_2 27$ (g) $\log_4 5$ (h) $\log_5 2$

2. On the same set of axes, sketch the graphs of the equations $y = \log_2 x$ and $y = \log_3 x$.
   (a) From these graphs determine the set $\{x \mid \log_2 x > \log_3 x\}$.
   (b) Use the graphs to solve the equation $\log_2 x = \log_3 \frac{1}{2}$.

3. If f is the logarithmic function with base 8, find the following numbers.
   (a) $f(64)$ (b) $f(\frac{1}{8})$ (c) $f(4)$ (d) $f(\sqrt[3]{2})$
   (e) $-\frac{1}{2}f(\frac{1}{2})$ (f) $|f(\frac{1}{4})|$ (g) $[\![f(1492)]\!]$ (h) $[\![f(\frac{1}{20})]\!]$

4. What is the base of the logarithmic function whose graph contains the point $(100, 2)$?

5. Discuss the graph of the equation $y = e^x$, where e is the logarithmic base we mentioned in the text.

6. Using the rules of logarithms, can you suggest a way to obtain
   (a) the graph of the equation $y = \log_2 2x$ from the graph of the equation $y = \log_2 x$?
   (b) the graph of the equation $y = \log_2 \sqrt{x}$ from the graph of the equation $y = \log_2 x$?

7. Find the domain of the function f defined by the given equation.
   (a) $f(x) = \log_2 x - \log_2 (x + 1)$ (b) $f(x) = \log_2 \left(\dfrac{x}{x+1}\right)$

8. Sketch the graphs of f and g if $f(x) = \exp_2 (\log_2 x)$ and $g(x) = \log_2 (\exp_2 (x))$.

9. Find $[\![\log_{10} N]\!]$ in the following cases.
   (a) $1 \leq N < 10$ (b) $10 \leq N < 100$ (c) $.1 \leq N < 1$

10. Let $f = \{(x, \log_e x)\}$. With the aid of Fig. 21-2, solve graphically the equation $2f(x) = 2f(e^2) - 3$.

11. Show that if $b > 1$, then $b^{[\![\log_b N]\!]} \leq N$.

12. Show that $f \subseteq g$ for each of the following pairs of functions.
    (a) $f(x) = \exp_{10} (\log x)$, $g(x) = \log \exp_{10} (x)$
    (b) $f(x) = 2 \log x$, $g(x) = \log x^2$
    (c) $f(x) = \log x$, $g(x) = \log |x|$
    (d) $f(x) = \log x + \log (x - 1)$, $g(x) = \log (x^2 - x)$

22
LOGARITHMS
TO THE
BASE 10

We will now make a rather detailed study of the system of logarithms to the base 10. Such logarithms are called **common logarithms,** and the subscript denoting the base is often omitted. Thus, in this book, **log N** means **$\log_{10} N$**.

The logarithm to the base 10 of a positive number N is the number x that satisfies the equation $10^x = N$. We can solve this equation easily only for certain values of N; for example, $N = 100 = 10^2$, $N = \frac{1}{10} = 10^{-1}$, $N = 10 = 10^1$, and so on. For other values of N we must look up the answer in a table, for example, in Table I in the back of this book.

Table I lists (to 4 decimal places) the logarithms of the numbers from 1.00 to 9.99, in steps of .01. The first two digits of each such number appear in the left-hand column (the one headed n), and the third digit is written at the top of the page. Thus to find the number log 4.53, we look down the left-hand column till we come to the row headed by 4.5 and we look along the top row till we find the column headed by 3. The row that begins with 4.5 and the column that begins with 3 intersect in the number .6561, and this number is the four-place decimal approximation to log 4.53. We may use the table in the same way to see that $\log 8.37 = .9227$, $\log 5.50 = .7404$, and so on. We may also read the table backwards to see that if $\log n = .3692$, then $n = 2.34$; if $\log n = .9763$, then $n = 9.47$, and so forth.

Although the table only lists the logarithms of numbers between 1 and 10, we can use it to find the logarithm of any positive number, whether it is in this range or not. As a result of the decimal notation we use to write numbers, any positive number N can be written in the form

(22-1)
$$N = n \times 10^c,$$

where n satisfies the inequalities $1 \leq n < 10$, and c is an integer (which may be positive or negative or zero). Thus if we write $238 = 2.38 \times 10^2$, we have $n = 2.38$ and $c = 2$. Similarly, $.00238 = 2.38 \times 10^{-3}$. Again, $n = 2.38$, but now $c = -3$. When we write a number N in the form shown in Equation 22-1, we say that N is expressed in **scientific notation.** The number n lists the digits used in writing N, and the integer c locates the decimal point. Scientific notation enables us to use Table I to find logarithms of numbers not listed there.

If we write a number N in scientific notation, as in Equation 22-1, then $\log N = \log n \times 10^c = \log n + \log 10^c$. Since $\log 10^c = c$, we see that

(22-2)
$$\log N = \log n + c.$$

Because the number n is between 1 and 10, its logarithm is listed in Table I, and so we have the following rule for finding the logarithm of a positive number. *Write the number in scientific notation, $n \times 10^c$, look up the number $\log n$ in Table I, and add c.*

EXAMPLE 22-1   Find log 7960.

Solution   In scientific notation, $7960 = 7.96 \times 10^3$. Here, $n = 7.96$, and Table I tells us that $\log 7.96 = .9009$. Since $c = 3$, we have

$$\log 7960 = .9009 + 3 = 3.9009.$$

EXAMPLE 22-2   Find log .0041.

    Solution   In scientific notation, $.0041 = 4.1 \times 10^{-3}$. Therefore, $\log .0041 = \log 4.1 + (-3) = .6128 - 3 = -2.3872$.

When we follow our rule for finding the logarithm of a given positive number N, we calculate two numbers, the numbers $\log n$ and c, and add them. Since n is a number between 1 and 10, the number $\log n$ is between 0 and 1, and c is an integer. Thus, if we write m for $\log n$, we have

(22-3) $$\log N = m + c,$$

where m satisfies the inequalities $0 \leq m < 1$, and c is an integer. When $\log N$ is written this way, we say that it is in **standard form.** The integer c (which may be positive, or negative, or zero) is called the **characteristic** of $\log N$, and the number m (the positive decimal part of $\log N$) is the **mantissa** of $\log N$.

We have just described how to find $\log N$, supposing N is given. How do we go the other way; that is, how do we find N if $\log N$ is given? We first write $\log N$ in standard form, Equation 22-3. Definition 19-1 tells us that this equation is equivalent to the equation

(22-4) $$N = 10^{m+c} = 10^m \times 10^c.$$

So we find N by finding 10^m and multiplying it by 10^c. Since c is an integer, this multiplication merely amounts to shifting the decimal point; so the key to finding N is finding 10^m. But 10^m is the number whose logarithm is m. Since m is a number between 0 and 1, it is listed in the body of Table I; hence we find 10^m simply by referring to Table I to find the number whose logarithm is m.

EXAMPLE 22-3   Find N if $\log N = 6.8904$.

    Solution   In standard form, $\log N = .8904 + 6$. From Table I we find that the number whose logarithm is .8904 is 7.77; that is, $10^{.8904} = 7.77$. Thus,

$$N = 7.77 \times 10^6 = 7,770,000.$$

EXAMPLE 22-4   Find N if $\log N = -1.6180$.

    Solution   Students frequently write $.6180 - 1$ as the standard form of this number, which, of course, is wrong, since -1.6180 is really $-.6180 - 1$. But this latter number is not in standard form, because the decimal expression is not positive. To put a given number x in standard form, we subtract and add $[[x]]$. For example, $[[-1.6180]] = -2$, so we have

$$-1.6180 = 2 - 1.6180 - 2 = (2 - 1.6180) - 2 = .3820 - 2.$$

Thus our problem is to find the number N when

$$\log N = .3820 - 2.$$

From Table I we find that $10^{.3820} = 2.41$, and so

$$N = 2.41 \times 10^{-2} = .0241.$$

EXAMPLE 22-5 Find N if $\log N = -\frac{1}{3}$.

Solution In terms of decimals, $\log N = -.3333$, and to write this number in standard form, we subtract and add $[\![-.3333]\!] = -1$:

$$\log N = 1 - .3333 - 1 = .6667 - 1.$$

The number .6667 is not in Table 1, but .6665 is, so let us say that $10^{.6667}$ is approximately 4.64, and hence N is approximately $4.64 \times 10^{-1} = .464$. We will discuss the problem of interpolating between tabular entries in the next section.

PROBLEMS
22

1. Evaluate from Table I.
   (a) $\log 72,900$ (b) $\log 36.2$ (c) $\log .0000912$
   (d) $1 - \log 4.61$ (e) $(\log 23.7)/(\log .237)$ (f) $\log \frac{3}{8}$

2. Solve for x.
   (a) $10^x = 8.91$ (b) $10^x = 273$ (c) $10^x = .0171$
   (d) $10^x = \frac{33}{2500}$ (e) $10^x = 4010$ (f) $10^x = 5^3$

3. Solve for x.
   (a) $\log x = 5.1903$ (b) $\log x = 3.9053$ (c) $\log x = .4014$
   (d) $\log x = 9.49$ (e) $\log x = -3.1593$ (f) $\log x = -.4001 + 2$

4. Compute the following numbers.
   (a) $10^{1.786}$ (b) $10^{-2.699}$ (c) $\sqrt[5]{104}$ (d) $1/\sqrt[5]{100}$

5. Solve the following equations for N.
   (a) $\log N = .6580 - 3$ (b) $\log N = -2.9309 - 1$
   (c) $\log N = -1.1232$ (d) $\log N = -4.5317$
   (e) $\log N = \frac{1}{2}(1.6702)$ (f) $\log N = 1.3145 + 2.3631$
   (g) $\log N = \frac{1}{2} + 3.1160$ (h) $\log N = \frac{9}{5} - 2.4478$

6. Using Table I, we can evaluate $\log \frac{5}{2}$ in two ways:
   (i) $\log \frac{5}{2} = \log 2.5 = .3979$
   (ii) $\log \frac{5}{2} = \log 5 - \log 2 = .6990 - .3010 = .3980$
   Why do we get two different answers?

7. Use the two procedures indicated in the preceding question to find the following logarithms and compare your answers.
   (a) $\log \frac{4}{5}$ (b) $\log \frac{9}{4}$ (c) $\log \frac{3}{8}$ (d) $\log \frac{13}{4}$
   (e) $\log \frac{61}{2}$ (f) $\log \frac{101}{2}$

8. Use Theorem 20-4 and Table I to find the following numbers (see Example 20-5).

(a) $\log_5 10$ (b) $\log_5 6$ (c) $\log_7 2$ (d) $\log_8 3.07$
(e) $\log_{3.14} 86.2$ (f) $\log_{.314} .862$

9. Use the rules of logarithms and Table I to solve for x.

(a) $10^x = \sqrt[3]{12.3}$ (b) $10^x = (1.42)(7.89)$ (c) $10^x = \sqrt[3]{.0123}$
(d) $10^{-x} = 14.2/.789$ (e) $10^{3x-5} = (.0063)^4$ (f) $10^{1/x} = 1/.123$

10. Sketch the graphs of the following equations.

(a) $y = x10^{-[\![\log x]\!]}$ (b) $y = \log x - [\![\log x]\!]$

11. Show that $[\![\log N]\!]$ is the characteristic of $\log N$ and $\log N - [\![\log N]\!]$ is its mantissa. The standard form of $\log N$ is therefore

$$\log N = (\log N - [\![\log N]\!]) + [\![\log N]\!].$$

12. Show that Equation 22-4 has the form $N = 10^{x-[\![x]\!]}10^{[\![x]\!]}$.

<h2>23
INTERPOLATION</h2>

From Table I, we can immediately find the logarithm of 2.34 and the number whose logarithm is .7168. But the logarithm of 2.347 and the number whose logarithm is .1234 are not listed in Table I, and we shall now see how to find such unlisted numbers.

The simplest way to find the logarithm of a number N that is not listed in Table I is to use the logarithm of the number that we get by "rounding off" N. If N is written in scientific form as $n \times 10^c$, then the number of digits in n is called the number of **significant digits** of N. To round off N to three significant digits, we replace n with the nearest three digit number. For example, instead of $\log 2.347$, we could find $\log 2.35$. When we are in doubt about which number to choose in the rounding off process, we adopt the convention that the rounded number should be even. For example, 1.415 is rounded to 1.42, and 7.465 is rounded to 7.46. If we are given a positive decimal m that is not listed in the body of Table I and are asked to find the number N for which $\log N = m$, we could choose the number whose logarithm is the number in Table I that is nearest to m. Thus we might say that the solution of the equation $\log N = .1234$ is $N = 1.33$, because $\log 1.32 = .1206$ and $\log 1.33 = .1239$, and .1239 is nearer to .1234 than .1206 is.

A more sophisticated (and more accurate) way to approximate non-listed values is by the method of **linear interpolation.** As the name suggests, in linear interpolation we proceed as if our given function were linear. It is easy to find intermediate values of a linear function f; let us see how. Suppose that $f(x) = mx + b$, and let r, s, and n be three given numbers. We wish to find an equation that expresses the value of f at n in terms of its values at r and s. In other words, we want to express the number $f(n)$ in terms of the numbers $f(r)$ and $f(s)$. In Section 16 we saw how the slope m of the line that is the graph of f can be calculated from the coordinates of any two points of the line. Thus,

$$\frac{f(n) - f(r)}{n - r} = \frac{f(s) - f(r)}{s - r},$$

since each side of this equation is equal to the slope of our line. When we solve this equation for $f(n)$, we get the interpolation formula

(23-1) $$f(n) = f(r) + \frac{n - r}{s - r} [f(s) - f(r)].$$

Since the logarithm function is not linear, it is not strictly correct to replace f with log in the last equation. However, if the points r, s, and n are fairly close together, such a replacement gives us a good approximation formula:

(23-2) $$\log n \approx \log r + \frac{n - r}{s - r} (\log s - \log r).$$

EXAMPLE 23-1   Approximate log 2.347.

Solution   Since the numbers log 2.34 and 2.35 can be found from Table I, we will set $r = 2.34$, $n = 2.347$, and $s = 2.35$ in Formula 23-2:

$$\log 2.347 \approx \log 2.34 + \frac{2.347 - 2.34}{2.35 - 2.34} (\log 2.35 - \log 2.34)$$

$$= .3692 + \frac{.007}{.01} (.3711 - .3692)$$

$$= .3692 + .7 \times .0019$$

$$= .3692 + .0013 = .3705.$$

After you become familiar with linear interpolation, you won't have to go through this complicated substitution routine every time you interpolate. The basic argument that we used in Example 23-1 can be summed up as follows: *Since 2.347 is $\frac{7}{10}$ of the way from 2.34 to 2.35, we will assume that* log 2.347 *is $\frac{7}{10}$ of the way from* log 2.34 *to* log 2.35.

EXAMPLE 23-2   Use linear interpolation to find log 5.723.

Solution   Since 5.723 is .3 of the way from 5.72 to 5.73, we argue that log 5.723 is approximately .3 of the way from log 5.72 to log 5.73:

$$\log 5.723 \approx \log 5.72 + .3(\log 5.73 - \log 5.72)$$

$$= .7574 + .3(.7582 - .7574) = .7576.$$

We usually write log 5.723 = .7576, rather than log 5.723 ≈ .7576. In context, it is clear that an approximation is meant.

The inverse procedure is similar. If we are asked, for example, to find the number m such that $\log m = .1234$, we first look in Table I to find logarithms that bracket .1234. Thus we find that $\log 1.32 = .1206$ and $\log 1.33 = .1239$, so we see that $\log m$ is

$$\frac{.1234 - .1206}{.1239 - .1206} = \frac{.0028}{.0033} = .8$$

of the way from $\log 1.32$ to $\log 1.33$. We therefore conclude that m is (approximately) .8 of the way from 1.32 to 1.33; that is, $m = 1.328$.

EXAMPLE 23-3   Find N if $\log N = 3.6129$.

Solution   As a first step, we write this number in standard form:

$$\log N = .6129 + 3.$$

Now we look in Table I and find that $\log 4.10 = .6128$ and $\log 4.11 = .6138$. Since .6129 is .1 of the way from the first of these numbers to the second, we assume that it is the logarithm of the number that is .1 of the way from 4.10 to 4.11; that is, $\log 4.101 = .6129$. Therefore, from our work in the last section, we see that

$$N = 4.101 \times 10^3 = 4{,}101.$$

Graphically speaking, linear interpolation amounts to replacing a small section of the graph of the logarithm function with a line segment. Thus Fig. 23-1 illustrates Example 23-1. The number we really want ($\log 2.347$) is the Y-coordinate of the point P. The number we found (.3705) is the Y-coordinate of the point Q. This number will be a good approximation to the desired number if the line segment is a good approximation to the logarithm curve. Our graph in Fig. 23-1 considerably exaggerates the curvature of the logarithm curve; actually, the graph is very nearly linear in this range. Our value of .3705 is really the value of $\log 2.347$, correct to four decimal places.

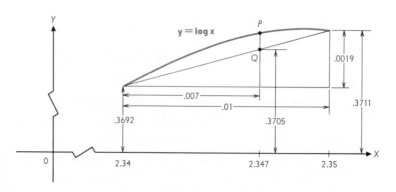

FIGURE 23-1

96

Of course, linear interpolation doesn't work so well when we are dealing with a function whose graph does not approximate a line for the values under consideration. Thus if $g = \{(x, x^2 - 1)\}$, then the graph of g is not even approximately linear for x in the interval from -1 to 1. If we use linear interpolation here to argue that since 0 is half-way between -1 and 1, then $g(0)$ is half-way between $g(-1) = 0$ and $g(1) = 0$, we would conclude that $g(0) = 0$, when in fact $g(0) = -1$. This simple example shows that we cannot use linear interpolation indiscriminately. We will use it later to approximate values of the trigonometric functions.

PROBLEMS
23

1. Write the following logarithms in standard form.
   (a) log 9.017
   (b) log .001372
   (c) log 2476
   (d) log (1.697) · 10⁴
   (e) log (3.726) · 10⁻⁶
   (f) log e
   (g) log 10$^\pi$
   (h) log $[\![10\pi]\!]$
   (i) log π^{10}

2. Find the number x to 4 significant digits.
   (a) log x = .5521
   (b) log x = 3.9286
   (c) log x = 4.7182 − 5
   (d) log x = −1.8347
   (e) log x = $\frac{1}{3}$
   (f) log x = $\sqrt{2}$

3. Find the number x.
   (a) 10^x = .3976
   (b) $10^x = e$
   (c) 10^x = 23.14
   (d) 10^x = .006723
   (e) $10^x = \frac{1}{3}$
   (f) $10^x = \sqrt{2}$

4. Write as a decimal number.
   (a) $10^{.395}$
   (b) $10^{.1}$
   (c) $\sqrt[5]{10}$
   (d) $\sqrt[4]{160}$
   (e) 10^{-e}
   (f) $10^{-.0013}$

5. We know that $\log_2 8 = 3$ and that $\log_2 16 = 4$. Use linear interpolation to find $\log_2 10$. Is your approximate value too large or too small?

6. Use the tabulated values of log 1 and log 1.2, and interpolate to find log 1.1. Interpolate to find log 9.9 by using the listed values of log 9.8 and log 10. Compare your answers with the tabulated values, and discuss your findings.

7. Using $\sqrt{16} = 4$ and $\sqrt{25} = 5$ and linear interpolation, find $\sqrt{18}$. Discuss the accuracy of your result.

8. If $f(x) = 2x - 3$, then $f(2) = 1$ and $f(3) = 3$. Use linear interpolation to find $f(2.5)$. Discuss the accuracy of your result.

9. Let $f(x) = 1/x^2$. Use linear interpolation to find $f(1.5)$ from the values of $f(1)$ and $f(2)$. Also use linear interpolation to find $f(4.5)$ from the values of $f(4)$ and $f(5)$. Discuss the accuracy of your results.

10. We know that $\exp_2 (0) = 1$ and $\exp_2 (1) = 2$. Use linear interpolation to find $\exp_2 (.5)$. Discuss the accuracy of your result.

At one time, logarithms were studied mainly because they were helpful in simplifying long numerical computations. Today the widespread use of computing machines has removed the necessity of doing complicated computations by hand. This development does not mean that logarithms are no longer important, only that they are important for different reasons. Basic logarithmic computations still have some practical value, and it will increase your mastery of the theory of logarithms if we work a few arithmetic problems with them. There are a number of conventions and short cuts that speed up logarithmic computations, but we won't consider them here. If you ever have to learn them, you can in a relatively short time.

EXAMPLE 24-1   Calculate the number $\dfrac{3480 \times 1265}{.00143}$.

Solution   One way to proceed is to write

$$\frac{3480 \times 1265}{.00143} = \frac{3.48 \times 10^3 \times 1.265 \times 10^3}{1.43 \times 10^{-3}} = \frac{3.48 \times 1.265 \times 10^9}{1.43}$$

and then to calculate the number $N = \dfrac{3.48 \times 1.265}{1.43}$.

According to the rules of logarithms,

$$\log N = \log 3.48 \times 1.265 - \log 1.43$$
$$= \log 3.48 + \log 1.265 - \log 1.43.$$

We use Table I to find the logarithms on the right-hand side of this equation:

$$\log N = .5416 + .1021 - .1553 = .4884.$$

Now we solve this equation, again using Table I, and we find that $N = 3.079$. Therefore,

$$\frac{3480 \times 1265}{.00143} = 3.079 \times 10^9.$$

We would have had no great difficulty in carrying out the calculation in Example 24-1 by ordinary arithmetic. We could not work the following examples, however, without using logarithms.

EXAMPLE 24-2   Find $\sqrt[5]{20}$.

Solution   Let $N = \sqrt[5]{20} = 20^{1/5}$. Then

$$\log N = \log 20^{1/5} = \tfrac{1}{5} \log 20 = \frac{1.3010}{5} = .2602.$$

Thus,

$$N = \sqrt[5]{20} = 1.82.$$

EXAMPLE 24-3   Find $\sqrt[5]{.2}$.

Solution   Let $N = (.2)^{1/5}$. Then

$$\log N = \log (.2)^{1/5} = \tfrac{1}{5} \log .2 = \frac{.3010 - 1}{5}.$$

If we proceed with the arithmetic at this point, we shall find that $\log N = \frac{-.6990}{5} = -.1398$. But this last number must be written in standard form before we can solve for N. It is easier to replace the number $(.3010 - 1)$ with its equivalent expression $(4.3010 - 5)$ before we divide by 5. Then

$$\log N = \frac{4.3010 - 5}{5} = .8602 - 1.$$

Here $\log N$ is expressed in standard form, and we can use Table I to find that $N = \sqrt[5]{.2} = .7248$.

Logarithms also provide us with a convenient way of obtaining an idea of the size of numbers that are quite large.

EXAMPLE 24-4   We mentioned earlier that under favorable conditions a single cell of the bacterium *Escherichia coli* divides into two about every 20 minutes. If this same rate of division is maintained for 10 hours, how many organisms will be produced from a single cell?

Solution   The 10-hour interval may be divided into 30 periods of 20 minutes each. At the end of the first period there are 2 bacteria, at the end of the second there are $2 \cdot 2 = 2^2$, at the end of the third there are 2^3, and so on. At the end of the 30th period there will be 2^{30} bacteria, which we will call N, the number we were seeking. Now $\log N = \log 2^{30} = 30 \log 2 = 30 \times .3010 = 9.0300$. From the table of logarithms, we see that $N = 1.072 \times 10^9 = 1,072,000,000$, so a single cell is potentially capable of producing about a billion organisms in a 10-hour period.

**PROBLEMS
24**

1. Use logarithms to make the following computations.

(a) $\dfrac{573 \times 6.11}{487 \times .0021}$

(b) $\dfrac{-413 \times 22}{816}$

(c) $21^{-1/5}$

(d) $(3^{1/2})^{1/4}$

(e) $10^{\pi}/\pi^{10}$

(f) $\sqrt[3]{71^4}$

(g) $\dfrac{1861}{1776} + \dfrac{1776}{1861}$

(h) $\dfrac{91.5 + .00318}{\sqrt[3]{.028}}$

(i) $\sqrt{\sqrt{\sqrt{.0236}}}$

(j) $-(18.3)^2 (.176)^4 / (258.6)^2$

(k) $(18.3)^2 + (.176)^4 - (258.6)^2$

(l) $(18.3)^{-1/3} + (.176)^{1/4} - (258.6)^{1/2}$

2. Solve for x.

(a) $10^x = \tfrac{123}{321}$

(b) $10^{-x} = \tfrac{123}{321}$

(c) $10^x = \pi$

(d) $100^x = \pi$

(e) $10^x = 2^{\pi}$

(f) $2 \cdot 10^x = 5^{.5}$

3. Determine which number is larger.

(a) $25^{7/6}$ or $28^{10/9}$ (b) 5^{π} or 157 (c) 3^{21} or 95^5 (d) $3^{\sqrt{2}}$ or $2^{\sqrt{3}}$

4. The radioactive isotope ^{11}C decomposes rapidly. If at one instant there is 1 microgram of ^{11}C, then t minutes later there will be A micrograms, where $A = (\frac{1}{2})^{t/20}$. Find the amount of ^{11}C that will be present at the end of 10, 20, and 30 minutes.

5. If you invest p dollars at 6% interest "compounded annually," then at the end of n years you will have accumulated a sum of A dollars. Express A in terms of p and n. If you invest \$5585, how much will you have in your account at the end of 10 years?

6. An angstrom (Å) is a unit of length equal to 10^{-10} meters (1 meter is 39.37 inches long). The wavelength of a certain gamma ray is .8973 Å. Use logarithms to compute the wavelength in inches.

7. The time required for a simple pendulum L meters long to make one complete oscillation is t seconds, where $t = 2\pi\sqrt{L/9.8}$. Use logarithms to find the length of a pendulum with a period of 2 seconds.

8. The volume of a spherical ball is given by the formula $V = \frac{4}{3}\pi r^3$. What is the radius of a ball that has a volume of 5 cubic meters?

9. Suppose $f(x) = \left(1 + \frac{1}{x}\right)^x$. Find $f(1)$, $f(10)$, and $f(100)$.

10. Let us assume that the radius of the earth is 3960 miles. What is the approximate volume in cubic feet of the earth? It is estimated that the earth weighs 5.883×10^{21} tons. How much does an average cubic foot weigh?

11. Originally, the meter was defined to be 10^{-7} times the distance from the earth's equator to a pole.
(a) What is the radius of the earth in kilometers?
(b) What is the earth's area in hectares (1 hectare is 10^4 square meters)?

12. Compute the slope of the chord that joins the points of the graph of the equation $y = 2^x$ whose X-coordinates are .01 and $-.01$. This slope approximates the slope of the "tangent line" to the graph of $y = 2^x$ at the point $(0, 1)$. Conclude that when x is close to zero, $2^x \approx 1 + .7x$.

25
EXPONENTIAL AND LOGARITHMIC EQUATIONS

We have already observed that Table I may be considered as a table of solutions for equations of the type $10^x = N$, where N is a given positive number. For example, the solution of the equation $10^x = 2$ is given in the table as $\log 2 = .3010$. The equation $10^x = 2$ is of the form $a^x = b$, where a and b are given positive numbers. Of course, the solution of the equation $a^x = b$ can be written as $x = \log_a b$, but we must go a little farther to get a more meaningful answer. To evaluate this quantity by using Table I, we must recall that $\log_a b = \log b/\log a$ (see Theorem 20-4).

Instead of memorizing the formula relating $\log_a b$ and $\log b$, we can solve the equation $a^x = b$ directly. Simply take logarithms (to the base 10) of both sides of the equation to obtain $\log a^x = \log b$, and therefore (Theorem 20-2) $x \log a = \log b$. It follows that $x = \log b / \log a$, as above.

EXAMPLE 25-1   Solve the equation $2^x = 7$ for x.

Solution   By taking logarithms of both sides of the equation, we obtain the equation $x \log 2 = \log 7$. From Table I, $\log 2 = .3010$ and $\log 7 = .8451$. So our equation becomes $.3010x = .8451$. Hence,

$$x = \frac{.8451}{.3010} = 2.808.$$

Remark 1. Since $2^2 = 4$ and $2^3 = 8$, it should be obvious at the start that the solution of the equation $2^x = 7$ is a number between 2 and 3.

Remark 2. Since $x \log 2 = \log 7$, it follows that $x = \log 7 / \log 2$. It should be emphasized that the expression $\log 7 / \log 2$ is a *quotient*. We do not evaluate this quotient by looking up $\log 2$ and $\log 7$ in the table and subtracting; we look up the two numbers and divide. We can divide with the aid of logarithms, but it still will be division.

EXAMPLE 25-2   Solve the equation $2^x = 3^{x+1}$ for x.

Solution   We take logarithms of both sides of the equation to obtain the equation

$$x \log 2 = (x + .1) \log 3.$$

Hence

$$x \log 2 - x \log 3 = \log 3,$$

or in other words,

$$x = \frac{\log 3}{\log 2 - \log 3}.$$

Table I gives

$$x = \frac{.4771}{.3010 - .4771} = \frac{.4771}{-.1761} = -2.709.$$

EXAMPLE 25-3   Solve the inequality $(.3)^x < \frac{4}{3}$.

Solution   We have already seen in Section 21 that if $M < N$, then $\log_b M < \log_b N$, and conversely. Thus $(.3)^x < \frac{4}{3}$ if, and only if, $\log (.3)^x < \log \frac{4}{3}$. We can then use the rules of logarithms and Table I to write the following chain of equivalent inequalities:

$$x \log (.3) < \log 4 - \log 3,$$
$$x(.4771 - 1) < .6021 - .4771,$$
$$-.5229x < .1250,$$
$$x > -.239.$$

Therefore, $\{x > -.239\}$ is the solution set of our given inequality.

EXAMPLE 25-4 Solve the equation $I = \dfrac{E}{R}(1 - e^{-Rt/L})$ for t.

Solution We can write the equation as

$$\frac{RI}{E} = 1 - e^{-Rt/L} \quad \text{or} \quad e^{-Rt/L} = 1 - \frac{RI}{E}.$$

Then

$$\log e^{-Rt/L} = \log\left(1 - \frac{RI}{E}\right),$$

$$\frac{-Rt}{L}\log e = \log\left(1 - \frac{RI}{E}\right),$$

$$\frac{-Rt}{L} = \frac{\log\left(1 - \dfrac{RI}{E}\right)}{\log e},$$

$$t = \frac{-L\log\left(1 - \dfrac{RI}{E}\right)}{R \log e}.$$

If we use logarithms to the base e, the expression for t takes the simpler form

$$t = \frac{-L}{R}\log_e\left(1 - \frac{RI}{E}\right).$$

EXAMPLE 25-5 Solve the equation $2 \log x - \log 10x = 0$.

Solution We can use the fundamental properties of logarithms to simplify the left-hand side of this equation:

$$2 \log x - \log 10x = 2 \log x - \log 10 - \log x = \log x - 1 = 0.$$

Therefore, $\{2 \log x - \log 10x = 0\} = \{\log x = 1\} = \{10\}$.

EXAMPLE 25-6 Solve the equation $x^{\log x} = 100x$.

Solution By taking logarithms of both sides of the equation, we obtain the equivalent equation

$$\log (x^{\log x}) = \log 100x.$$

But

$$\log (x^{\log x}) = (\log x)(\log x) = (\log x)^2,$$

and

$$\log 100x = \log 100 + \log x = 2 + \log x.$$

We can therefore write our equation as

$$(\log x)^2 = 2 + \log x,$$

and so it is equivalent to the equation

$$(\log x)^2 - \log x - 2 = (\log x - 2)(\log x + 1) = 0.$$

Now

$$\{(\log x - 2)(\log x + 1) = 0\} = \{x \mid \log x = 2 \text{ or } \log x = -1\}$$
$$= \{\log x = 2\} \cup \{\log x = -1\}$$
$$= \{100\} \cup \{\tfrac{1}{10}\}$$
$$= \{100, \tfrac{1}{10}\},$$

and this is the set of numbers we were to find.

PROBLEMS 25

1. Solve for x.
   (a) $3^x = 2$ (b) $4^x = \frac{5}{3}$ (c) $5^x = 17$ (d) $(.01)^x = 5$
   (e) $7^{-x} = 8$ (f) $7^{1/x} = 8$ (g) $x^7 = 8$ (h) $x^{-7} = 8$
   (i) $[\![x^7]\!] = 8$ (j) $[\![7^x]\!] = 8$ (k) $\exp_2 (x^2) = 5$

2. Solve for x.
   (a) $3^{1-x} = 2$ (b) $5^{x+2} = 7^{x-1}$ (c) $10^{x^2} = 5$
   (d) $(10^x)^2 = 5$ (e) $5^{2x}7^{-3x} = 6$ (f) $(2x)^5(7x)^{-3} = 6$
   (g) $3^{5x-2}/2^{3x+1} = 51$ (h) $6 \cdot 5^{2x} = 7 \cdot 2^{5x}$ (i) $5^{2x+3} = 7^{x+3}$

3. Solve for x.
   (a) $\log (x + 1) - \log x = \frac{1}{2}$ (b) $\log (x^2 - 1) - \log (x + 1) = 2$
   (c) $5^{\log x} = 5x$ (d) $\log |x + 1| + \log |x - 1| = 2$
   (e) $\log |x + 1| + \log |x - 1| = -2$ (f) $\log (x + 1)/\log |x - 1| = 2$

4. Sketch the graph of the equation $5^{x+y} = 10$.

5. Show that $\{x \mid \log (x - 4) - \log (x + 1) = \log 6\} = \varnothing$.
   What is the set $\{x \mid \log |x - 4| - \log |x + 1| = \log 6\}$?

6. Use the data in Example 24-4 to determine how long it would take a single cell of *E. coli* to produce 500,000 organisms.

7. What are the numbers N and k if the graph of the function $\{(x, N \cdot 10^{kx})\}$ contains the points $(0, 7)$ and $(1, 14)$?

8. What are the numbers N and k if the graph of the function $\{(x, N \log kx)\}$ contains the points $(5, 7)$ and $(.5, 0)$?

9. The number of milligrams of radium present at the end of t years is given by the formula $A = A_0 10^{-.000174t}$. What is the initial amount of radium? How long will it take a given sample to reduce to $\frac{1}{2}$ its original size; that is, what is the half-life period?

10. At what interest compounded annually must we invest $100 if we want it to double in 12 years?

11. The intensity I of an x-ray beam after passing through x centimeters of a certain material is given by $I = I_0 e^{-kx}$, where I_0 is the intensity as the beam enters the material, and k is a constant called the linear coefficient of absorption. Calculate k if the material absorbs $\frac{1}{2}$ the x-ray beam after penetrating 5 centimeters.

12. Solve the equation $[\![\log x]\!] = \log [\![x]\!]$.

You should be able to answer the following questions without referring back to the text.

1. Let $f = \exp_{10}$ and $g = \log_{10}$. What is the domain of f, the domain of g, the range of f, and the range of g? Which of the following equations are true?
   (a) $f(x + y) = f(x)f(y)$ (b) $f(xy) = f(x) + f(y)$ (c) $g(x + y) = g(x)g(y)$
   (d) $g(xy) = g(x) + g(y)$ (e) $g(x^p) = pg(x)$ (f) $f(g(x)) = x$
   (g) $g(f(x)) = x$ (h) $f(0) = g(10)$

2. Solve for x.
   (a) $(\log x)^2 = \log x^2$ (b) $\log [\log (\log x)] = 0$ (c) $|\log x| = \log |x|$

3. Which number is larger, e^π or π^e? (Take $e = 2.718$ and $\pi = 3.142$.)

4. Solve the following inequalities.
   (a) $\exp_{10} (x^2 + x - 2) > 1$ (b) $\log (x^2 + x - 2) > 1$

5. What changes have to be made in Table I to obtain a table of values of $\log_{100}$?

6. A *light year* is the distance traveled by light in 1 year. If the speed of light is 186,000 miles per second, find the approximate length in miles of 1 light year.

7. Solve the following equations.
   (a) $\log (x + 1) = \log x + 1$ (b) $10^{x+1} = 10^x + 1$

8. How many digits are there in the number 2^{30}?

9. Sketch the graphs of the equations $\exp_{10} (x) = \exp_{.1} (y)$ and $\log_{10} x = \log_{.1} y$.

10. Convince yourself graphically that the equation $10^x + \log x = 0$ has a solution r between 0 and 1. Is r greater than or less than $\frac{1}{10}$?

MISCELLANEOUS PROBLEMS, CHAPTER THREE

These problems are designed to test your ability to apply your knowledge of exponential and logarithmic functions to somewhat more difficult problems.

1. If u and v are positive numbers and b is a logarithmic base, show that $u^{\log_b v} = v^{\log_b u}$.

2. Find the solution sets of the equations $(\log x)^{-1} = \log x^{-1}$ and $|\log x|^{-1} = |\log x^{-1}|$.

3. Suppose that $f(px) = (f(x))^p$ for every real number p and each real number x. Show that $f = \exp_b$ for some number b.

4. Show that $[\![x + \frac{1}{2}]\!]$ is the nearest integer to x. Show that the formula $.01[\![100x + \frac{1}{2}]\!]$ rounds off x to two decimal places.

5. Find all the pairs of *integers* that satisfy the equation $\log (x + y) = \log x + \log y$.

6. Sketch the graph of the equation $y = \log_x 3$.

7. Solve the inequality $\log_x 5 < \log_5 x$.

8. Can you convince yourself graphically that the equation $\log^{-1} x = (\log x)^{-1}$ has a solution r? Is r greater than or less than .5?

9. Let $f(x) = |\log |x||$.
   (a) Sketch the graph of f.
   (b) Show that $f(xy) \leq f(x) + f(y)$.
   (c) Show that $f(x^p) = |p| f(x)$ for each real number p.

10. Suppose that $x \geq 1$.
    (a) Explain why $10^{[\![\log x]\!]} \leq 10^{\log x} = x$.
    (b) From the fact that the left-hand side of the above inequality is an integer, infer that $10^{[\![\log x]\!]} \leq [\![x]\!]$.
    (c) Now infer that $[\![\log x]\!] \leq \log [\![x]\!]$.
    (d) Are there choices of $x(>1)$ for which $\leq$ may be replaced by $=$?
    (e) Can you use the same type of argument to show that $[\![\sqrt{x}]\!] \leq \sqrt{[\![x]\!]}$?

11. If a and b are positive numbers, show that

$$\log (a + b + |a - b|) = \log (a + b - |a - b|) + |\log (a/b)|.$$

The
Circular
Functions

The circular functions are very closely related to functions that you probably know as the trigonometric functions. The word "trigonometry" stems from the Greek words for "triangle measurement," and the subject was originally developed to solve geometric problems involving triangles. Such problems called for functions that were defined by pairing with an acute angle a number, the ratio of the lengths of certain sides of a right triangle containing the angle. Thus, the domain of a trigonometric function, as it was originally defined, consisted of acute angles, not numbers. But the trigonometric functions have since been extended so that they are also useful in situations in which angles are not involved. The importance of these applications justifies a non-geometric approach to the subject of trigonometry.

The functions we introduce in this chapter are functions of the type we have been studying all along; that is, they are sets of pairs of real numbers. If you have already studied trigonometry in terms of angles, you will recognize that the circular functions we are talking about have the same names and properties as the trigonometric functions you already know. But their domains are different. The domain of a circular function is a set of numbers, and the domain of a trigonometric function is a set of angles. This distinction may seem like a minor technicality to you. Nevertheless, you should try to adopt the numerical point of view, because it is the basis for many of the present applications of trigonometry.

26
THE
CIRCULAR
POINT

We call the circle with a radius of 1 unit, and whose center is the origin of a cartesian coordinate system, the **unit circle.** The circular functions are based on a function P whose domain is the set of real numbers and whose range is the set of points of the unit circle. We will now give the rule by which this function pairs with a given number t a point $P(t)$ of the unit circle. Starting at the point $(1, 0)$, move $|t|$ units along the circle, counterclockwise if t is positive and clockwise if t is negative (Fig. 26-1). You will

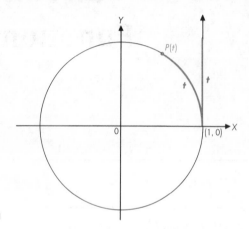

FIGURE 26-1

arrive at a point of the unit circle that we call the **circular point $P(t)$.** Thus to find $P(1)$, we start at the point $(1, 0)$ and move one unit along the circumference of the unit circle in the counter-clockwise direction, arriving at the point $P(1)$. To find $P(-1)$, we move from $(1, 0)$ one unit along the unit circle in the clockwise direction. To find $P(0)$, we move 0 units from $(1, 0)$, and so on. Assuming that we know what $|t|$ units along the unit circle means, then with each real number t there is paired exactly one trigonometric point $P(t)$, so the function P is well-defined.

The function P is sometimes called the **winding function.** This name comes from the fact that its rule of correspondence amounts to winding a number scale around the unit circle. Suppose (as shown in Fig. 26-1) that we place a vertical number scale so that its zero point coincides with the point $(1, 0)$. To find the circular point $P(t)$ that corresponds to a given point t of this number scale, we simply wind the number scale around the circle. The point t will fall on $P(t)$.

It is easy to find some values of the winding function P and hard to find others. For example, we said that $P(0)$ is the point we arrive at when we move 0 units from $(1, 0)$, so $P(0) = (1, 0)$. On the other hand, to find $P(1)$ we must measure 1 unit along the arc of the unit circle—a difficult task without a curved ruler. It is necessary to use tables (which we will discuss later) to locate most circular points, in particular, $P(1)$. But we can use our knowledge of the geometry of the circle to locate certain points directly.

The circumference of the unit circle is 2π units, and so one-half of the circumference is π units, and one-quarter of the circumference is $\frac{1}{2}\pi$ units. Therefore, to find the circular points $P(2\pi)$, $P(\pi)$, and $P(\frac{1}{2}\pi)$, we proceed once around the circle, halfway around, and one-quarter of the way around.

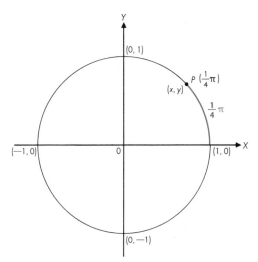

FIGURE 26-2

Thus (look at Fig. 26-2), $P(2\pi) = (1, 0)$; $P(\pi) = (-1, 0)$; and $P(\frac{1}{2}\pi) = (0, 1)$. The point $P(\frac{3}{2}\pi)$ is three-quarters of the way around the unit circle in the positive (counter-clockwise) direction, and the point $P(-\frac{1}{2}\pi)$ is one-quarter of the way around in the negative direction. Hence it follows that $P(\frac{3}{2}\pi) = P(-\frac{1}{2}\pi) = (0, -1)$. It should be easy for you to locate the point $P(t)$ if t is any integral multiple of $\frac{1}{2}\pi$, for example, $\frac{7}{2}\pi$, $-\frac{13}{2}\pi$, and so forth.

Suppose that t is a given number and that (x, y) are the coordinates of the circular point $P(t)$. Since this point belongs to the unit circle, it is one unit from the origin, and hence the distance formula tells us that

(26-1) $$x^2 + y^2 = 1.$$

Thus, *if we know one of the coordinates of $P(t)$, then we can use Equation 26-1 to determine the other (except for sign).*

Let us now find the coordinates (x, y) of the circular point $P(\frac{1}{4}\pi)$. Since $\frac{1}{4}\pi$ is one-eighth of the circumference of the unit circle, the point $P(\frac{1}{4}\pi)$ is the midpoint of the shorter arc that joins the points $(1, 0)$ and $(0, 1)$

(see Fig. 26-2). Therefore the point $P(\frac{1}{4}\pi)$ is equidistant from the X- and Y-axes, and hence $x = y$. Thus Equation 26-1 becomes

$$x^2 + x^2 = 2x^2 = 1;$$

that is, $x^2 = \frac{1}{2}$. Since x is clearly positive, we have $x = \sqrt{\frac{1}{2}} = \frac{1}{2}\sqrt{2} = y$. Therefore, the coordinates of $P(\frac{1}{4}\pi)$ are $(\frac{1}{2}\sqrt{2}, \frac{1}{2}\sqrt{2})$. You should be able to find the coordinates of circular points such as $P(\frac{3}{4}\pi)$, $P(\frac{5}{4}\pi)$, and so forth, by locating them on the unit circle and observing that they are symmetrically placed with respect to $P(\frac{1}{4}\pi)$.

EXAMPLE 26-1 Find the coordinates of the points $P(\frac{3}{4}\pi)$ and $P(-\frac{3}{4}\pi)$.

Solution We use the obvious symmetry of Fig. 26-3 and the known coordinates of $P(\frac{1}{4}\pi)$ to find that $P(\frac{3}{4}\pi)$ is the point $(-\frac{1}{2}\sqrt{2}, \frac{1}{2}\sqrt{2})$ and that $P(-\frac{3}{4}\pi)$ is the point $(-\frac{1}{2}\sqrt{2}, -\frac{1}{2}\sqrt{2})$.

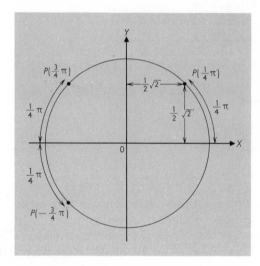

FIGURE 26-3

The geometric reasoning that we use to find the coordinates (x, y) of the circular point $P(\frac{1}{3}\pi)$ is only a little more complicated. We notice that the points $P(\frac{1}{3}\pi)$ and $P(\frac{2}{3}\pi)$ are symmetrically located with respect to the Y-axis (Fig. 26-4). Therefore, the coordinates of $P(\frac{2}{3}\pi)$ are $(-x, y)$. The chord $P(\frac{2}{3}\pi)P(\frac{1}{3}\pi)$ is $2x$ units long, and the distance formula tells us that the chord $P(\frac{1}{3}\pi)P(0)$ is $\sqrt{(x-1)^2 + y^2}$ units long. These two chords are equal, since the corresponding arcs are equal, and so we see that $2x = \sqrt{(x-1)^2 + y^2}$. Let us square both sides of this equation to remove the radical sign:

$$4x^2 = (x-1)^2 + y^2 = x^2 - 2x + 1 + y^2.$$

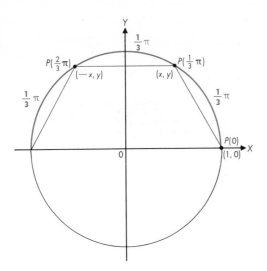

FIGURE 26-4

If we use Equation 26-1 to simplify this last equation, we obtain the equation $4x^2 = 2 - 2x$, and so

$$4x^2 + 2x - 2 = 2(2x - 1)(x + 1) = 0.$$

The solution set of this equation is $\{-1, \frac{1}{2}\}$, and a glance at Fig. 26-4 shows that $\frac{1}{2}$ is the solution we are seeking.

Now from Equation 26-1 we find

$$y^2 = 1 - x^2 = 1 - \tfrac{1}{4} = \tfrac{3}{4},$$

and hence $y = \frac{1}{2}\sqrt{3}$, since Fig. 26-4 shows that y is positive. Thus, $(\frac{1}{2}, \frac{1}{2}\sqrt{3})$ are the coordinates of the circular point $P(\frac{1}{3}\pi)$.

The coordinates of $P(\frac{1}{6}\pi)$ can be found by an algebraic method like the one we have just used, but it is easier to use the known coordinates of $P(\frac{1}{3}\pi)$ as follows. Figure 26-5 shows the points $P(\frac{1}{3}\pi)$ and $P(\frac{1}{6}\pi)$ of the unit circle. The arc between the X-axis and $P(\frac{1}{6}\pi)$ and the arc between the Y-axis and $P(\frac{1}{3}\pi)$ are both $\frac{1}{6}\pi$ units long. Therefore, the distance between $P(\frac{1}{6}\pi)$ and the X-axis is equal to the distance between $P(\frac{1}{3}\pi)$ and the Y-axis. In other words, the Y-coordinate of $P(\frac{1}{6}\pi)$ is equal to the X-coordinate of $P(\frac{1}{3}\pi)$. Similarly, the X-coordinate of $P(\frac{1}{6}\pi)$ is equal to the Y-coordinate of $P(\frac{1}{3}\pi)$. Since the coordinates of $P(\frac{1}{3}\pi)$ are $(\frac{1}{2}, \frac{1}{2}\sqrt{3})$, it follows that the coordinates of $P(\frac{1}{6}\pi)$ are $(\frac{1}{2}\sqrt{3}, \frac{1}{2})$. You should now be able to find the coordinates of circular points such as $P(\frac{5}{6}\pi)$, $P(-\frac{1}{6}\pi)$, and so forth, by locating them on the unit circle and observing that they are symmetrically placed with respect to $P(\frac{1}{3}\pi)$ and $P(\frac{1}{6}\pi)$. You will find it worthwhile to memorize the coordinates of the circular points $P(\frac{1}{3}\pi)$, $P(\frac{1}{4}\pi)$, and $P(\frac{1}{6}\pi)$.

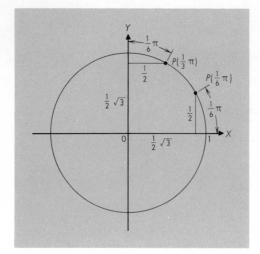

FIGURE 26-5

PROBLEMS
26

1. Determine the coordinates of each of the following circular points.
   (a) $P(3\pi)$ (b) $P(-2\pi)$ (c) $P(\frac{11}{2}\pi)$
   (d) $P(-\frac{7}{2}\pi)$ (e) $P(1861\pi)$ (f) $P(\frac{1861}{2}\pi)$

2. Determine the coordinates of each of the following circular points.
   (a) $P(\frac{5}{4}\pi)$ (b) $P(-\frac{1}{4}\pi)$ (c) $P(\frac{9}{4}\pi)$
   (d) $P(-\frac{7}{4}\pi)$ (e) $P(\frac{1861}{4}\pi)$ (f) $P(-\frac{1861}{4}\pi)$

3. Determine the coordinates of each of the following circular points.
   (a) $P(-\frac{1}{3}\pi)$ (b) $P(\frac{2}{3}\pi)$ (c) $P(\frac{5}{3}\pi)$
   (d) $P(-\frac{5}{3}\pi)$ (e) $P(\frac{1492}{3}\pi)$ (f) $P(-\frac{1492}{3}\pi)$

4. Determine the coordinates of each of the following circular points.
   (a) $P(\frac{5}{6}\pi)$ (b) $P(-\frac{5}{6}\pi)$ (c) $P(\frac{11}{6}\pi)$
   (d) $P(-\frac{7}{6}\pi)$ (e) $P(\frac{1861}{6}\pi)$ (f) $P(-\frac{1861}{6}\pi)$

5. Make a careful drawing of the unit circle. By rotating a ruler or using a piece of string, estimate from your figure the coordinates of $P(1)$.

6. If you multiply the coordinates of the given circular point, will you get a positive or a negative number?
   (a) $P(2)$ (b) $P(3)$ (c) $P(1.5)$ (d) $P(4)$
   (e) $P(5)$ (f) $P(6)$ (g) $P(\frac{22}{7})$ (h) $P(6.2832)$

7. Suppose that $P(t) = (x, y)$. Can you convince yourself (draw lots of pictures) that the following equations are true?
   (a) $P(-t) = (x, -y)$ (b) $P(t + \pi) = (-x, -y)$
   (c) $P(t + 200\pi) = (x, y)$ (d) $P(\frac{1}{2}\pi - t) = (y, x)$
   (e) $P(t + \frac{1}{2}\pi) = (-y, x)$ (f) $P(t - \frac{1}{2}\pi) = (y, -x)$

8. Solve the following equations.
   (a) $P(t) = P(3)$
   (b) the Y-coordinate of $P(t)$ = the Y-coordinate of $P(3)$
   (c) the X-coordinate of $P(t)$ = the Y-coordinate of $P(2)$
   (d) $P(-t) = P(1)$

9. Let $\overline{P(s)P(t)}$ denote the distance between the circular points $P(s)$ and $P(t)$. Find the following numbers.
   (a) $\overline{P(0)P(\frac{1}{2}\pi)}$ (b) $\overline{P(\pi)P(2\pi)}$
   (c) $\overline{P(\frac{1}{6}\pi)P(\frac{5}{3}\pi)}$ (d) $\overline{P(\frac{27}{4}\pi)P(-\frac{13}{3}\pi)}$

10. Let $P(t_1) = (x_1, y_1)$ and $P(t_2) = (x_2, y_2)$ be two circular points. Show that $\overline{P(t_1)P(t_2)} = \sqrt{2 - 2(x_1x_2 + y_1y_2)}$.

11. Find the equation of the line determined by the given pair of circular points.
   (a) $P(0)$ and $P(\frac{1}{2}\pi)$ (b) $P(0)$ and $P(\pi)$
   (c) $P(\frac{1}{4}\pi)$ and $P(\frac{3}{4}\pi)$ (d) $P(-\frac{1}{3}\pi)$ and $P(\frac{5}{6}\pi)$

12. What is the area of the circular sector $P(1)OP(3)$?

27
THE CIRCULAR FUNCTIONS

With each real number t the winding function P that we introduced in the last section associates a point $P(t)$ and hence a pair (x, y) of real numbers, the coordinates of $P(t)$. For example, with the number 0 the function P associates the pair of numbers $(1, 0)$, with the number $\frac{1}{4}\pi$ it associates the pair $(\frac{1}{2}\sqrt{2}, \frac{1}{2}\sqrt{2})$, and so on. Now we will use this association to construct the six circular functions, the **cosine** function, the **sine** function, the **tangent** function, the **cotangent** function, the **secant** function, and the **cosecant** function. The values of these functions at a point t are denoted by **cos t, sin t, tan t, cot t, sec t,** and **csc t.** They are expressed in terms of the coordinates (x, y) of the circular point $P(t)$ by means of the following defining equations.

Definition 27-1 *If the coordinates of the circular point $P(t)$ are (x, y), then*

$$\cos t = x, \qquad\qquad \cot t = \frac{x}{y} \ (\text{if } y \neq 0),$$

(27-1) $\qquad\qquad \sin t = y, \qquad\qquad \sec t = \frac{1}{x} \ (\text{if } x \neq 0),$

$$\tan t = \frac{y}{x} \ (\text{if } x \neq 0), \qquad \csc t = \frac{1}{y} \ (\text{if } y \neq 0).$$

If t is a number for which the X-coordinate of $P(t)$ is 0, then tan t and sec t are not defined; that is, t is not in the domain of either the tangent or the secant function. For example, the X-coordinate of $P(\frac{1}{2}\pi)$ is 0, so $\frac{1}{2}\pi$ is not in the domain of either the tangent or the secant function. Similarly, some real numbers are not in the domains of the cotangent and cosecant functions, but the domain of both the cosine and sine functions is the set of all real numbers.

EXAMPLE 27-1   Calculate the values of the six circular functions at the point $\frac{1}{3}\pi$.

Solution   In the preceding section, we saw that the coordinates of the circular point $P(\frac{1}{3}\pi)$ are $(\frac{1}{2}, \frac{1}{2}\sqrt{3})$. Hence, according to Equations 27-1:

$$\cos \tfrac{1}{3}\pi = \tfrac{1}{2}, \qquad \sec \tfrac{1}{3}\pi = 2,$$
$$\sin \tfrac{1}{3}\pi = \tfrac{1}{2}\sqrt{3}, \qquad \csc \tfrac{1}{3}\pi = \tfrac{2}{3}\sqrt{3},$$
$$\tan \tfrac{1}{3}\pi = \sqrt{3}, \qquad \cot \tfrac{1}{3}\pi = \tfrac{1}{3}\sqrt{3}.$$

EXAMPLE 27-2   Calculate the values of the six circular functions at the point -3π.

Solution   To locate the circular point $P(-3\pi)$, we proceed 3π units along the unit circle in a clockwise direction, and we find that the coordinates of $P(-3\pi)$ are $(-1, 0)$. Hence, from Equations 27-1:

$\cos(-3\pi) = -1$, $\tan(-3\pi) = 0$, $\sin(-3\pi) = 0$, and $\sec(-3\pi) = -1$.

Since the Y-coordinate of $P(-3\pi)$ is 0, the symbols $\cot(-3\pi)$ and $\csc(-3\pi)$ are not defined.

We can establish certain relations among the various circular functions directly from their definitions. If (x, y) are the coordinates of any circular point $P(t)$, then $x^2 + y^2 = 1$ (Equation 26-1). But according to Equations 27-1, $x = \cos t$ and $y = \sin t$, and therefore we have the basic relation

(27-2) $\cos^2 t + \sin^2 t = 1$ for any number t.

In Equation 27-2 we used a shorthand notation that avoids the use of parentheses. For example, we write $\sin^2 t$ instead of $(\sin t)^2$. We use this convention with all *positive* exponents and all circular functions. But we can't (and don't) write $\sin^{-1} t$ to mean $1/\sin t$, since according to the notation we introduced in Section 17, $\sin^{-1}$ would denote the inverse of the sine function.

Other relations that we can derive from Equations 27-1 are

(27-3) $$\tan t = \frac{\sin t}{\cos t},$$

(27-4) $$\cot t = \frac{\cos t}{\sin t},$$

(27-5) $$\cot t = \frac{1}{\tan t},$$

(27-6)
$$\sec t = \frac{1}{\cos t},$$

(27-7)
$$\csc t = \frac{1}{\sin t}.$$

We can verify these equations very quickly. For example, $\tan t$ is the quotient y/x, where (x, y) are the coordinates of $P(t)$. But $y = \sin t$ and $x = \cos t$, and so

$$\tan t = \frac{y}{x} = \frac{\sin t}{\cos t}.$$

The other equations are verified in the same way. These five equations, together with Equation 27-2, are called the **elementary identities.** Each is true for every number t for which both sides of the equation are defined, and you should know these elementary identities by heart.

EXAMPLE 27-3   Show that $\tan t + \cot t = \csc t \sec t$.

Solution   We have

$$\tan t + \cot t = \frac{\sin t}{\cos t} + \frac{\cos t}{\sin t} \qquad \text{(Equations 27-3 and 27-4)}$$

$$= \frac{\sin^2 t + \cos^2 t}{(\sin t)(\cos t)} \qquad \text{(algebra)}$$

$$= \frac{1}{(\sin t)(\cos t)} \qquad \text{(Equation 27-2)}$$

$$= \csc t \sec t \qquad \text{(Equations 27-6 and 27-7)}.$$

EXAMPLE 27-4   Show that $\tan^2 t + 1 = \sec^2 t$.

Solution   This equation is meaningless if the X-coordinate of $P(t)$ is 0, and so we tacitly assume that it is not. In other words, $\cos t \neq 0$ for each number t in which we are interested. Thus we may divide both sides of the elementary identity $\sin^2 t + \cos^2 t = 1$ by $\cos^2 t$ and use other elementary identities to obtain, successively, the equations

$$\frac{\sin^2 t}{\cos^2 t} + 1 = \frac{1}{\cos^2 t},$$

and

(27-8)
$$\tan^2 t + 1 = \sec^2 t.$$

Since every circular point $P(t)$ belongs to the unit circle, neither of its coordinates can be greater than 1 in absolute value; that is, both $|x| \leq 1$ and $|y| \leq 1$. Therefore,

(27-9) $|\cos t| \leq 1$ and $|\sin t| \leq 1,$ for every number t.

It then follows from Equations 27-6 and 27-7 that

(27-10) $|\sec t| \geq 1$ and $|\csc t| \geq 1$,   for every number t
in the domains of these functions.

In set notation, Inequalities 27-9 and 27-10 tell us that the range of the sine
and cosine functions is the set $\{|y| \leq 1\}$, and the range of the secant and
cosecant functions is the set $\{|y| \geq 1\}$. The range of the tangent and co-
tangent functions is R^1.

PROBLEMS
27

1. Find the values of the six circular functions at a number whose circular point
    has the given coordinates.
    (a) $(\frac{3}{5}, -\frac{4}{5})$
    (b) $(1/\sqrt{3}, \sqrt{\frac{2}{3}})$
    (c) $(-\frac{1}{3}, \frac{2}{3}\sqrt{2})$
    (d) $(.24, .97)$
    (e) $(-.6, .8)$
    (f) $(.4447, -.8957)$

2. Complete the following table.

t	$\cos t$	$\sin t$	$\tan t$	$\cot t$	$\csc t$	$\sec t$
0	1	0	0	$\cdots$	$\cdots$	1
$\frac{1}{6}\pi$						
$\frac{1}{4}\pi$						
$\frac{1}{3}\pi$	$\frac{1}{2}$ .500	$\frac{1}{2}\sqrt{3}$ .866	$\sqrt{3}$ 1.732	$\frac{1}{3}\sqrt{3}$ .577	$\frac{2}{3}\sqrt{3}$ 1.154	2
$\frac{1}{2}\pi$	0	1	$\cdots$	0	1	$\cdots$
$\frac{2}{3}\pi$						
$\frac{3}{4}\pi$	$-\frac{1}{2}\sqrt{2}$ $-.707$	$\frac{1}{2}\sqrt{2}$ .707	-1	-1	$\sqrt{2}$ 1.414	$-\sqrt{2}$ -1.414
$\frac{5}{6}\pi$						
π	-1	0	0	$\cdots$	$\cdots$	-1
$\frac{7}{6}\pi$						
$\frac{5}{4}\pi$						
$\frac{4}{3}\pi$						
$\frac{3}{2}\pi$	0	-1	$\cdots$	0	-1	$\cdots$
$\frac{5}{3}\pi$						
$\frac{7}{4}\pi$						
$\frac{11}{6}\pi$						

3. Use a sketch of the unit circle to estimate the following numbers.
   (a) sin 3
   (b) cos 1
   (c) cot 4.7
   (d) sec 6.2
   (e) tan 2
   (f) tan (−2)

4. Which of the following numbers are positive?
   (a) sin 2
   (b) tan 3
   (c) cos 1.5
   (d) csc 4
   (e) cos 2 sin 2
   (f) cos 2 + sin 2

5. Use linear interpolation and the values of the sine function at $\frac{1}{4}\pi$ and $\frac{1}{3}\pi$ to estimate sin 1. Is your estimate too large or too smal?

6. Which number is larger?
   (a) sin (cos 2) or cos (sin 2)
   (b) sin (cos 3) or cos (sin 3)
   (c) sin (cos 0) or cos (sin 0)
   (d) sin (cos 4) or cos (sin 4)

7. Prove that sin $(u+v)$ and sin $u +$ sin v need not be the same number. What about tan $2u$ and 2 tan u?

8. Show that $1 + \cot^2 t = \csc^2 t$ for any number t in the domain of the cotangent function (see Example 27-4).

9. Show that each of the following expressions has the value 1 for each number t for which the expression is defined.
   (a) $\frac{1}{2}(\tan t \csc t \cos t + \cot t \sec t \sin t)$
   (b) $\dfrac{\sec t \csc t}{\tan t + \cot t}$
   (c) $\dfrac{\cos t + \sin^2 t \sec t}{\sec t}$
   (d) $(\sec t)^{-2} + (\csc t)^{-2}$
   (e) $\cos^2 (t^2 + 1) + \sin^2 (t^2 + 1)$
   (f) $[\![\cos^2 t]\!] + [\![\sin^2 t]\!] - [\![\sin t]\!] - [\![-\sin t]\!]$

10. Solve the following equations.
    (a) $|\cos t| = 1$
    (b) $|\sin t| = 1$
    (c) $[\![\sin t]\!] = -1$
    (d) $[\![\tan t]\!] = 0$

28
TABLES

We used geometric reasoning to locate the circular point $P(t)$ for numbers such as $t = 0$, $\frac{3}{2}\pi$, $-\pi$, $\frac{1}{3}\pi$, and so forth. Since the values of the circular functions are expressed in terms of the coordinates of these circular points, we are therefore able to calculate such numbers as cos π and sec $\frac{1}{3}\pi$. But most of the time we must evaluate circular functions by referring to tables. Table II in the back of the book lists values of the circular functions for numbers that range from 0 to 1.60 in steps of .01. We read this table in much the same way that we read the table of logarithms. In particular, we can use linear interpolation to find values of the circular functions for numbers (between 0 and 1.60) not listed in Table II. The numbers in this table are actually four-place decimal approximations of the values of the circular functions, and when we write an equation such as sin .82 = .7311, we really mean that .73105 ≤ sin .82 ≤ .73115.

EXAMPLE 28-1   Find sin .827.

Solution   Since .827 is .7 of the way from .82 to .83, we suppose that sin .827 is .7 of the way from sin .82 to sin .83; that is,

$$\sin .827 = \sin .82 + .7(\sin .83 - \sin .82).$$

From Table II we see that sin .82 = .7311 and sin .83 = .7379, so we have

$$\sin .827 = .7311 + .7(.7379 - .7311) = .7311 + .0048 = .7359.$$

EXAMPLE 28-2   Find cos 1.262.

Solution   Here we assume (see Equation 23-1) that

$$\cos 1.262 = \cos 1.26 + .2(\cos 1.27 - \cos 1.26)$$
$$= .3058 + .2(.2963 - .3058)$$
$$= .3058 - .0019 = .3039.$$

You will notice that .0019 is subtracted, not added, because cos 1.27 is smaller than cos 1.26. If you think about our definition of the cosine function, you will understand why its values decrease as we go down the list in Table II. You will want to keep this fact in mind as you interpolate in the table, but it doesn't affect the interpolation procedure.

EXAMPLE 28-3   Find the coordinates of the circular point $P(.73)$.

Solution   If (x, y) denote the desired coordinates, then $x = \cos .73$ and $y = \sin .73$. We look these numbers up in Table II, and we find that the coordinates of $P(.73)$ are (.7452, .6669).

The solution set of the equation log $x = .4771$ contains only one number, 3. But the solution set of the equation sin $t = 0$ contains many numbers, for example, 0, π, $-\pi$, 2π, and so on. Such behavior is typical of equations involving circular functions, and in a later section we shall discuss the question of finding the solution sets of such equations. Here, however, when we are just beginning our study of the circular functions, we shall be content to find only one or two members of the solution set of an equation that involves a circular function.

EXAMPLE 28-4   Find one number in the solution set of the equation tan $t = \sqrt{3}$.

Solution   You should remember that $(\frac{1}{2}, \frac{1}{2}\sqrt{3})$ are the coordinates of the circular point $P(\frac{1}{3}\pi)$, and therefore tan $\frac{1}{3}\pi = \sqrt{3}$. Hence $\frac{1}{3}\pi \in \{\tan t = \sqrt{3}\}$. Can you show that $\frac{4}{3}\pi$ also belongs to this set?

EXAMPLE 28-5   Find a solution of the equation cos $t = .6241$.

Solution   From Table II we see that

$$\cos .89 = .6294,$$
$$\cos t = .6241,$$
$$\cos .90 = .6216.$$

Since .6241 is $\frac{53}{78} = .7$ of the distance from .6294 to .6216, we conclude that a suitable choice for t is .897.

We end this section with an example that can be solved much more easily by using elementary identities than by using tables.

EXAMPLE 28-6   If $0 < t < \frac{1}{2}\pi$, and $\tan t = \frac{1}{2}\sqrt{5}$, find $\cos t$.

Solution   One way to find $\cos t$ is to express $\frac{1}{2}\sqrt{5}$ as a decimal, use Table II to solve the equation $\tan t = \frac{1}{2}\sqrt{5}$ for t, and then again use the table to find $\cos t$. But it is much easier to use some elementary identities to solve this problem. From Equation 27-8,

$$\sec^2 t = 1 + \tan^2 t = 1 + \tfrac{5}{4} = \tfrac{9}{4}.$$

Thus,

$$\cos^2 t = 1/\sec^2 t = \tfrac{4}{9}.$$

Since $0 < t < \frac{1}{2}\pi$, we know that $\cos t > 0$, and hence it follows from the last equation that $\cos t = \frac{2}{3}$.

PROBLEMS
28

1. Use Table II to find the following numbers.
   (a) $\tan .237$ (b) $\sec .409$ (c) $\csc 1.103$
   (d) $\cot 1.234$ (e) $\sin \frac{3}{8}$ (f) $\cos \frac{7}{8}$

2. Calculate the following numbers.
   (a) $\tan 1.6 - 2 \tan .8$ (b) $\sin .47 - \sin .27 - \sin .2$
   (c) $\sec \frac{1}{8} - \frac{1}{8} \sec 1$ (d) $1 + \tan^2 .245$
   (e) $\sin (\cos 1) - \cos (\sin 1)$ (f) $\sec (\cos^2 1)$

3. Find a member of the solution set of each of the following equations.
   (a) $\sin t = \frac{3}{8}$ (b) $\cos t = \frac{3}{8}$ (c) $\tan t = \frac{8}{3}$
   (d) $\cot t = \frac{1}{4}\pi$ (e) $\cot \frac{1}{4}\pi = t$ (f) $\pi \cos t = 1$

4. Use the values of $\cos .82$ and $\sin .82$ listed in Table II to test the identity $\cos^2 t + \sin^2 t = 1$.

5. With the aid of Table II, find the value of each of the following expressions. (Use elementary identities to simplify the expression first.)
   (a) $\sqrt{1 - \cos^2 .56}$ (b) $\sqrt{\sec^2 .79 - 1}$
   (c) $(\tan 1.4 \cos 1.4 - \csc 1.4) \csc 1.4$ (d) $\sec .32/(\tan .32 + \cot .32)$

6. With what circular functions may C be replaced to make the inequality $C(.666) > C(\frac{2}{3})$ a true statement?

119

7. Explain why the values of the cosine, tangent, cotangent, and secant functions at the numbers 1.58, 1.59, and 1.60 are listed as negative in Table II.

8. The approximation $\cos t \approx 1 - \frac{1}{2}t^2 + \frac{1}{24}t^4$ is quite good if t is a relatively small number. Try it for $t = .1, .3$, and 1.

9. Find 3 members of each of the following sets.
    (a) $\{[\![\tan t]\!] = 3\}$ (b) $\{[\![\csc t]\!] = 7\}$
    (c) $\{2 \le \tan t \le 2.1\}$ (d) $\{9 \le \csc t \le 10\}$

10. How do the identities $\cot t = 1/\tan t$, $\csc t = 1/\sin t$, and $\sec t = 1/\cos t$ show up in Table II?

11. Later, we will derive the formulas $\cos t = \sin(\frac{1}{2}\pi - t)$, $\cot t = \tan(\frac{1}{2}\pi - t)$, and $\csc t = \sec(\frac{1}{2}\pi - t)$. How do these formulas show up in Table II? (Recall that $\frac{1}{2}\pi = 1.57$.)

12. Use Table II to convince yourself that if t is a very small positive number, then $(\sin t)/t$ is very close to 1.

29
FUNCTIONAL VALUES AT ANY NUMBER t

Just as we are able to find the logarithm of any positive number by using only a table of logarithms of numbers between 1 and 10, so we can evaluate circular functions by using only a table of values (such as Table II) that correspond to numbers between 0 and $\frac{1}{2}\pi$ ($= 1.5708$). The values in such a table are computed from coordinates of circular points that lie in the first quadrant, and we can use symmetry arguments to express the coordinates of a point in another quadrant in terms of the coordinates of a first quadrant point. An example will show us how.

EXAMPLE 29-1 Find $\cos 2$.

Solution The number $\cos 2$ is the X-coordinate of the circular point $P(2)$ that is located 2 units along the unit circle from the point $(1, 0)$, as shown in Fig. 29-1. Now observe that $P(2)$ is $\pi - 2 = 3.14 - 2 = 1.14$ (ap-

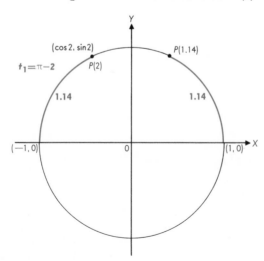

FIGURE 29-1

120

proximately) units along the unit circle from the point $(-1, 0)$. It is therefore obvious from the figure that the point $P(1.14)$ in the first quadrant is symmetric to $P(2)$ with respect to the Y-axis. Hence, the X-coordinate of $P(2)$ (the number we are looking for) is the negative of the X-coordinate of $P(1.14)$. But this number is simply $\cos 1.14$, and hence it is listed in Table II. Thus, $\cos 2 = -\cos 1.14 = -.4176$.

The method we followed in Example 29-1 illustrates the general technique of finding the values of circular functions at a number t that is not listed in Table II. First we locate the point $P(t)$ of the unit circle and calculate the length of the shortest arc of the unit circle that joins $P(t)$ to one of the points $(1, 0)$ or $(-1, 0)$. Figure 29-2 shows the possible configurations.

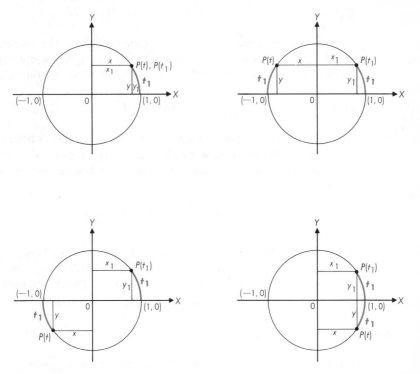

FIGURE 29-2

We denote this length by t_1 and call it the **reference number** associated with t. In Example 29-1 we saw that the reference number associated with 2 is the number 1.14. Figure 29-3 shows that the reference numbers associated with 4 and 5 are $4 - \pi = .86$ and $2\pi - 5 = 1.28$.

Because the circular point $P(t)$ cannot be more than one-quarter of a circle away from one or the other of the points $(1, 0)$ or $(-1, 0)$, the reference number t_1 associated with t will always be between 0 and $\frac{1}{2}\pi$. Therefore, the circular point $P(t_1)$ will lie in the first quadrant. In order to find the relationship between the coordinates (x_1, y_1) of $P(t_1)$ and the coordinates (x, y) of $P(t)$, we examine the drawings in Fig. 29-2. You can easily convince your-

121

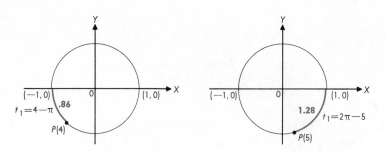

FIGURE 29-3

self that in every case either $x = x_1$ or $x = -x_1$, and that either $y = y_1$ or $y = -y_1$. In other words,

(29-1) $$|x| = x_1 \quad \text{and} \quad |y| = y_1.$$

These relations between the coordinates of the circular points $P(t)$ and $P(t_1)$ lead to similar relations between the values of the circular functions at the numbers t and t_1. Thus, *if C is any circular function, then*

(29-2) $$|C(t)| = C(t_1).$$

There are really six possible versions of Equation 29-2, obtained by replacing C with cos, sin, tan, and so on. Let us verify just one of them, for example, the equation $|\tan t| = \tan t_1$. According to the definition of the tangent function,

$$|\tan t| = \left|\frac{y}{x}\right| = \frac{|y|}{|x|} = \frac{y_1}{x_1} = \tan t_1.$$

The values of the circular functions at t_1 are listed in Table II, and therefore we can determine the absolute values of the circular functions at any number t as soon as we find the reference number t_1. To find the actual values, we need only notice the quadrant in which $P(t)$ lies and affix the proper sign.

EXAMPLE 29-2   Find sin 4.

Solution   The circular point $P(4)$ lies in the third quadrant (see Fig. 29-3), so sin $4 < 0$. The associated reference number is $t_1 = 4 - \pi = .86$. According to Equation 29-2 and Table II,

$$|\sin 4| = \sin .86 = .7578.$$

Therefore, sin $4 = -.7578$.

EXAMPLE 29-3   Find sec $(-\frac{31}{4}\pi)$.

Solution   We obtain the point $P(-\frac{31}{4}\pi)$ by proceeding $\frac{31}{4}\pi = 7\pi + \frac{3}{4}\pi$ units around the unit circle from the point $(1, 0)$ in a clockwise direction (Fig. 29-4). The point is in the first quadrant, and the reference number is $t_1 = \frac{1}{4}\pi$. Hence, sec $(-\frac{31}{4}\pi) = \sec\frac{1}{4}\pi = \sqrt{2}$.

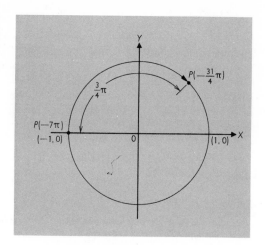

FIGURE 29-4

Our method for finding the values of the circular functions at any number t consists of four steps:

(i) Determine the quadrant in which $P(t)$ lies;
(ii) Find the reference number t_1 associated with t;
(iii) Use Table II to find $C(t_1) = |C(t)|$;
(iv) Affix the proper algebraic sign.

If t is a number between -2π and 2π, it is easy to locate $P(t)$. If t is not a number between -2π and 2π, and if the location of $P(t)$ is not obvious, we might proceed in the manner illustrated by the following two examples.

EXAMPLE 29-4 Locate $P(60)$, find the reference number t_1 associated with $t = 60$, and find the value of cos 60.

Solution To locate the circular point $P(60)$ we must proceed 60 units around the unit circle in a counter-clockwise direction, starting from the point $(1, 0)$. Since the circumference of the unit circle is $2\pi = 6.283$ units, it is apparent that we will traverse the entire circumference a number of times. To discover the exact number of complete revolutions, we divide 60 by 2π, using the process of "long division":

$$\begin{array}{r} 9. \\ 6.283\overline{\smash{)}60.000} \\ 56.547 \\ \hline 3.453 \end{array}$$

Thus we may write $60 = 9 \times 6.283 + 3.453$. This result means that to locate the point $P(60)$ we must proceed nine times around the unit circle and then 3.453 units in a counter-clockwise direction. It follows that the point $P(3.453)$ is the same as the point $P(60)$. It is clear that $P(3.453)$ is in the third quadrant, so $t_1 = 3.453 - \pi = 3.453 - 3.142 = .311$. With the aid of Table II, we find that $\cos 60 = -\cos .311 = -.9520$.

123

EXAMPLE 29-5   Find tan (−26).

Solution   The first step in locating $P(-26)$ is to use "long division" to write $-26 = -4 \times 6.283 - .868$. This result means that we are to proceed four times around the unit circle and then .868 units in a *clockwise* direction. The point $P(-.868)$ is therefore the same as the point $P(-26)$. Since $-\frac{1}{2}\pi = -1.571$, we see that $P(-.868)$ is in the fourth quadrant and that $t_1 = .868$. Thus, we have tan $(-26) = -\tan .868 = -1.180$.

PROBLEMS 29

1. Evaluate the following numbers.
   (a) $\sin \frac{41}{4}\pi$ (b) $\cos \left(-\frac{27}{4}\pi\right)$ (c) $\sec \frac{37}{6}\pi$ (d) $\tan \frac{19}{3}\pi$
   (e) $\sin \frac{13}{3}\pi$ (f) $\cos \frac{17}{6}\pi$ (g) $\tan \frac{22}{3}\pi$ (h) $\sin \left(-\frac{35}{6}\pi\right)$

2. Use Table II (with $\pi = 3.14$) to evaluate the following numbers.
   (a) $\cos 3.56$ (b) $\sin 2.79$ (c) $\cot (-5.41)$ (d) $\tan 4.75$
   (e) $\csc (-3.95)$ (f) $\sec 6.19$ (g) $\cos 4.37$ (h) $\tan 4.19$

3. Use Table II (with $\pi = 3.142$ and $2\pi = 6.283$) to find the coordinates of each of the following circular points.
   (a) $P(2.341)$ (b) $P(3.241)$ (c) $P(4.231)$ (d) $P(6)$
   (e) $P(20)$ (f) $P(-\frac{1}{8})$ (g) $P(\sqrt{8})$ (h) $P(\sqrt{27})$

4. Evaluate the following numbers.
   (a) $\sin \pi^2$ (b) $\tan \frac{22}{7}\pi$ (c) $\cos \frac{20}{\pi}$ (d) $\sec \frac{\pi}{20}$

5. Evaluate the following numbers (use $\pi = 3.142$ and $2\pi = 6.283$).
   (a) $\sin 40$ (b) $\cos 50$ (c) $\tan 35$ (d) $\cot 25$

6. Find the smallest number t that is larger than π and that also satisfies the given equation.
   (a) $\sin t = \frac{1}{2}$ (b) $\cos t = \frac{1}{2}\sqrt{3}$ (c) $\tan t = 1$
   (d) $\sec t = 2$ (e) $\csc t = 2$ (f) $\cos t = .7379$

7. Let u and v be two real numbers with associated reference numbers u_1 and v_1. Show that if $|\sin u| = |\sin v|$, then $u_1 = v_1$.

8. One end of a shaft is fastened to a piston that moves vertically. The other end is connected by prongs to a peg P on the rim of the wheel, as shown in Fig. 29-5. The wheel has a radius of 1 foot, and the shaft is k feet long. The wheel rotates counter-clockwise at the rate of 1 revolution per second; that is, P moves 2π feet per second around a unit circle. A coordinate system is introduced as shown in the figure.
   (a) Find a formula for the distance d feet between the bottom of the piston and the X-axis T seconds after P is at the point $(1, 0)$.
   (b) Locate the piston when T is $\frac{7}{8}$, $\frac{10}{3}$, and $13/\pi$.

124

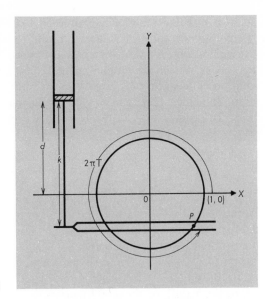

FIGURE 29-5

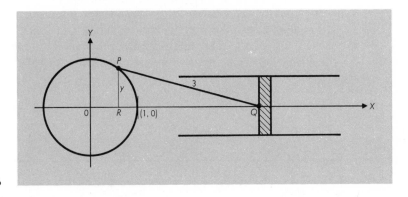

FIGURE 29-6

9. A piston is connected to the rim of a wheel as shown in Fig. 29-6. The radius of the wheel is 1 foot and the connecting rod PQ is 3 feet long. The wheel rotates counter-clockwise at the rate of 1 revolution per second. A coordinate system is introduced as shown in the figure.

(a) Show that T seconds after the point P has coordinates $(1, 0)$, the X-coordinate of Q is given by the equation

$$x = \sqrt{9 - \sin^2 2\pi T} + \cos 2\pi T.$$

(b) Where is the point Q when T is $\frac{1}{3}$, $\frac{11}{3}$, and $\frac{4}{5}$?

10. If a and b are positive integers, the process of long division produces integers q and r, where $0 \le r < b$, such that $a = bq + r$. Can you convince yourself that $\left| C\left(\frac{a}{b}\pi\right) \right| = \left| C\left(\frac{r}{b}\pi\right) \right|$ for each circular function C?

125

To draw the graph of a circular function we proceed as we do with any function, we plot the points of a table of values and join them. It will make our job a lot easier if we first observe that the circular functions are periodic; that is, their values repeat. Since the circular point $P(t+2\pi)$ is 2π units (one complete revolution) along the unit circle from the point $P(t)$, we see that these two points coincide. Thus the values of the circular functions repeat every 2π units. In other words, for each number t and each circular function C, we have

(30-1) $$C(t+2\pi) = C(t).$$

For example, $\sin(2+2\pi) = \sin 2$ and $\tan(4\pi - 3) = \tan(2\pi - 3) = \tan(-3)$. From Equation 30-1 it follows that the graph of a circular function need only be plotted for values of t from any interval 2π units long; the complete graph consists of duplications of this portion.

To plot the graph of the sine function we shall first choose values of t between and including 0 and 2π. (Any other interval of length 2π would do just as well.) From the values of the sine function that we calculated in Problem 27-2, we can find 17 points of this portion of the graph. If we need additional points, we can use the methods we discussed in the preceding section to find them. We may plot these points and join them smoothly to form one cycle of the sine curve. The remainder of the graph of the sine function (Fig. 30-1) consists of duplications of this part of the graph.

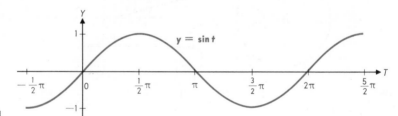

FIGURE 30-1

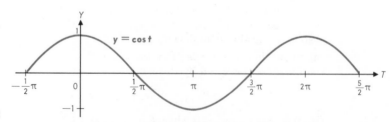

FIGURE 30-2

We can plot the graphs of the other circular functions in the same way. The graph of the cosine function is shown in Fig. 30-2, and the graph of the cosecant function in Fig. 30-3. Since $\csc t = 1/\sin t$, we have drawn a black sine curve in Fig. 30-3 to show how it is related to the cosecant curve.

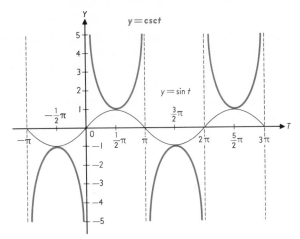

FIGURE 30-3

Not only does the graph of the tangent function repeat every 2π units, it repeats every π units as well. Thus, $\tan(t + \pi) = \tan t$ for every number t in the domain of the tangent function. This periodicity equation is a consequence of the fact that the coordinates of the circular point $P(t + \pi)$ are the negatives of the coordinates of $P(t)$. It tells us that we need only plot the portion of the tangent curve that corresponds to values of t between $-\frac{1}{2}\pi$ and $\frac{1}{2}\pi$ and then duplicate it to the left and to the right to get the entire curve (Fig. 30-4).

Up to this point, we have used the letter t rather than the letter x to denote a number in the domain of a circular function, in order to avoid confusion with the X-coordinate of the circular point $P(t)$. When there is no danger of such confusion, we often use the letter x to denote a number in

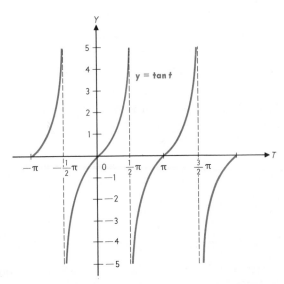

FIGURE 30-4

127

the domain. In particular, we frequently use the letter x instead of t when dealing with graphs, since we usually label the horizontal coordinate axis as the X-axis rather than the T-axis.

The graphs display certain properties of the circular functions that we shall later study analytically. It is clear, for example, that the cosine and sine curves are the "same" except for their positions relative to the Y-axis. For example, if we "shift" the sine curve $\frac{1}{2}\pi$ units to the left, it will become the cosine curve.

A graph is **symmetric with respect to the Y-axis** if for each point (x, y) of the graph the point $(-x, y)$ also belongs to the graph. In geometric terms this statement means that for every point P of the graph there is a point Q of the graph such that the Y-axis is the perpendicular bisector of the segment PQ. When a graph is symmetric with respect to the Y-axis, we say that the part of the graph to the left of the Y-axis is the **reflection** about the Y-axis of the part of the graph to the right. Figure 30-2 shows at a glance that the graph of the cosine function is symmetric with respect to the Y-axis.

A graph is **symmetric with respect to the origin** if for each point (x, y) of the graph the point $(-x, -y)$ also belongs to the graph. Geometrically, this sentence means that for every point P of the graph there is a point Q of the graph such that the origin is the midpoint of the segment PQ. Figures 30-1 and 30-4 show clearly that the graphs of both the sine and tangent functions are symmetric with respect to the origin.

Figures 30-1 and 30-4 also illustrate some properties of the circular functions that are developed in a course in calculus. You should become familiar with the graphs shown in these figures and be able to make a quick sketch of the graph of any circular function.

PROBLEMS
30

1. Sketch graphs of the secant and cotangent functions.

2. Discuss the symmetry of the graphs of the cotangent, secant, and cosecant functions.

3. Use the figures in this section to find the following numbers.
    (a) $\sin 5$ (b) $\cos 3$ (c) $\tan (-2)$
    (d) $\cos (-3)$ (e) $\sin 167$ (f) $\csc (-4)$

4. Use the figures in this section to determine *how many* solutions of the given equation belong to the interval $\{-1 \le t \le 6\}$.
    (a) $\sin t = .1$ (b) $\cos t = .1$ (c) $\tan t = .1$
    (d) $\csc t = 3$ (e) $\cos t = -.1$ (f) $\tan t = -.1$

5. Explain how you could use a ruler and compass to find $\sin (\sin \frac{3}{4}\pi)$ from Fig. 30-1. Try it, and check your result against the tables.

6. If you are told that the graph of the equation $y = mt + b$ does not intersect the cosecant curve, what can you say about the numbers m and b?

7. We remarked that if we shift the sine curve $\frac{1}{2}\pi$ units to the left, it becomes the cosine curve. What happens if we shift it $\frac{1}{2}\pi$ units to the right and "turn it over"? What if we shift it π units to the right and turn it over?

8. Sketch the graphs of the equations $y = x$ and $y = \cos x$ in the same cartesian plane. Find an approximate solution of the equation $x = \cos x$ by finding the X-coordinate of the point of intersection of the two graphs. Check your answer with Table II. How many solutions does the equation have? How many solutions does the equation $x = \sin x$ have? The equation $x = \tan x$?

9. Use the figures in this section to help you sketch the graphs of the following equations.
   (a) $y = |\cos x|$ (b) $y = \cos |x|$ (c) $y = [\![\sin x]\!]$
   (d) $y = \sin [\![x]\!]$ (e) $y = [\![\tan x]\!]$ (f) $y = \tan [\![x]\!]$

10. Sketch the graphs of the following equations in the interval $\{0 < x < \pi\}$.
    (a) $y = \sin 2x$ (b) $y = 2 \sin x$ (c) $y = \cot 2x$ (d) $y = 2 \cot x$

11. How are the numbers $f(x)$ and $f(-x)$ related if
    (a) the graph of the equation $y = f(x)$ is symmetric with respect to the Y-axis?
    (b) the graph of the equation $y = f(x)$ is symmetric with respect to the origin?

12. Use a graphical argument to show that $\sin t \geq 2t/\pi$ and $\cos t \geq 1 - 2t/\pi$ if if $0 \leq t \leq \frac{1}{2}\pi$.

31
ADDITION FORMULAS

A formula relating the function values $f(u)$, $f(v)$, and $f(u+v)$ is called an **addition formula.** For example, if f is defined by the direct variation equation $f(x) = mx$, we have the addition formula $f(u+v) = f(u) + f(v)$. If f is an exponential function $\exp_b$, we have the addition formula $f(u+v) = f(u)f(v)$. In this section, we start developing addition formulas for the circular functions.

Let u and v be any two numbers, and $P(u)$ and $P(v)$ be the corresponding circular points. Figure 31-1 is drawn for particular numbers u and v, but the following argument is applicable to any pair of numbers. With the number $u - v$ is associated the circular point $P(u - v)$, and it is clear that the arc joining the points $P(0)$ and $P(u - v)$ has the same length as the arc joining $P(u)$ and $P(v)$—namely, $|u - v|$ units. We know from geometry that the corresponding chords are therefore also of equal length. The lengths of these chords may be expressed in terms of the coordinates of their endpoints as follows. According to the definitions of the sine and cosine functions, the coordinates of any circular point $P(t)$ are $(\cos t, \sin t)$. Therefore the coordinates of $P(u)$ are $(\cos u, \sin u)$, the coordinates of $P(v)$ are $(\cos v, \sin v)$, and the coordinates of $P(u - v)$ are $(\cos (u - v), \sin (u - v))$. The coordinates of $P(0)$ are, of course, $(1, 0)$. Then according to the distance formula,

$$\overline{P(u)P(v)}^2 = (\cos u - \cos v)^2 + (\sin u - \sin v)^2$$
$$= \cos^2 u - 2 \cos u \cos v + \cos^2 v + \sin^2 u$$
$$- 2 \sin u \sin v + \sin^2 v.$$

129

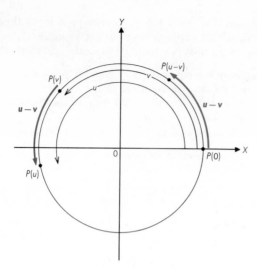

FIGURE 31-1

Two applications of the fundamental identity $\cos^2 t + \sin^2 t = 1$ reduce this equation to

(31-1) $$\overline{P(u)P(v)}^2 = 2 - 2\cos u \cos v - 2\sin u \sin v.$$

Similarly,

$$\overline{P(0)P(u-v)}^2 = [\cos(u-v) - 1]^2 + \sin^2(u-v)$$
$$= \cos^2(u-v) - 2\cos(u-v)$$
$$+ 1 + \sin^2(u-v),$$

and so

(31-2) $$\overline{P(0)P(u-v)}^2 = 2 - 2\cos(u-v).$$

Since the two chords have the same length, we can equate the right-hand sides of Equations 31-1 and 31-2 to obtain the equation

$$2 - 2\cos(u-v) = 2 - 2\cos u \cos v - 2\sin u \sin v.$$

Now a slight simplification leads to the identity

(31-3) $$\cos(u-v) = \cos u \cos v + \sin u \sin v.$$

Notice, in particular, that $\cos(u-v)$ is *not* the same as $\cos u - \cos v$.

130

EXAMPLE 31-1 Find $\cos \frac{1}{12}\pi$.

Solution We can write $\frac{1}{12}\pi = \frac{1}{3}\pi - \frac{1}{4}\pi$ and therefore can apply Equation 31-3 to obtain

$$\cos \tfrac{1}{12}\pi = \cos (\tfrac{1}{3}\pi - \tfrac{1}{4}\pi)$$
$$= \cos \tfrac{1}{3}\pi \cos \tfrac{1}{4}\pi + \sin \tfrac{1}{3}\pi \sin \tfrac{1}{4}\pi$$
$$= \tfrac{1}{2} \cdot \tfrac{1}{2}\sqrt{2} + \tfrac{1}{2}\sqrt{3} \cdot \tfrac{1}{2}\sqrt{2}$$
$$= \tfrac{1}{4}\sqrt{2}(1 + \sqrt{3}).$$

Equation 31-3 is the fundamental formula from which we will derive a great deal of information about the circular functions, including further addition formulas. For example, since the formula must be valid for any two numbers u and v, it must be valid if $u = \frac{1}{2}\pi$. Thus,

$$\cos (\tfrac{1}{2}\pi - v) = \cos \tfrac{1}{2}\pi \cos v + \sin \tfrac{1}{2}\pi \sin v.$$

We already know that $\cos \frac{1}{2}\pi = 0$ and $\sin \frac{1}{2}\pi = 1$, so it follows that

(31-4) $$\cos (\tfrac{1}{2}\pi - v) = \sin v,$$

and *this equation is valid for any number v.*

A similar identity is valid for the sine function. Let u be any number, and replace v in Equation 31-4 with $\frac{1}{2}\pi - u$. In this way, we obtain the equation $\cos (\frac{1}{2}\pi - (\frac{1}{2}\pi - u)) = \sin (\frac{1}{2}\pi - u)$, which simplifies to

(31-5) $$\sin (\tfrac{1}{2}\pi - u) = \cos u.$$

This equation is valid for any number u.

EXAMPLE 31-2 If u and v are two numbers such that $u + v = \frac{1}{2}\pi$, show that $\sin^2 u + \sin^2 v = 1$.

Solution Since $u + v = \frac{1}{2}\pi$, $v = \frac{1}{2}\pi - u$. Therefore, $\sin v = \cos u$ (Equation 31-5), and

$$\sin^2 u + \sin^2 v = \sin^2 u + \cos^2 u = 1.$$

We have seen sufficiently many examples of functions to know that for a given function f, the numbers $f(-x)$ and $-f(x)$ are not necessarily the same. In the case of the sine function, however, they are:

(31-6) $$\sin (-v) = -\sin v \quad \text{for every number } v.$$

The proof of this assertion follows readily from the identities we have just derived:

$$\sin (-v) = \cos (\tfrac{1}{2}\pi - (-v)) \qquad \text{(Equation 31-4)}$$
$$= \cos (v - (-\tfrac{1}{2}\pi)) \qquad \text{(algebra)}$$
$$= \cos v \cos (-\tfrac{1}{2}\pi) + \sin v \sin (-\tfrac{1}{2}\pi) \qquad \text{(Equation 31-3)}.$$

But since $\cos (-\tfrac{1}{2}\pi) = 0$ and $\sin (-\tfrac{1}{2}\pi) = -1$, this last equation reduces to Equation 31-6.

In Section 30 we observed that the graph of the sine function is symmetric with respect to the origin. Equation 31-6 is the analytic statement of this fact. For suppose that the point (x, y) belongs to the graph of the sine function; that is, that $y = \sin x$. Then, since $\sin (-x) = -\sin x = -y$, we see that the point $(-x, -y)$ also belongs to the graph, and so the graph is symmetric with respect to the origin.

But $\cos (-v)$ and $-\cos v$ are in general not the same number; indeed,

(31-7) $\qquad\qquad \boldsymbol{\cos (-v) = \cos v} \quad \textit{for every number } v.$

This identity is established by setting $u = 0$ in Equation 31-3 to obtain

$$\cos (0 - v) = \cos (-v) = \cos 0 \cos v + \sin 0 \sin v.$$

Since $\cos 0 = 1$ and $\sin 0 = 0$, this last equation reduces to Equation 31-7.

In geometric terms, Equation 31-7 tells us that the graph of the cosine function is symmetric with respect to the Y-axis. For if the point (x, y) belongs to the graph of the equation $y = \cos x$, then the point $(-x, y)$ also belongs to the graph, since $\cos (-x) = \cos x = y$.

EXAMPLE 31-3 Show that

$$\boldsymbol{\tan (-v) = -\tan v} \quad \textit{for every number } v \textit{ in the domain of the tangent}$$
$$\textit{function}.$$

Solution We have

$$\tan (-v) = \frac{\sin (-v)}{\cos (-v)} = \frac{-\sin v}{\cos v} = -\tan v.$$

PROBLEMS

31

1. Use the equation $\tfrac{1}{12}\pi = \tfrac{1}{4}\pi - \tfrac{1}{6}\pi$ to find $\cos \tfrac{1}{12}\pi$, and compare the result with the number we obtained in Example 31-1.

2. What equation do we get when we let $v = u$ in Equation 31-3?

3. Show that the following equations are identities; they are true for any number t.

   (a) $\cos (\pi - t) = -\cos t$ \qquad\qquad (b) $\cos (2\pi - t) = \cos t$

   (c) $\cos (\tfrac{1}{2}\pi + t) = -\sin t$ \qquad\qquad (d) $\cos (\tfrac{3}{2}\pi - t) = -\sin t$

4. Show that $\csc u = \sec (\frac{1}{2}\pi - u)$ and $\cot u = \tan (\frac{1}{2}\pi - u)$, thus demonstrating that co-$C(u) = C(\frac{1}{2}\pi - u)$ for any circular function C and number u in the domain of C.

5. How are the following pairs of numbers related?
    (a) $\cot t$ and $\cot (-t)$ (b) $\sec t$ and $\sec (-t)$ (c) $\csc t$ and $\csc (-t)$

6. Find $P(\frac{7}{12}\pi)$.

7. Show that $|C(\pi - t)| = |C(t)|$ for any circular function C and number t in its domain.

8. (a) Show that $\cos (x - \frac{1}{2}\pi) = \sin x$ for each number x and explain why this identity tells us that the sine curve is simply the cosine curve shifted $\frac{1}{2}\pi$ units to the right.
    (b) Show that $\sin (x - \frac{1}{2}\pi) = -\cos x$, and explain why this identity tells us that the cosine curve is the sine curve shifted $\frac{1}{2}\pi$ units to the right and "turned over."

9. Find a member of the solution set of each of the following equations.
    (a) $\cos x = \sin \frac{2}{3}\pi$ (b) $\cos x = \sin \frac{2}{3}$
    (c) $\cos (3x - 2) = \sin (2 - 2x)$ (d) $\cos (\frac{1}{2}\pi - t) = \cos \frac{1}{2}\pi - \cos t$

10. Find at least one solution of each of the following equations.
    (a) $\sin (x - 3) + \sin (5 - 2x) = 0$ (b) $\cos (3x + 1) - \cos (2x - 9) = 0$
    (c) $\tan (x + 4) + \tan (5 - 4x) = 0$ (d) $\cot (5x - 4) + \tan (4x - 4) = 0$

11. If $f(-x) = f(x)$ for every number x in the domain of f, then f is called an **even function.** For example, the cosine function is an even function. Determine which of the following equations define even functions.
    (a) $f(x) = \csc x$ (b) $f(x) = \sec x$ (c) $f(x) = |x|$
    (d) $f(x) = x^{52}$ (e) $f(x) = \csc x \cot x$ (f) $f(x) = x \sin x$
    (g) $f(x) = x + \sin x$ (h) $f(x) = [\![\cos x]\!]$

12. If $f(-x) = -f(x)$ for every number x in the domain of f, then f is called an **odd function.** For example, the sine function is an odd function. Determine which of the following equations define odd functions.
    (a) $f(x) = \cot x$ (b) $f(x) = \csc x^2$ (c) $f(x) = x^{25}$
    (d) $f(x) = x + 1/x$ (e) $f(x) = x \sin x$ (f) $f(x) = x + \sin x$
    (g) $f(x) = \cos (\sin x)$ (h) $f(x) = [\![\sin x]\!]$

13. What can you say about the symmetry of the graph of an even function? An odd function?

32

ADDITION
FORMULAS
FOR THE
CIRCULAR
FUNCTIONS

We can use the identities we found in the last section to derive addition formulas for the circular functions. Since $\cos (u + v) = \cos (u - (-v))$, we may apply Equation 31-3 to this latter expression and obtain the equation

$$\cos (u + v) = \cos (u - (-v)) = \cos u \cos (-v) + \sin u \sin (-v).$$

We already know that $\cos(-v) = \cos v$ and $\sin(-v) = -\sin v$, so

(32-1) $$\cos(u+v) = \cos u \cos v - \sin u \sin v.$$

To find our addition formula for $\sin(u+v)$, we write

$$
\begin{aligned}
\sin(u+v) &= \cos(\tfrac{1}{2}\pi - (u+v)) && \text{(Equation 31-4)}\\
&= \cos((\tfrac{1}{2}\pi - u) - v) && \text{(algebra)}\\
&= \cos(\tfrac{1}{2}\pi - u)\cos v + \sin(\tfrac{1}{2}\pi - u)\sin v && \text{(Equation 31-3)}\\
&= \sin u \cos v + \cos u \sin v && \text{(Equations 31-4}\\
& && \text{and 31-5).}
\end{aligned}
$$

Thus we obtain the identity

(32-2) $$\sin(u+v) = \sin u \cos v + \cos u \sin v.$$

We find our formula for $\sin(u-v)$ by replacing v with $-v$ in this equation:

$$\sin(u-v) = \sin u \cos(-v) + \cos u \sin(-v).$$

Now we replace $\cos(-v)$ with $\cos v$ and $\sin(-v)$ with $-\sin v$, and our identity becomes

(32-3) $$\sin(u-v) = \sin u \cos v - \cos u \sin v.$$

There are so many identities of the type we have developed in this section and preceding ones that it is difficult to keep them all in mind. The key facts to memorize are the formulas for $\sin(u+v)$ and $\cos(u+v)$, plus the equations $\sin(-v) = -\sin v$ and $\cos(-v) = \cos v$. With this information, and a knowledge of the values of the circular functions at $t = 0$, $\tfrac{1}{2}\pi$, and so on, it should be easy for you to derive the other formulas we have discussed and many more.

EXAMPLE 32-1 Show that $\sin(\tfrac{1}{2}\pi + t) = \cos t$ for every number t.

Solution According to Equation 32-2, $\sin(\tfrac{1}{2}\pi + t) = \sin\tfrac{1}{2}\pi \cos t + \cos\tfrac{1}{2}\pi \sin t$. But $\sin\tfrac{1}{2}\pi = 1$ and $\cos\tfrac{1}{2}\pi = 0$, and therefore $\sin(\tfrac{1}{2}\pi + t) = \cos t$.

EXAMPLE 32-2 Find an expression for $\tan(u+v)$.

Solution According to an elementary identity, $\tan(u+v) = \dfrac{\sin(u+v)}{\cos(u+v)}$. When we expand the numerator and denominator we get

$$\tan(u+v) = \frac{\sin u \cos v + \cos u \sin v}{\cos u \cos v - \sin u \sin v}.$$

134

If neither $\cos u = 0$ nor $\cos v = 0$, we can divide both the numerator and the denominator of this fraction by the product $\cos u \cos v$ to obtain a formula that involves only the tangent function:

$$\tan (u+v) = \frac{\dfrac{\sin u \cos v}{\cos u \cos v} + \dfrac{\cos u \sin v}{\cos u \cos v}}{\dfrac{\cos u \cos v}{\cos u \cos v} - \dfrac{\sin u \sin v}{\cos u \cos v}},$$

so that

$$\tan (u+v) = \frac{\tan u + \tan v}{1 - \tan u \tan v}.$$

EXAMPLE 32-3   Find $\sec \frac{7}{12}\pi$.

Solution   Since $\frac{7}{12}\pi = \frac{1}{3}\pi + \frac{1}{4}\pi$, we have

$$\sec \tfrac{7}{12}\pi = \frac{1}{\cos \frac{7}{12}\pi} = \frac{1}{\cos \left(\frac{1}{3}\pi + \frac{1}{4}\pi\right)}$$

$$= \frac{1}{\cos \frac{1}{3}\pi \cos \frac{1}{4}\pi - \sin \frac{1}{3}\pi \sin \frac{1}{4}\pi}$$

$$= \frac{1}{\frac{1}{2} \cdot \frac{1}{2}\sqrt{2} - \frac{1}{2}\sqrt{3} \cdot \frac{1}{2}\sqrt{2}} = \frac{2\sqrt{2}}{1 - \sqrt{3}}$$

$$= -\sqrt{2}(1 + \sqrt{3}).$$

PROBLEMS
32

1. Derive a formula for $\cot (u+v)$.

2. Derive a formula for $\tan (u-v)$.

3. Verify the following identities.
    (a) $\sin (\pi + t) = -\sin t$
    (b) $\tan (\pi - u) = -\tan u$
    (c) $\sin (\frac{3}{2}\pi - t) = -\cos t$
    (d) $\sec (t + \frac{3}{2}\pi) = \csc t$
    (e) $\csc (\pi - x) = \csc x$
    (f) $\tan (\frac{1}{2}\pi + t) = -\cot t$

4. Evaluate the following numbers.
    (a) $\sin \frac{1}{12}\pi$
    (b) $\sin \frac{7}{12}\pi$
    (c) $\cos \frac{13}{12}\pi$
    (d) $\tan \frac{7}{12}\pi$
    (e) $\sin \frac{1775}{12}\pi$
    (f) $\cos \left(-\frac{271}{12}\pi\right)$

5. Use Table II and the addition formulas to evaluate the following numbers.
    (a) $\sin \frac{1}{2} \cos 1 + \cos \frac{1}{2} \sin 1$
    (b) $\cos .3 \cos .1 - \sin .3 \sin .1$
    (c) $\sin 3 \cos 2 - \cos 3 \sin 2$
    (d) $\sin 6 \sin 7 - \cos 6 \cos 7$

6. If $\sin u = \frac{3}{5}$, $\sin v = \frac{4}{5}$, and $0 < u < v < \frac{1}{2}\pi$, evaluate the following numbers.
    (a) $\sin (u+v)$
    (b) $\sin (u-v)$
    (c) $\cos (u+v)$
    (d) $\tan (u+v)$

7. Simplify each of the following expressions.
    (a) $\cos (u+v) \cos v + \sin (u+v) \sin v$
    (b) $\sin (u-v) \cos v + \cos (u-v) \sin v$
    (c) $\frac{1}{2}[\sin (x+y) + \sin (x-y)] \csc x$
    (d) $\frac{1}{2}[\cos (x+y) + \cos (x-y)] \sec x$

8. Derive a formula for sin $(x + y + z)$.

9. If $x + y + z = \pi$, show that $\cos z + \cos x \cos y - \sin x \sin y = 0$.

10. Find an expression for sin $(\sin x + \sin y)$.

11. Set $v = \pi$ in the formula for tan $(u + v)$ to show that tan $(u + \pi) = \tan u$.

12. Let $f(x) = \sin \log x$. What is the domain of f? What is its range? Show that $f(10^{\pi/2}x) = \cos \log x$. Show that

$$f(xy) = f(x)f(10^{\pi/2}y) + f(10^{\pi/2}x)f(y).$$

33

VALUES OF THE CIRCULAR FUNCTIONS AT MULTIPLES OF t

It is easy to see that the numbers sin $2t$ and $2 \sin t$ are not the same for every number t. If $t = \frac{1}{2}\pi$, for example, we see that sin $2t = \sin \pi = 0$, while $2 \sin t = 2 \sin \frac{1}{2}\pi = 2$. To find an expression for sin $2t$, we need only replace each of the numbers u and v in Equation 32-2 with the same number t. We then obtain the identity

$$\sin (t + t) = \sin t \cos t + \cos t \sin t,$$

which reduces to the equation

(33-1) $$\textbf{sin } 2t = 2 \textbf{ sin } t \textbf{ cos } t.$$

We can derive a formula for cos $2t$ in a similar manner. From Equation 32-1 we have

$$\cos (t + t) = \cos t \cos t - \sin t \sin t,$$

and hence,

(33-2) $$\textbf{cos } 2t = \textbf{cos}^2 t - \textbf{sin}^2 t.$$

We can write this equation in different forms by using the identity $\cos^2 t + \sin^2 t = 1$:

(33-3) $$\textbf{cos } 2t = 1 - 2 \textbf{ sin}^2 t$$

and

(33-4) $$\textbf{cos } 2t = 2 \textbf{ cos}^2 t - 1.$$

The identities that result when these last two equations are solved for $\sin^2 t$ and $\cos^2 t$ are particularly useful for certain operations in calculus. They are

(33-5) $$\textbf{sin}^2 t = \frac{1 - \cos 2t}{2}$$

and

(33-6) $$\textbf{cos}^2 t = \frac{1 + \cos 2t}{2}.$$

136

EXAMPLE 33-1   Find $\sin \frac{1}{8}\pi$.

Solution   We have not discussed the values of the circular functions at $t = \frac{1}{8}\pi$, but we do know the values at $2t = \frac{1}{4}\pi$. So we replace t by $\frac{1}{8}\pi$ in Equation 33-5 to obtain

$$\sin^2 \tfrac{1}{8}\pi = \frac{1 - \cos \tfrac{1}{4}\pi}{2} = \tfrac{1}{2}(1 - \tfrac{1}{2}\sqrt{2}) = \tfrac{1}{4}(2 - \sqrt{2}).$$

The number $\sin \frac{1}{8}\pi$ is positive, and hence $\sin \frac{1}{8}\pi = \frac{1}{2}\sqrt{2 - \sqrt{2}}$.

Equations 33-5 and 33-6 give only the squares of the numbers $\sin t$ and $\cos t$. If we want to find the numbers themselves, we must, by locating the quadrant in which the circular point $P(t)$ lies, determine the correct sign. A corresponding formula for $\tan t$ is more complete. If we multiply the numerator and denominator of the right-hand member of the identity $\tan t = \sin t / \cos t$ by $2 \cos t$, we obtain

$$\tan t = \frac{2 \sin t \cos t}{2 \cos^2 t}.$$

From Equation 33-1 we see that the numerator is $\sin 2t$, and from Equation 33-6 we find that the denominator is $1 + \cos 2t$. Hence,

(33-7)
$$\tan t = \frac{\sin 2t}{1 + \cos 2t}.$$

EXAMPLE 33-2   Use Equation 33-7 to find $\tan \frac{5}{8}\pi$.

Solution   By setting $t = \frac{5}{8}\pi$ in Equation 33-7 we obtain

$$\tan \tfrac{5}{8}\pi = \frac{\sin \tfrac{5}{4}\pi}{1 + \cos \tfrac{5}{4}\pi} = \frac{-\tfrac{1}{2}\sqrt{2}}{1 - \tfrac{1}{2}\sqrt{2}} = \frac{-\sqrt{2}}{2 - \sqrt{2}}$$

$$= \frac{-\sqrt{2}(2 + \sqrt{2})}{(2 - \sqrt{2})(2 + \sqrt{2})} = \frac{-\sqrt{2}}{2}(2 + \sqrt{2}) = -(\sqrt{2} + 1).$$

EXAMPLE 33-3   Find the maximum value of the product $\sin t \cos t$.

Solution   According to Equation 33-1, $\sin t \cos t = \frac{1}{2} \sin 2t$ The maximum value assumed by $\sin 2t$ is 1. Hence the maximum value of the product $\sin t \cos t$ is $\frac{1}{2}$. (The maximum value occurs when $2t = \frac{1}{2}\pi$ or $t = \frac{1}{4}\pi$, for instance.)

PROBLEMS 33

1. Derive formulas for (a) $\tan 2t$ and (b) $\cot 2t$.

2. Derive a formula for $\tan^2 t$ by "dividing Equation 33-5 by Equation 33-6."

3. Find a number that is in the set $\{\sin 2x = \sin x\}$ but not in the set $\{\sin x = 0\}$.

4. Simplify the following expressions.
   (a) $(\sin x + \cos x)^2 - \sin 2x$
   (b) $\sin^2 2t/(1 + \cos 2t)^2 + 1$
   (c) $\sin^2 2t - \cos^2 2t$
   (d) $(\sin 2t)(\cos 2t)$

5. With the aid of Table II and the formulas given in this section, evaluate the following numbers.
   (a) $\sin \frac{1}{2} \cos \frac{1}{2}$
   (b) $2 \cos^2 .3$
   (c) $\cos^2 \frac{2}{5} - \cos^2 (\frac{1}{2}\pi - \frac{2}{5})$
   (d) $(1 + \cos 2) \csc 2$

6. If t satisfies the inequalities $0 < t < \frac{1}{2}\pi$, and if $\sin t = \frac{3}{5}$, find each of the following.
   (a) $\sin 2t$
   (b) $\cos 2t$
   (c) $\sin 3t$
   (d) $\cos 3t$

7. Find the following numbers.
   (a) $\cos \frac{1}{8}\pi$
   (b) $\sin \frac{5}{8}\pi$
   (c) $\tan \frac{1}{8}\pi$
   (d) $\cos \frac{5}{8}\pi$
   (e) $\tan \frac{5}{8}\pi$
   (f) $\sec \frac{99}{8}\pi$
   (g) $\csc (-\frac{1865}{8}\pi)$
   (h) $\sin \frac{1492}{8}\pi$

8. Use Equation 33-6 and the equation $2(\frac{1}{12}\pi) = \frac{1}{6}\pi$ to find $\cos \frac{1}{12}\pi$. Compare your answer with the result of Example 31-1.

9. Derive an identity for $\sin 3t$.

10. Find the following numbers.
    (a) $\sin \frac{1}{16}\pi$
    (b) $\cos \frac{1}{16}\pi$
    (c) $\tan \frac{1}{16}\pi$
    (d) $\sin \frac{17}{16}\pi$

11. Show that $[\![\sin t \cos t]\!] \leq [\![\sin t]\!][\![\cos t]\!]$ for each $t \in R^1$.

12. Show that $\{\sin 2t \leq 2 \sin t\} = \{0 \leq \sin t\}$.

34
SUMMARY OF IDENTITIES

We have developed many relationships among the circular functions in the last few sections. Let us bring them together here.

First we list eight elementary identities that we can obtain directly from the definitions of the circular functions:

(34-1)

$$\csc t = \frac{1}{\sin t}, \qquad \cot t = \frac{\cos t}{\sin t},$$

$$\sec t = \frac{1}{\cos t}, \qquad \sin^2 t + \cos^2 t = 1.$$

$$\cot t = \frac{1}{\tan t}, \qquad \tan^2 t + 1 = \sec^2 t,$$

$$\tan t = \frac{\sin t}{\cos t}, \qquad \cot^2 t + 1 = \csc^2 t.$$

Some of the following equations are marked with an asterisk (*). You should be able to obtain the remaining relations by using these equations and the elementary identities.

The basic addition formulas are:

$$(*) \ \sin (u \pm v) = \sin u \cos v \pm \cos u \sin v,$$

(34-2) $\quad (*) \ \cos (u \pm v) = \cos u \cos v \mp \sin u \sin v,$

$$\tan (u \pm v) = \frac{\tan u \pm \tan v}{1 \mp \tan u \tan v} \cdot$$

The basic relations involving the values of the circular functions at $-t$ are:

$$(*) \ \sin (-t) = -\sin t,$$

(34-3) $\quad (*) \ \cos (-t) = \cos t,$

$$\tan (-t) = -\tan t.$$

The following equations state the basic relations involving the values of the circular functions at $2t$:

$$\sin 2t = 2 \sin t \cos t,$$

(34-4) $\quad \cos 2t = \cos^2 t - \sin^2 t = 1 - 2 \sin^2 t = 2 \cos^2 t - 1,$

$$\tan 2t = \frac{2 \tan t}{1 - \tan^2 t} \cdot$$

You will frequently find it useful to express Relations 34-4 in the following form:

$$\sin^2 t = \frac{1}{2} (1 - \cos 2t),$$

(34-5) $\quad \cos^2 t = \frac{1}{2} (1 + \cos 2t),$

$$\tan t = \frac{\sin 2t}{1 + \cos 2t} = \frac{1 - \cos 2t}{\sin 2t} \cdot$$

All of the above relations are "identities" in the sense that they are valid for any numbers t, u, and v in the domains of the functions involved. Such identities are used principally to change the form of an expression involving values of circular functions to a simpler or more useful form.

EXAMPLE 34-1 Simplify the expression $\sin t \left(\dfrac{\cot t}{\sec t} + \csc t \right) \cdot$

Solution   We can use the elementary identities to simplify the given expression in the following manner:

$$\sin t \left(\frac{\cot t}{\sec t} + \csc t \right) = \sin t \cdot \frac{\cos t}{\sin t} \cdot \cos t + 1$$

$$= \cos^2 t + 1.$$

EXAMPLE 34-2   Write the expression $(\sin 2t / \sin t) - (\cos 2t / \cos t)$ in a form that involves only values of circular functions at t.

Solution

$$\frac{\sin 2t}{\sin t} - \frac{\cos 2t}{\cos t} = \frac{2 \sin t \cos t}{\sin t} - \frac{2 \cos^2 t - 1}{\cos t}$$

$$= 2 \cos t - 2 \cos t + \sec t = \sec t.$$

EXAMPLE 34-3   Simplify the expression $\csc 2t + \cot 2t$.

Solution

$$\csc 2t + \cot 2t = \frac{1}{\sin 2t} + \frac{\cos 2t}{\sin 2t}$$

$$= \frac{1 + \cos 2t}{\sin 2t}$$

$$= \frac{2 \cos^2 t}{2 \sin t \cos t}$$

$$= \frac{\cos t}{\sin t} = \cot t.$$

EXAMPLE 34-4   If $0 < x < 2\pi$, show that $\sin \frac{1}{2} x = \sqrt{\frac{1}{2}(1 - \cos x)}$.

Solution   We substitute $x = 2t$ in the formula $\sin^2 t = \frac{1}{2}(1 - \cos 2t)$ to obtain the equation $\sin^2 \frac{1}{2} x = \frac{1}{2}(1 - \cos x)$. Thus we find that either $\sin \frac{1}{2} x = \sqrt{\frac{1}{2}(1 - \cos x)}$, or $\sin \frac{1}{2} x = -\sqrt{\frac{1}{2}(1 - \cos x)}$. From the inequalities $0 < x < 2\pi$, it follows that $0 < \frac{1}{2} x < \pi$, so that $\sin \frac{1}{2} x > 0$, and the equation we are seeking then follows.

The identities in the next set are frequently useful, so we list them here. They follow directly from Equations 34-2, and we will leave their verification to you in the problems. Since these identities express sums as products, they are called the **factoring identities:**

(34-6)
$$\sin (u + v) + \sin (u - v) = 2 \sin u \cos v,$$
$$\sin (u + v) - \sin (u - v) = 2 \cos u \sin v,$$
$$\cos (u - v) + \cos (u + v) = 2 \cos u \cos v,$$
$$\cos (u - v) - \cos (u + v) = 2 \sin u \sin v.$$

140

If we substitute $x = u + v$ and $y = u - v$ (and hence $u = \frac{1}{2}(x + y)$ and $v = \frac{1}{2}(x - y)$) in Equations 34-6, we obtain the identities

$$\sin x + \sin y = 2 \sin \frac{1}{2}(x + y) \cos \frac{1}{2}(x - y),$$

$$\sin x - \sin y = 2 \cos \frac{1}{2}(x + y) \sin \frac{1}{2}(x - y),$$

(34-7)

$$\cos x + \cos y = 2 \cos \frac{1}{2}(x + y) \cos \frac{1}{2}(x - y),$$

$$\cos x - \cos y = -2 \sin \frac{1}{2}(x + y) \sin \frac{1}{2}(x - y).$$

PROBLEMS 34

1. Simplify the following expressions.
   (a) $\csc t - \cot t \cos t$ (b) $\tan x \sin x + \cos x$
   (c) $\sin^2 u / (1 - \cos u) - 1$ (d) $\tan t (1 + \cot^2 t)/(1 + \tan^2 t)$
   (e) $\tan t + \cot t$ (f) $1 + \cos t / (2 + \sec t)$

2. Let t satisfy the inequalities $0 < t < \frac{1}{2}\pi$. Express each of the following numbers in terms of $\sin t$.
   (a) $\sec t$ (b) $\tan t$ (c) $\cot t$ (d) $\csc 2t$

3. Simplify the following expressions.
   (a) $[(2 \sin^2 t - 1)/\sin t \cos t] + \cot 2t$
   (b) $(\sec t + \tan t)/(\cos t - \tan t - \sec t)$
   (c) $\sin t \cot t (\sec t - 1)/(1 - \cos t)$
   (d) $\sin x / (\sec x + 1) + \sin x / (\sec x - 1)$
   (e) $\csc^2 x - 2 + \sin^2 x$ (f) $\cos 2t / (\sin t + \cos t)$
   (g) $\cos 2t + \sin^4 t$ (h) $\cot t \csc 2t - \frac{1}{2}$

4. Show that $\sqrt{1 + \sin 2x} = |\sin x + \cos x|$.

5. Find a point of the solution set of the equation $\tan^2 x + \sec^2 x = 3$.

6. Write each of the following expressions in terms of values of circular functions at $2t$.
   (a) $(\sin^3 t - \cos^3 t)/(\sin t - \cos t)$
   (b) $(\cos t - 1)^2 - (\sin t - 1)^2 + 2(\cos t - \sin t)$
   (c) $(\tan t + \cot t)/(\cot t - \tan t)$
   (d) $(\sec 2t - 1)(1 - 2 \sin^2 t)$

7. Write a formula for $\tan \frac{1}{2}x$ (see Equation 34-5).

8. Verify the factoring identities (Equations 34-6).

9. Verify the following identities.
   (a) $2 \sin 2x \cos 3x = \sin 5x - \sin x$ (b) $\cos 4x = 2 \sin 5x \sin x + \cos 6x$

10. Find at least 3 numbers that belong to the set $\{0 \leq x \leq \frac{1}{2}\pi\}$ and also to the solution set of the given equation.

(a) $\sin 3x + \sin 5x = 0$ (b) $\sin 3x - \sin 5x = 0$

(c) $\cos 3x + \cos 5x = 0$ (d) $\cos 3x - \cos 5x = 0$

11. Let $f(x) = \sin (2x + 1) + \sin (2x - 3)$. What is the maximum value of f? The minimum value?

12. Let $f(x) = \cos \log x$. Show that $f(x)f(y) = \frac{1}{2}[f(x/y) + f(xy)]$.

13. Verify the following identities.

(a) $\dfrac{\sin 2x + \sin 2y}{\cos 2x + \cos 2y} = \tan (x + y)$ (b) $\dfrac{\cos x + \sin x}{\cos x - \sin x} = \tan 2x + \sec 2x$

35

THE EQUATION $y = A \sin (ax + b)$

The voltage drop across the terminals of an ordinary electric outlet varies with time. Thus suppose we start to measure the drop at an instant when it is 0 and increasing. Then the formula that expresses the voltage drop t seconds later is

(35-1)
$$E(t) = 170 \sin 120\pi t.$$

With our knowledge of circular functions, we can read a lot of information from this formula. For example, suppose we compare the voltage at a certain time t with the voltage $\frac{1}{60}$ of a second later, at time $t + \frac{1}{60}$. In other words, we are to compare the numbers $E(t)$ and $E(t + \frac{1}{60})$. According to our formula, $E(t + \frac{1}{60}) = 170 \sin 120\pi(t + \frac{1}{60}) = 170 \sin (120\pi t + 2\pi)$. But

$$\sin (120\pi t + 2\pi) = \sin 120\pi t,$$

so we see that $E(t + \frac{1}{60}) = E(t)$. Therefore, the values of the function E that is defined by Equation 35-1 repeat every $\frac{1}{60}$ of a second; we say that E is **periodic** with a period of $\frac{1}{60}$. The reciprocal of the period is called the **frequency** of the function. It represents the number of cycles (or the fraction of a cycle) completed in one second. In our example, we see that the frequency of ordinary household voltage is 60 cycles per second. The values of E vary between -170 and 170; the number 170 is the **amplitude** of the function. Graphically speaking, the amplitude represents the maximum displacement of the graph of the function from the horizontal axis. We speak of "120-volt current," and the number 120 is also found from Equation 35-1. It is an "average value" (computed by calculus) of our function E.

It is easy to generalize these results to the function that is defined by the equation

(35-2)
$$y = A \sin ax,$$

where A and a are given numbers (and a is positive). The values of this

142

function vary between $-|A|$ and $|A|$, and so $|A|$ is its amplitude. Furthermore, since $\sin a(x + 2\pi/a) = \sin(ax + 2\pi) = \sin ax$, it follows that the period of the function defined by Equation 35-2 is $2\pi/a$. Its frequency is therefore $a/2\pi$. Basically, the graph of Equation 35-2 is a "sine-type" curve, so compressed or elongated in the horizontal direction that an entire cycle is completed in an interval of length $2\pi/a$. The effect of the amplitude factor A is to magnify the curve in the vertical direction by a factor of $|A|$, and reflect it about the X-axis if A is negative.

EXAMPLE 35-1   Sketch the graph of the equation $y = -3 \sin \pi x$.

Solution   To show the effects of the amplitude factor -3 and the frequency factor π separately, we will first plot the graph of the equation $y = \sin \pi x$ and then modify it to get the required curve. The equation $y = \sin \pi x$ defines a function whose period is $2\pi/\pi = 2$, so we need only plot the points of its graph with X-coordinates between 0 and 2 and then duplicate this portion to obtain the entire graph. The five "key points" of this portion are $(0, 0)$, $(\frac{1}{2}, 1)$, $(1, 0)$, $(\frac{3}{2}, -1)$, and $(2, 0)$. We plot them and join them with the "compressed sine curve" shown in black in Fig. 35-1. The amplitude factor -3 expands this curve three-fold and reflects it. The resulting colored curve in Fig. 35-1 is the graph of the equation $y = -3 \sin \pi x$.

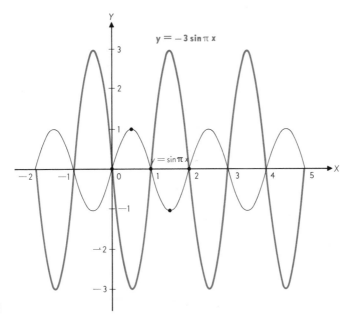

FIGURE 35-1

We said that Equation 35-1 gives the voltage across the terminals of an electric outlet t seconds after an initial instant at which the voltage is 0 and increasing. If our intial measurement were made at some other point of the cycle, then Equation 35-1 would have to be replaced by an equation of the form $y = 170 \sin (120\pi t + b)$, where b is a number that depends on the point in the cycle at which the initial measurement is made. This equation is of the type

(35-3) $$y = A \sin (ax + b),$$

where A, a, and b are given numbers (a positive), and it is such equations that we wish to study in this section.

The graph of Equation 35-3 is the same as the graph of the equation $y = A \sin ax$, except that it is shifted along the X-axis. It is a sine-type curve with an amplitude of A and a period of $2\pi/a$. Notice that $y = 0$ if $ax + b = 0$; that is, if $x = -b/a$. The number $-b/a$ is called the **phase shift**. The phase shift represents the number of units that the graph of the equation $y = A \sin ax$ must be shifted along the X-axis in order to coincide with the graph of the equation $y = A \sin (ax + b)$. This shift is to the left when the phase shift is negative and to the right when the phase shift is positive.

EXAMPLE 35-2   Sketch the graph of the equation $y = 2 \sin (2x - \pi)$.

Solution   This graph is a sine-type curve with an amplitude of 2 and a period of $2\pi/2 = \pi$. Notice that $y = 0$ when $2x - \pi = 0$; that is, when $x = \frac{1}{2}\pi$. Thus the graph of our given equation is the same as the graph of the equation $y = 2 \sin 2x$, except that we "start" it at $\frac{1}{2}\pi$ instead of at 0. When we shift the graph of the equation $y = 2 \sin 2x$ to the right $\frac{1}{2}\pi$ units, we obtain the graph of the equation $y = 2 \sin (2x - \pi)$ shown in Fig. 35-2.

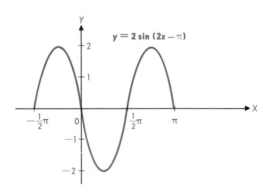

FIGURE 35-2

In Section 30, where we first studied the graphs of the circular functions, we learned that the sine and cosine curves were the same in the sense that they can be made to coincide by shifting one of them along the X-axis. We now see exactly what this statement means, for (see Example 32-1) $\cos x = \sin (x + \frac{1}{2}\pi)$, and so the graph of the cosine function is the same as the graph of the equation $y = \sin(x + \frac{1}{2}\pi)$. But the graph of this last equation is merely a sine curve with a phase shift of $-\frac{1}{2}\pi$ units; that is, a sine curve shifted $\frac{1}{2}\pi$ units to the left.

We can use the addition formula we developed for the sine function to write $A \sin(ax + b) = A \sin ax \cos b + A \cos ax \sin b$. When we introduce the abbreviations

$$(35\text{-}4) \qquad B = A \cos b \quad \text{and} \quad C = A \sin b,$$

this equation takes the form

$$(35\text{-}5) \qquad A \sin(ax + b) = B \sin ax + C \cos ax.$$

If we are given the numbers A and b, then of course it is easy to use Equations 35-4 to find B and C. But in the applications of mathematics, we often start with B and C and try to determine A and b; that is, we proceed from the right-hand side of Equation 35-5 to the left.

If we are given B and C, we can use the following procedure to find a positive number A and a number b that satisfy Equations 35-4. We choose $A = \sqrt{B^2 + C^2}$. Then $A^2 = B^2 + C^2$, and so $(B/A)^2 + (C/A)^2 = 1$. Therefore, $(B/A, C/A)$ is a point of the unit circle (Fig. 35-3). We take b to be

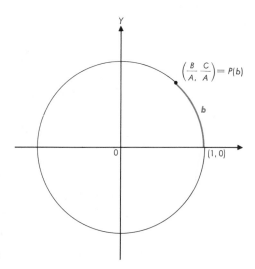

FIGURE 35-3

any number such that this point is the circular point $P(b)$; for example, we could choose b as the distance measured along the unit circle in the counterclockwise direction from the point $(1, 0)$ to the point $(B/A, C/A)$. According to the definition of the circular functions, $\sin b = C/A$ and $\cos b = B/A$, so Equations 35-4 are satisfied.

EXAMPLE 35-3 Two alternating current generators produce currents that are given, in terms of the time t, by the equations

$$I_1 = \sqrt{3} \sin 120\pi t \quad \text{and} \quad I_2 = -\cos 120\pi t.$$

If the output of the second is added to the output of the first, find the maximum current output, when it occurs, and determine the phase shift produced.

Solution The combined current is given by the equation

$$I = \sqrt{3} \sin 120\pi t - \cos 120\pi t.$$

In this case, $A = \sqrt{3+1} = 2$. The point $P(b)$ has coordinates $(\frac{1}{2}\sqrt{3}, -\frac{1}{2})$. Thus, $\sin b = -\frac{1}{2}$ and $\cos b = \frac{1}{2}\sqrt{3}$, so b can be chosen as $-\frac{1}{6}\pi$. The total current can therefore be represented by the equation

$$I = 2 \sin (120\pi t - \tfrac{1}{6}\pi).$$

It follows that the maximum current is 2 and that the current experiences a phase shift of $\frac{1}{720}$ units of time. The maximum value of I occurs when $120\pi t - \frac{1}{6}\pi = \frac{1}{2}\pi + k \cdot 2\pi$ (k can be any integer), and hence when $t = \frac{1}{180} + \frac{1}{60}k$.

PROBLEMS 35

1. Sketch the graph of each of the following equations.
   (a) $y = \sin (x + \frac{3}{2}\pi)$ (b) $y = \tan (x + \frac{1}{2}\pi)$
   (c) $y = \cos (x - \frac{1}{2}\pi)$ (d) $y = \sin (x + \frac{1}{6}\pi)$

2. Sketch the graph of each of the following equations. In each case determine the frequency and phase shift.
   (a) $y = \sin (2\pi x + \pi)$ (b) $y = \sin (\frac{1}{2}x + 3)$
   (c) $y = \cos (2\pi x - \pi)$ (d) $y = \tan (\pi x - 4)$

3. Sketch the graph of each of the following equations. In each case determine the frequency, phase shift, and amplitude.
   (a) $y = 4 \sin (3x - \frac{1}{2}\pi)$ (b) $y = \dfrac{\sin (\frac{1}{2}\pi x + 1)}{2}$
   (c) $y = 3 \cos (2\pi x - 4\pi)$ (d) $y = -\frac{1}{4} \cos (\pi x - 3\pi)$

4. Sketch the graph of the equation $y = |2 \sin (\pi x - \frac{1}{2}\pi)|$, and determine the frequency and phase shift.

5. Write the following expressions in the form $A \sin (ax + b)$, where A and a are positive.
   (a) $-2 \sin (2\pi x + 1)$ (b) $3 \sin (-5x + 2)$
   (c) $-\sin (-4x + 3)$ (d) $2 \cos (2\pi x + \pi)$

6. Show that if A and a are not 0, then $A \sin (ax + b) = |A| \sin (|a| x + b_1)$, where

$$b_1 = \frac{a}{|a|} b + \frac{1}{2} \left(1 - \frac{aA}{|aA|} \right) \pi.$$

7. Write the following expressions in the form $A \sin (ax + b)$.
   (a) $\sin x + \cos x$ (b) $4 \sin 2x + 4\sqrt{3} \cos 2x$
   (c) $3 \sin \pi x + 4 \cos \pi x$ (d) $\sin 2x - \cos 2x$

8. Sketch the graphs of the following equations for x in the interval $\{0 \le x \le 6\}$.
   (a) $y = [\![\sin \pi x]\!]$ (b) $y = \sin \pi [\![x]\!]$
   (c) $y = \sin \pi [\![x]\!] x$ (d) $y = [\![x]\!] \sin \pi x$
   (e) $y = \sin (\pi x - \frac{1}{3}\pi [\![x]\!])$ (f) $y = \sin [\![\pi]\!] x$

9. The graph of the equation $y = A \sin (ax + b) + K$ is a sine-type curve translated K units in the Y-direction. Sketch the graphs of the following equations.
   (a) $y = \sin x + 1$ (b) $y = \sin 2x - 2$
   (c) $y = \frac{1}{2}(1 + \cos 2x)$ (d) $y = \frac{1}{2}(1 - \cos 2x)$

10. Sketch the graphs of the equations $y = \sin^2 x$ and $y = \cos^2 x$. Make use of Equations 34-5 and parts (c) and (d) of the preceding problem.

36

THE PRINCIPAL SINE AND COSINE FUNCTIONS AND THEIR INVERSES

No circular function has an inverse. For if C is a circular function and y is a given number of its range, then to each solution of the equation $y = C(x)$ we can add 2π, 4π, -2π, and so on, and obtain other solutions. Thus, the equation $y = C(x)$ does not have *just one* solution for each choice of y in the range of C, and so C does not have an inverse. But we can construct, by "restricting the domain" of the circular functions, new functions that do have inverses.

The domain of the sine function is the entire set of real numbers, and its rule of correspondence is described in Section 27. Now we want to consider a function with the same rule of correspondence but whose domain is only the interval $\{-\frac{1}{2}\pi \le x \le \frac{1}{2}\pi\}$. This new function is the **principal sine function,** and we name it **Sin** (the capital letter S distinguishes it from the ordinary sine function). Thus, for example, $\text{Sin} \frac{1}{6}\pi = \sin \frac{1}{6}\pi = \frac{1}{2}$, but $\text{Sin} \frac{7}{6}\pi$ is a meaningless symbol, since $\frac{7}{6}\pi$ is not in the domain of the principal sine function. The colored curve in Fig. 36-1 is the graph of the principal sine function. We obtain it simply by removing the black portion of the sine curve.

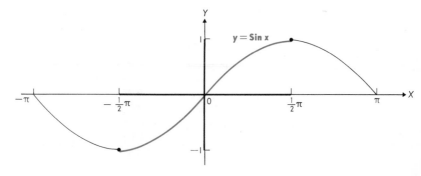

FIGURE 36-1

Of course, we chose our definition of the principal sine function so that it would have an inverse, and a glance at Fig. 36-1 shows that we have indeed produced an invertible function. This inverse function is the **Arcsine function.** The range of the principal sine function is the interval $\{-1 \leq y \leq 1\}$, so this interval serves as the domain of the Arcsine function. The number that the Arcsine function associates with a given number y of its domain is commonly denoted either by Arcsin y or by $\text{Sin}^{-1} y$. (Notice that $\text{Sin}^{-1} y$ *does not mean* $1/\text{Sin } y$.) If we think of our functions as sets of pairs of numbers, then $\text{Sin} = \{(x, y) \mid -\tfrac{1}{2}\pi \leq x \leq \tfrac{1}{2}\pi, y = \sin x\}$. We obtain its inverse by interchanging the numbers in each pair; that is,

$$\text{Arcsin} = \{(y, x) \mid -\tfrac{1}{2}\pi \leq x \leq \tfrac{1}{2}\pi, y = \sin x\}.$$

Our formal definition of the Arcsine function states the conditions under which a pair belongs to this set.

Definition 36-1   *The equation*

$$x = \textbf{Arcsin } y = \textbf{Sin}^{-1}\, y$$

is equivalent to the two statements

(i) $y = \sin x$ *and*   (ii) $-\tfrac{1}{2}\pi \leq x \leq \tfrac{1}{2}\pi$.

EXAMPLE 36-1   Find Arcsin $\tfrac{1}{2}$.

Solution   We know from our study of circular functions that $\tfrac{1}{2} = \sin \tfrac{1}{6}\pi$. Furthermore, $-\tfrac{1}{2}\pi \leq \tfrac{1}{6}\pi \leq \tfrac{1}{2}\pi$, so both Conditions (i) and (ii) of Definition 36-1 are satisfied, and hence $\tfrac{1}{6}\pi = \text{Arcsin } \tfrac{1}{2}$.

EXAMPLE 36-2   Find Arcsin $(-.7895)$.

Solution   We find from Table II that $\sin .91 = .7895$, so $\sin(-.91) = -.7895$. Since $-\tfrac{1}{2}\pi \leq -.91 \leq \tfrac{1}{2}\pi$, we see that Arcsin $(-.7895) = -.91$.

From our discussion of the graphs of inverse functions in Section 17, we see that we can obtain the graph of the Arcsine function by reflecting the graph of the principal sine function about the line $y = x$. The resulting graph of the equation $y = \text{Sin}^{-1} x$ is shown in Fig. 36-2.

We treat the cosine function in the same way. To obtain a function that has an inverse, we first define the **principal cosine function** as the function whose domain is the interval $\{0 \leq x \leq \pi\}$ and whose rule of correspondence is expressed by the equation $y = \cos x$. This function's name is Cos, and its graph is the colored curve in Fig. 36-3. The graph shows that

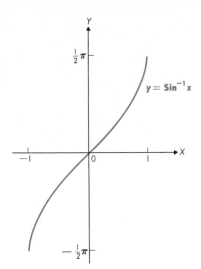

FIGURE 36-2

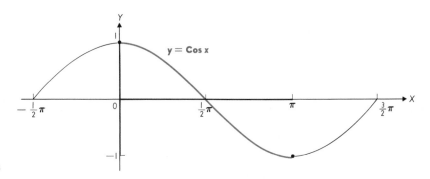

FIGURE 36-3

the principal cosine function has an inverse whose domain is the interval $\{-1 \le y \le 1\}$. We call this inverse the **Arccosine function,** and the number corresponding to a given number y in the domain of the Arccosine function is denoted either by **Arccos** y or by **Cos**$^{-1}$ y. Since $\text{Cos} = \{(x, y) \mid 0 \le x \le \pi, y = \cos x\}$, we have $\text{Arccos} = \{(y, x) \mid 0 \le x \le \pi, y = \cos x\}$. Our formal definition of the Arccosine function states the conditions under which a pair belongs to this set.

Definition 36-2   *The equation*

$$x = \textbf{Arccos } y = \textbf{Cos}^{-1} y$$

is equivalent to the two statements

(i) $y = \cos x$ and   (ii) $0 \le x \le \pi.$

The graph of the equation $y = \text{Cos}^{-1} x$ is shown in Fig. 36-4.

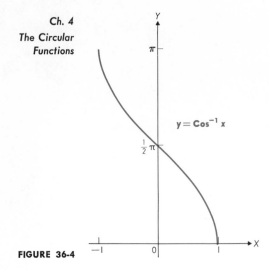

$$y = \text{Cos}^{-1} x$$

FIGURE 36-4

EXAMPLE 36-3   Calculate $\sin (\text{Cos}^{-1} \frac{3}{5})$.

Solution   Let $t = \text{Cos}^{-1} \frac{3}{5}$. According to Definition 36-2, $\cos t = \frac{3}{5}$, and $0 \leq t \leq \pi$. In fact, since $\cos t > 0$, we know that $0 \leq t \leq \frac{1}{2}\pi$. From the fundamental identity $\cos^2 t + \sin^2 t = 1$, we obtain the equation

$$\sin^2 t = 1 - (\tfrac{3}{5})^2 = \tfrac{16}{25}.$$

Since $0 \leq t \leq \frac{1}{2}\pi$, we know that $\sin t \geq 0$, and therefore $\sin t = \frac{4}{5}$. Thus, we have $\sin (\text{Cos}^{-1} \frac{3}{5}) = \frac{4}{5}$.

EXAMPLE 36-4   Show that $\text{Arcsin } x + \text{Arccos } x = \frac{1}{2}\pi$ for any number x such that $-1 \leq x \leq 1$.

Solution   Let $u = \text{Arcsin } x$. Then $\sin u = x$, and $-\frac{1}{2}\pi \leq u \leq \frac{1}{2}\pi$. Therefore, if we write $v = \frac{1}{2}\pi - u$, it follows that $0 \leq v \leq \pi$, and $\cos v = \cos (\frac{1}{2}\pi - u) = \sin u = x$. Since $0 \leq v \leq \pi$, and $\cos v = x$, we see from Definition 36-2 that $v = \text{Arccos } x$. Thus,

$$\text{Arcsin } x + \text{Arccos } x = u + v = \tfrac{1}{2}\pi.$$

**PROBLEMS
36**

1.   Evaluate the following expressions.
   (a) $\text{Arcsin } \frac{1}{2}\sqrt{3}$ (b) $\text{Arccos } (-\frac{1}{2})$ (c) $\text{Sin}^{-1} .8573$
   (d) $\text{Cos}^{-1} .4536$ (e) $\text{Sin}^{-1} 1 + \text{Cos}^{-1} 1$ (f) $\text{Sin}^{-1} \frac{1}{2} + \text{Cos}^{-1} (-\frac{1}{2})$
   (g) $\text{Arccos } (\text{Sin}^{-1} 0)$ (h) $\text{Arcsin } (\text{Cos}^{-1} 1)$

2.   Find the following numbers.
   (a) $\cos (\text{Sin}^{-1} .8)$ (b) $\sin (\text{Arccos } \frac{12}{13})$
   (c) $\sin (\text{Arccos } .28)$ (d) $\cos (\text{Sin}^{-1} .96)$

Sec. 37

The Principal
Tangent and
Cotangent
Functions and
Their Inverses

3. Find the following numbers.
   (a) $\sin (2 \operatorname{Arccos} \frac{1}{2})$
   (b) $\cos (2 \operatorname{Sin}^{-1} \frac{1}{2})$
   (c) $\operatorname{Cos}^{-1} (2 \sin \frac{1}{6}\pi)$
   (d) $\operatorname{Sin}^{-1} (2 \cos \frac{2}{3}\pi)$

4. Sketch the graphs of the following equations.
   (a) $y = \sin (\operatorname{Sin}^{-1} x)$
   (b) $y = \operatorname{Sin}^{-1} (\sin x)$
   (c) $y = \operatorname{Sin}^{-1} (\cos x)$
   (d) $y = \cos (\operatorname{Sin}^{-1} x)$

5. Verify the following statements.
   (a) $\operatorname{Sin}^{-1} 2x \neq 2 \operatorname{Sin}^{-1} x$
   (b) $\operatorname{Cos}^{-1} u + \operatorname{Cos}^{-1} v \neq \operatorname{Cos}^{-1} (u + v)$

6. Verify the following identities.
   (a) $\cos (\operatorname{Arcsin} a + \operatorname{Arccos} b) = b\sqrt{1 - a^2} - a\sqrt{1 - b^2}$
   (b) $\sin (2 \operatorname{Arcsin} t) = 2t\sqrt{1 - t^2}$
   (c) $\sin (\operatorname{Cos}^{-1} y) = \cos (\operatorname{Sin}^{-1} y)$
   (d) $\cos (2 \operatorname{Cos}^{-1} y) = 2y^2 - 1$

7. Solve the following equations.
   (a) $\operatorname{Cos}^{-1} \frac{3}{5} - \operatorname{Sin}^{-1} \frac{4}{5} = \operatorname{Cos}^{-1} t$ (b) $\operatorname{Sin}^{-1} (-\frac{5}{13}) + \operatorname{Cos}^{-1} \frac{4}{5} = \operatorname{Sin}^{-1} t$

8. Verify the following identities.
   (a) $\operatorname{Sin}^{-1} (-y) = -\operatorname{Sin}^{-1} y$
   (b) $\operatorname{Cos}^{-1} (-y) = \pi - \operatorname{Cos}^{-1} y$

9. Let $-1 \leq u < v \leq 1$. Determine which of the following numbers is larger.
   (a) $\operatorname{Sin}^{-1} u$ or $\operatorname{Sin}^{-1} v$
   (b) $\operatorname{Cos}^{-1} u$ or $\operatorname{Cos}^{-1} v$
   (c) $\operatorname{Sin}^{-1} (u - v)$ or $\operatorname{Sin}^{-1} (v - u)$
   (d) $\operatorname{Cos}^{-1} (u - v)$ or $\operatorname{Cos}^{-1} (v - u)$

10. A chord divides a circular disk into two regions, and in calculus we learn that the area of the smaller region can be expressed in terms of the radius r of the disk and the distance d of the chord from the center by means of the equation $f(r) - f(d)$, where

$$f(x) = x\sqrt{r^2 - x^2} + r^2 \operatorname{Sin}^{-1} \frac{x}{r}.$$

Suppose we cut a disk of radius 1 by a chord that is halfway between the center and the boundary. What is the area of the smaller region?

11. What are the following sets?
   (a) $\cos \cap \operatorname{Cos}$
   (b) $\sin \cup \operatorname{Sin}$
   (c) $\operatorname{Sin} \cap \operatorname{Cos}$
   (d) $\sin \cap \operatorname{Cos}$

37

THE PRINCIPAL TANGENT AND COTANGENT FUNCTIONS AND THEIR INVERSES

Although no circular function has an inverse, we have just seen how to restrict the domains of the sine and cosine functions to produce new functions that do have inverses. So let us apply the same procedure to the tangent and cotangent functions.

We restrict the domain of the tangent function to the interval $\{-\frac{1}{2}\pi < x < \frac{1}{2}\pi\}$, thereby obtaining the **principal tangent function.** Thus, by definition,

$$\operatorname{Tan} = \{(x, y) \mid -\frac{1}{2}\pi < x < \frac{1}{2}\pi, y = \tan x\}.$$

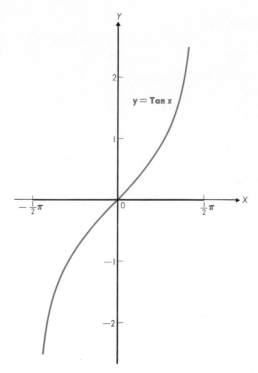

$y = \text{Tan } x$

FIGURE 37-1

We have drawn the graph of this function in Fig. 37-1, and from this graph we see that the principal tangent function has an inverse whose domain is the set of all real numbers. This inverse function is called the **Arctangent** function, and the number that it associates with a given real number y is denoted by Arctan y or by $\text{Tan}^{-1} y$. From our description of Tan as a set of pairs of numbers, we see that

$$\text{Arctan} = \{(y, x) \mid -\tfrac{1}{2}\pi < x < \tfrac{1}{2}\pi, \ y = \tan x\}.$$

Our formal definition of the Arctangent function states the conditions under which a pair belongs to this set.

Definition 37-1 *The equation*

$$x = \textbf{Arctan } y = \textbf{Tan}^{-1} y$$

is equivalent to the two statements

(i) $y = \tan x$ *and*   (ii) $-\tfrac{1}{2}\pi < x < \tfrac{1}{2}\pi$.

We obtain the graph of the Arctangent function (Fig. 37-2) by reflecting the graph of the principal tangent function about the line $y = x$. It is important that you know what this graph looks like.

Sec. 37

The Principal
Tangent and
Cotangent
Functions and
Their Inverses

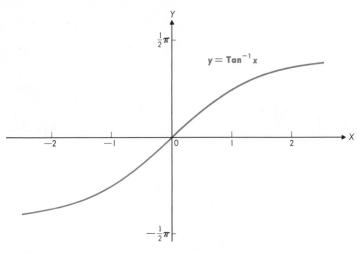

FIGURE 37-2

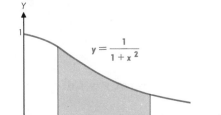

FIGURE 37-3

EXAMPLE **37**-1   Figure 37-3 shows the graph of the equation $y = 1/(1 + x^2)$ for values of $x \geq 0$. When you study calculus, you will find that the area of the region bounded by this curve, the X-axis, and vertical lines at the points a and b (the shaded region in the figure) is given by the formula $A =$ Arctan b — Arctan a. What is this area if $a = 0$ and $b = 1$?

Solution   According to the formula we have just stated, $A =$ Arctan 1 — Arctan 0. You can easily show that Arctan $1 = \frac{1}{4}\pi$ and Arctan $0 = 0$; therefore, $A = \frac{1}{4}\pi$. (Inspect Fig. 37-3. Does this answer seem reasonable?)

We treat the cotangent function in a similar manner. First, we restrict its domain to the interval $\{0 < x < \pi\}$, thereby obtaining the **principal cotangent function,** whose graph appears in Fig. 37-4. The inverse of this function is the **Arccotangent** function, which pairs with a given real number y the number Arccot y or $\mathrm{Cot}^{-1} y$. Thus, by definition,

$$\mathrm{Cot} = \{(x, y) \mid 0 < x < \pi, y = \cot x\},$$

and therefore

$$\mathrm{Arccot} = \{(y, x) \mid 0 < x < \pi, y = \cot x\}.$$

Our formal definition of the Arccotangent function states the conditions under which a pair belongs to this set.

153

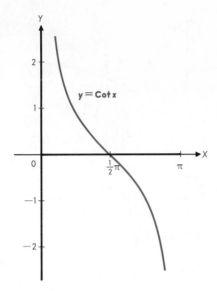

FIGURE 37-4

Definition 37-2   *The equation*

$$x = \mathbf{Arccot}\ y = \mathbf{Cot}^{-1}\ y$$

is equivalent to the two statements

(i) $y = \cot x$ *and*   (ii) $0 < x < \pi$.

The graph of the Arccotangent function is shown in Fig. 37-5.

EXAMPLE 37-2   Find Arctan $1 +$ Arccot 5.

Solution   By interpolating in Table II, we find that cot .197 = 5. Since Arctan 1 = $\frac{1}{4}\pi$ = .785, we have Arctan 1 + Arctan 5 = .785 + .197 = .982.

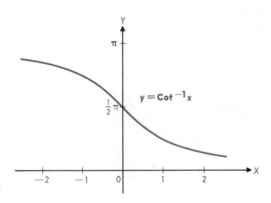

FIGURE 37-5

154

EXAMPLE 37-3   Show that if $x > 0$, then Arctan $x = $ Arccot $\dfrac{1}{x}$.

Solution   Let $u = $ Arctan x. Then tan $u = x$, and $-\frac{1}{2}\pi < u < \frac{1}{2}\pi$. In fact, since $x > 0$, we know that $0 < u < \frac{1}{2}\pi$. Since

$$\frac{1}{x} = \frac{1}{\tan u} = \cot u,$$

and $0 < u < \frac{1}{2}\pi$, we have $u = $ Arccot $(1/x)$; that is, Arctan $x = $ Arccot $(1/x)$.

EXAMPLE 37-4   Find sin (Arctan x), where x may be any real number.

Solution   Let $t = $ Arctan x. Then tan $t = x$, and $-\frac{1}{2}\pi < t < \frac{1}{2}\pi$. Also (see Fig. 37-2), t and x have the same sign. We wish to express sin t in terms of x. Since tan $t = x$, we have

$$\sin t = x \cos t,$$
$$\sin^2 t = x^2 \cos^2 t,$$
$$\sin^2 t = x^2(1 - \sin^2 t).$$

When we solve this last equation for $\sin^2 t$ we get

$$\sin^2 t = \frac{x^2}{1 + x^2}.$$

Therefore, either

$$\sin t = \frac{x}{\sqrt{1 + x^2}} \quad \text{or} \quad \sin t = \frac{-x}{\sqrt{1 + x^2}}.$$

Since t is between $-\frac{1}{2}\pi$ and $\frac{1}{2}\pi$, sin t and t have the same sign. We have already seen that t and x have the same sign. Now $x/\sqrt{1 + x^2}$ has the same sign as x, while $-x/\sqrt{1 + x^2}$ has the opposite sign, so we conclude that

$$\sin t = \sin (\text{Arctan } x) = \frac{x}{\sqrt{1 + x^2}}.$$

We shall not discuss inverse functions for the cosecant and secant functions. They may be defined in the same way that the inverse functions for the other circular functions are defined. For example, we could introduce a principal secant function and its inverse. If you master the concepts in this and the preceding section, you will have no trouble defining inverse functions for the secant and cosecant functions when you need to.

PROBLEMS 37

1. Calculate the following numbers.

    (a) Arctan $\sqrt{3}$ (b) Tan^{-1} (-1) (c) Arccot (-1)
    (d) Cot^{-1} $3^{-1/2}$ (e) Tan^{-1} .2553   (f) Arctan (-1.138)
    (g) Arccot 19.983   (h) Cot^{-1} $(-.3323)$

2. Find the following numbers (without using Table II).

(a) $\cos (\text{Tan}^{-1} .75)$ (b) $\sin \text{Arctan} (-\frac{4}{3})$

(c) $\sin (\text{Arccot} \frac{7}{24})$ (d) $\cos \text{Cot}^{-1} (-\frac{24}{7})$

3. Write the following without using trigonometric expressions.

(a) $\tan (\text{Tan}^{-1} x)$ (b) $\cot (\text{Cot}^{-1} x)$ (c) $\sin (\text{Cos}^{-1} x)$

(d) $\sec^2 (\text{Tan}^{-1} x)$ (e) $\cos (\text{Tan}^{-1} x)$ (f) $\sin (\text{Cot}^{-1} x)$

(g) $\cos (\text{Cot}^{-1} x)$ (h) $\tan (\text{Cot}^{-1} 1/x)$

4. Write the following without using trigonometric expressions.

(a) $\tan (2 \text{Tan}^{-1} x)$ (b) $\cot (2 \text{Cot}^{-1} x)$

(c) $\sin (2 \text{Arctan} x)$ (d) $\cos (2 \text{Arccot} x)$

5. Verify the following identities.

(a) $\text{Tan}^{-1} (-x) = -\text{Tan}^{-1} x$ (b) $\text{Cot}^{-1} (-x) = \pi - \text{Cot}^{-1} x$

6. Find the equation that relates $\text{Arctan} x$ and $\text{Arccot} 1/x$ if $x < 0$ (see Example 37-3). What is the equation that relates $\text{Arctan} x$ and $\text{Arccot} x$?

7. Solve the following equations.

(a) $\text{Arctan} \frac{1}{3} + \text{Arctan} \frac{1}{2} = \text{Arcsin} t$ (b) $\text{Arccot} \frac{1}{3} + \text{Arccot} \frac{1}{2} = \text{Arccos} t$

8. Solve the equation $\tan (2 \text{Arcsin} x) = \dfrac{2\sqrt{1 - x^2}}{x(1 - x^2)}$.

9. Verify the following identities.

(a) $\tan (\frac{1}{2} \text{Cos}^{-1} a) = \sqrt{\dfrac{1 - a}{1 + a}}$

(b) $\tan (\text{Tan}^{-1} a + \text{Tan}^{-1} 1) = \dfrac{1 + a}{1 - a}$

(c) $\tan (\text{Tan}^{-1} a - \text{Tan}^{-1} b) = \dfrac{a - b}{1 + ab}$

(d) $\tan (\text{Cos}^{-1} a + \text{Sin}^{-1} b) = \dfrac{\sqrt{(1 - a^2)(1 - b^2)} + ab}{a\sqrt{1 - b^2} - b\sqrt{1 - a^2}}$

10. Sketch the graphs of the following equations.

(a) $y = \tan (\text{Tan}^{-1} x)$ (b) $y = \text{Tan}^{-1} (\tan x)$

(c) $y = \sin (\text{Tan}^{-1} x)$ (d) $y = \text{Tan}^{-1} (\sin x)$

38
EQUATIONS INVOLVING CIRCULAR FUNCTIONS

If y is any number in the range of the sine function, the solution set of the equation $\sin x = y$ has infinitely many members. For example, the equation $\sin x = \frac{1}{2}$ has the number $\frac{1}{6}\pi$ as one solution. Since we can obtain new solutions by adding integral multiples of 2π to old ones, we see that all the members of the set $\{\frac{1}{6}\pi + 2k\pi\}$ are solutions of our equation. (In this set notation, we are assuming that k is an integer—positive, negative, or 0. Written out more fully, our set can be expressed as $\{x \mid x = \frac{1}{6}\pi + 2k\pi, k$ is an integer$\}$.) But this set is only a subset of the solution set of the equation

$\sin x = \frac{1}{2}$. For the number $\frac{5}{6}\pi$ is also a solution, and hence $\{\frac{5}{6}\pi + 2k\pi\}$ is also a subset of the solution set. The solution set of our equation is, as you can readily see, the union of these two sets. In other words,

$$\{\sin x = \tfrac{1}{2}\} = \{\tfrac{1}{6}\pi + 2k\pi\} \cup \{\tfrac{5}{6}\pi + 2k\pi\}.$$

Many equations involving circular functions have solutions that are typified by the example we have just studied. Their solution sets have infinitely many numbers, and we have to use care to be sure that we do not miss solutions.

EXAMPLE 38-1   Solve the equation $\sin^2 x = 1$.

Solution   We have
$$\{\sin^2 x = 1\} = \{\sin x = 1\} \cup \{\sin x = -1\}$$
$$= \{\tfrac{1}{2}\pi + 2k\pi\} \cup \{-\tfrac{1}{2}\pi + 2k\pi\}.$$

EXAMPLE 38-2   Solve the equation $\sin^2 x - 4\sin x + 3 = 0$.

Solution   In factored form, this equation reads
$$(\sin x - 3)(\sin x - 1) = 0,$$
and so our desired solution set is the union
$$\{\sin x = 3\} \cup \{\sin x = 1\}.$$

The first of these sets is the empty set (Why?), and the second is the set $\{\tfrac{1}{2}\pi + 2k\pi\}$. This latter set is therefore the soluton set of our given equation.

EXAMPLE 38-3   Solve the equation $2\sin^2 x + \cos^2 x - 4\sin x + 2 = 0$.

Solution   We replace $\cos^2 x$ with $1 - \sin^2 x$ and simplify to obtain an equivalent equation:
$$2\sin^2 x + 1 - \sin^2 x - 4\sin x + 2 = 0,$$
$$\sin^2 x - 4\sin x + 3 = 0.$$

This last equation was solved in Example 38-2.

EXAMPLE 38-4   Solve the equation $\cos 2x + 3\sin^2 x - 4\sin x + 2 = 0$.

Solution   We may use the identity $\cos 2x = 1 - 2\sin^2 x$ to transform this equation into the equivalent equation that we solved in Example 38-2.

As in the case of algebraic equations, solutions may be "lost" or extraneous "solutions" may be introduced as you simplify a given equation involving circular functions. Thus, it is always wise to check your solutions by substituting in the given equation.

EXAMPLE 38-5   Solve the equation $\sin x + \cos x = 1$.

Solution   If we square both sides of this equation, we get

$$\sin^2 x + 2 \sin x \cos x + \cos^2 x = 1.$$

Since $\sin^2 x + \cos^2 x = 1$, and $2 \sin x \cos x = \sin 2x$, this last equation reduces to

$$\sin 2x = 0.$$

Now x is a solution of this last equation if $2x \in \{k\pi\}$; that is, if $x \in \{\tfrac{1}{2}k\pi\}$. But not every number of this latter set satisfies the original equation. The numbers $\tfrac{3}{2}\pi$ and π, for example, satisfy the equaton $\sin 2x = 0$, but not the equation $\sin x + \cos x = 1$. In other words, the solution set of our given equation is merely a subset of the set $\{\tfrac{1}{2}k\pi\}$. To find this subset we need only check the numbers 0, $\tfrac{1}{2}\pi$, π, and $\tfrac{3}{2}\pi$ (Why?). While the numbers π and $\tfrac{3}{2}\pi$ do not satisfy the original equation, you can easily check and see that 0 and $\tfrac{1}{2}\pi$ do satisfy the equation $\sin x + \cos x = 1$. Thus, the solution set we seek is $\{2k\pi\} \cup \{\tfrac{1}{2}\pi + 2k\pi\}$.

The equation in Example 38-5 has the form

(38-1) $$B \sin x + C \cos x = r.$$

In Section 35 we showed how to find numbers A and b such that

$$B \sin x + C \cos x = A \sin (x + b).$$

Therefore, Equation 38-1 is equivalent to the equation

$$A \sin (x + b) = r.$$

Hence, to solve the equation of Example 38-5 we could have observed that

$$\sin x + \cos x = \sqrt{2} \sin (x + \tfrac{1}{4}\pi).$$

Our given equation can therefore be written as $\sin (x + \tfrac{1}{4}\pi) = \tfrac{1}{2}\sqrt{2}$, and it is easy to see that

$$\{\sin (x + \tfrac{1}{4}\pi) = \tfrac{1}{2}\sqrt{2}\} = \{2k\pi\} \cup \{\tfrac{1}{2}\pi + 2k\pi\}.$$

Most equations involving circular functions are difficult to solve explicitly by analytic methods. Often we use a computer and some kind of approximation technique. Usually the computer uses an iterative procedure to generate a sequence $\{r_1, r_2, \ldots\}$ of numbers, each term of which is a better approximation to a given solution of the equation than its predecessor was. To use the computer in this fashion, we must give it a starting number. We often use graphs to find this starting approximation. Furthermore, a graphical analysis of a problem can tell us many other things about its solutions, as the following examples illustrate.

158

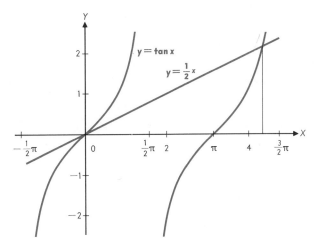

FIGURE 38-1

EXAMPLE 38-6   Solve the equation $x - 2 \tan x = 0$.

Solution   If we write this equation in the form $\frac{1}{2}x = \tan x$, we see that its solutions are the X-coordinates of the points of intersection of the graphs of the equations $y = \frac{1}{2}x$ and $y = \tan x$. We have drawn these curves in Fig. 38-1, and the figure shows that two solutions of our equation are 0 and 4.3 (approximately). It is clear that the equation has infinitely many positive solutions and infinitely many negative solutions. If we move far to the right (or to the left) of the Y-axis, the X-coordinates of the points of intersection are approximately odd multiples of $\frac{1}{2}\pi$. Thus, for example, if $\{p_1, p_2, \ldots\}$ is the sequence of *positive* solutions of our equation, then the figure suggests that

$$p_{1776} \approx \tfrac{1}{2}(2 \cdot 1776 + 1)\pi = \tfrac{3553}{2}\pi.$$

EXAMPLE 38-7   Solve the equation $x \sin x = 1$.

Solution   We first observe, as we could have in the last example, that if t is a solution, so is $-t$. For $(-t) \sin(-t) = t \sin t$, and if the right-hand side of this equation is 1, so is the left-hand side. Therefore, we need only look for positive solutions. We write the given equation as $\sin x = 1/x$, and sketch the graphs of the equations $y = \sin x$ and $y = 1/x$ for $x > 0$ (Fig. 38-2). The figure shows us that the given equation has infinitely many solutions, two of which are approximately 1.1 and 2.8. It is clear from Fig. 38-2 that those solutions whose absolute values are large are approximately integral multiples of π. This result was to be expected, since $1/x$ is small when x is large, and so solutions of our equation are almost solutions of the equation $\sin x = 0$.

Even though we can explicitly solve a problem analytically, it is often helpful to look at it graphically.

159

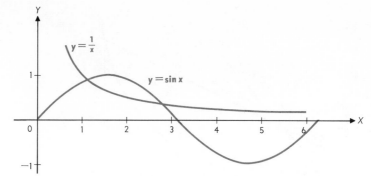

FIGURE 38-2

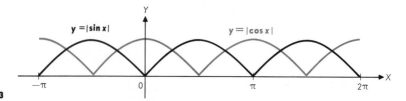

FIGURE 38-3

EXAMPLE 38-8   Solve the inequality $|\cos x| > |\sin x|$.

Solution   In Fig. 38-3 we have drawn a few cycles of the graphs of the equations $y = |\sin x|$ (in black) and $y = |\cos x|$ (in color). From the figure, it is clear that

$$\{|\cos x| > |\sin x|\} = \bigcup_{k \text{ an integer}} \{-\tfrac{1}{4}\pi + k\pi < x < \tfrac{1}{4}\pi + k\pi\}.$$

(Here we mean that we are to take the union of all sets of the type that follow the symbol $\bigcup\limits_{k \text{ an integer}}$ in which k is an integer.) It would be a good exercise for you to use analytic methods to solve this inequality.

PROBLEMS 38

1. Solve the following equations.
   (a) $\sqrt{3} \sin x + 2 = 0$
   (b) $2 \cos x - 1 = 0$
   (c) $\sqrt{3} \tan x + 1 = 0$
   (d) $\cot (x + 1) - \sqrt{3} = 0$
   (e) $[\![\sin x]\!] = 0$
   (f) $[\![\tan x]\!] = 0$
   (g) $4 \sin^2 t - 1 = 0$
   (h) $6 \cos^2 t + 5 \cos t + 1 = 0$
   (i) $2 \tan t - 2 \cot t + 3 = 0$
   (j) $2 \sec^2 t + 3 \sec t - 2 = 0$

2. Solve the following equations.
   (a) $\cos 2x + \sin x = 1$
   (b) $\sin 2x + \cos x = 0$
   (c) $4 \tan x + \sin 2x = 0$
   (d) $\tan 2x = \cos x$
   (e) $\sin t = \sin (2t - \pi)$
   (f) $\cos t = \sin (\pi - 3t)$
   (g) $\cot t = \tan (2t - 3\pi)$
   (h) $\sec t = \csc (t + \tfrac{1}{6}\pi)$
   (i) $\sqrt{3} \cos x - \sin x = 1$
   (j) $\sin x - 2 \cos x = 1$
   (k) $\sin x - \cos x = 2$
   (l) $\cos x + \sqrt{3} \sin x = \sqrt{2}$

160

3. Solve the following systems of equations.
   (a) $\sin x + \cos y = 1$
       $\sin 2x = 2 \sin y$
   (b) $y = 7 \sin x + 3 \cos x$
       $y = 7 \cos x + 3 \sin x$
   (c) $y \cos x = 4$
       $y \sin x = 4\sqrt{3}$
   (d) $\tan x = \cot y$
       $y = 3x + 2$

4. Solve the following equations.
   (a) $3 \operatorname{Sin}^{-1} x = \frac{1}{2}\pi$
   (b) $\operatorname{Arctan}(x - 1) = \frac{1}{3}\pi$
   (c) $\operatorname{Arcsin}(2x - x^2) = \operatorname{Arcsin}\frac{1}{2}$
   (d) $\operatorname{Arccos}(x^2 - 2x) = \operatorname{Arccos} 1$
   (e) $\operatorname{Arctan} x = 2 \operatorname{Sin}^{-1}\frac{1}{2}\pi$
   (f) $\sin(\operatorname{Tan}^{-1} x) = \tan(\operatorname{Sin}^{-1} x)$

5. (a) Show that if $\sin a = \sin b$ and $\cos a = \cos b$, then $a = b + 2k\pi$ for some integer k.
   (b) Show that if $\sin a = \cos b$ and $\cos a = \sin b$, then $a + b = \frac{1}{2}\pi + 2k\pi$ for some integer k.

6. Find a triple (r, θ, ϕ) such that $r > 0$, $0 \le \theta \le \pi$, $0 \le \phi \le 2\pi$, and the triple is a solution of the system of equations

$$r \sin \theta \cos \phi = 1$$
$$r \sin \theta \sin \phi = 1$$
$$r \cos \theta = 2.$$

7. Use graphical methods to find approximations to some of the solutions of each of the following equations.
   (a) $\sin 2x - \tan x = 0$
   (b) $x \sin 2x = 1$
   (c) $2x \sin \pi x = 1$
   (d) $\tan \pi x = x + 1$
   (e) $\pi x - 3 \sin \pi x = 0$
   (f) $\sin 2\pi x + \cos 2\pi x = \pi x$
   (g) $\sin(3x - 1) = x$
   (h) $x \tan \pi x = 1$
   (i) $x^2 = \cos 2\pi x$
   (j) $x - \operatorname{Arcsin} x = 0$
   (k) $x - \operatorname{Arccos} x = 0$
   (l) $x + \operatorname{Tan}^{-1} x = 1$
   (m) $2x = \sin \pi x$
   (n) $\log x = \sin \pi x$

8. Solve the equation $\operatorname{Tan}^{-1} x = (\operatorname{Tan} x)^{-1}$ graphically.

REVIEW PROBLEMS, CHAPTER FOUR

You should be able to answer the following questions without referring back to the text.

1. Suppose that $\tan t = \frac{3}{4}$. Can you find the following numbers?
   (a) $\sin t$
   (b) $|\sin t|$
   (c) $\sin 2t$
   (d) $\cos 2t$

2. If n is an integer, find the following numbers.
   (a) $\sin n\pi$
   (b) $\cos n\pi$
   (c) $\sin \frac{1}{2}(2n + 1)\pi$
   (d) $\cos \frac{1}{2}(2n + 1)\pi$

3. Verify the identity $\cos 2t = \cos^4 t - \sin^4 t$.

4. Sketch the graph of the equation $\sin(y - x) = 0$.

5. Convince yourself that none of the following numbers is negative.

   (a) $\sin (\text{Cos}^{-1} y)$ (b) $\cos (\text{Sin}^{-1} y)$

   (c) $y \tan (\text{Arcsin } y)$ (d) $y \sin (\text{Arctan } y)$

6. What are the domains of the functions defined by the following equations?

   (a) $f(x) = \sin (\log x)$ (b) $g(x) = \log (\sin x)$

7. Can you tell which number is larger without using tables?

   (a) $\sin (\cos 1)$ or $\cos (\sin 1)$ (b) $\cos (\log 1)$ or $\log (\cos 1)$

   (c) $\tan (\log 1)$ or $\log (\tan 1)$ (d) $[\![\cos \sqrt{5}]\!]$ or $\cos [\![\sqrt{5}]\!]$

8. Solve the following equations.

   (a) $\sin (\sin x) = 0$ (b) $\cos (\cos x) = 0$

   (c) $\text{Sin}^{-1} (\text{Sin}^{-1} x) = 0$ (d) $\text{Cos}^{-1} (\text{Cos}^{-1} x) = 0$

9. Find the following sets of numbers.

   (a) $\{x \mid |\sin x| = |\cos x|\}$ (b) $\{x \mid \sin x < \cos x\}$

   (c) $\{x \mid [\![\sin x]\!] = [\![\cos x]\!]\}$

10. Show that the following relations are valid for every number t that belongs to the domains of the functions involved.

    (a) $|\tan t + \cot t| \geq 2$ (b) $|\sin t + \cos t| \leq \sqrt{2}$

    (c) $(\sin t + \csc t)^2 \geq 4$ (d) $|\sin (\cos t)| < \frac{1}{2}\sqrt{3}$

MISCELLANEOUS PROBLEMS, CHAPTER FOUR

These problems are designed to test your ability to apply your knowledge of the circular functions to somewhat more difficult problems.

1. Show how to find (in terms of a and b) numbers p and q so that

$$\sin at + \sin bt = 2 \sin pt \cos qt.$$

   Use your result to discuss the graph of the equation $y = \sin at + \sin bt$, where a and b are large numbers of nearly equal size.

2. Show that if $\sin c = r$, then $\{\sin x = r\} = \{c + 2k\pi\} \cup \{(2k - 1)\pi - c\}$.

3. Let F be a function with an inverse, and suppose that F satisfies the addition formula $F(u + v) = F(u) + F(v)$. Show that the addition formula for F^{-1} is $F^{-1}(u + v) = F^{-1}(u) + F^{-1}(v)$.

4. Use graphical methods to solve the equation $|\sin \pi x| = x - [\![x]\!]$.

5. Show that t_1, the reference number associated with t, is given by the formula

$$t_1 = |t - [\![t/\pi + \tfrac{1}{2}]\!]\pi|.$$

6. Let m be a positive integer, and let $N(m)$ be the number of solutions of the equation $\sin x = \frac{1}{2}$ that belong to the set $\{0 < x < m\pi\}$. Express $N(m)$ in terms of m.

7. Use graphs to find two solutions of the system of equations

$$y = \tan x$$
$$x = \tan y.$$

8. The area of the smaller region between a unit circle and a chord d units from the center of the circle is $\frac{1}{4}$ of the area of the whole circular disk. Use the formula in Problem 36-10 and graphs to find d.

9. Sketch the graphs of the following equations.

   (a) $\sin \pi x \sin \pi y = 0$ (b) $|\sin \pi x| + |\sin \pi y| = 0$

   (c) $[\![\sin \pi x \sin \pi y]\!] = 0$ (d) $[\![\sin \pi x]\!] + [\![\sin \pi y]\!] = 0$

10. Let x be the solution of the equation $\sin x = 1/x$ that lies between 10π and $\frac{21}{2}\pi$. Show that $x - 10\pi < \frac{1}{20}$.

The
Trigonometric
Functions

five

A function is a set of pairs, and the functions we have been dealing with up to now have been, for the most part, sets of pairs of numbers. For example, the first two columns of Table I list some of the pairs, such as (1.0, .0000) and (2.5, .3979), that comprise the logarithmic function with base 10. The first two columns of Table II list some of the pairs that make up the sine function, and so on. But the pairs that constitute a function don't have to be pairs of numbers. When men first studied trigonometry they were interested in functions in which the first member of each pair was an *angle*, not a number, although the second member of the pair was a number. Thus, with each *angle* θ, the sine function paired a *number* sin θ, the cosine function paired a *number* cos θ, and so on. We will call these functions, whose domains are sets of angles rather than numbers, *trigonometric functions* to distinguish them from the circular functions that we defined earlier. From a strictly formal point of view, it is important to distinguish between the circular and the trigonometric functions. In practice, however, it is always clear from the context which type of function we are dealing with, so we don't have to be too careful about the distinction.

Our first task in this chapter will be to gather together some facts about angles and circles. Next, we will define the trigonometric functions in terms of the circular functions we discussed in the preceding chapter. Then, since trigonometric functions are used in solving geometric problems, we will devote the last few sections of this chapter to problems in a geometric setting.

A point P of a line separates it into two components; if we adjoin P to one of them, we have a **half-line** with **endpoint** P, as shown in color in Fig. 39-1.

FIGURE 39-1

If Q is the endpoint of another half-line, we see that the intersection of these two half-lines is the line segment that has P and Q as endpoints. Thus, we can think of a line segment as the intersection of two half-lines.

Now let us move up a dimension and consider the analogous situation in the plane. A line l of the plane separates it into two components; if we adjoin l to one of them, we have a **half-plane** with **edge** l, as shown in color on the left-hand side of Fig. 39-2. On the right-hand side of the figure we

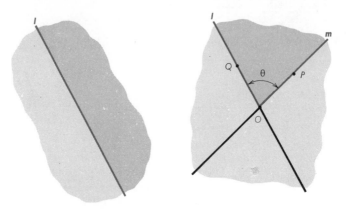

FIGURE 39-2

show the intersection of two half-planes whose edges are the lines l and m. This set of points is called a **wedge,** and you see that a wedge with its two boundary half-lines is the two-dimensional analogue of a line segment and its two boundary points. We call the boundary of a wedge an **angle.** The boundary half-lines are the **sides** of the angle, and their common endpoint is its **vertex.** Just as a line segment consists of two points of a line and the points between them, a wedge consists of two half-lines and the half-lines "between" them. The intersection of two half-planes with parallel edges is just a half-plane, and we say that its edge is a **straight angle.**

We will label angles with Greek letters, such as θ, ϕ, α, β, and γ. Thus the wedge in Fig. 39-2 is bounded by the angle θ whose vertex is O. If P and Q are points (different from O) of the two sides of the angle, we also speak of the angle POQ. Just as we introduce units of length to measure line

166

segments, we introduce units of width to measure wedges. The length of a line segment is the distance between its boundary points, and in like manner the width of a wedge is called the width of the angle that is its boundary.

To measure the width of a given angle, we can set up a coordinate system in which one side of the angle is the positive X-axis and the other side lies in the upper half-plane (Fig. 39-3). In this position, the vertex of

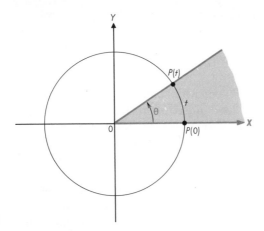

FIGURE 39-3

the angle is the origin of our coordinate system, and the two sides of the angle intersect the unit circle in two circular points $P(0)$ and $P(t)$, where t is a number betweee 0 and π. We may use this number t as a measure of the size of the angle. Thus, the angle shown in Fig. 39-3 is t units wide. The unit of angular measurement that we are using here is the **radian.** If we apply this procedure to a right angle, for example, one side will be the positive X-axis and the other side the positive Y-axis of our coordinate system. These sides intersect the unit circle in the circular points $P(0)$ and $P(\frac{1}{2}\pi)$, so a right angle is $\frac{1}{2}\pi$ radians wide. Similarly, a straight angle is π radians wide, and so on. The angle $P(0)OP(1)$ is 1 radian wide; that is, *the angle subtended by a unit arc of a unit circle is* 1 *radian wide.*

Just as distances are measured in different units, such as feet, yards, and meters, so angles are also measured in different units. Another common unit of angular measurement, besides the radian, is the degree (°). You are surely familiar with the division of a circle into 360°, and you know that a right angle is 90° wide, a straight angle is 180° wide, and so on. Since a straight angle is both 180° and π radians wide, we see that $1° = \frac{1}{180}\pi$ radians, and 1 radian $= (180/\pi)°$. Therefore, we can express the relationship between degree and radian measurement by means of a ratio. *If an angle θ is d degrees wide and also t radians wide, then the numbers d and t are related by the equation*

(39-1)
$$\frac{d}{180} = \frac{t}{\pi}.$$

To remember this formula, you need only recall that π radians $= 180°$.

EXAMPLE 39-1   If θ is an angle of 30°, what is its width in radians?

Solution   If our angle is t radians wide, then Equation 39-1 tells us that

$$\frac{30}{180} = \frac{t}{\pi} ; \quad \text{that is,} \quad t = \tfrac{1}{6}\pi.$$

Just as a line segment whose boundary points coincide is said to have zero length, we say that a wedge whose boundary half-lines coincide has zero width, and we speak of zero angles in this case. Thus, geometric angles can vary in size from 0° to 180°.

The introduction of a coordinate system enables us to use the terminology of angles in a somewhat broader sense, too. The angle θ shown in Fig. 39-3 is in **standard position** relative to the coordinate system, and we think of θ as being formed by rotating the positive X-axis in a counterclockwise direction until it contains the point $P(t)$. We say that we have rotated the axis through an angle of t radians. The axis forms the **initial side** of the angle, and the final position of the rotating half-line is the **terminal side** of the angle. The positive direction of rotation is the counterclockwise direction. This way of looking at angles frees us of any restriction on their size. Thus, we can let the positive X-axis make one complete revolution and generate an angle of 360° (2π radians). Or we can even make several circuits. An angle of $\tfrac{7}{2}\pi$ radians, for example, is obtained by rotating the positive X-axis through $1\tfrac{3}{4}$ revolutions in the counter-clockwise direction. Rotating the positive X-axis in the clockwise direction produces negative angles. For example, the initial side of an angle of $-\tfrac{1}{2}\pi$ radians is the positive X-axis, and its terminal side is the negative Y-axis, and so on.

We will use the word "angle" in both the geometric and rotational senses, and we will use the geometric terminology of "width" in both cases. But you see, of course, that in a geometric sense, an angle that is 20° wide is exactly as "wide" as an angle that is 380° wide.

EXAMPLE 39-2   If θ is an angle of 4 radians, what is its width in degrees?

Solution   Here Equation 39-1 becomes

$$\frac{d}{180} = \frac{4}{\pi}, \quad \text{and hence} \quad d = \frac{720}{\pi} \approx 229°.$$

We will sometimes talk about the sum of two angles or of the product of an angle by a number. Figure 39-4 illustrates what we mean. Thus, for example, to form $\theta + \phi$ we use the terminal side of θ as the initial side of ϕ. If θ and ϕ are angles of t_1 and t_2 radians, then $\theta + \phi$ is $t_1 + t_2$ radians wide, and $k\theta$ is kt_1 radians wide.

The unit circle has a radius of 1, and its center is the origin. Now suppose we look at a circle whose center is also the origin, but whose radius

168

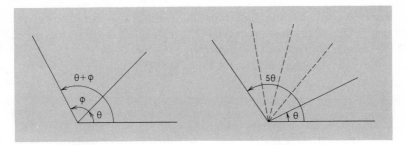

FIGURE 39-4

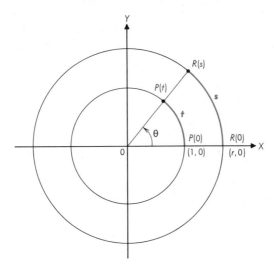

FIGURE 39-5

is a positive number r not necessarily equal to 1 (see Fig. 39-5). Just as we considered the point $(1, 0)$ as an "initial" point of the unit circle, here we consider the corresponding point $(r, 0)$ as an "initial" point. If s is any real number, we determine the point $R(s)$ by starting at the point $(r, 0)$ and proceeding $|s|$ units along the circumference of the circle, in the counter-clockwise direction if s is positive, and clockwise if s is negative. The process of locating the point $R(s)$ can also be described as follows. We rotate the positive X-axis about the origin, counter-clockwise or clockwise depending on whether s is positive or negative, until the point $(r, 0)$ has moved $|s|$ units to the point $R(s)$. During this process the point $(1, 0)$ will be moved $|t|$ units around the unit circle to a point $P(t)$. Let us find the relation between the numbers s and t.

The arcs $\overparen{R(0)R(s)}$ and $\overparen{P(0)P(t)}$ determine the same central angle θ. From plane geometry we know that the lengths of these arcs are therefore in the same ratio as the radii of the two circles, so that $|s|/|t| = r/1$ or $|s| = r|t|$. Since s and t both have the same sign, we have

(39-2) $$s = rt.$$

The number t in Equation 39-2 is the width in radians of the central angle θ. Thus Equation 39-2 says, *"The arclength s along a circle of radius r subtended by a central angle t radians wide is equal to rt."*

We may also write Equation 39-2 as

$$t = \frac{s}{r}.$$

In this form the equation says, *"The radian measure t of a central angle θ is the quotient of the subtended arclength s and the radius of the circle."*

The length $|s|$ of the arc joining the points $R(0)$ and $R(s)$ is $|s|/2\pi r$ times the circumference of the circle. Therefore, the area of the sector $R(0)OR(s)$ (Fig. 39-6) is $|s|/2\pi r$ times the area of the entire circular disk.

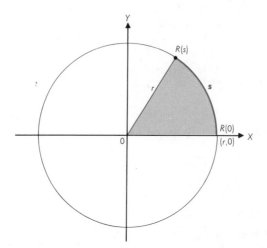

FIGURE 39-6

If the area of this sector is K square units, then $K = \dfrac{|s|}{2\pi r}\, \pi r^2$; that is,

(39-3)
$$K = \frac{1}{2} r\, |s|.$$

Notice the similarity between Equation 39-3 and the formula for the area of a triangular region (area equals one-half the base times the altitude). Since $s = rt$, Formula 39-3 may also be written as

(39-4)
$$K = \frac{1}{2} r^2\, |t|.$$

Formulas 39-3 and 39-4 actually give the total area of the sector that is "swept out" by the line segment $OR(0)$ as $R(0)$ is moved $|s|$ units along the circle of radius r. For example, if $s = 3\pi r$—that is, $1\frac{1}{2}$ times the circumference—Equation 39-3 gives $K = \frac{3}{2}\pi r^2$ or $1\frac{1}{2}$ times the area of the disk.

If we drop the absolute value signs in Equations 39-3 and 39-4, the new formulas state that the area of the sector that is swept out by the segment $OR(0)$ is positive if $OR(0)$ rotates in a counter-clockwise direction and negative if $OR(0)$ rotates in a clockwise direction. It is sometimes convenient to adopt this convention.

If a point moves along the arc of a circle of radius r at a constant velocity v, then this velocity is given by the equation $v = s/T$, where T is the number of units of time required to traverse s units of distance. The quotient s/T measures the rate at which distance changes with time. If, for example, s is measured in feet and T in seconds, then v is measured in feet per second. As the point moves around the circle, the angle between the radius through the point and any fixed radius will change. If this angle is initially 0 units wide and is t units wide at time T, then the **angular velocity** ω of the point is given by the equation $\omega = t/T$. When t is the radian measure of the angle, then angular velocity is measured in radians per unit of time. We can easily obtain the relation between the linear velocity v and the angular velocity ω (measured in radians per unit of time). Since $s = rt$, we have $v = s/T = rt/T = r(t/T) = r\omega$, and hence

(39-5) $$v = r\omega.$$

You should notice that v will be *negative* when the motion is clockwise.

EXAMPLE 39-3   A fly sits on the tip of the minute hand of a clock. If the hand is 4 inches long, how fast is the fly moving in miles per hour, assuming that the clock keeps perfect time?

Solution   The angular velocity of the minute hand is -2π radians per hour (ω is negative since the minute hand obviously moves in a clockwise direction). Thus the linear velocity of the fly is $v = -8\pi$ inches per hour or $v = -8\pi/(12 \cdot 5280) = -.0003967$ miles per hour.

PROBLEMS 39

1. Complete the following table.

Width of θ in radians	0	$\frac{1}{6}\pi$	$\frac{1}{4}\pi$		$\frac{1}{2}\pi$	$\frac{2}{3}\pi$			π
Width of θ in degrees	0°			60°			135°	150°	

2. Find the width in degrees of an angle whose width in radians is given.
   (a) $\frac{1}{12}\pi$ (b) $-\frac{5}{6}\pi$ (c) 3   (d) $\frac{1}{2}$

3. Angles are also measured in **minutes** (′) and **seconds** (″). These units are related to the degree unit by the equations $1° = 60'$ and $1' = 60''$. Find the degree-minute-second width (to the nearest second) of angles of the following widths.
   (a) 27.2°   (b) 163.36°   (c) 1 radian   (d) .612 radian

171

4. Find the widths in radians of angles of the following widths.

(a) 1° (b) 1′ (c) 1″ (d) 34°22′
(e) 148°37′ (f) 25°31′14″

5. Suppose that the circumference of the unit circle is divided into 7 equal parts by 7 equally spaced points. If two successive division points are joined to the origin, what is the degree-minute-second measure of the resulting angle?

6. A circle with a radius of 2 contains a central angle θ that intercepts an arc 5 units long. How wide is θ in radians? In degrees?

7. The latitude of Columbus, Ohio, is 40° north. Assuming that the earth is a sphere with a radius of 3960 miles, calculate the distance from Columbus to the North Pole.

8. The outside diameter of the wheel of a car is 20 inches. Find the number of revolutions the wheel makes in 1 second if the car is traveling 60 miles per hour.

9. Draw a sketch to show that the arc and chord subtended by a small central angle θ are almost the same length. Use this fact to calculate the sun's diameter, assuming the sun to be 93 million miles from the earth and as seen from the earth to subtend an angle of 32′.

10. The radius of the front wheel of a child's tricycle is 10 inches, and the rear wheels each have a radius of 6 inches. The pedals are fastened to the front wheels by arms that are 7 inches long. How far does a pedal travel when the rear wheels make 1 revolution?

11. At a distance of 10 feet, the least speed that the average eye can detect is about .005 feet per second. How long must the minute hand of a clock be made so that the average person will see it move when he is standing 10 feet from the clock?

12. A piston is connected to the rim of a wheel as shown in Fig. 29-6. The radius of the wheel is r feet, the length of the connecting rod is L feet, and the angular velocity of the wheel is ω radians per second. A coordinate system is introduced as shown in the figure. Find an expression for the X-coordinate of the point Q if the coordinates of P are $(r, 0)$ when $T = 0$.

13. The hour hand of a certain clock is 3 inches long What is the area of the region that it "sweeps out" during the time interval from 12 o'clock till the next time that the minute and hour hands are together?

40
THE TRIGONOMETRIC FUNCTIONS

Strictly speaking, trigonometric functions and circular functions are different, since the domains of the former are sets of angles and the domains of the latter are sets of numbers. But their similarities are so great that if you know about one of these classes of functions, you know about the other. Since we have already studied circular functions, we will reverse the historical order and define the trigonometric functions in terms of them.

With each circular function we associate a trigonometric function of the same name. Thus, to define the trigonometric sine function, we must state a rule for pairing with a given *angle* θ a *number* $\sin \theta$. We accomplish this pairing by first assigning to θ a number, its width t in radians, and then using the circular sine function to pair with t the number $\sin t$. In other words, we define $\sin \theta = \sin t$, where t is the radian measure of the angle θ. We follow the same procedure for the other trigonometric functions.

Definition 40-1   *If C is any one of the six circular functions, there is a corresponding trigonometric function T of the same name whose domain is a set of angles. If θ is an angle that is t radians wide, then*

$$T(\theta) = C(t).$$

EXAMPLE 40-1   Find $\cos \theta$ if θ is an angle of 1.3 radians.

Solution   By definition, $\cos \theta = \cos 1.3$. From Table II we find that $\cos 1.3 = .2675$, and therefore $\cos \theta = .2675$.

Sometimes it is convenient to replace an angle θ with its measure. For example, we will write $\theta = 30°$ to mean that θ is $30°$ wide, and we write $\tan 30°$ for the tangent of this angle. This convention (together with Definition 40-1) tells us that in the expression $\tan \theta$ we can replace θ with either its radian measure or its degree measure and the resulting expression stands for the same number.

EXAMPLE 40-2   Find $\tan 30°$.

Solution   An angle of $30°$ is $\frac{30}{180}\pi = \frac{1}{6}\pi$ radians wide, and hence $\tan 30° = \tan \frac{1}{6}\pi = 1/\sqrt{3} = .5774$.

We don't have to go back to Definition 40-1 every time we want to find $\sin \theta$ or $\tan \theta$ for a given angle θ. The angle will be measured either in radians or in degrees. If θ is measured in radians, we can use Table II to find corresponding values of the trigonometric functions, and if θ is measured in degrees, the values of the trigonometric functions are listed directly in Table III. You can use this table, for example, to check our result in Example 40-2. The symmetry of the trigonometric functions allows us to print Table III in compact form. We read the table in the natural way (that is, from top to bottom) for angles between $0°$ and $45°$, but we "invert it" for angles between $45°$ and $90°$. Thus, for example, $\sin 58° = .8480$, and $\cot 79° = .1944$.

Table III lists values of the trigonometric functions only at angles between $0°$ and $90°$. To find the values at other angles, we modify the procedure we used in Section 29 to find the values of trigonometric functions at numbers not listed in Table II. Thus, let T be a trigonometric function,

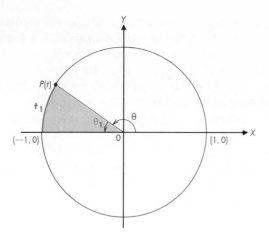

FIGURE 40-1

and let θ be an angle in standard position (Fig. 40-1). Then according to Definition 40-1, $T(\theta) = C(t)$, where t is the radian measure of θ and C is the circular function that corresponds to T. To find $C(t)$, we first must find the reference number t_1 associated with t. This number measures the distance along the unit circle between the circular point $P(t)$ and the nearest point of the X-axis. In terms of angles, the quantity that corresponds to the reference number t_1 is the **reference angle** θ_1. *This angle is the positive acute angle between the terminal side of θ and the X-axis.* Its radian measure is t_1, and so $T(\theta_1) = C(t_1)$. But $C(t_1) = |C(t)| = |T(\theta)|$, and hence we see that $|T(\theta)| = T(\theta_1)$. We look up $T(\theta_1)$ in Table II if θ_1 is measured in radians and in Table III if θ_1 is measured in degrees. Thus we know $T(\theta)$ except for sign, and the sign of $T(\theta)$ is determined by the quadrant in which the terminal side of θ lies.

The steps in finding $T(\theta)$ for a given angle θ may be summarized as follows.

(i) Locate the quadrant in which the terminal side of θ lies.

(ii) Calculate the reference angle θ_1.

(iii) Use Table II or Table III to find $T(\theta_1) = |T(\theta)|$.

(iv) Affix the proper sign to obtain $T(\theta)$.

EXAMPLE 40-3 Find cos 1059°.

Solution Since $1059 = 2 \cdot 360 + 339$, we see that an angle of 1059° is generated by rotating the positive X-axis (in a counter-clockwise direction) two complete revolutions, and then through an angle of 339°. The terminal side of the angle therefore lies in the fourth quadrant, and the reference angle $\theta_1 = 360° - 339° = 21°$. Hence, from Table III, $|\cos 1059°| = \cos 21° = .9336$. The cosine of an angle whose terminal side lies in the fourth quadrant is positive, so we finally have $\cos 1059° = .9336$.

EXAMPLE 40-4   Find csc $(-130°)$.

Solution   The terminal side of an angle of $-130°$ lies in the third quadrant, and the reference angle $\theta_1 = 180° - 130° = 50°$. We therefore see that

$$\text{csc } (-130°) = -\text{csc } 50°.$$

The number csc $50°$ is not listed in Table III, but we know that csc $50° = 1/\sin 50°$, and hence csc $(-130°) = -1/\sin 50° = -1/.7660 = -1.305$.

All the identities and formulas we have previously derived are still valid when numbers such as t, u, and so forth, are replaced by angles θ, ϕ, and the like. For example, if θ and ϕ are any two angles, then

$$\sin (\theta + \phi) = \sin \theta \cos \phi + \cos \theta \sin \phi.$$

We see from Fig. 40-1 that the terminal side of an angle θ in standard position intersects the unit circle in the point $P(t)$, where t is the radian measure of θ. The coordinates of $P(t)$ are $(\cos t, \sin t)$. Therefore, since $\cos t = \cos \theta$ and $\sin t = \sin \theta$, *the terminal side of an angle θ in standard position intersects the unit circle in the point* $(\cos \theta, \sin \theta)$.

Now let us extend this result. Again, let θ be an angle in standard position (Fig. 40-2), and suppose that P with coordinates (x, y) is any point

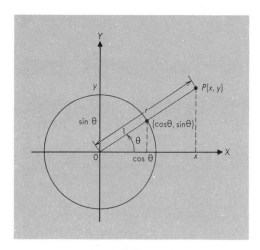

FIGURE 40-2

of the terminal side of θ. The distance $\overline{OP}$ is $r = \sqrt{x^2 + y^2}$. If we drop perpendiculars from the points $(\cos \theta, \sin \theta)$ and (x, y) to the X-axis, two similar triangles are formed, from which we see that $x/r = \cos \theta/1$ and $y/r = \sin \theta/1$. Thus, $x = r \cos \theta$ and $y = r \sin \theta$. These remarks are summarized in the following useful theorem.

175

Theorem 40-1   *If P is a point r units from the origin along the terminal side of an angle θ in standard position, then the coordinates of P are given by the equations*

(40-1) $x = r \cos \theta$ and $y = r \sin \theta.$

EXAMPLE 40-5   A regular pentagon is inscribed in the circle whose radius is 5 and whose center is the origin in such a way that one vertex is the point $(5, 0)$. Find two other vertices.

Solution   If you make a sketch of this problem and use Theorem 40-1, you will find that two other vertices are the points $(5 \cos 72°, 5 \sin 72°)$ and $(5 \cos (-72°),\ 5 \sin (-72°))$. According to Table III, these points can also be written as $(1.545, 4.7555)$ and $(1.545, -4.7555)$.

The next theorem gives another interpretation of the results we summarized in Theorem 40-1.

Theorem 40-2   *If P is a point r units from the origin along the terminal side of an angle θ in standard position, and if the coordinates of P are (x, y), then the values of the trigonometric functions at the angle θ are given by the equations*

(40-2)
$$\sin \theta = \frac{y}{r}, \qquad \cot \theta = \frac{x}{y},$$

$$\cos \theta = \frac{x}{r}, \qquad \sec \theta = \frac{r}{x},$$

$$\tan \theta = \frac{y}{x}, \qquad \csc \theta = \frac{r}{y}.$$

Proof   The first two equations are simply restatements of Equations 40-1. We can establish the other equations by using the first two equations and the elementary trigonometric identities. For example, $\tan \theta = \sin \theta / \cos \theta = (y/r)/(x/r) = y/x$.

The graph of the equation $y = mx + b$ is a line with slope m and Y-intercept b. The line that is parallel to this line and that contains the origin is the graph of the equation $y = mx$ (Fig. 40-3). This latter line and the positive X-axis determine an angle α between $0°$ and $180°$, which is called the **angle of inclination** of a line of slope m. The line $y = mx$ intersects the unit circle in the point $(\cos \alpha, \sin \alpha)$. Therefore, the coordinates of this point satisfy the equation of the line; that is, $\sin \alpha = m \cos \alpha$. Thus, $m = \tan \alpha$, and so we see that *the slope of a line is the tangent of its angle of inclination.*

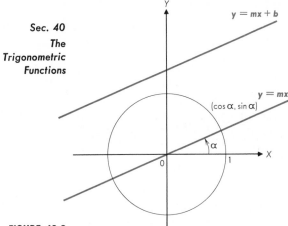

$y = mx + b$

$y = mx$

$(\cos \alpha, \sin \alpha)$

α

FIGURE 40-3

As we said earlier, the subject of trigonometry was originally developed to solve geometric problems involving triangles. In particular, the values of the trigonometric functions were defined in terms of the ratios of the lengths of sides of a right triangle. Let us denote the two acute angles of a right triangle by α and β, label the lengths of the sides opposite these angles a and b respectively, and let r be the length of the hypotenuse of the triangle. Take the vertex of the angle α to be the origin of a cartesian coordinate system whose positive X-axis is an extension of the side of length b, so that the triangle appears as in Fig. 40-4. The coordinates of the vertex of β are

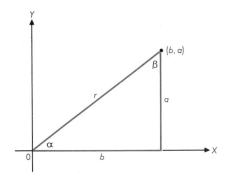

(b, a)

β

r

a

α

b

FIGURE 40-4

therefore (b, a). Since the angle α is in standard position, we may use Theorem 40-2 to write the equations

$$\sin \alpha = \frac{a}{r}, \qquad \csc \alpha = \frac{r}{a},$$

(40-3)
$$\cos \alpha = \frac{b}{r}, \qquad \sec \alpha = \frac{r}{b},$$

$$\tan \alpha = \frac{a}{b}, \qquad \cot \alpha = \frac{b}{a}.$$

177

Using words instead of letters may help you remember these equations. *If α is an acute angle of a right triangle, then*

$$\sin \alpha = \frac{\text{side opposite } \alpha}{\text{hypotenuse}}, \qquad \csc \alpha = \frac{\text{hypotenuse}}{\text{side opposite } \alpha},$$

(40-4) $$\cos \alpha = \frac{\text{side adjacent } \alpha}{\text{hypotenuse}}, \qquad \sec \alpha = \frac{\text{hypotenuse}}{\text{side adjacent } \alpha},$$

$$\tan \alpha = \frac{\text{side opposite } \alpha}{\text{side adjacent } \alpha}, \qquad \cot \alpha = \frac{\text{side adjacent } \alpha}{\text{side opposite } \alpha}.$$

It is easy to verify that $T(\alpha) = $ co-$T(\beta)$, where T may be replaced by any trigonometric function and co-T by the corresponding co-function. For example,

$$\sin \alpha = \frac{\text{side opposite } \alpha}{\text{hypotenuse}} = \frac{\text{side adjacent } \beta}{\text{hypotenuse}} = \cos \beta.$$

This equality and the fact that α and β are complementary angles accounts for the terminology of co-functions. The word "cosine," for example, is an abbreviation for "complement's sine."

Equations 40-4 are the original definitions of the trigonometric functions. They are extremely useful for solving certain problems that can be stated in geometric terms, so you should not neglect them. Traditionally, the study of trigonometry begins with a study of the trigonometric functions of acute angles as defined by Equations 40-4. Then the functions are defined for angles other than acute angles, identities are established, and so on.

The right-triangle definitions also provide us with a convenient memory device for obtaining the values of the trigonometric functions at certain angles—namely, 30°, 45°, and 60°. On the left-hand side of Fig. 40-5 we have

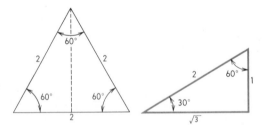

FIGURE 40-5

drawn an equilateral triangle whose sides are two units long. You will recall from geometry that its angles are also equal. Each is one-third of a straight angle, that is, each is an angle of 60°. A perpendicular from a vertex to the opposite side will bisect the vertical angle and the side, thus splitting the original into two right triangles, one of which is shown on the right-hand side

178

of Fig. 40-5. (We used the theorem of Pythagoras to find that the third side of this triangle is $\sqrt{3}$ units long.) This right triangle, together with Equations 40-4, helps us remember the values of the trigonometric functions at 30° and at 60°. For example, we see that $\tan 60° = \sqrt{3}/1 = \sqrt{3}$, and so on. In the problems we ask you to find a right triangle that will help you remember the values of the trigonometric functions at 45°.

Many practical triangle problems can be solved by right-triangle trigonometry. A typical example of such a problem is the following.

EXAMPLE 40-6   From a certain point on a level plain at the foot of a mountain, the angle of elevation of the peak is 45°. From a point 2000 feet farther away, the angle of elevation of the peak is 30°. How high above the plain is the peak of the mountain?

Solution   The situation is illustrated in Fig. 40-6. Triangles APC and BPC are right triangles, so Equations 40-4 give us the two equations

$$\tan 45° = \frac{y}{x},$$

and

$$\tan 30° = \frac{y}{x + 2000}.$$

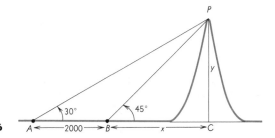

FIGURE 40-6

Since $\tan 45° = 1$ and $\tan 30° = 1/\sqrt{3}$, these equations become

$$y = x,$$

and

$$x + 2000 = \sqrt{3}y.$$

When we solve these last two equations for y, we get

$$y = \frac{2000}{\sqrt{3} - 1} = 2732 \text{ feet.}$$

EXAMPLE 40-7   If t is a number such that $0 < t < \frac{1}{2}\pi$, show that

$$\cos t < \frac{t}{\sin t} < \frac{1}{\cos t}.$$

Solution   Figure 40-7 shows the distance t laid off on the unit circle. The central angle θ is t radians wide so $T(\theta) = T(t)$ for any trigonometric function T. The point P is the circular point of t, and R is the foot of the perpendicular

179

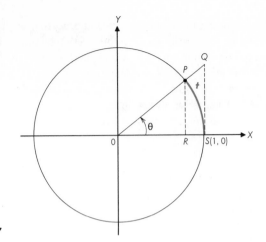

FIGURE 40-7

dropped from P to the X-axis. The coordinates of the point S are $(1, 0)$, and the segment QS is drawn perpendicular to the X-axis. Comparing the area of the circular sector POS to the area of the triangular regions POR and QOS, we see that

(40-5) $Area\ of\ POR < Area\ of\ POS < Area\ of\ QOS,$

and we shall now find expressions for each of these areas. Since $\cos \theta = \overline{OR}/1$, the base of the triangle POR is $OR = \cos \theta = \cos t$. Similarly, the altitude of triangle POR is $\overline{PR} = \sin t$. The area of POR is therefore $\frac{1}{2} \sin t \cos t$. According to Equation 39-3, the area of the circular sector POS is $\frac{1}{2}t$. Triangle QOS is a right triangle with base $\overline{OS} = 1$. Since $\tan \theta = \overline{QS}/1$, the altitude is $\overline{QS} = \tan \theta = \tan t$. The area of QOS is therefore $\frac{1}{2} \tan t$. We may now write Inequalities 40-5 as

$$\tfrac{1}{2} \sin t \cos t < \tfrac{1}{2}t < \tfrac{1}{2} \tan t.$$

If we divide each term of these inequalities by $\frac{1}{2} \sin t$, we get the result we are seeking. You may encounter these inequalities again when you study calculus.

PROBLEMS
40

1. Use Table III to evaluate each of the following.
   (a) $\sin 40.6°$ (b) $\cos 125°$ (c) $\tan 95°$ (d) $\cot -261.4°$
   (e) $\sin 3°21'$ (f) $\cos 698°$ (g) $\sec -21°$ (h) $\sin 37°1'1''$
   (i) $\csc -705°$

2. Determine which is the larger number.
   (a) $\cos 3°$ or $\cos 3$ (b) $\cot 1°$ or $\cot 1$ (c) $\sin 7°$ or $\sin 7$ (d) $\tan 4°$ or $\tan 4$

3. Find a right triangle (as in Fig. 40-5) which will help you remember the values of the trigonometric functions at 45°.

4. Complete the following table. (You should be able to work this problem without referring to the text.)

θ	$\sin \theta$	$\cos \theta$	$\tan \theta$
0			
30°			
45°			
60°			
90°			
120°			
135°			
150°			
180°			

5. Make a sketch of the unit circle, showing a small central angle θ that is t radians wide. Explain why the figure suggests that $\sin \theta = t$ (approximately). Does $\sin 1° = 1$ (approximately)?

6. Use an addition formula and the equation $15° = 45° - 30°$ to calculate $\cos 15°$. Compare your result with the value listed in Table III.

7. Use Theorem 40-1 to calculate the coordinates of a point P of a circle of radius 5 if P belongs to the terminal side of an angle of the following width.
   (a) 35°   (b) 122°   (c) 192°   (d) 303°

8. Find the widths of the angles of a right triangle if the lengths of two of the sides are as given.
   (a) $a = 3, b = 4$ (b) $a = 5, b = 12$
   (c) $a = 7, r = 25$ (d) $a = 8, r = 17$

9. Find the lengths of the sides of a right triangle if the width of one angle and the length of one side are as given.
   (a) $a = 5, \beta = 40°$ (b) $a = 3, \alpha = 10°$
   (c) $b = 8.21, \alpha = 26°31'$ (d) $r = 7, \alpha = 64°17'$

10. At a time when the angle of elevation of the sun (measured from the horizontal) measures 50°, the length of the shadow of a tree is 75 feet. What is the height of the tree?

11. Find the width of an angle α of a right triangle if it satisfies the following equations.
   (a) $\sin \alpha = \cos (\alpha + 45°)$ (b) $\tan \alpha = \cot 2\alpha$
   (c) $\sec \frac{1}{2}\alpha = \csc 2\alpha$ (d) $\sin (\alpha + 25°) = \cos (\alpha - 25°)$

12. Find the width in degrees of the angle of inclination of the following lines.
   (a) $y = 2x - 1$ (b) $2y = x + 1$
   (c) $y - 2 = 3(x - 4)$ (d) $y = x \sin 3 + 1$

13. A given positive integer m can be written as $m = 180q + r$, where q is a non-negative integer, and $0 \leq r < 180$. Convince yourself that $\sin m° = \sin (90 - |90 - r|)°$.

The lengths of the sides and the values of the trigonometric functions at the angles of a triangle satisfy certain relations that are useful in solving geometric problems. One of these relations, called the **Law of Sines,** can be viewed as a generalization of the equation $\tan \alpha = a/b$ of Equations 40-3. If we write that equation as $\dfrac{\sin \alpha}{a} = \dfrac{\cos \alpha}{b}$ and realize that $\cos \alpha = \sin \beta$, we see that *the ratio of the sine of an angle to the length of the opposite side is the same for each of the acute angles of a right triangle.* This italicized statement remains valid if we delete the words "acute" and "right."

Theorem 41-1   *(Law of Sines). Let α, β, and γ be the angles of a triangle, and a, b, and c be the lengths of the opposite sides. Then*

$$\frac{\sin \alpha}{a} = \frac{\sin \beta}{b} = \frac{\sin \gamma}{c}.$$

Proof   We will show that $(\sin \alpha)/a = (\sin \beta)/b$. Consider two ways of placing the triangle with respect to a cartesian coordinate system. First, take the vertex of the angle α as the origin, with the side of length c extending along the positive X-axis (Fig. 41-1). Then turn the triangle over so that the vertex of β becomes the origin and the side of length c again extends along the positive X-axis (Fig. 41-2). In the first case, the vertex of γ will be a point P and in the second case a point Q. The essential thing to observe is that *the points P and Q have the same Y-coordinate*—namely, the altitude of the triangle. In Fig. 41-1 the angle α is in standard position, and in Fig. 41-2 the angle β is in standard position. According to Theorem 40-1, the Y-coordinate of P is $b \sin \alpha$, and the Y-coordinate of Q is $a \sin \beta$. Hence, $b \sin \alpha = a \sin \beta$, and so $(\sin \alpha)/a = (\sin \beta)/b$. Since we used no special properties of the angles α and β in establishing this equation, we have really shown that the ratio of the sine of an angle of a triangle to the length of the opposite side is the same for all angles of the triangle. We have therefore proved the Law of Sines.

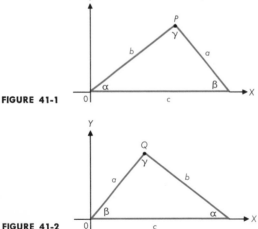

FIGURE 41-1

FIGURE 41-2

The lengths of the sides of a triangle and the widths of its angles cannot be assigned arbitrarily. For example, the sum of the angles of a triangle must measure 180°, and therefore if we know the widths of two angles, we can find the width of the third. If, in addition to knowing the widths of two angles of a triangle, we also know the length of one side, then we can use the Law of Sines to find the lengths of the other two sides. This procedure is illustrated in the following example.

EXAMPLE 41-1 Let $\alpha = 45°$, $\beta = 75°$, and $c = 10$. Find a, b, and γ.

Solution Since $\alpha + \beta + \gamma = 180°$, we see that $\gamma = 180° - 120° = 60°$. According to the Law of Sines, $(\sin \alpha)/a = (\sin \gamma)/c$, so

$$a = \frac{c \sin \alpha}{\sin \gamma} = \frac{10 \sin 45°}{\sin 60°} = \frac{10 \cdot \frac{1}{2}\sqrt{2}}{\frac{1}{2}\sqrt{3}} = 10\sqrt{\tfrac{2}{3}}.$$

Similarly,

$$b = \frac{c \sin \beta}{\sin \gamma} = \frac{10 \sin 75°}{\sin 60°}.$$

We find the number $\sin 75°$ by writing $\sin 75° = \sin (30° + 45°)$ and using an addition formula:

$$\sin 75° = \frac{1 + \sqrt{3}}{2\sqrt{2}}.$$

Thus,

$$b = \frac{10(1 + \sqrt{3})/2\sqrt{2}}{\sqrt{3}/2} = \frac{10(1 + \sqrt{3})}{\sqrt{6}}.$$

If we know the lengths of two sides of a triangle and the width of the angle opposite one of the sides, then we can determine the remaining side and angles by using the Law of Sines.

EXAMPLE 41-2 Let $\alpha = 30°$, $a = 10$, and $c = 15$. Find β, γ, and b.

Solution Since $(\sin \gamma)/c = (\sin \alpha)/a$, we have $\sin \gamma = c(\sin\alpha)/a$. Thus,

$$\sin \gamma = \frac{15 \sin 30°}{10} = \frac{15}{20} = \frac{3}{4}.$$

There are two angles between 0° and 180° for which $\sin \gamma = \frac{3}{4}$, the angles 48°36″ and 131°24′. If $\gamma = 48°36′$, then $\beta = 101°24′$ and $b = a \sin \beta/\sin \alpha = 20(.98) = 19.6$. In case $\gamma = 131°24′$, then $\beta = 18°36′$ and $b = 20(.32) = 6.4$. These two possibilities are shown in Fig. 41-3.

If two numbers a and c and an angle α are given, there may be one, two, or no triangles that have one side of length a, another of length c, and the angle α opposite the side of length a. We should therefore use care in solving problems such as Example 41-2. Frequently a sketch will immediately

183

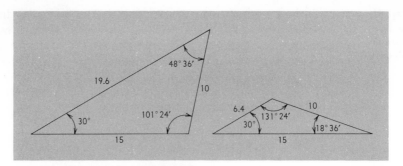

FIGURE 41-3

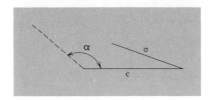

FIGURE 41-4

make the situation clear. For example, Fig. 41-4 shows that if $\alpha \geq 90°$, then it is impossible to construct a triangle in which $a \leq c$. If $a > c$, there is exactly one triangle. If α is acute, the problem is more complicated. There is exactly one triangle if $a \geq c$, but if $a < c$, several things can happen. Figure 41-5 illustrates the possibilities of one, two, or no solutions. We find

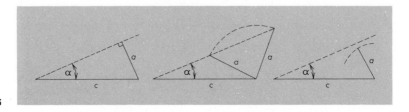

FIGURE 41-5

which of these situations exists when we calculate $\sin \gamma = c(\sin \alpha)/a$. If $\sin \gamma = 1$, then γ is a right angle, and there is just one suitable triangle. If $\sin \gamma < 1$, then there are two possible choices for γ, as there were in Example 41-2. But if it should appear that $\sin \gamma$ is greater than 1, then there are no triangles that satisfy the given conditions. You shouldn't take the trouble to memorize all the possibilities in this so-called *ambiguous case*. Work each problem as it comes, but keep your eyes open.

The **Law of Cosines** is a generalization of the Pythagorean Theorem to triangles that do not necessarily contain a right angle.

Theorem 41-2   (Law of Cosines). *Let a, b, and c be the lengths of the sides of a triangle, and label the angle opposite the side of length a by α. Then*

(41-1) $$a^2 = b^2 + c^2 - 2bc \cos \alpha.$$

184

Written out, the Law of Cosines states that the square of the length of a given side of a triangle is equal to the sum of the squares of the lengths of the other two sides minus twice the product of the lengths of the other sides and the cosine of the angle between them.

Proof We place the triangle so that the vertex of the angle α is the origin of a cartesian coordinate system, and the side of length c lies along the positive X-axis (see Fig. 41-6). It is clear that the coordinates of the vertex of the

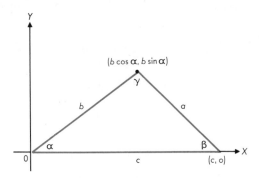

FIGURE 41-6

angle β are $(c, 0)$. Since the angle α is in standard position, Theorem 40-1 tells us that the coordinates of the vertex of γ are $(b \cos \alpha, b \sin \alpha)$. The distance between the vertices of β and γ is a, and according to the distance formula,

$$a^2 = (b \cos \alpha - c)^2 + (b \sin \alpha)^2$$
$$= b^2 \cos^2 \alpha + b^2 \sin^2 \alpha + c^2 - 2bc \cos \alpha.$$

Since

$$b^2 \cos^2 \alpha + b^2 \sin^2 \alpha = b^2 (\cos^2 \alpha + \sin^2 \alpha) = b^2 \cdot 1 = b^2,$$

our equation reduces to Equation 41-1, and the Law of Cosines is established.

If α is a right angle, then $\cos \alpha = 0$, and Equation 41-1 simply re-states the Theorem of Pythagoras.

If we know the lengths of two sides of a triangle and the width of the angle between them, then we can use the Law of Cosines to find the length of the remaining side and the widths of the other angles.

EXAMPLE 41-3 If $\gamma = 60°$, $a = 12$, and $b = 5$, find c, α, and β.

Solution According to the Law of Cosines,

$$c^2 = 12^2 + 5^2 - 2 \cdot 12 \cdot 5 \cdot \cos 60°.$$

Since $\cos 60° = \frac{1}{2}$, we have

$$c^2 = 144 + 25 - 2 \cdot 12 \cdot 5 \cdot \tfrac{1}{2} = 109,$$

185

and hence $c = \sqrt{109}$. We can also use the Law of Cosines (or the Law of Sines) to find the angles α and β. For example, from Equation 41-1

$$2bc \cos \alpha = b^2 + c^2 - a^2,$$

so that

$$\cos \alpha = \frac{b^2 + c^2 - a^2}{2bc} = \frac{134 - 144}{10\sqrt{109}} = \frac{-1}{\sqrt{109}} = -.0958.$$

From Table III we find that $\cos 84°30' = .0958$, and therefore

$$\alpha = 180° - 84°30' = 95°30'. \text{ Then}$$
$$\beta = 180° - (95°30' + 60°) = 24°30'.$$

EXAMPLE 41-4 Find the angles of a triangle if the lengths of the sides are 5, 6 and 8.

Solution Let $a = 5$, $b = 6$, and $c = 8$. Then

$$\cos \alpha = \frac{6^2 + 8^2 - 5^2}{2 \cdot 6 \cdot 8} = \frac{25}{32} = .7812$$

$$\cos \beta = \frac{5^2 + 8^2 - 6^2}{2 \cdot 5 \cdot 8} = \frac{53}{80} = .6625$$

$$\cos \gamma = \frac{5^2 + 6^2 - 8^2}{2 \cdot 5 \cdot 6} = \frac{-1}{20} = -.0500$$

With the aid of Table III we find $\alpha = 38°37'$, $\beta = 48°30'$, and $\gamma = 92°52'$. Notice that $\alpha + \beta + \gamma = 179°59'$, rather than $180°$, but this result is not surprising since we did some "rounding off" in our caculations.

EXAMPLE 41-5 If a, b, and c are the lengths of the sides of a triangle, show that α is an acute angle if, and only if, $b^2 + c^2 > a^2$.

Solution Since α is an angle of a triangle, it must measure between $0°$ and $180°$. Such an angle is acute if, and only if, its cosine is positive. According to to the Law of Cosines, $\cos \alpha = (b^2 + c^2 - a^2)/2bc$, and this quotient is positive if, and only if, $b^2 + c^2 - a^2 > 0$—that is, if, and only if, $b^2 + c^2 > a^2$.

**PROBLEMS
41**

1. Find the remaining parts of a triangle with the following measurements.
   (a) $\gamma = 75°$, $\beta = 30°$, $a = 5$ (b) $\alpha = 42°$, $\gamma = 76°$, $b = 18$
   (c) $\alpha = 33°30'$, $\beta = 50°48'$, $c = 12.4$ (d) $\beta = 2\alpha$, $\gamma = 2\beta$, $a = 1.6$

2. Either show that no triangle exists or find the missing parts of all triangles with the following measurements.
   (a) $a = 4$, $b = 5$, $\alpha = 30°$ (b) $a = 12$, $c = 6$, $\gamma = 67°$
   (c) $b = 20$, $c = 26$, $\beta = 148°$ (d) $a = 23.8$, $b = 31.4$, $\alpha = 23°40'$

3. Find the remaining parts of a triangle with the following measurements.
   (a) $a = 3$, $b = 2$, $\gamma = 60°$ (b) $b = 4$, $c = \sqrt{3}$, $\alpha = 30°$
   (c) $b = 7$, $c = \sqrt{2}$, $\alpha = 135°$ (d) $b = 6$, $c = 9$, $\alpha = 53°$
   (e) $a = 9$, $b = 4$, $\gamma = 93°15'$ (f) $a = 2$, $c = 2$, $\beta = 45°$

4. Determine the angles of a triangle with sides of the following lengths.
   (a) 5, 7, and 8   (b) 15, 25, and 30
   (c) 28, 45, and 53   (d) 4, 5, and 10

5. City A is directly south of city B, but there are no direct airline flights from A to B. Planes first fly 143 miles from city A to city C, which is 51° east of north from A, and then fly 212 miles to city B. What is the straight-line distance between A and B?

6. Observers in cities A and B, which are 5 miles apart, see an object in the sky above the line on the ground joining the cities. The angle of elevation at A is 23° and the angle of elevation at B is 31°. What is the altitude of the object?

7. The longer diagonal of a certain parallelogram is 10 inches long. At one end the diagonal makes angles of 33° and 25° with the sides of the parallelogram. Find the lengths of the sides of the parallelogram.

8. In order to measure the distance between two points A and B on opposite sides of a canyon, a third point C is chosen so that the following measurements can be made: $\overline{CA} = 175$ feet, and $\overline{CB} = 212$ feet. At B, the angle ABC is 47° wide. Compute the distance $\overline{AB}$.

9. Four ships, A, B, C, and D, are at sea in the following relative positions. B is on a line between A and C, B is due north of D, and D is due west of C. The distance between B and D is 2 miles. From D, the angle BDA is 40° wide, and from C the angle BCD is 25° wide. What is the distance between A and D?

10. Let D be a point of the side c of a triangle ABC such that CD bisects the angle γ. Use the Law of Sines of prove that $\overline{AD}/\overline{DB} = b/a$.

11. In order to measure the distance between two points A and B on opposite sides of a building, a third point C is chosen such that the following measurements can be made: $\overline{CA} = 215$ feet, $\overline{CB} = 371$ feet, and the angle ACB measures 56° at C. What is the distance between A and B?

12. A man is at a point directly south of factory A and on a line southwest from factory B. He hears the noon whistle of factory B 5 seconds after noon, and the noon whistle of factory A 7 seconds after noon. Assuming the velocity of sound to be 1100 feet per second, what is the distance between the factories?

13. A pilot intends to fly from city A to city B, a distance of 150 miles. He starts 15° off his course, and proceeds 50 miles before discovering his error. How much should he alter his course and how far must he fly to reach B?

42
AREA FORMULAS

In our terminology, a triangle is a one-dimensional figure consisting of three vertices and the line segments that join them. Therefore, its area is 0. But everyone knows that when we speak of the "area of a triangle," we mean the area of the *triangular region* consisting of the triangle itself and the interior points it bounds. So even though it may not be strictly correct, we will continue to use this language as we study area formulas in this section. If a triangle is placed with the vertex of the angle α as the origin of a cartesian coordinate system and with the side of length c extending along the positive X-axis as shown in Fig. 42-1, then α is in standard position and according to

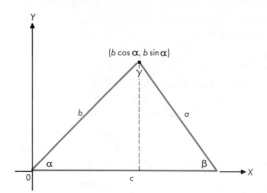

FIGURE 42-1

Theorem 40-1, the Y-coordinate of the vertex of γ is $b \sin \alpha$. This number is the length of the altitude of the triangle. Its base is b units long, so the area A of the triangle in Fig. 42-1 is given by the equation

(42-1)
$$A = \frac{1}{2} bc \sin \alpha.$$

We have therefore shown that *the area of a triangle is one-half the product of the lengths of any two sides and the sine of the included angle.*

By using the relation $b = (c \sin \beta)/\sin \gamma$ from the Law of Sines, we can write Equation 42-1 as

(42-2)
$$A = c^2 \frac{\sin \alpha \sin \beta}{2 \sin \gamma}.$$

Equation 42-2 expresses the area of a triangle in terms of the sines of all three angles and the length of one side.

If we square both sides of Equation 42-1, we get the equations

(42-3)
$$A^2 = \frac{b^2 c^2}{4} \sin^2 \alpha = \frac{b^2 c^2}{4} (1 - \cos^2 \alpha).$$

Now, from the Law of Cosines,

$$\cos \alpha = \frac{b^2 + c^2 - a^2}{2bc},$$

so

$$1 - \cos^2 \alpha = 1 - \frac{(b^2 + c^2 - a^2)^2}{4b^2c^2}$$

$$= \frac{4b^2c^2 - (b^2 + c^2 - a^2)^2}{4b^2c^2}.$$

The numerator of this fraction is the difference of two squares, and if we use the identity $x^2 - y^2 = (x + y)(x - y)$, with $x = 2bc$ and $y = b^2 + c^2 - a^2$, we have

$$1 - \cos^2 \alpha = \frac{(2bc + b^2 + c^2 - a^2)(2bc - b^2 - c^2 + a^2)}{4b^2c^2}$$

$$= \frac{[(b + c)^2 - a^2][a^2 - (b - c)^2]}{4b^2c^2}$$

$$= \frac{(b + c + a)(b + c - a)(a + b - c)(a - b + c)}{4b^2c^2}.$$

When we substitute this expression in Equation 42-3, we obtain the equation

(42-4) $$A^2 = \frac{(a + b + c)(a + b - c)(b + c - a)(c + a - b)}{16}.$$

This equation expresses *the area of a triangle in terms of the lengths of its sides.* We can put it in a simpler form by introducing the **semiperimeter** $s = \frac{1}{2}(a + b + c)$ of the triangle. You can easily verify the following equations:

$$2s = a + b + c, \qquad\qquad 2(s - b) = c + a - b,$$

(42-5)

$$2(s - a) = b + c - a, \qquad 2(s - c) = a + b - c.$$

When we introduce these quantities into Equation 42-4, it becomes

(42-6) $$A = \sqrt{s(s - a)(s - b)(s - c)}.$$

Let us now inscribe a circle in a triangle (Fig. 42-2). Suppose that the point P is the center of the circle and that r is its radius. We divide the

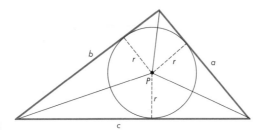

FIGURE 42-2

triangle into three small triangles by joining P to each of the vertices; then we find the area of the large triangle by adding the areas of the small triangles. Since the radius of the inscribed circle is perpendicular to a side of the large triangle at its point of contact, the altitude of each of these small triangles is r. We therefore see that

$$A = \tfrac{1}{2}ra + \tfrac{1}{2}rb + \tfrac{1}{2}rc = r\tfrac{1}{2}(a+b+c), \quad \text{so}$$

(42-7)
$$A = rs.$$

By equating the right-hand sides of Equations 42-6 and 42-7, we find that the radius of the inscribed circle is given by the formula

(42-8)
$$r = \sqrt{\frac{(s-a)(s-b)(s-c)}{s}}.$$

If you search through other trigonometry books, particularly older ones, you will uncover many other formulas that connect the sides and the values of the trigonometric functions at the angles of a triangle.

PROBLEMS
42

1. Find the area of a triangle with the following measurements.
   (a) $b = 14$, $c = 5$, $\alpha = 30°$ (b) $a = 10$, $b = 20$, $\gamma = 179°$
   (c) $a = 5$, $c = 7.2$, $\beta = 71°$ (d) $b = 1.5$, $c = 4$, $\alpha = \tfrac{5}{7}\pi$ radians

2. Find the area of a triangle with the following measurements.
   (a) $c = 3$, $\beta = 40°$, $\gamma = 60°$ (b) $a = 2$, $\alpha = 60°$, $\beta = 75°$
   (c) $b = 11$, $\alpha = 1$ radian, $\beta = 1.5$ radians
   (d) $c = 3.2$, $\alpha = 16°$, $\beta = 31°45'$

3. Find the area of a triangle and the radius of the inscribed circle if the triangle has the following measurements.
   (a) $a = 2$, $b = 3$, $c = 4$ (b) $a = 2$, $b = 7$, $c = 8$
   (c) $a = \log 2$, $b = \log 3$, $c = \log 5$ (d) $a = \sqrt{2}$, $b = 2$, $c = 3$

4. If two triangles are similar, the ratio of the length of any side of one triangle to the length of the corresponding side of the other is a number k. How are the areas related? How are the radii of the inscribed circles related?

5. Equation 42-6 expresses the area of a triangle in terms of the lengths of its sides (angles are not involved). Is it possible to express the area of a triangle in terms of its angles only?

6. Find the length of the longest side of a triangle whose area is 10 and whose angles measure 30°, 50°, and 100°.

7. A field is in the shape of a parallelogram. The lengths of two of the sides are 5 rods and 7 rods. The length of the longer diagonal is 10 rods. Find the area of the field.

8. Let a circle of radius r be inscribed in a triangle ABC and let x, y, and z denote the lengths of the segments from the points of contact of the inscribed circle to the vertices A, B, and C. Show that the area of the triangle is given by the equation $A = r(x + y + z)$. From this equation deduce the equation $A = r^2(\cot \frac{1}{2}\alpha + \cot \frac{1}{2}\beta + \cot \frac{1}{2}\gamma)$.

9. Show that the area of an isosceles triangle is a maximum if the angle between the two equal sides is a right angle.

10. Let B and C be fixed points, l a line parallel to BC, and A a point of l. Show that the product

$$[(\overline{AC} + \overline{AB})^2 - \overline{BC}^2][\overline{BC}^2 - (\overline{AC} - \overline{BA})^2]$$

is the same for every choice of the point A.

11. Let P be the point of contact between the inscribed circle of radius r and the side of length b of a triangle ABC. Show that $\overline{AP} = s - a$, and thus deduce the equation $\tan \frac{1}{2}\alpha = r/(s - a)$. This equation can be used to find the angles of a triangle when the lengths of the three sides are given.

REVIEW PROBLEMS, CHAPTER FIVE

You should be able to answer the following questions without referring back to the text.

1. Show that $T(45° + \alpha) = \text{co-}T(45° - \alpha)$ for any trigonometric function T and angle α.

2. A bicycle odometer counts the number of revolutions of the front wheel. For a 26-inch (diameter) wheel, how many revolutions measure a mile of travel?

3. Let Q be the point $(-1, 0)$ and let P be any other point of the unit circle. If we denote by ϕ the angle PQO, express the coordinates of P in terms of ϕ.

4. Determine the angles of a triangle if its vertices are the following points.
    (a) $(0, 0)$, $(3, 0)$, $(3, 4)$ (b) $(0, 0)$, $(2, 0)$, $(2 + \sqrt{3}, 1)$
    (c) $(-1, 0)$, $(1, 0)$, $(\frac{1}{2}, \frac{1}{2}\sqrt{3})$ (d) $(1, 2)$, $(-2, 1)$, $(-1, -2)$

5. Columbus, Ohio, and Boulder, Colorado, are both at 40°N latitude, and their respective longitudes are 83°W and 105°W. How far is it from Columbus to Boulder along the 40° parallel if the radius of the earth is 3960 miles?

6. A ship is steaming on a straight course and at a steady speed. At 12 o'clock the captain spots a lighthouse at an angle of 45° with the ship's course, and at 1 o'clock he finds that the angle has increased to 60°. When will the lighthouse be abeam?

7. Two concentric circles have radii r and R inches $(r < R)$. Two half-lines drawn from the center determine a central angle that is t radians wide. Find a formula for the area of the region between the inner and outer circles and bounded by the two half-lines. Can you express this area entirely in terms of the lengths of the edges of the region?

8. Suppose that α and β are two angles of a triangle and a and b are the lengths of the opposite sides. Show that if $\beta = 2\alpha$, then $\cos \alpha = b/2a$.

9. Use the Law of Cosines to prove that there is no triangle whose side lengths would satisfy the inequality $a > b + c$.

10. In a certain triangle the sides opposite the angles α, β, and γ are a, b, and c units long; so $a + b > c$. Use the Law of Sines and this inequality to show that, therefore, $\sin \alpha + \sin \beta > \sin \gamma$. But $\sin \gamma = \sin (\alpha + \beta)$ (Why?). Thus, we have shown that $\sin \alpha + \sin \beta > \sin (\alpha + \beta)$ if α and β are two angles of a triangle. Can you think of another simple proof of this inequality?

MISCELLANEOUS PROBLEMS, CHAPTER FIVE

These problems are designed to test your ability to apply your knowledge of trigonometry to somewhat more difficult problems.

1. (a) If the lengths of the sides of an isosceles triangle are a, b, and b, use the Law of Cosines to show that $a^2 = 2b^2(1 - \cos \alpha)$.
   (b) A perpendicular dropped from the vertex A of angle α to the side of length a bisects the angle α and also the side of length a. Thus, from right-angle trigonometry, $a = 2b \sin \frac{1}{2}\alpha$. Show that the two equations for a are consistent.

2. Show that the altitude from the vertex C of triangle ABC is ab/c if, and only if, γ is a right angle.

3. The angles of an isosceles triangle are α, α, and β, and the opposite sides are a, a, and b units long. Express b in terms of a and α.

4. The diagonals of a parallelogram divide its obtuse interior angle into the sum $\alpha + \beta$ and its acute interior angle into the sum $\gamma + \delta$. Show that the labels α, β, γ, and δ can be assigned so that $\dfrac{\sin \alpha}{\sin \beta} = \dfrac{\sin \gamma}{\sin \delta}$.

5. Use the Law of Cosines to show that the sum of the squares of the diagonals of a parallelogram is the sum of the squares of its sides.

6. Two circular disks of radius r are situated so that the center of one is a point of the circumference of the other.
   (a) Find a formula for the area of the intersection of the disks.
   (b) Calculate this area if the radius of each disk is 2.

7. Show that the area of a regular polygon with n sides each of length a is $\frac{1}{4}na^2 \cot (\pi/n)$.

8. Let α, β, and γ be the angles of a triangle. Show that if

$$\tan (\alpha - \beta) + \tan (\beta - \gamma) + \tan (\gamma - \alpha) = 0,$$

then the triangle is isosceles.

9. Determine the length of an uncrossed belt running around two pulleys whose radii are 4 inches and 8 inches and whose centers are 24 inches apart. How long is the belt if it is crossed once?

10. A picture 5 feet high is hung on the wall of an art gallery with its bottom edge 9 feet above the floor. A man whose eye level is 5 feet above the floor stands x feet from the wall. Show that the vertical angle θ subtended by the picture at the man's eye can be found from the equation $\tan \theta = 5x/(x^2 + 36)$. Show that the man has the best view when he stands 6 feet from the wall. (*Hint:* Show that $5x/(x^2 + 36) \leq \frac{5}{12}$.)

11. A regular five-pointed star is inscribed in a circle of radius r. Find a formula for the area of the region inside the star.

12. A man on the bridge of a ship at sea is h feet above the water. Show that the radius of his field of vision is about $\sqrt{3h/2}$ miles (assuming that the radius of the earth is 3960 miles).

13. Discuss the problem of constructing a triangle with angles α, β, and γ and opposite sides a, b, and c units long such that angle α is a radians wide and angle β is b radians wide.

The
Complex
Numbers

six

In Chapter One we discussed briefly three systems of numbers: the integers, the rational numbers, and the real numbers. The system of *integers* is adequate for solving problems involving addition and subtraction, but not for solving certain problems involving multiplication and division. For example, there are no integers in the solution set of the equation $2x + 1 = 0$. To solve such an equation, we must use the system of *rational numbers*. But the rational numbers are not completely satisfactory either, since, for example, they do not provide us with a number that can be used to designate the length of the diagonal of a unit square. In other words, there is no rational number in the set $\{x^2 - 2 = 0\}$. So we must consider a still larger system— the *real number system*. However, the real numbers, which are satisfactory for many purposes, fail to provide a solution of the equation $x^2 + 1 = 0$. In order to solve this equation and more complicated equations that we shall encounter in the next chapter, we introduce here a still broader class of numbers—the *complex number system*. Complex numbers are extremely useful for describing many physical phenomena, and no physical scientist or engineer can afford to be ignorant of them. In this chapter we will study only the very basic ideas relating to complex numbers. If you pursue your study of mathematics long enough, you will study the "theory of the functions of a complex variable," one of the most beautiful and useful branches of mathematical analysis.

43
THE COMPLEX NUMBERS

In our earlier discussion of number systems we concentrated on three main points:

(i) how to recognize that a number belongs to the system,

(ii) how the system can be related to other number systems, and

(iii) how various mathematical operations are performed with the numbers of the system.

Let us illustrate these points by briefly reviewing our discussion of the system of rational numbers. We recognize a rational number as one that can be written in the form a/b, where a and b are integers and $b \neq 0$. A rational number of the form $a/1$ is regarded as the same as the integer a; hence, the integers can be considered part of the system of rational numbers. The product of two rational numbers a/b and c/d is, by definition, the rational number ac/bd. Notice that this definition assumes that we already know how to compute the products ac and bd; that is, that we know how to multiply integers. The other mathematical operations with the rational numbers are also defined in terms of mathematical operations with the integers. The definitions of the operations with the rational numbers are motivated by a desire to preserve certain laws of arithmetic, such as $a(b + c) = ab + ac$, which are valid for operations on the integers.

The development of the complex numbers from the real numbers follows a similar pattern. Just as the rational numbers are determined by pairs of integers, so *a complex number z is determined by a pair of real numbers* (a, b). And just as we find it more convenient to denote a given rational number by a/b rather than (a, b), we will write $z = a + bi$ instead of (a, b) Two examples of complex numbers are $3 + 5i$ and $17 + \pi i$. For the moment we can consider the letter i as an indicator to show that the number b is the second member of the pair (a, b). We shall soon see that the notation $a + bi$ makes it easy for us to remember the rules of arithmetic for complex numbers.

The real number a and the complex number $a + 0i$ *are regarded as the same.* In particular, 0 is written for $0 + 0i$. For this reason, we say that the real numbers are part of the complex number system. It is also convenient to write bi instead of $0 + bi$. Thus, $2i$ really means $0 + 2i$. Other notational simplifications are $i = 1i = 0 + 1i$ and $a - bi = a + (-b)i$.

Now that we know what complex numbers look like, we will list some of their rules of arithmetic. *If* $u = a + bi$ *and* $v = c + di$ *are two complex numbers that are expressed in terms of the real numbers a, b, c, and d, then*

(43-1) $u = v$ means $a = c$ and $b = d$ *(Definition of Equality)*,

(43-2) $u + v = (a + c) + (b + d)i$ *(Definition of Addition)*,

(43-3) $u \cdot v = (ac - bd) + (bc + ad)i$ *(Definition of Multiplication).*

The associative, commutative, and distributive laws we studied in Section 1 remain valid for the system of complex numbers, which means that the arithmetic of complex numbers is similar to the arithmetic of real numbers. Ordinarily, the symbol $<$ is not used with complex numbers; *symbols such as $1 < i$ are not defined.*

EXAMPLE 43-1   Calculate i^2.

Solution   We first write i in formal notation: $i = 0 + 1i$. Now we can use Equation 43-3:

$$(0 + 1i)(0 + 1i) = (0 - 1) + (0 + 0)i = -1 + 0i.$$

Hence, in the simplified notation we discussed above, $i^2 = -1$.

If, in adding or multiplying two complex numbers $a + bi$ and $c + di$, we treat all terms in the "ordinary" way (that is, as if they were real numbers), except that i^2 is replaced by -1, then we obtain

$$(a + bi) + (c + di) = a + c + bi + di$$
$$= (a + c) + (b + d)i,$$

and

$$(a + bi)(c + di) = ac + bci + adi + bdi^2$$
$$= (ac - bd) + (bc + ad)i.$$

These results agree with Equations 43-2 and 43-3. The notational form $a + bi$ thus provides an easy way to remember the rules for adding and multiplying complex numbers. *We operate with complex numbers just as if we were dealing with real numbers, except that we replace i^2 by -1.*

EXAMPLE 43-2   Compute the sum and product of the complex numbers $3 + 2i$ and $1 - 3i$.

Solution $$(3 + 2i) + (1 - 3i) = 3 + 1 + 2i - 3i = 4 - i.$$
$$(3 + 2i)(1 - 3i) = 3 - 7i - 6i^2 = 3 - 7i + 6 = 9 - 7i.$$

According to the definition of equality, *one equation involving complex numbers is equivalent to two equations involving real numbers.*

EXAMPLE 43-3   Find the real numbers a and b such that
$$(a + bi) + (2 - 3i) = 2(-2 + i).$$

Solution   We see that $a + bi = (-4 + 2i) - (2 - 3i)$
$$= -6 + 5i,$$

and hence $a = -6$ and $b = 5$.

197

We mentioned in the introduction to this chapter that the system of real numbers fails to provide a solution of the equation $x^2 + 1 = 0$. Since $i^2 = -1$, the complex number i is a solution of the equation $x^2 + 1 = 0$. In problem 43-5 we ask you to show that $\{-i, i\}$ is the solution set of this equation.

EXAMPLE 43-4 Solve the equation $x^2 = 2i$.

Solution We seek a complex number $x = a + bi$, where a and b are real numbers, such that $(a + bi)^2 = 2i$. Therefore, according to the multiplication rule, $a^2 - b^2 + 2abi = 0 + 2i$. From the definition of equality, this equation is equivalent to two equations involving the real numbers a and b,

$$a^2 - b^2 = 0 \quad \text{and} \quad 2ab = 2.$$

Replacing b in the first of these equations with the expression $b = 1/a$ that we obtain from the second, gives us the equation $a^4 = 1$. The only real numbers that satisfy this last equation are 1 and -1. If $a = 1$, then $b = 1/a = 1$ and so $x = 1 + i$. If $a = -1$, then $b = -1$ and so $x = -1 - i$. Thus we see that $\{x^2 = 2i\} = \{1 + i, -1 - i\}$.

Some of the terminology used with complex numbers is rather unfortunate. In the complex number $z = a + bi$, the real number a is called the **real part,** and the real number b is called the **imaginary part** of z. There is a historical reason behind this choice of words that harks back to the days when complex numbers were considered in some sense not as "real" as real numbers. It would perhaps be better to identify these numbers merely as the "non-i part" and the "i part" of the complex number, but the other terminology is almost always used.

1. Perform the following additions and subtractions.
   (a) $(4 + 3i) + (2 - 6i)$
   (b) $(2 - 2i) - [4i + (2 - 3i)]$
   (c) $2 - 6i - 7 + i - 3$
   (d) $2 - [(1 + i) - (3i - 2)]$

2. Perform the following multiplications.
   (a) $(6 + 2i)(1 - i)$
   (b) $2(1 + 3i)(4 - i)$
   (c) $i(3 + i)(3 - i)$
   (d) $(1 + 2i)(1 + 3i)(1 + i)$

3. Find the set $\{i^n \mid n = 2, 5, 7, 9, 83\}$.

4. Perform the indicated operations, and write your answer in the form $a + bi$.
   (a) $(3 - 4i) - (6 + 2i)(1 + i)$
   (b) $(2 - i)^3$
   (c) $(\sqrt{3} + 2i)(\sqrt{3} - 2i)$
   (d) $(\sqrt[3]{2} + i)(\sqrt[3]{4} - i)$
   (e) $i[i(3 - i) + (1 - i)(2 - i)]$
   (f) $(\frac{2}{3} + \frac{1}{4}i) - i(\frac{1}{6}i - \frac{3}{2}i)$
   (g) $(-1 - \sqrt{3}i)^3$
   (h) $(\frac{1}{2}\sqrt{2} - \frac{1}{2}\sqrt{2}i)^{20}$

5. (a) Show that $-i \in \{x^2 + 1 = 0\}$.
   (b) Show that $\{-i, i\} = \{x^2 + 1 = 0\}$. (Use the method of Example 43-4.)

6. Follow the procedure in Example 43-4 to solve the equation $x^2 + 8i = 0$.

7. Each of the following equations defines a function f. What is its domain in each case? Find the numbers $f(i)$ and $f(1 - i)$.

(a) $f(z) = (2 - i)z + 3 + 4i$ (b) $f(z) = z^2 - iz + 2 - i$

8. The equation $f(z) =$ the imaginary part of z defines a function f.
   (a) What is $f(3 + 4i)$?
   (b) What is $f(a + bi)$?
   (c) Is $f(3z) = 3f(z)$ for each complex number z in the domain of f?
   (d) What is the domain of f?
   (e) What is the range of f?

9. If instead of denoting a complex number as $a + bi$ we merely write it as (a, b), then the rule for addition is $(a, b) + (c, d) = (a + c, b + d)$. Use this notation to write the rule for multiplication.

10. Find complex numbers u and v such that $2u + 3v = 3 + i$ and $3u - 2v = -2 + 8i$.

11. Show that $\{n \mid i^n = -1\} = \{n \mid n = 4k + 2, k \text{ an integer}\}$.

12. Can you convince yourself that $i^n = \cos\frac{1}{2}n\pi + i\sin\frac{1}{2}n\pi$ for each positive integer n? (Try some examples.)

13. Solve the equation $\log a + 2^b i = 2 + 5i$ for real numbers a and b.

14. Can you use Definitions 43-1, 43-2, and 43-3 to *prove* that the associative, commutative, and distributive laws are valid in the system of complex numbers?

44

THE CONJUGATE OF A COMPLEX NUMBER

We have just seen that the way to add, subtract, and multiply with complex numbers is to proceed as if we were dealing with real numbers, except that we substitute -1 for i^2. Division cannot be handled this way. As in the case of real numbers, the number $x + yi$ is the quotient $(a + bi)/(c + di)$ if, and only if, $a + bi = (c + di)(x + yi)$. (Division by zero is again excluded.) To find the quotient, we must solve this last equation for the real numbers x and y. An easy way to find these numbers depends on the notion of the *conjugate* of a complex number, which we will now introduce.

Definition 44-1   The **conjugate** of the complex number $z = a + bi$ (a and b are real numbers) is the complex number $\bar{z} = \overline{a + bi} = a - bi$.

For example, $\overline{3 + 4i} = 3 - 4i$ and $\overline{\pi - 3i} = \pi + 3i$. It is also fairly common to see the symbol z^* denoting the conjugate of z. In this notation, then, $(3 + 4i)^* = 3 - 4i$.

We summarize a number of simple results concerning complex conjugates in the theorems that follow.

Theorem 44-1   *Suppose that* $u = a + bi$ *and* $v = c + di$. *Then*

$$\overline{u + v} = \overline{u} + \overline{v} \quad and \quad \overline{uv} = \overline{u}\,\overline{v}.$$

Written out, the theorem states that the result will be the same complex number whether we first add two complex numbers and then take the conjugate of the sum, or first take the conjugate of the two numbers and then add. The same is true for multiplication.

Proof   According to the definition of addition, $u + v = (a + c) + (b + d)i$, and from the definition of conjugation,

$$\overline{u + v} = (a + c) - (b + d)i.$$

On the other hand, $\overline{u} = a - bi$ and $\overline{v} = c - di$, so

$$\overline{u} + \overline{v} = (a + c) - (b + d)i,$$

which is the same as $\overline{u + v}$. The proof for multiplication is similar.

Theorem 44-2   *The equation* $\overline{z} = z$ *is valid if, and only if, z is a real number.*

Proof   If z is a real number, then it may be written in the form $z = a + 0i$. Therefore, $\overline{z} = a - 0i$, so $\overline{z} = z$. Conversely, if $\overline{z} = z$, where $z = a + bi$, then $a - bi = a + bi$. Hence, $-b = b$, so $2b = 0$. Therefore, $b = 0$ and $z = a + 0i$; that is, z is a real number.

EXAMPLE 44-1   Show that if $z \in \{x^2 = 2x - 2\}$, then $\overline{z} \in \{x^2 = 2x - 2\}$.

Solution   The statement that z belongs to the solution set of the equation $x^2 = 2x - 2$ is merely a complicated way of saying that $z^2 = 2z - 2$. We take the conjugate of both sides of this equation and obtain the equation

(44-1)
$$\overline{z^2} = \overline{2z - 2}.$$

According to Theorem 44-1, $\overline{z^2} = \overline{zz} = \overline{z}\,\overline{z} = \overline{z}^2$. Also according to Theorem 44-1, $\overline{2z - 2} = \overline{2z} - \overline{2}$. Since $\overline{2} = 2$ (Theorem 44-2), we have $\overline{2z - 2} = 2\overline{z} - 2$. Therefore, Equation 44-1 may be written $\overline{z}^2 = 2\overline{z} - 2$, which says that $\overline{z}$ is a member of the solution set of the equation $x^2 = 2x - 2$. We will soon be able to show that $\{x^2 = 2x - 2\} = \{1 + i, 1 - i\}$.

The easy method of division that we are looking for is based on the observation that if $z = x + yi$, then $z\overline{z} = (x + yi)(x - yi) = x^2 + y^2$, which is a positive number unless $z = 0$. To calculate $(a + bi)/(c + di)$ we need only multiply both the numerator and the denominator of this fraction by $\overline{c + di} = c - di$. The result is

$$\frac{(a + bi)(c - di)}{c^2 + d^2} = \frac{(ac + bd)}{c^2 + d^2} + \frac{(bc - ad)}{c^2 + d^2}i.$$

(We are assuming, of course, that the complex number $c + di \neq 0$.)

EXAMPLE 44-2 Find $(10 + 5i)/(3 + 4i)$.

Solution Multiply the numerator and denominator by $\overline{3 + 4i} = 3 - 4i$ to obtain

$$\frac{(10 + 5i)(3 - 4i)}{(3 + 4i)(3 - 4i)} = \frac{50 - 25i}{25} = 2 - i.$$

You might check to see that this number really is the quotient of the given numbers by showing that $10 + 5i = (2 - i)(3 + 4i)$.

The product of two real numbers a and b is 0 if, and only if, either $a = 0$ or $b = 0$ (or both). This property is one of the most important that a number system can possess. From it we can conclude, for example, that if $(x - 1)(x - 3) = 0$, then either $x = 1$ or $x = 3$. There are systems in which a multiplication is defined and in which the above property does not hold, but the next example shows that the complex number system is not one of them.

EXAMPLE 44-3 If $u = a + bi$ and $v = c + di$ are complex numbers, show that $uv = 0$ if, and only if, $u = 0$ or $v = 0$ (or both).

Solution If $u = 0$ or $v = 0$ (or both), it follows from direct multiplication that $uv = 0$. Conversely, if we know that $uv = 0$, then $uv\overline{u}\overline{v} = u\overline{u}v\overline{v} = 0$. Since $u\overline{u} = a^2 + b^2$ and $v\overline{v} = c^2 + d^2$, the equation $u\overline{u}v\overline{v} = 0$ becomes

$$(a^2 + b^2)(c^2 + d^2) = 0.$$

We now have a product of two real numbers, and such a product can be 0 only if at least one of the factors is 0. If $a^2 + b^2 = 0$, then $a = b = 0$ and hence $u = 0$. Similarly, if $c^2 + d^2 = 0$, then $v = 0$. Therefore, at least one of the numbers u or v must be 0.

PROBLEMS
44

1. Show by direct computation that the following equations are correct.
   (a) $\overline{(1 + 2i) + (2 - 3i)} = \overline{(1 + 2i)} + \overline{(2 - 3i)}$
   (b) $\overline{(1 + 2i)(2 - 3i)} = \overline{(1 + 2i)}\,\overline{(2 - 3i)}$

2. Express the following quotients in the form $a + bi$.
   (a) $\dfrac{6 + 2i}{3 + 2i}$ (b) $\dfrac{i}{1 + i}$ (c) $\dfrac{2 - i}{i}$ (d) $\dfrac{\sqrt{2} + \sqrt{6}i}{\sqrt{3} + 2i}$

3. Perform the indicated operations, and write your answers in the form $a + bi$.
   (a) $3 - i - \dfrac{(2 - i)^2}{1 + i}$ (b) $\dfrac{6 + i}{2 - i} - \dfrac{1 + i}{4 - i}$
   (c) $\dfrac{(3 + i)^2}{(2 - i)^4}$ (d) $(1 + 2i)^{-1}(2 - i)^2$

4. Show that $\overline{\overline{z}} = z$.

201

5. Show that the real part of z is the number $\frac{1}{2}(z + \bar{z})$, and the imaginary part of z is the number $\frac{1}{2}i(\bar{z} - z)$.

6. Find $(\cos t + i \sin t)^{-1}$, where t is any real number.

7. Show that $i^{-n} = (-1)^n i^n$ for each integer n.

8. Solve the following equations.

(a) $2iz - 3 = 0$ (b) $(1 + i)z + 2 = 0$

(c) $(3 + i)z - 1 - i = 0$ (d) $\dfrac{4 + 2i}{1 - i}z + \dfrac{3 + i}{7 + 2i} = 0$

(e) $(3iz + 1)(2z - 3 + 2i) = 0$ (f) $z^2 + 4 = 0$

9. The equation $f(z) = \bar{z}$ defines a function f.

(a) What is the domain of f?
(b) What is the range of f?
(c) Is $f(u + v) = f(u) + f(v)$ for each pair u, v of complex numbers?
(d) Is $f(uv) = f(u)f(v)$ for each pair u, v of complex numbers?
(e) Is $f(u/v) = f(u)/f(v)$ for each pair u, v of complex numbers?
(f) What is $f(f(z))$?

10. Solve the following equations.

(a) $3z + 2\bar{z} = 10 - i$ (b) $z + 4\bar{z} = 5$

11. What is the relation between $\overline{z^{1492}}$ and $\bar{z}^{1492}$?

12. Explain why the set equation

$$\{iz^3 + (\pi + i)z^2 + 7 = 0\} \cup \{z^2 + z + i = 0\}$$
$$= \{(iz^3 + (\pi + i)z^2 + 7)(z^2 + z + i) = 0\}$$

is a consequence of the multiplication property proved in Example 44-3.

45
GRAPHICAL REPRESENTATION OF COMPLEX NUMBERS

Geometrically, we think of real numbers as points of a line. Since a complex number is determined by a *pair of real numbers*, and since we think of pairs of real numbers as points of a plane, it is natural to represent complex numbers as points of a plane. *We therefore associate with each complex number $z = a + bi$, where a and b are real numbers, the point (a, b) of a cartesian plane* (see Fig. 45-1). The number $\bar{z} = a - bi$ may then be plotted as the point $(a, -b)$. Let r denote the distance between the origin and the point that represents z. Then $r^2 = a^2 + b^2 = z\bar{z}$, and $r = 0$ only when $z = 0$. If $r \neq 0$, let θ be an angle in standard position whose terminal side contains the point z. (We say *an* angle because there are really infinitely many possibilities, all differing by $360°$.) From Theorem 40-1, we see that $a = r \cos \theta$ and $b = r \sin \theta$. Thus,

(45-1) $$z = a + bi = r(\cos \theta + i \sin \theta).$$

The expression $r(\cos \theta + i \sin \theta)$ is called the **trigonometric form** of the complex number z. The number r is termed the **absolute value** of z, and the

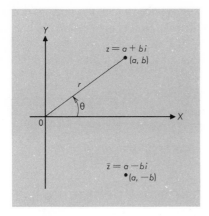

FIGURE 45-1

angle θ is called an **argument** of z. Notice that the absolute value of a complex number z is the distance between the origin and the point z, just as the absolute value of a real number x is the distance between the origin and the point x. We use the symbol $|z|$ to denote the absolute value of z, so that $|z| = r$. Since $r^2 = z\bar{z}$, we see that $|z| = \sqrt{z\bar{z}}$.

EXAMPLE 45-1   Express the complex number $1 - i$ in trigonometric form.

Solution   We plot the point $(1, -1)$ and notice (Fig. 45-2) that $r = \sqrt{2}$. The angle

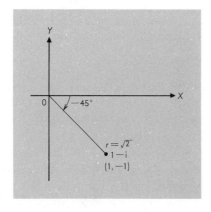

FIGURE 45-2

θ is determined from the equations $1 = \sqrt{2} \cos \theta$ and $-1 = \sqrt{2} \sin \theta$. In this case we may take our angle to be one of $-\frac{1}{4}\pi$ radians. Hence, $z = \sqrt{2} [\cos (-\frac{1}{4}\pi) + i \sin (-\frac{1}{4}\pi)] = \sqrt{2}(\cos \frac{1}{4}\pi - i \sin \frac{1}{4}\pi)$. Or we could use degree measurement for our argument of z and write

$$z = \sqrt{2}(\cos 45° - i \sin 45°).$$

Trigonometric form is particularly useful for multiplying or dividing complex numbers. Suppose that u and v are two complex numbers whose

203

trigonometric forms are $u = r(\cos \theta + i \sin \theta)$ and $v = s(\cos \phi + i \sin \phi)$. Then by straightforward multiplication their product is

$$uv = rs[(\cos \theta \cos \phi - \sin \theta \sin \phi) + i(\sin \theta \cos \phi + \cos \theta \sin \phi)].$$

From the addition Formulas 34-2 we have

$$\cos \theta \cos \phi - \sin \theta \sin \phi = \cos (\theta + \phi), \quad \text{and}$$

$$\sin \theta \cos \phi + \cos \theta \sin \phi = \sin (\theta + \phi).$$

Hence,

(45-2) $$\boldsymbol{uv = rs[\cos (\theta + \phi) + i \sin (\theta + \phi)].}$$

Written out, Equation 45-2 states that *the product of two complex numbers is the complex number whose absolute value is the product of the absolute values of the factors, and with an argument that is the sum of arguments of the factors.*

EXAMPLE 45-2   Let z be a complex number. Find the trigonometric form of the product iz.

Solution   Suppose the trigonometric form of the number z is $z = r(\cos \theta + i \sin \theta)$.

Since $i = 1(\cos 90° + i \sin 90°)$, Equation 45-2 gives us

$$iz = r \cos (\theta + 90°) + i \sin (\theta + 90°).$$

Geometrically, we obtain the complex number iz from the complex number z by rotating the line segment joining z and the origin $90°$ in a counterclockwise direction.

EXAMPLE 45-3   If $z = r(\cos \theta + i \sin \theta)$, express z^2 in trigonometric form.

Solution   The absolute value of z^2 is r^2, and the argument of z^2 is $\theta + \theta = 2\theta$. Hence, $z^2 = r^2(\cos 2\theta + i \sin 2\theta)$.

The quotient u/v of the complex numbers $u = r(\cos \theta + i \sin \theta)$ and $v = s(\cos \phi + i \sin \phi)$, where $s \neq 0$, is the complex number $w = t(\cos \psi + i \sin \psi)$ such that $u = vw$. Hence, Equation 45-2 tells us that t and ψ should be chosen so that

(45-3) $$r(\cos \theta + i \sin \theta) = st[\cos (\phi + \psi) + i \sin (\phi + \psi)].$$

It is clear that *two complex numbers in trigonometric form are equal if, and only if,*

(i) their absolute values are equal and
(ii) their arguments are equal or differ by a multiple of $360°$.

Thus, we must have

$$r = st$$

$$\theta = \phi + \psi + k \cdot 360° \quad (k \text{ may be any integer}).$$

Finally then, $t = r/s$ and $\psi = \theta - \phi - k \cdot 360°$. Since $T(\theta - \phi - k \cdot 360°) = T(\theta - \phi)$ for any trigonometric function T, our quotient $w = u/v$ is

(45-4)
$$\frac{u}{v} = \frac{r}{s}[\cos (\theta - \phi) + i \sin (\theta - \phi)].$$

EXAMPLE 45-4   Find the trigonometric form of the reciprocal of the complex number $z = r(\cos \theta + i \sin \theta)$.

Solution   We are to calculate the quotient $1/z$. Since 1 may be written as $1 = 1(\cos 0° + i \sin 0°)$, it follows that

$$\frac{1}{z} = \frac{1}{r}[\cos (0° - \theta) + i \sin (0° - \theta)]$$

$$= \frac{1}{r}(\cos \theta - i \sin \theta).$$

**PROBLEMS
45**

1. Find the absolute value and an argument of each of the following complex numbers, and plot it as a point of the plane.
   (a) $\sqrt{3} - i$ (b) i (c) -1
   (d) $-2 - 2i$ (e) $-1 + \sqrt{3}i$ (f) $1 + 2i$

2. Write each of the following numbers in non-trigonometric form $a + bi$ and plot as a point of the plane.
   (a) $2(\cos \frac{1}{6}\pi + i \sin \frac{1}{6}\pi)$ (b) $5(\cos \pi + i \sin \pi)$
   (c) $2(\cos \frac{1}{4}\pi - i \sin \frac{1}{4}\pi)$ (d) $10(\cos 1 + i \sin 1)$
   (e) $7(\cos 200° - i \sin 200°)$ (f) $4(\cos 3 + i \sin 3)$

3. Show that $|uv| = |u| \cdot |v|$ and that $|u/v| = |u|/|v|$.

4. Show that $|z| = |-z| = |\bar{z}| = |iz| = |z(\cos \theta + i \sin \theta)|$.

5. Show that $|z| = 0$ if, and only if, $z = 0$.

6. Let $u = a + bi$ and $v = c + di$. Show that $|u - v|$ is the distance between the points that represent u and v. Use this fact to show that $|u + v| \leq |u| + |v|$. (See Inequality 8-3.)

7. Put Equation 45-4 into words.

8. Find the product uv and the quotient $\dfrac{u}{v}$ if $u = 12(\cos \frac{1}{3}\pi + i \sin \frac{1}{3}\pi)$ and if:

   (a) $v = 4(\cos \frac{1}{4}\pi + i \sin \frac{1}{4}\pi)$ (b) $v = 3(\cos \pi + i \sin \pi)$
   (c) $v = .5(\cos 20° + i \sin 20°)$ (d) $v = 1.5(\cos 40° - i \sin 40°)$

9. Perform the following operations both algebraically and by first writing the numbers in trigonometric form.

(a) $(1-i)(2-2i)(3-3i)$

(b) $\dfrac{1+i}{3-i}$

(c) $(1-i)(1-i\sqrt{3})(\frac{1}{2}+\frac{1}{2}i)(\frac{1}{2}+\frac{1}{2}i\sqrt{3})$

(d) $\dfrac{(\sqrt{3}-i)(-1+i)}{(1-i\sqrt{3})(2\sqrt{3}+2i)}$

10. Prove that two complex numbers in trigonometric form are equal if, and only if, (a) their absolute values are equal, and (b) their arguments are equal or differ by a multiple of 360° or 2π radians.

11. Describe the following sets of complex numbers geometrically as subsets of the cartesian plane.

(a) $\{|z| = 2\}$

(b) $\{z + \bar{z} = 2i\}$

(c) $\{|z| < 1\}$

(d) $\{z + \bar{z} > 2\}$

(e) $\{$an argument of $z = 45°\}$

(f) $\{z^2 = \bar{z}^2\}$

12. Explain why the equation

$$x + iy = \sqrt{x^2+y^2}\left[\cos\left(\text{Tan}^{-1}\frac{y}{x}\right) + i\sin\left(\text{Tan}^{-1}\frac{y}{x}\right)\right]$$

is true for $x > 0$ but false for $x \le 0$.

46
ROOTS OF COMPLEX NUMBERS

In Example 45-3 we showed that if $z = r(\cos\theta + i\sin\theta)$, then $z^2 = r^2(\cos 2\theta + i\sin 2\theta)$. Now $z^3 = z^2 \cdot z$, so another application of the rule for multiplying complex numbers in trigonometric form (Equation 45-2) shows that

$$z^3 = r^3[\cos(2\theta + \theta) + i\sin(2\theta + \theta)] = r^3(\cos 3\theta + i\sin 3\theta).$$

If we repeat the process once more, we get

$$z^4 = z^3 \cdot z = r^4[\cos(3\theta + \theta) + i\sin(3\theta + \theta)] = r^4(\cos 4\theta + i\sin 4\theta).$$

It should now be clear what the general formula is, and in Section 74 (Example 74-4) we will use the process of "mathematical induction" to show that for each positive integer n,

(46-1) $$z^n = r^n(\cos n\theta + i\sin n\theta).$$

EXAMPLE 46-1 If θ is any angle and n is any positive integer, show that

(46-2) $$(\cos\theta + i\sin\theta)^n = \cos n\theta + i\sin n\theta.$$

Solution We simply let z be the complex number $z = \cos\theta + i\sin\theta$. Then Equation 46-1 becomes $(\cos\theta + i\sin\theta)^n = 1^n \cdot (\cos n\theta + i\sin n\theta)$, which is the equation to be verified. Equation 46-2 is known as **DeMoivre's Theorem**.

206

Let n be a positive integer and u be a complex number. *A complex number z is called an **nth root of u** if*

(46-3)
$$z^n = u.$$

The numbers 1, -1, i, and $-i$, for example, are 4th roots of the number 1, since $(1)^4 = (-1)^4 = (i)^4 = (-i)^4 = 1$. We can easily solve Equation 46-3 if u and z are expressed in trigonometric form. Suppose that

$$u = s(\cos \phi + i \sin \phi)$$

and that z is an nth root of u expressed in the trigonometric form $z = r(\cos \theta + i \sin \theta)$. Let us suppose that $u \neq 0$, and therefore $s > 0$. Then, using Equation 46-1, we may write Equation 46-3 in the form

$$r^n(\cos n\theta + i \sin n\theta) = s(\cos \phi + i \sin \phi).$$

Our problem is to find a positive number r and an angle θ for which this last equation is valid. Two complex numbers can be equal only if their absolute values are equal, and hence we must choose r so that $r > 0$ and $r^n = s$. Therefore,

(46-4)
$$r = \sqrt[n]{s}.$$

Our equation then reduces to the two equations

$$\cos n\theta = \cos \phi \quad \text{and} \quad \sin n\theta = \sin \phi.$$

It is clear from these equations that the angles $n\theta$ and ϕ can differ only by a multiple of $360°$ or 2π radians (see Problem 45-10). More precisely, $n\theta = \phi + k \cdot 360°$, where k is an integer—positive, negative, or 0. Therefore,

(46-5)
$$\theta = \frac{\phi}{n} + k\frac{360°}{n}.$$

We can also write this equation as

(46-6)
$$\theta = \frac{\phi}{n} + k\frac{2\pi}{n} \text{ radians.}$$

EXAMPLE 46-2   Find the cube roots of the number $u = 8(\cos 60° + i \sin 60°)$.

Solution   According to Equations 46-4 and 46-5, $z = r(\cos \theta + i \sin \theta)$ will be a cube root of u if $r = \sqrt[3]{8} = 2$ and θ is an angle of the form $(60/3)° + k(360/3)° = 20° + k120°$, where k is any integer. The cube roots we are seeking are therefore given by the formula

$$z_k = 2[\cos (20° + k\ 120°) + i \sin (20° + k\ 120°)].$$

207

We now observe that even though there are an infinite number of possibilities for the integer k, there are really only three values of z_k. To
illustrate this statement, we will list a few values of z_k for specific choices
of the integer k:

$$z_0 = 2(\cos 20° + i \sin 20°),$$
$$z_1 = 2(\cos 140° + i \sin 140°),$$
$$z_2 = 2(\cos 260° + i \sin 260°),$$
$$z_3 = 2(\cos 380° + i \sin 380°),$$
$$z_4 = 2(\cos 500° + i \sin 500°).$$

But it is clear that $z_3 = z_0$ and $z_4 = z_1$, since $T(380°) = T(20°)$ and
$T(500°) = T(140°)$ for any trigonometric function T. You should be able
to convince yourself that any choice of k will lead to one of the numbers
z_0, z_1, or z_2.

The results of our work with the roots of complex numbers suggest the
following theorem.

Theorem 46-1   *A non-zero complex number $u = s(\cos \phi + i \sin \phi)$ has, for any given
positive integer n, exactly n distinct nth roots given by the formula*

(46-7) $$z_k = \sqrt[n]{s}\left[\cos\left(\frac{\phi}{n} + k \cdot \frac{360°}{n}\right) + i \sin\left(\frac{\phi}{n} + k \cdot \frac{360°}{n}\right)\right],$$

$k = 0, 1, 2, \ldots, (n-1)$. *If we use set notation and radian measure
for our angles, we can write this result as*

$$\{z \mid z^n = s(\cos \phi + i \sin \phi)\} =$$

$$\left\{\sqrt[n]{s}\left[\cos\left(\frac{\phi}{n} + k\frac{2\pi}{n}\right) + i \sin\left(\frac{\phi}{n} + k\frac{2\pi}{n}\right)\right] \middle| k = 0, 1, \ldots, n-1\right\}.$$

EXAMPLE 46-3   Find the 5th roots of 32.

Solution   Using radian measure for angles, we can write $32 = 32(\cos 0 + i \sin 0)$.
Hence, the 5th roots of 32 are the five numbers of the set

$$\{\sqrt[5]{32}[\cos(0 + \tfrac{2}{5}k\pi) + i \sin(0 + \tfrac{2}{5}k\pi)] \mid k = 0, 1, 2, 3, 4\}$$
$$= \{2(\cos \tfrac{2}{5}k\pi + i \sin \tfrac{2}{5}k\pi) \mid k = 0, 1, 2, 3, 4\}.$$

It is interesting to plot the 5th roots of 32 as points of the plane. We see
that all the 5th roots of 32 belong to a circle with a radius of 2 and are
equally spaced about the circumference (one of them is a point of the
X-axis). The geometric picture of the 5th roots of 32 appears in Fig. 46-1.
In general, the nth roots of a complex number z divide a circle of radius
$|z|^{1/n}$ into n arcs of equal length.

If $u \geq 0$, we reserve the symbol $\sqrt{u}$ to mean the non-negative number
whose square is u. This usage is standard; wherever you see the symbol $\sqrt{9}$,

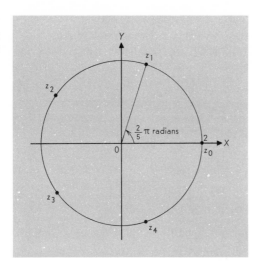

FIGURE 46-1

for example, it always means 3, not -3. If u is not a positive number, the symbol $\sqrt{u}$ is ordinarily not used without some explanation of its meaning. Since we shall have occasion to use this symbol later, we will agree that *if u is not a negative real number, then the symbol $\sqrt{u}$ will denote the solution of the equation $z^2 = u$ whose real part is not negative. If $u < 0$, then $\sqrt{u} = i\sqrt{|u|}$.* These statements do not represent fundamental mathematical truths. They are just a means of selecting which of the two numbers whose square is u we shall designate by the symbol $\sqrt{u}$. A different selection may be used in other books.

EXAMPLE 46-4 Find $\sqrt{2i}$ and $\sqrt{-4}$.

Solution In Example 43-4 we found that the two numbers whose square is $2i$ are the numbers $1 + i$ and $-1 - i$. The real part of the first is the positive number 1, so according to our agreement, we write $\sqrt{2i} = 1 + i$. Since $-4 < 0$, we have agreed that $\sqrt{-4} = i\sqrt{|-4|} = i\sqrt{4} = 2i$.

EXAMPLE 46-5 Find $\sqrt{-1 + \sqrt{3}i}$.

Solution To find $\sqrt{-1 + \sqrt{3}i}$ we must find that solution of the equation $z^2 = -1 + \sqrt{3}i$ whose real part is non-negative. In trigonometric form,

$$-1 + \sqrt{3}i = 2(\cos 120° + i \sin 120°),$$

so

$$z_1 = \sqrt{2}(\cos 60° + i \sin 60°) = \tfrac{1}{2}\sqrt{2}(1 + \sqrt{3}i) \quad \text{and}$$
$$z_2 = \sqrt{2}(\cos 240° + i \sin 240°) = -\tfrac{1}{2}\sqrt{2}(1 + \sqrt{3}i).$$

Since the real part of z_1 is $\tfrac{1}{2}\sqrt{2} > 0$, we write

$$\sqrt{-1 + \sqrt{3}i} = \tfrac{1}{2}\sqrt{2}(1 + \sqrt{3}i).$$

209

1. Use Equation 46-1 to calculate the following numbers.
   (a) $(\cos 60° + i \sin 60°)^6$ (b) $(1 - i)^{12}$ (c) $(1 + \sqrt{3}i)^{12}$
   (d) $(2 + 2i)^6$ (e) $(\cos 1 + i \sin 1)^5$ (f) $(.9820 + .1889i)^3$

2. Find three members of the set $\{z \mid z^3 = u\}$ for each of the following choices of u.
   (a) $u = 8$ (b) $u = 8i$ (c) $u = -8$ (d) $u = -8i$

3. Write the following radicals in the form $a + bi$.
   (a) $\sqrt{-64}$ (b) $\sqrt{16i}$ (c) $\sqrt{2 + 2i}$ (d) $\sqrt{\sqrt{3} + i}$
   (e) $\sqrt{i^3}$ (f) $\sqrt{.1736 + .9848i}$

4. Show that if $u^5 = 1$, then $\bar{u}^5 = 1$.

5. Explain why the two square roots of a non-zero complex number z can be written as $\sqrt{z}$ and $-\sqrt{z}$.

6. Plot the fourth roots of u as points of the plane without figuring them out numerically.
   (a) $u = 16$ (b) $u = i$ (c) $u = -1$ (d) $u = 1 + i$

7. Explain why $|z|^n = |z^n|$.

8. Solve the following equations.
   (a) $z^2 - 2i = 0$ (b) $3z^4 + 1 = 0$ (c) $iz^3 + 1 = 0$
   (d) $(z - i)^4 - 16 = 0$

9. DeMoivre's Theorem says that $(\cos \theta + i \sin \theta)^3 = \cos 3\theta + i \sin 3\theta$. Multiply out the left-hand side of this equation and equate real and imaginary parts of the result to obtain formulas for $\cos 3\theta$ and $\sin 3\theta$ in terms of $\cos \theta$ and $\sin \theta$.

10. Let n be a positive integer. Use DeMoivre's Theorem to show that if $z = r(\cos \theta + i \sin \theta)$, then $z^{-n} = r^{-n}[\cos (-n\theta) + i \sin (-n\theta)]$.

11. Show that our definition of the symbol $\sqrt{}$ does not preserve the rule $\sqrt{ab} = \sqrt{a}\sqrt{b}$.

12. Suppose that $\cos p \cdot 45° + i \sin p \cdot 45° = \overline{\cos q \cdot 45° + i \sin q \cdot 45°}$, where p and q are different members of the set $\{0, 1, \ldots, 7\}$. What is the relation between p and q?

REVIEW PROBLEMS, CHAPTER SIX

You should be able to answer the following questions without referring back to the text.

1. If u and v are complex numbers, show that $|u + v|^2 + |u - v|^2 = 2|u|^2 + 2|v|^2$. (*Hint:* Recall that $|z|^2 = z\bar{z}$.)

2. Show that $u + v$ is a real number if, and only if, $v - \bar{u}$ is real.

3. The equation $f(z) = z - \bar{z}$ defines a function f.
    (a) What is the domain of f?
    (b) What is the range of f?
    (c) Does $f(u+v) = f(u) + f(v)$?
    (d) Does $f(3u) = 3f(u)$?
    (e) Does $f(uv) = f(u)f(v)$?

4. Describe geometrically the result of multiplying a number by $\sqrt{i}$.

5. Which of the following equations are always true?
    (a) $|z|^2 = |z^2|$ (b) $|u+v| = |u| + |v|$
    (c) $|3z| = 3|z|$ (d) $\sqrt{|z|} = |\sqrt{z}|$

6. Consider the set of complex numbers $A = \{3t + 4ti \mid t \in R^1\}$.
    (a) Describe A geometrically.
    (b) What does the set $B = \{z \mid \bar{z} \in A\}$ look like graphically?
    (c) What is $A \cap B$?

7. Show that $\cos \frac{1}{2}n\pi = \frac{1}{2}(i^n + i^{-n})$ for each integer n.

8. Describe the set $\{n \mid i^n = -1\}$.

9. There are three cube roots of 1, one of which is the number 1 itself. Let ω denote either of the others, and show that the set of cube roots of 1 is $\{\omega, \omega^2, \omega^3\}$.

10. Let $C_{\text{rat}} = \{x + iy \mid x \text{ and } y \text{ are rational numbers}\}$, and suppose that u and v are members of C_{rat}. Which of the following numbers are certainly in C_{rat}?
    (a) $u + v$ (b) uv (c) $\bar{u}$ (d) $|u|$ (e) $|u|^2$ (f) u/v

---

MISCELLANEOUS PROBLEMS, CHAPTER SIX

These problems are designed to test your ability to apply your knowledge of complex numbers to somewhat more difficult problems.

1. Give a geometric description of the following sets of complex numbers.
    (a) $\{|z - 2| = 3\}$ (b) $\{|z - \bar{z}| = 2\}$
    (c) $\{(3 + 4i)z \text{ is a non-negative real number}\}$
    (d) $\{1 < |z - 3| < 4\}$

2. The equation $E(t) = \cos t + i \sin t$ defines a function E whose domain is the set of real numbers.
    (a) What is the range of E?
    (b) Show that $E(s+t) = E(s)E(t)$.
    (c) Show that $E(t)^n = E(nt)$ for each integer n.
    (d) By replacing t with t/n in part (c), show that $E(t)^{1/n} = E(t/n)$.
    (e) From parts (c) and (d), show that $E^{p/q}(t) = E(pt/q)$ (p and q assumed to be integers).

3. Solve the equation $(z + \bar{z})z = 2 + 4i$.

4. (a) Can you convince yourself that if two complex numbers u and v are plotted as points of the plane, their sum is represented by the vertex that must be added to complete the parallelogram that has Ou and Ov as two of its sides?

(b) Now try to convince yourself that the other diagonal of this parallelogram is $|u - v|$ units long. Hence, the equation that you verified in Review Problem 1 says that the sum of the squares of the lengths of the diagonals of a parallelogram equals the sum of the squares of the lengths of its sides.

5. Show that $x + iy = -\sqrt{x^2 + y^2}[\cos (\text{Tan}^{-1} y/x) + i \sin (\text{Tan}^{-1} y/x)]$ if $x < 0$.

6. Let f be a function whose domain is the set $\{|z| \leq 1\}$. Describe geometrically the range of f if $f(z)$ is given by the following formula.

(a) $f(z) = 2z$ (b) $f(z) = z^2$

7. Show that if $z \neq 1$, then

$$1 + z + z^2 + z^3 + z^4 + z^5 = (1 - z^6)/(1 - z).$$

Use this relation and DeMoivre's Theorem to derive the formula

$$1 + \cos \theta + \cos 2\theta + \cos 3\theta + \cos 4\theta + \cos 5\theta = \frac{1}{2} + \frac{\sin \frac{11}{2}\theta}{2 \sin \frac{1}{2}\theta}.$$

8. Suppose we say that $u > v$ if the real part of u is greater than the real part of v. Show that Properties 4-2 and 4-3 are valid, but that Property 4-4 fails.

9. If $z = x + iy$ is a complex number, let us write $[\![z]\!] = [\![x]\!] + i[\![y]\!]$. Find the set $\{[\![z]\!] \mid z^4 = 81i\}$.

10. Discuss the problems that arise when we try to define the symbols $(-1)^{p/q}$, where p and q are integers. Think about the problem of defining $(-1)^{\pi}$.

The Theory
of
Equations

seven

Anyone who knows how to add, subtract, and multiply can show that when we substitute 2 for x in the expression $2x^3 - 3x + 7$, we get the number 17. But it is more difficult to *start* with the number 17 and then solve the equation $2x^3 - 3x + 7 = 17$. In this chapter we will touch on some of the high spots of the theory of such algebraic equations. In a sense, the material of this chapter is also an introduction to some of the attitudes and techniques of "higher algebra." We have tried to keep matters as down to earth as possible, although some of the topics in this chapter are basically pretty abstract.

47
POLYNOMIALS   If n is a positive integer or 0, and if $a_0, a_1, \ldots, a_n$ are $n + 1$ numbers, where $a_n \neq 0$, then the expression

$$a_n x^n + a_{n-1} x^{n-1} + \cdots + a_1 x + a_0$$

is called a **polynomial of degree n in x.** We shall use symbols such as $P(x), Q(x), R(x)$, and $D(x)$ to denote polynomials. The numbers $a_0, a_1, \ldots, a_n$ are the **coefficients** of the polynomial. Notice that in a polynomial of degree n, the coefficient a_n is not 0, but some of the other coefficients may be. Each expression such as $a_j x^j$ is called a **term** of the polynomial. The number 0 is also regarded as a polynomial, but no degree is assigned to it. Examples of polynomials are:

(i) $P(x) = 3x^4 - \pi x^2 + x - 2$. We use some natural abbreviations here; formally, you should think of this polynomial as $3x^4 + 0x^3 + (-\pi)x^2 + 1x + (-2)$. Therefore, its degree is 4, and its coefficients are $a_4 = 3$, $a_3 = 0$, $a_2 = -\pi$, $a_1 = 1$, and $a_0 = -2$.

(ii) $Q(x) = ix^2 - 3x + 17 - 2i$. In this example, $n = 2$, $a_2 = i$, $a_1 = -3$, and $a_0 = 17 - 2i$.

(iii) $R(x) = 4$. The degree of this polynomial is 0, and its only coefficient is $a_0 = 4$.

A polynomial in x consists of a sum of one or more terms; each term is either a number or consists of the product of a number and a *positive integral power* of x. Expressions such as $3x^2 + 2\sqrt{x} - 5$, $2^x + i$, $2x + 3 - 1/x$, and $2x + \sin x + 3$ are *not* polynomials in x. If $P(x)$ is a polynomial in x, and if r is a given number, then we use the symbol $P(r)$ to denote the number that results from replacing x in $P(x)$ by r. Thus, if $P(x) = x^2 - 3x + 1$, then $P(2) = 2^2 - 3(2) + 1 = -1$. At times we shall confine our attention to polynomials whose coefficients are real numbers, or rational numbers, or integers; but unless otherwise indicated, the following theory is applicable when the coefficients are any complex numbers.

Two polynomials are **equal** if, and only if, the coefficients of the corresponding powers of x are equal. Thus $ax^2 + bx + c = -x^2 + 2$, if, and only if, $a = -1$, $b = 0$, and $c = 2$. If $P(x)$ and $Q(x)$ are equal polynomials, then the numbers $P(r)$ and $Q(r)$ are equal every time a number r is substituted for x. The algebraic operations of addition, subtraction, and multiplication of polynomials should be familiar to you from elementary algebra. Thus, for example,

$$(3x^2 - 2x + 1) + (x^3 - 5) = x^3 + 3x^2 - 2x - 4,$$

and

$$(3x^2 - 2x + 1)(x^3 - 5) = 3x^5 - 2x^4 + x^3 - 15x^2 + 10x - 5.$$

But perhaps we should say a word or two about division. In Section 1, we defined the quotient a/b of a by b as the number q that satisfies the equation $a = qb$. We then noticed that if the numbers a, b, and q were limited to the integers, a given pair of numbers need not have a quotient. If $a = 7$ and $b = 3$, for example, then there is no integer q such that $7 = 3q$. To cope with this situation, we then introduced the rational numbers and found that there was a rational number that satisfies the equation $7 = 3q$. There is, however, another way to avoid the difficulty—namely, to devise a somewhat different definition of division. Suppose that we do not demand that $a = qb$, but that $a = qb + r$, where the integer r is in some sense the smallest possible choice. For example, we could write $7 = 2 \cdot 3 + 1$. The number q is then called the *quotient* and the number r is called the *remainder* in the division of a by b. The ordinary process of "long division" of a positive integer a by a positive integer b is simply a method of finding the numbers q and r such that $a = qb + r$, and $0 \leq r < b$.

EXAMPLE 47-1 If $a = 217$ and $b = 23$, find the quotient q and the remainder r when a is divided by b.

Solution By long division,

$$
\begin{array}{r}
9 = q \\
23\overline{)217} \\
207 \\
\hline
10 = r.
\end{array}
$$

Thus, $217 = 9 \cdot 23 + 10$, and $0 \leq 10 < 23$.

Division of polynomials works the same way. We state the result as a theorem, but omit the proof.

Theorem 47-1 *Let $P(x)$ and $D(x)$ be polynomials (and $D(x)$ is not the polynomial 0). Then there are polynomials $Q(x)$ and $R(x)$, where $R(x)$ is either the number 0 or is of lower degree than $D(x)$, such that*

(47-1) $$P(x) = Q(x)D(x) + R(x).$$

The dividend $P(x)$ and the divisor $D(x)$ determine the **quotient** $Q(x)$ *and the* **remainder** $R(x)$ *uniquely.*

Finding the quotient and remainder when polynomials $P(x)$ and $D(x)$ are given is merely a matter of "long division."

EXAMPLE 47-2 Let $P(x) = 2x^4 - x^3 + 7x^2 - 2x - 2$ and $D(x) = 2x^2 - x + 1$. Find $Q(x)$ and $R(x)$.

Solution

$$
D(x) = 2x^2 - x + 1 \overline{\smash{\big)}\,}
\begin{array}{l}
x^2 + 3 = Q(x) \\
2x^4 - x^3 + 7x^2 - 2x - 2 = P(x) \\
\underline{2x^4 - x^3 + x^2} \\
\qquad\qquad 6x^2 - 2x - 2 \\
\qquad\qquad \underline{6x^2 - 3x + 3} \\
\qquad\qquad\qquad\quad x - 5 = R(x).
\end{array}
$$

Thus, $Q(x) = x^2 + 3$ and $R(x) = x - 5$.

Example 47-2 illustrates the following useful fact. If $P(x) = a_n x^n + \cdots + a_0$ and $D(x) = b_m x^m + \cdots + b_0$ (and if $m \leq n$), then the first term of the quotient polynomial $Q(x)$ is $(a_n/b_m)x^{n-m}$.

If $D(x)$ is of the form $x - d$, the process of long division can be shortened considerably by a method known as **synthetic division.** We can explain how it works by showing you an example.

EXAMPLE 47-3   Divide $2x^4 + x^3 - 3x^2 + x - 10$ by $x - 2$.

Solution

$$
\begin{array}{r}
2x^3 + 5x^2 + 7x + 15 = Q(x) \\
x - 2 \overline{\smash{\big)}\ 2x^4 + x^3 - 3x^2 + x - 10} \\
\underline{2x^4 - 4x^3} \\
5x^3 - 3x^2 \\
\underline{5x^3 - 10x^2} \\
7x^2 + x \\
\underline{7x^2 - 14x} \\
15x - 10 \\
\underline{15x - 30} \\
20 = R.
\end{array}
$$

The remainder R will always be simply a number, maybe 0, since the degree of the divisor polynomial is 1. We have put the number R and the coefficients of the quotient $Q(x)$ in bold-faced type where they first appear in the division. The problem of division amounts to finding these bold-faced numbers. When we inspect our division, we see that *each coefficient of the quotient results from multiplying the coefficient that precedes it by* (-2) *and then subtracting the result from the corresponding coefficient of the dividend.* For example, the second coefficient, 5, is $1 - (-2)2 = 5$. Since the first coefficient of the quotient is the first coefficient of the dividend, we can start with this information and determine all the coefficients (and the remainder) successively. Instead of multiplying by (-2) and then subtracting, we multiply by $+2$ and add. The result in schematic form is

$$
\begin{array}{r|rrrrr}
 & 2 & 1 & -3 & 1 & -10 \\
 & & 4 & 10 & 14 & 30 \\
\hline
2 & 2 & 5 & 7 & 15 & 20.
\end{array}
$$

The first row in the array consists of the coefficients of the dividend. We write the number 2 at the lower left because we are dividing by $x - 2$; if we were dividing by $x + 2$, the number would be -2. The arrows, usually omitted, indicate that the number at the "tail," when multiplied by 2, gives the number at the head. The number 20 at the lower right is the remainder.

EXAMPLE 47-4   Use synthetic division to find $Q(x)$ and R if $P(x) = 2x^5 - 30x^3 + x - 1$ and $D(x) = x + 4$.

Solution The coefficients of the terms in x^4 and x^2 of $P(x)$ are 0; that is, $P(x) = 2x^5 + 0x^4 - 30x^3 + 0x^2 + x - 1$. Therefore, the starting point of our synthetic division is the array

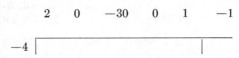

$$2 \quad 0 \quad -30 \quad 0 \quad 1 \quad -1$$

$$-4 \, \lfloor$$

The coefficients of $Q(x)$ and R will appear in the last line. To carry out the synthetic division, we bring down the first coefficient, 2, and then follow the procedure outlined above:

$$
\begin{array}{r|rrrrrr}
 & 2 & 0 & -30 & 0 & 1 & -1 \\
 & & -8 & 32 & -8 & 32 & -132 \\
\hline
-4 & 2 & -8 & 2 & -8 & 33 & -133.
\end{array}
$$

We therefore have $Q(x) = 2x^4 - 8x^3 + 2x^2 - 8x + 33$, and $R = -133$.

1. Write each of the following expressions in the form

$$a_n x^n + a_{n-1} x^{n-1} + \cdots + a_0.$$

(a) $(x - 3)(x + 5) + x(x^3 - 1)$ (b) $(x^2 - 9)(2 - x) - 3(x - 1)$
(c) $(x + 3)(x^2 - 2)(x^3 + 1)$
(d) $(i - x)(2 - 4ix + 3x^2) - (x^3 - 4 + i)$

2. Let $A(x) = 2x^2 - 3x + 1$, $B(x) = x^3 - 5$, and $C(x) = x + 2$. Find the following polynomials.

(a) $A(x)B(x) + A(x)C(x)$ (b) $A(x)^2 - B(x)^2$
(c) $xA(x) + 5B(x) - 5C(x)$ (d) $B(x) + 2[C(x) - 3]^2$

3. If $P(x)$ is a polynomial of degree 1492 and $Q(x)$ is a polynomial of degree 1776, what is the degree of each of the following polynomials?

(a) $P(x)Q(x)$ (b) $P(x) + Q(x)$ (c) $P(x)^3$

4. (a) If $A(x) = x^2 + 2x - 4$, $B(x) = x^3 + 24$, $P(x) = -x^2 + 2x - 8$, and $Q(x) = x$, show that $P(x)A(x) + Q(x)B(x) = 32$.
    (b) Now find polynomials $R(x)$ and $S(x)$ such that $R(x)A(x) + S(x)B(x) = 1$.

5. Find non-negative integers q and r such that $a = qb + r$, $0 \le r < b$, for the following choices of a and b.

(a) $a = 29$, $b = 3$ (b) $a = 1492$, $b = 1776$
(c) $a = 229$, $b = 17$ (d) $a = 1084$, $b = 16$

6. Suppose that a and b are positive integers and q and r are the quotient and remainder that we obtain by dividing a by b. Convince yourself that $q = [\![a/b]\!]$ and $r = a - [\![a/b]\!]b$.

7. Find the quotient $Q(x)$ and remainder $R(x)$ such that $P(x) = Q(x)D(x) + R(x)$ for the following choices of $P(x)$ and $D(x)$.

(a) $P(x) = x^4 + x^2 + x + 1$, $D(x) = x^2 + x + 1$

(b) $P(x) = 3x^3 + x - 1$, $D(x) = 2x^2 - 1$

(c) $P(x) = x^4 - x^2 - 1$, $D(x) = x^4 + x^2 + 1$

(d) $P(x) = 3x^2 - x + 2$, $D(x) = 2x^3 + 5x^2 - x + 7$

(e) $P(x) = x^3 - 5ix^2 + 2x - 3$, $D(x) = ix + 2$

(f) $P(x) = ix^3 - 5x^2 + 2x - 3$, $D(x) = x^2 + x - 2$

8. Use synthetic division to find the quotient $Q(x)$ and remainder R such that $P(x) = Q(x)D(x) + R$ for the following choices of $P(x)$ and $D(x)$.

(a) $P(x) = x^3 - 2x^2 + 3x - 2$, $D(x) = x - 1$

(b) $P(x) = x^5 + 4x^3 - 5x^2 + 5$, $D(x) = x + 1$

(c) $P(x) = x^7 + 1$, $D(x) = x + 1$

(d) $P(x) = 3x^4 + 22x^3 - 19x^2 - 22x + 16$, $D(x) = x + 8$

(e) $P(x) = 3x^3 - ix^2 + 2x - 3i$, $D(x) = x - i$

(f) $P(x) = ix^3 + 3x^2 - 2ix + 1$, $D(x) = x + 2i$

9. (a) Let a be any number and n be a positive integer. Use synthetic division to convince yourself that the quotient obtained by dividing $x^n - a^n$ by $x - a$ is $x^{n-1} + ax^{n-2} + a^2x^{n-3} + \cdots + a^{n-1}$.

(b) Can you convince yourself that if you multiply the quotient we obtained in part (a) by $x - a$ you do indeed obtain $x^n - a^n$?

10. If $P(x) = a_nx^n + \cdots + a_0$ is a polynomial with complex coefficients, we may define $\overline{P}(x)$ to be the polynomial $\bar{a}_nx^n + \cdots + \bar{a}_0$. Can you convince yourself that if $P(x) + Q(x) = R(x)$ and $P(x)Q(x) = S(x)$, then $\overline{P}(x) + \overline{Q}(x) = \overline{R}(x)$ and $\overline{P}(x)\overline{Q}(x) = \overline{S}(x)$?

11. Suppose we say that a polynomial $a_nx^n + \cdots + a_0$ of degree n (with real coefficients) is *greater than* 0 if $a_n > 0$, and *less than* 0 if $a_n < 0$. We will then say that $P(x) > Q(x)$ if $P(x) - Q(x) > 0$. The polynomial 0, and only the polynomial 0, equals 0. Which of Properties 4-1 to 4-5 [with a, b, and c replaced by $P(x)$, $Q(x)$, and $R(x)$, of course] remain valid?

12. A non-zero polynomial in x is said to be **even** if all the coefficients of odd powers of x are 0, and **odd** if all the coefficients of even powers of x are 0. Suppose that $P(x)$ is even and $Q(x)$ is odd. Can you convince yourself that

(a) $P(x)Q(x)$ is odd?

(b) $Q(x)^2$ is even?

(c) $P(x) + Q(x)$ is neither even nor odd?

48
POLYNOMIALS OF THE FIRST DEGREE

A polynomial of the first degree has the form $P(x) = ax + b$, where $a \neq 0$. Of course, nothing could be simpler than such a polynomial, so it is a happy fact that polynomials of the first degree are useful in themselves and that a study of them helps us to understand general polynomials better.

One of the first questions we ask about a given polynomial $P(x)$ is, "What are its zeros?" That is, we want to find the set $\{r \mid P(r) = 0\}$. This question is hard to answer for a general polynomial, but it is easy to find the zeros of a polynomial of degree 1:

$$\{r \mid ar + b = 0\} = \{-b/a\}.$$

In other words, the set of zeros of the polynomial $P(x) = ax + b$ contains just the single member $-b/a$.

Anticipating some of our future discussion, we now state a theorem that is extremely simple, but whose generalization to polynomials of higher degree is extremely important.

Theorem 48-1 *If $P(x) = ax + b$ and r is its zero, then $P(x) = a(x - r)$.*

Proof Since $r = -b/a$, we have

$$a(x - r) = a\left[x - \left(-\frac{b}{a}\right)\right] = ax + b = P(x).$$

Many algebraic equations may be reduced to the form $ax + b = 0$; let us look at some examples.

EXAMPLE 48-1 Solve the equation $\frac{3}{2}x - \frac{2}{3} = 2x + 1$.

Solution We write the given equation in ever simpler, but equivalent, forms:

$$\frac{3}{2}x - 2x = 1 + \frac{2}{3},$$

$$-\frac{1}{2}x = \frac{5}{3},$$

$$x = -\frac{10}{3}.$$

Thus the solution set of our given equation is just the set $\{-\frac{10}{3}\}$.

EXAMPLE 48-2 Solve the equation

$$\frac{5}{x - 1} + \frac{1}{4 - 3x} = \frac{3}{6x - 8}.$$

Solution This equation may be written as

$$\frac{5}{x - 1} + \frac{1}{4 - 3x} + \frac{3}{2(4 - 3x)} = 0.$$

Now we multiply both sides of this last equation by $2(x - 1)(4 - 3x)$ to obtain the equivalent equation

$$10(4 - 3x) + 2(x - 1) + 3(x - 1) = 0.$$

This equation simplifies to

$$-5x + 7 = 0,$$

and so we see that

$$\left\{\frac{5}{x - 1} + \frac{1}{4 - 3x} = \frac{3}{6x - 8}\right\} = \left\{\frac{7}{5}\right\}.$$

EXAMPLE 48-3 Seven engineering firms agree to contribute equally to the cost of a joint technical library. If three more firms join the plan, the cost to each member of the group would be reduced by $600. Find the cost of the library.

Solution Suppose the library will cost c dollars. Therefore, $\frac{1}{7}c$ is each firm's share when there are 7 firms involved, and $\frac{1}{10}c$ is each firm's share when there are 10 firms. We are told that the latter share is 600 less than the former, in symbols,

$$\tfrac{1}{10}c = \tfrac{1}{7}c - 600.$$

We leave it to you to solve this equation and thus find that the library costs $14,000.

We know that a real (domain and range are subsets of R^1) linear function f is defined by an equation $f(x) = mx + b$, where m and b are given real numbers. Thus the expression that defines a linear function is a polynomial of degree one if $m \neq 0$. First-degree polynomials are also known as **linear polynomials,** and an equation $mx + b = 0$ is a linear equation. We saw in Section 16 that the graph of any real linear function is a line; that is, the graph of an equation $y = mx + b$, where m and b are real numbers, is a line. The question of whether or not every line is the graph of such an equation is answered by the following theorem.

Theorem 48-2 *A line not parallel to the Y-axis is the graph of an equation $y = mx + b$. A line parallel to the Y-axis is the graph of an equation $x = a$.*

Proof First, let us consider the case of a line that is parallel to the Y-axis. A line parallel to the Y-axis must intersect the X-axis in some point $(a, 0)$. Then the X-coordinate of every point of the line must be $x = a$, which is the equation of the line.

If a line is not parallel to the Y-axis, then it must intersect that axis and every line parallel to that axis. Suppose such a line intersects the Y-axis in the point $(0, b)$ and the line $x = 1$ in the point $(1, c)$, and let $m = c - b$. We shall show that with these choices of m and b, the equation of our given line is $y = mx + b$. In the first place, we know that this equation is the equation of *some* line. To show that it is the line we started with, we need only show that it contains two points of the given line. The obvious candidates for testing are the points $(0, b)$ and $(1, c)$, and when we substitute these coordinates in the equation $y = mx + b$, we obtain the correct equations $b = b$ and $c = m + b$.

The following theorem completes our discussion of lines for now. We will have more to say about them in Chapter Eleven.

Theorem 48-3 *If A, B, and C are three real numbers such that A and B are not both 0, then the graph of the equation $Ax + By + C = 0$ is a line.*

Proof (i) Suppose $B \neq 0$ and (x, y) is a pair of numbers such that $Ax + By + C = 0$. Then $y = (-A/B)x - C/B$. Thus every pair of coordinates that satisfies the given equation satisfies the equation $y = mx + b$, where $m = -A/B$ and $b = -C/B$. We know this latter equation is the equation of a line, so the points whose coordinates satisfy the given equation lie in a line. Notice that the line has slope $-A/B$ and Y-intercept $-C/B$.

220

(ii) If $B = 0$, then $A \neq 0$, since we have assumed that A and B are not both 0. The equation $Ax + By + C = 0$ then reduces to the equation $x = -C/A$, which we know is the equation of a line parallel to the Y-axis. In this case the slope is undefined.

PROBLEMS
48

1. The following equations are linear in x; solve them. State any conditions that are necessary for your solution to be valid; for example, $b \neq 0$ in Problem 1(d).

(a) $2x - 4(2 - 3x) = 20$

(b) $\dfrac{x-1}{2} + \dfrac{x-2}{3} = \dfrac{2+x}{10} + \dfrac{5}{6}$

(c) $4ix - 1 = 3x - i$

(d) $2bx - 5 = d$

(e) $b^2 + bx = a^2 - ax$

(f) $xy^2 + 6 = 3y + 2xy$

(g) $S = \dfrac{x - xr^n}{1 - r}$

(h) $\dfrac{x}{r} + r = rx - 1$

2. The circumference of the outer of two concentric circles is 5 inches longer than the circumference of the inner. What is the difference in their radii?

3. TV station A broadcasts a commercial every 2 minutes, while commercials come every 3 minutes on station B. During the course of a broadcasting day (which is the same length for both stations) the two stations combined put out 700 commercials. How many are broadcast by station A? How long is the broadcasting day?

4. It takes Farmer Jones 7 minutes to milk a cow, while Farmer Brown can do the job in 5 minutes. How fast can they drain old Boss if they work together?

5. (a) How much whiskey that is 40% alcohol should a distiller add to 100 gallons of whiskey that is 50% alcohol to obtain whiskey that is 43% alcohol? (b) Suppose that our distiller wants to end up with 100 gallons of 43% whiskey, so he sells a certain number of gallons of his 50% whiskey and replaces them with 40% whiskey. How many gallons should he sell?

6. At what time between 8 and 9 o'clock will the minute hand of a watch be directly over the hour hand?

7. Solve the following equations.

(a) $3 \log x - 2 = 7$

(b) $3(\log x + 3) = 2(2 - \log x)$

(c) $\frac{1}{3} 10^x = 7 - 2(10^x + 1)$

(d) $\frac{1}{2}(10^x - 1) = 10^{x-1}$

(e) $6[\![x]\!] - 4 = [\![x]\!] + 11$

(f) $4[\![x]\!] - 6 = [\![x]\!] + 11$

(g) $3(2 - \tan x) = 5 \tan x - 2$

(h) $7 \sin x - 6 = 5 \sin x - 7$

8. If we multiply both sides of the equation of Example 48-2 by

$$(x - 1)(4 - 3x)(6x - 8)$$

and simplify, we obtain the equation $15x^2 - 41x + 28 = (3x - 4)(5x - 7) = 0$. Thus, it would seem that the solution set of the given equation is $\{\frac{4}{3}, \frac{7}{5}\}$. Explain.

9. Sketch the graphs of the following equations:

(a) $x = 3$ (b) $2x - 3y = 1$
(c) $x + 2y - 4 = 0$ (d) $5x + 2 = 3(y - 4)$

10. Show that the graphs of the equations $x - 2y + 3 = 0$ and $4y - 2x - 6 = 0$ are the same line.

11. Suppose sugar is 10¢ a pound and shortening is 40¢ a pound. A mixture containing x pounds of sugar and y pounds of shortening is worth 50¢. Display graphically the set of points that represent possible values of x and y.

12. Sketch the graphs of the following equations.

(a) $|x| + |y| = 1$ (b) $|x + y| = 1$ (c) $|x + 1| = y$ (d) $x + 1 = |y|$

49
POLYNOMIALS OF THE SECOND DEGREE

A polynomial of the second degree is called a **quadratic polynomial** and has the form $P(x) = ax^2 + bx + c$, where $a \neq 0$. As with polynomials of the first degree, we shall be interested in finding the zeros of quadratic polynomials.

EXAMPLE 49-1 Find the zeros of the polynomial $P(x) = x^2 + x - 6$.

Solution Since $x^2 + x - 6 = (x + 3)(x - 2)$, our problem is to find the set

$$\{(x + 3)(x - 2) = 0\}.$$

The product of two numbers is 0 if, and only if, at least one of the numbers is 0, and so

$$\{(x + 3)(x - 2) = 0\} = \{x + 3 = 0\} \cup \{x - 2 = 0\} = \{-3, 2\}.$$

Finding the zeros of a quadratic polynomial by factoring, as we did in Example 49-1, works well if we can easily recognize the factors. If we cannot recognize the factors, the following device, known as **completing the square,** will always work. To illustrate this method, we will find the zeros of a general quadratic polynomial; that is, we will solve the general quadratic equation

(49-1) $$ax^2 + bx + c = 0.$$

We first subtract c from both sides of the equation, and then divide both sides by a (which is not 0 by hypothesis) to obtain the equivalent equation

$$x^2 + \frac{b}{a} x = -\frac{c}{a}.$$

Now we add the quantity $b^2/4a^2$ to both sides of this equation to get

(49-2) $$x^2 + \frac{b}{a} x + \frac{b^2}{4a^2} = \frac{b^2 - 4ac}{4a^2}.$$

222

It is now apparent that the left-hand side of Equation 49-2 is simply $(x + b/2a)^2$. We added $b^2/4a^2$ to both sides in order to "complete" the left-hand side in such a way as to make it a perfect square. Thus we have

(49-3)
$$\left(x + \frac{b}{2a}\right)^2 = \frac{b^2 - 4ac}{4a^2}.$$

By writing the numerator of the right-hand side of Equation 49-3 as the square of its own square root and transposing the result to the left-hand side of the equation, we obtain the equivalent equation

$$\left(x + \frac{b}{2a}\right)^2 - \left(\frac{\sqrt{b^2 - 4ac}}{2a}\right)^2 = 0.$$

The left-hand side of this equation is the difference of two squares, so we use the factoring formula $A^2 - B^2 = (A - B)(A + B)$ to write the equivalent equation

$$\left(x + \frac{b}{2a} - \frac{\sqrt{b^2 - 4ac}}{2a}\right)\left(x + \frac{b}{2a} + \frac{\sqrt{b^2 - 4ac}}{2a}\right) = 0.$$

From this equation it is apparent that the solution set of Equation 49-1 is

$$\left\{ \frac{-b + \sqrt{b^2 - 4ac}}{2a}, \quad \frac{-b - \sqrt{b^2 - 4ac}}{2a} \right\}.$$

We state this important result as a theorem.

Theorem 49-1 *The polynomial $P(x) = ax^2 + bx + c$ has two zeros (which may be equal) given by the formulas*

$$r_1 = \frac{-b + \sqrt{b^2 - 4ac}}{2a} \quad \text{and} \quad r_2 = \frac{-b - \sqrt{b^2 - 4ac}}{2a}.$$

These formulas are called the **Quadratic Formulas,** and you should memorize them.

EXAMPLE 49-2 Find the zeros of the polynomial $P(x) = 2x^2 - 2x + 1$.

Solution The quadratic formulas yield

$$r_1 = \frac{2 + \sqrt{4 - 8}}{4} = \frac{2 + 2i}{4} = \frac{1 + i}{2} \quad \text{and}$$

$$r_2 = \frac{2 - \sqrt{4 - 8}}{4} = \frac{1 - i}{2}.$$

Therefore, $\{\frac{1}{2}(1 + i), \frac{1}{2}(1 - i)\}$ is the set of zeros of our given quadratic polynomial.

Two interesting facts emerge from Example 49-2:

(i) Although the coefficients of $P(x)$ are real numbers, indeed integers, the zeros of $P(x)$ are not real.

(ii) The relation between the numbers r_1 and r_2 is $r_1 = \bar{r}_2$.

It shouldn't surprise you that a quadratic polynomial with real coefficients may have zeros that are not real numbers; look at the Quadratic Formulas. To find the zeros, we take the square root of the number $b^2 - 4ac$, and if this number is negative, the zeros will not be real. The number $b^2 - 4ac$ is called the **discriminant** of the quadratic polynomial $ax^2 + bx + c$. Table 49-4 summarizes the relationship between the sign of the discriminant and the character of the zeros of a quadratic polynomial with real coefficients. You can easily verify these facts by referring to the Quadratic Formulas.

$$P(x) = ax^2 + bx + c; \ a, \ b, \ \text{and} \ c \text{ real}$$

(49-4)

$b^2 - 4ac$	zeros of $P(x)$
positive	real, unequal
zero	real, equal
negative	complex conjugates

EXAMPLE 49-3   Solve the equation $x^2 + 2ix - 2 = 0$.

Solution   According to the Quadratic Formulas,

$$r_1 = \frac{-2i + \sqrt{-4 + 8}}{2} = \frac{-2i + 2}{2} = 1 - i \quad \text{and}$$

$$r_2 = \frac{-2i - \sqrt{-4 + 8}}{2} = \frac{-2i - 2}{2} = -1 - i, \quad \text{so}$$

$$\{x^2 + 2ix - 2 = 0\} = \{1 - i, -1 - i\}.$$

Notice that $b^2 - 4ac > 0$, but r_1 and r_2 are not real numbers. Why doesn't this fact contradict Table 49-4?

The method of completing the square has many uses in addition to solving quadratic equations, so it is a procedure worth remembering. In order to improve our understanding of this method, let us solve a quadratic equation by completing the square, rather than by using the Quadratic Formulas. You will use this procedure again in Chapter 11.

EXAMPLE 49-4   Solve the equation $2x^2 + 8x + 4 = 0$ by completing the square.

Solution   To solve this equation by completing the square, we go through the steps we followed in deriving the Quadratic Formulas, replacing the letters a, *b,* and *c* by the numbers 2, 8, and 4. Thus, we construct the following chain of equivalent equations:

$$x^2 + 4x = -2,$$
$$x^2 + 4x + 4 = 2,$$
$$(x + 2)^2 = 2.$$
$$(x + 2)^2 - (\sqrt{2})^2 = 0,$$
$$(x + 2 + \sqrt{2})(x + 2 - \sqrt{2}) = 0.$$

From this last equation it is clear that

$$\{2x^2 + 8x + 4 = 0\} = \{-2 - \sqrt{2}, -2 + \sqrt{2}\}.$$

The *graph* of a polynomial $P(x)$ with real coefficients is the graph of the equation $y = P(x)$. The graph of every quadratic polynomial is similar in appearance either to the graph in Fig. 49-1 or to the graph in Fig. 49-2.

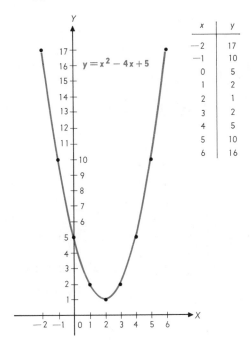

x	y
−2	17
−1	10
0	5
1	2
2	1
3	2
4	5
5	10
6	16

FIGURE 49-1

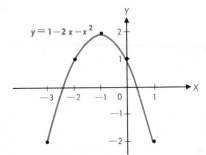

FIGURE 49-2

225

EXAMPLE 49-5   Sketch the graph of the polynomial $x^2 - 4x + 5$.

Solution   In order to sketch the graph of the equation $y = x^2 - 4x + 5$, we construct a table that shows corresponding values of x and y. If we plot the points whose coordinates are given in this table and join them by a smooth curve, we obtain the graph shown in Fig. 49-1. The graph does not intersect the X-axis. We could have predicted that it would not, because the discriminant of the polynomial $x^2 - 4x + 5$ is -4, which means that this polynomial does not have any real zeros.

In Chapter 1 we devoted some time to the solution of inequalities that involve quadratic polynomials. The next example illustrates another method by which we can solve such inequalities.

EXAMPLE 49-6   Solve the inequality $1 - 2x - x^2 > 0$.

Solution   We first sketch the graph of the equation $y = -x^2 - 2x + 1$ (Fig. 49-2). Our problem is to determine those values of x for which $y > 0$. It is clear from the figure that $y > 0$ at each point x that lies *between* the two points in which the graph intersects the X-axis. In other words, the solution set of the inequality is $\{r_1 < x < r_2\}$, where r_1 and r_2 are the zeros of the polynomial $-x^2 - 2x + 1$. By using the Quadratic Formulas, we find that the zeros of this polynomial are $-1 - \sqrt{2}$ and $-1 + \sqrt{2}$. Thus,

$$\{1 - 2x - x^2 > 0\} = \{-1 - \sqrt{2} < x < -1 + \sqrt{2}\}.$$

**PROBLEMS
49**

1. Solve the following equations by factoring.
    (a) $3x^2 + 2x = 0$ (b) $z^2 + 2z - 3 = 0$
    (c) $y^2 + 2y - 15 = 0$ (d) $6m^2 - 13m + 6 = 0$

2. Solve the following equations by completing the square.
    (a) $x^2 + 4x - 4 = 0$ (b) $x^2 - 3x + 2 = 0$
    (c) $x^2 - 8x + 20 = 0$ (d) $3x^2 + 2x + 1 = 0$

3. Use the Quadratic Formulas to solve the following equations.
    (a) $x^2 - 5x - 5 = 0$ (b) $x^2 + x + 1 = 0$
    (c) $(x + 1)(x - 2) = 3$ (d) $4x^2 + 4 = 5x$
    (e) $x^2 + 2\sqrt{3}x + 1 = 0$ (f) $\sqrt{5}x^2 + 2x + \sqrt{5} = 0$

4. (a) If $ax^2 + 2bx + c = 0$, solve for x.
    (b) If $Lm^2 + Rm + 1/C = 0$, solve for m.
    (c) If $s = v_0 t - \frac{1}{2}gt^2$, solve for t.
    (d) If $ay^2 + by + c = by^2 + cy + a$, solve for y.

5. Calculate the sign of the discriminant to determine the character of the zeros of $P(x)$.
    (a) $P(x) = 32x^2 - 20x + 3$ (b) $P(x) = 36x^2 - 60x - 25$
    (c) $P(x) = x^2 + (\sin 1)x + \sin \frac{1}{4}$ (d) $P(x) = x^2 + 2\sqrt{.3294}x + \log 2.135$
    (e) $P(x) = x^2 + .6x + \mathrm{Sin}^{-1} .09$ (f) $P(x) = x^2 + .6x + \mathrm{Tan}^{-1} .09$

6. Solve the following equations.
   (a) $2x^2 + ix - 1 = 0$ (b) $3x^2 - i\sqrt{2}x - \frac{1}{4} = 0$
   (c) $x^2 - 2ix - (1 + 2i) = 0$ (d) $x^2 - 2x + (1 - 8i) = 0$

7. In how many points does the graph of f intersect the X-axis if $f(x)$ is given by the following formula?
   (a) $f(x) = x^2 - 3x + 1$ (b) $f(x) = x^2 - 2\sqrt{2}x + 2$
   (c) $f(x) = 2 + x - 3x^2$ (d) $f(x) = x^2 + x + 1$

8. Solve the following equations for x.
   (a) $x^2 + 2xy - y^2 = 0$ (b) $x^2 - x + xy - y = 0$
   (c) $x^2 - xy - xz + yz = 0$ (d) $(x + y)^2 = x$

9. Sketch the graphs of the following equations.
   (a) $y = 4 - x^2$ (b) $y = x^2 + x$
   (c) $y = 2x^2 + 4x - 1$ (d) $y = x + 2 - x^2$

10. Solve the following inequalities.
    (a) $x^2 - 2x - 2 < 0$ (b) $x^2 - 4x - 7 > 0$
    (c) $3 + 4x < x^2$ (d) $x(3 - x) < 3$
    (e) $x(x - 2) > 2x(x - 4)$ (f) $x^3 < x^2 - 2x$

11. For each of the following equations, determine the values of k for which the solutions are real numbers.
    (a) $x^2 + 2kx + 5 = 0$ (b) $x^2 + 4x - k = 0$
    (c) $kx^2 - 4x + 7 = 0$ (d) $kx^2 - 2x + 7k = 0$

12. A target consists of a white circle surrounded by a black ring 5 inches wide. The white and black areas are equal. What is the radius of the white circle?

13. A plane takes $\frac{1}{2}$ hour to fly 150 miles with the wind and 65 miles back against the wind. The speed of the wind is 30 miles per hour. What is the plane's airspeed?

14. The outside dimensions of a rectangular picture frame are 2 feet by 3 feet. The area of the frame is equal to the area of the picture inside the frame. What is the width of the frame?

15. Suppose $P(x) = ax^2 + bx + c$, where $a > 0$ and b and c are real numbers. Under what conditions are the two zeros of $P(x)$ both positive? Both negative? Of opposite signs?

16. If $f(x) = x^2 + 2px + q$, we can complete the square to derive the equation $f(x) = (x + p)^2 - p^2 + q$. Use this formula to determine the minimum value of f.

50
EXPRESSIONS IN QUADRATIC FORM

We may sometimes use the methods described in Section 49 to solve equations which at first glance do not appear to be quadratic. The best way to make this point is via examples, which can range from the simple to the quite complicated.

EXAMPLE 50-1   Find the zeros of the polynomial $P(x) = x^4 - x^2 - 6$.

Solution   If we let $y = x^2$, the problem is reduced to solving the equation $y^2 - y - 6 = 0$. Since $y^2 - y - 6 = (y - 3)(y + 2)$, we see that

$$\{y^2 - y - 6 = 0\} = \{y = 3\} \cup \{y = -2\}.$$

Now we replace y with x^2, and we have

$$\{x^4 - x^2 - 6 = 0\} = \{x^2 = 3\} \cup \{x^2 = -2\}$$
$$= \{\sqrt{3}, -\sqrt{3}\} \cup \{i\sqrt{2}, -i\sqrt{2}\}$$
$$= \{\sqrt{3}, -\sqrt{3}, i\sqrt{2}, -i\sqrt{2}\}.$$

As Example 50-1 illustrates, if we can write an equation in the form $ax^2 + bx + c = 0$ (using a substitution if necessary), we may solve it by the techniques we use to solve quadratic equations.

EXAMPLE 50-2   Find the zeros of the function f that is defined by the equation $f(t) = \sin^2 t - 3 \cos t - 2$ which lie between 0 and π.

Solution   If we use the identity $\sin^2 t = 1 - \cos^2 t$, the problem is reduced to solving the equation $\cos^2 t + 3 \cos t + 1 = 0$. Now we let $x = \cos t$ and obtain the quadratic equation $x^2 + 3x + 1 = 0$. The Quadratic Formulas tell us that

$$\{x^2 + 3x + 1 = 0\} = \{x = \tfrac{1}{2}(\sqrt{5} - 3)\} \cup \{x = -\tfrac{1}{2}(\sqrt{5} + 3)\}$$

and when we replace x with $\cos t$, we have

$$\{t \mid f(t) = 0\} = \{\cos t = \tfrac{1}{2}(\sqrt{5} - 3)\} \cup \{\cos t = -\tfrac{1}{2}(\sqrt{5} + 3)\}.$$

Since $-\tfrac{1}{2}(3 + \sqrt{5})$ is less than -1 and hence is not in the range of the cosine function, we see that $\{\cos t = -\tfrac{1}{2}(3 + \sqrt{5})\} = \varnothing$. Therefore, taking 2.24 as the approximate value of $\sqrt{5}$, we have found that $\{t \mid \cos t = -.38\}$ is the set of zeros of our function f. There is only one member of this set that lies between 0 and π, and from Table II we find that it is $\pi - 1.18 = 1.96$.

EXAMPLE 50-3   Solve the equation $2x^{2/3} + 5x^{1/3} - 3 = 0$.

Solution   Let $y = x^{1/3}$. Then $y^2 = x^{2/3}$, and the equation to be solved becomes $2y^2 + 5y - 3 = (2y - 1)(y + 3) = 0$. Thus the solution of our original equation is

$$\{x^{1/3} = \tfrac{1}{2}\} \cup \{x^{1/3} = -3\} = \{x = \tfrac{1}{8}\} \cup \{x = -27\} = \{\tfrac{1}{8}, -27\}.$$

In the process of solving an equation we often perform various algebraic operations to obtain a new equation. If the solution set of the resulting equation is the same as the solution set of the original equation, then the two equations are **equivalent.** But if the solution set of the second equation turns out to be larger or smaller than the solution set of the original equation, we have a problem. When the solution set of the new equation has

members that are not solutions of the original equation, we call them **extraneous solutions.** For example, suppose we square both sides of the
linear equation $2x = x + 1$ and simplify the result. We obtain the equation
$3x^2 - 2x - 1 = 0$. Since $3x^2 - 2x - 1 = (3x + 1)(x - 1)$, it is clear that
$\{3x^2 - 2x - 1 = 0\} = \{-\frac{1}{3}, 1\}$. Of these solutions, only the number 1
satisfies the original equation; the solution $-\frac{1}{3}$ is extraneous. When we
squared both sides of the original equation, we really argued as follows:
If x is a number such that $2x = x + 1$, then *x is a number such that $3x^2 - 2x - 1 = 0$.* This argument shows that the solution set of the original
equation is a subset of the solution set of the new equation, but it *does not*
show that the two sets are equal. To demonstrate equality, we would also
have to show that the solution set of the new equation is a subset of the
solution set of the original equation.

Because certain algebraic operations can yield extraneous solutions,
and also simply to see that you didn't make any computational mistakes,
you should check every solution you find by substituting it into the original
equation to be solved.

EXAMPLE 50-4   Solve the equation

(50-1)
$$\sqrt{x + 7} = \sqrt{2x + 9}.$$

Solution   If we square both sides of the given equation, we obtain the equation
$x + 7 = 2x + 9$, which obviously has the single solution -2. If we replace x with -2 in Equation 50-1, we get $\sqrt{-2 + 7} = \sqrt{-4 + 9}$; that
is, $\sqrt{5} = \sqrt{5}$. Thus, -2 is the solution of Equation 50-1.

EXAMPLE 50-5   Solve the equation

(50-2)
$$\sqrt{x + 5} = x - 1.$$

Solution   Squaring both sides of Equation 50-2 produces the equation

$$x + 5 = x^2 - 2x + 1,$$

which may be simplified to

$$x^2 - 3x - 4 = (x - 4)(x + 1) = 0.$$

We therefore see that $\{-1, 4\}$ is the solution set of our new equation, so
the solution set of Equation 50-2 is a subset of $\{-1, 4\}$. If we replace x
with -1 in both sides of Equation 50-2, we obtain the unequal numbers
$\sqrt{4}$ and -2; thus, -1 is an extraneous solution. You can easily verify
that 4 is the solution of Equation 50-2.

The appearance of extraneous solutions may be a little bothersome,
but it does not present any serious difficulties. These come when we "lose"
a solution in the process of trying to solve an equation. It may happen that
we perform some operations on the terms of our given equation that produce
an equation whose solution set is smaller than the solution set of the equation
we started with. We have lost some solutions. For example, if we try to

229

solve the equation $x^2 = x$ by improperly dividing both sides by x to obtain the equation $x = 1$, we "lose" the solution 0. We need not memorize an elaborate set of rules in order to deal with the introduction of extraneous solutions, or the possible loss of solutions. All we have to do is carefully check the logical steps in our work. Solving equations consists of a series of steps in which one equation, say Equation A, is replaced by another, say Equation B. In each step we must ask ourselves two questions:

(i) If x is a solution of Equation A, is x necessarily a solution of Equation B?

(ii) If x is a solution of Equation B, is x necessarily a solution of Equation A?

If the answer to either question is "No," then Equations A and B are not equivalent, and we may either have introduced an extraneous solution or have lost a solution.

PROBLEMS
50

1. Solve the following equations.
   (a) $x^4 - 3x^2 - 4 = 0$
   (b) $x^2 - 8 + 15x^{-2} = 0$
   (c) $x^4 - 4x^2 + 4 = 0$
   (d) $x^4 - (4 + 2i)x^2 + 8i = 0$
   (e) $36x^4 - 5x^2 - 1 = 0$
   (f) $(x^2 - 1)^2 - 3(x^2 - 1) = 40$

2. Solve the following equations.
   (a) $x^{2/5} + x^{1/5} - 2 = 0$
   (b) $x^{1/3} + 1 = 6x^{-1/3}$
   (c) $x - 2\sqrt{x} - 3 = 0$
   (d) $4x^{1/2} - 4x^{1/4} = 3$
   (e) $x^{\cdot 2} = 1$
   (f) $\sqrt{x} - 2 = 1/\sqrt{x}$

3. Solve the following equations.
   (a) $4^x - 9 \cdot 2^x + 8 = 0$
   (b) $10^{2x} + 10^x - 6 = 0$
   (c) $(\log x)^2 + \log x - \log 100 = 0$
   (d) $15(\log x)^2 - \log x^2 = 1$
   (e) $\sin^2 t + 5 \sin t - \cos^2 t = 2$
   (f) $2 \sec^2 t - \tan t - \tan^2 t = 4$

4. Solve the following equations.
   (a) $\sqrt{x+5} = \sqrt{3x-1}$
   (b) $\sqrt{x-1} = \sqrt{x} - 1$
   (c) $\sqrt{x+1} = \sqrt{x} + 1$
   (d) $\sqrt{4x^2 + 6x + 6} - 2\sqrt{x^2 + x - 1} = 2$

5. Explain carefully why we lose the solution 0 when we divide both sides of the equation $x^2 = x$ by x.

6. Show that $\{4(\sin^2 t - \cos t) + 11 = 0\} = \emptyset$.

7. Solve the equation $x^6 - 7x^3 - 8 = 0$.

8. A target consists of a white center circle with a 6-inch radius surrounded by a black band of the same area. How wide is the band?

9. Find two numbers whose sum is 1 and such that the sum of their reciprocals is 1.

10. If you increase the sides of a certain equilateral triangle by 1 inch, you double its area. How long are the sides of the given triangle?

11. Suppose $A(x)$, $B(x)$, and $C(x)$ are algebraic expressions. Which of the following relations are certainly true if the symbol $*$ is replaced by $=$? By $\subseteq$? By $\supseteq$?

(a) $\{A(x)^2 = 0\} * \{A(x) = 0\}$
(b) $\{|A(x)| = |B(x)|\} * \{A(x) = B(x)\}$
(c) $\{A(x)C(x) = B(x)C(x)\} * \{A(x) = B(x)\}$
(d) $\{[\![A(x)]\!] = [\![B(x)]\!]\} * \{A(x) = B(x)\}$
(e) $\{A(x)C(x) < B(x)C(x)\} * \{A(x) < B(x)\}$
(f) $\{A(x)C(x) \neq B(x)C(x)\} * \{A(x) \neq B(x)\}$

12. Solve the following equations.

(a) $2x|x| - 5x - 3 = 0$ (b) $2x^2 - 5|x| - 3 = 0$
(c) $2[\![x]\!]^2 - 5[\![x]\!] - 3 = 0$ (d) $[\![2x^2 - 5x - 3]\!] = 0$

51
FACTORING POLYNOMIALS OF THE SECOND DEGREE

Factoring a polynomial and finding its zeros are essentially the same problem. We will prove this statement for a general polynomial in Section 53. Here, we show how our knowledge of the zeros of a quadratic polynomial enables us to factor it.

Our argument hinges on two formulas relating the zeros of a quadratic polynomial to its coefficients. If r_1 and r_2 are the zeros of the quadratic polynomial $P(x) = ax^2 + bx + c$, then

(51-1) $$r_1 + r_2 = -\frac{b}{a} \quad \text{and} \quad r_1 r_2 = \frac{c}{a}.$$

These equations stem directly from the Quadratic Formulas, for

$$r_1 + r_2 = \frac{-b + \sqrt{b^2 - 4ac}}{2a} + \frac{-b - \sqrt{b^2 - 4ac}}{2a} = -\frac{b}{a}$$

and

$$r_1 r_2 = \left(\frac{-b + \sqrt{b^2 - 4ac}}{2a}\right)\left(\frac{-b - \sqrt{b^2 - 4ac}}{2a}\right)$$

$$= \frac{b^2 - (b^2 - 4ac)}{4a^2} = \frac{c}{a}.$$

It is now an easy matter to show that every second-degree polynomial may be written as the product of first-degree polynomials.

Theorem 51-1   *If $P(x) = ax^2 + bx + c$ is a polynomial of the second degree with zeros r_1 and r_2, then*

(51-2) $$P(x) = a(x - r_1)(x - r_2).$$

Proof   We have

$$a(x - r_1)(x - r_2) = a[x^2 - (r_1 + r_2)x + r_1r_2] \quad \text{(algebra)}$$
$$= a[x^2 + bx/a + c/a] \quad \text{(Equations 51-1)}$$
$$= ax^2 + bx + c = P(x) \quad \text{(algebra)}.$$

EXAMPLE 51-1   Factor the polynomial $P(x) = 2x^2 + 8x + 4$.

Solution   In Example 49-4 we found that $\{-2 + \sqrt{2}, -2 - \sqrt{2}\}$ is the set of zeros of this polynomial. It therefore follows from Theorem 51-1 that

$$2x^2 + 8x + 4 = 2[x - (-2 + \sqrt{2})][x - (-2 - \sqrt{2})]$$
$$= 2(x + 2 - \sqrt{2})(x + 2 + \sqrt{2}).$$

EXAMPLE 51-2   Factor the polynomial $P(x) = 2x^2 - 2x + 1$.

Solution   In Example 49-2 we used the Quadratic Formulas to find that $\{\frac{1}{2}(1 + i), \frac{1}{2}(1 - i)\}$ is the set of zeros of $P(x)$. Therefore,

$$P(x) = 2[x - \tfrac{1}{2}(1 + i)][x - \tfrac{1}{2}(1 - i)]$$
$$= \tfrac{1}{2}(2x - 1 - i)(2x - 1 + i).$$

Notice that even though the coefficients of the original polynomial are real numbers, the coefficients of its factors are not.

EXAMPLE 51-3   Factor the polynomial $P(x) = x^2 + 2ix - 2$.

Solution   From Example 49-3 we have $r_1 = 1 - i$ and $r_2 = -1 - i$. Thus

$$x^2 + 2ix - 2 = [x - (1 - i)][x - (-1 - i)]$$
$$= (x - 1 + i)(x + 1 + i).$$

**PROBLEMS
51**

1. The sum of the zeros of a certain quadratic polynomial $P(x)$ is -1, and their product is -6.
    (a) Find b and c if $P(x)$ has the form $x^2 + bx + c$.
    (b) Find the zeros of $P(x)$.
    (c) Write $P(x)$ in factored form.

2. Use Equations 51-1 to find k if
    (a) $P(x) = 4x^2 - 4x + k$, and the zeros of $P(x)$ are equal.
    (b) $P(x) = 3x^2 - 10x + k$, and the zeros of $P(x)$ are reciprocals of each other.
    (c) $P(x) = x^2 - 6x + k$, and one zero of $P(x)$ is the square of the other.

3. Find a polynomial of the form $x^2 + bx + c$ that has the given set of zeros.
    (a) $\{-1, 3\}$ (b) $\{2 + \sqrt{3}, 2 - \sqrt{3}\}$ (c) $\{1 - 2i, 1 + 2i\}$
    (d) $\{i, 1 - i\}$ (e) $\{\sqrt{2} - i, -\sqrt{2} + i\}$ (f) $\{1 - \sqrt{i}, 1 + \sqrt{i}\}$

4. Factor the following quadratic polynomials.
    (a) $4x^2 + 3$ (b) $x^2 - 2x + 5$
    (c) $x^2 + 2ix - 10$ (d) $\tfrac{1}{4}x^2 - x + \tfrac{1}{4}$

232

5. Factor the following polynomials in x.

(a) $x^2 - 2x + \cos^2 t$
(b) $(\cos t)x^2 - 2(\sin t)x - \cos t$
(c) $(\sec t)x^2 - 2(\tan t)x + \sec t$
(d) $(\tan t)x^2 - 2(\sec t)x + \tan t$
(e) $(\sin t)x^2 - 2i(\cos t)x + \sin t$
(f) $(\cos t)x^2 + 8i(\sin t)x + 16 \cos t$

6. Factor the following expressions into products of linear polynomials in x.

(a) $x^2 - 4xy + y^2$
(b) $x^2 + y^2$
(c) $kx^2 - (k^2 - 1)x - k$
(d) $x^2 - 4x - y^2 + 2y + 3$
(e) $(\sin 2t)x^2 - 2x + \sin 2t$
(f) $x^2 + \log (1/uv)x + \log u^{\log v}$

7. One zero of a certain quadratic polynomial is 3, and the sum of its zeros is twice their product. If the coefficient of x^2 is 5, what is the polynomial?

8. Suppose that $P(x)$ is a quadratic polynomial with real coefficients and whose zeros are not real. Show that there are real numbers a, p, and q such that $P(x) = a(x^2 - 2px + p^2 + q^2)$.

9. Suppose that $P(x) = ax^2 + bx + c$. Find a quadratic polynomial, whose coefficients are expressed in terms of a, b, and c, the zeros of which may be obtained from the zeros of $P(x)$ by adding 1.

10. Suppose that the polynomial $ax^2 + bx + c$ has integer coefficients. Under what conditions does it have linear factors with rational coefficients?

11. Factor $ax^2 + bx + c$ by first writing

$$ax^2 + bx + c = a[(x + b/2a)^2 - (\sqrt{b^2 - 4ac}/2a)^2].$$

52
THE
REMAINDER
AND
FACTOR
THEOREMS

In Section 47 we discussed the problem of dividing a polynomial $P(x) = a_nx^n + a_{n-1}x^{n-1} + \cdots + a_0$ by a polynomial $D(x) = d_mx^m + d_{m-1}x^{m-1} + \cdots + d_0$. We mentioned that the process of long division produces two polynomials $Q(x)$ and $R(x)$, where $R(x)$ is either the number 0 or a polynomial of lower degree than $D(x)$, such that

(52-1) $$P(x) = Q(x)D(x) + R(x).$$

Now suppose that $P(x) = a_nx^n + \cdots + a_0$ and $D(x) = x - r$, where r is a given number. The polynomial $R(x)$ in Equation 52-1 is either the number 0 or a polynomial of lower degree than $D(x)$. Since $D(x)$ is a first-degree polynomial, $R(x)$ is 0 or its degree is 0. In either case, $R(x)$ is merely a number. These facts are summarized in the following theorem.

Theorem 52-1   *If $P(x) = a_nx^n + \cdots + a_0$ is a given polynomial and r is a given number, then there is a polynomial $Q(x)$ and a number R (which may be 0) such that*

(52-2) $$P(x) = Q(x)(x - r) + R.$$

It is apparent (if $n > 0$) that the first term of $Q(x)$ is a_nx^{n-1}, and therefore $Q(x)$ has the form $a_nx^{n-1} + \cdots + q_0$. For our purposes, it is unnecessary to find the general expressions for the coefficients of the terms of $Q(x)$ other than the first; they can be easily computed in any specific example.

EXAMPLE 52-1   Let $P(x) = 3x^3 - 2x^2 + x - 2$ and $r = 2$. Find $Q(x)$ and R.

Solution   We find $Q(x)$ and R by dividing $P(x)$ by $(x - 2)$, using synthetic division to shorten our labor:

$$
\begin{array}{r|rrrr}
 & 3 & -2 & 1 & -2 \\
 &   & 6 & 8 & 18 \\
\hline
2 & 3 & 4 & 9 & 16.
\end{array}
$$

We therefore see that $Q(x) = 3x^2 + 4x + 9$ and $R = 16$, so

$$3x^3 - 2x^2 + x - 2 = (3x^2 + 4x + 9)(x - 2) + 16.$$

A number of important theorems are immediate consequences of Theorem 52-1.

Theorem 52-2   (The Remainder Theorem). *If R is the remainder when $P(x)$ is divided by $x - r$, then $R = P(r)$.*

Proof   Equation 52-2 remains valid if x is replaced by r, and hence we have

$$P(r) = Q(r)(r - r) + R = 0 + R = R.$$

EXAMPLE 52-2   If $P(x) = 3x^3 - 2x^2 + x - 2$, find $P(2)$ by using the Remainder Theorem.

Solution   If we divide $P(x)$ by $x - 2$, then $R = 16$ (Example 52-1). Therefore, according to the Remainder Theorem, $P(2) = 16$. To check this result, we observe that $P(2) = 3(2)^3 - 2(2)^2 + (2) - 2 = 24 - 8 + 2 - 2 = 16$.

From the Remainder Theorem and Equation 52-2, we see that if $P(x)$ is any polynomial and r is any number, then there is a polynomial $Q(x)$ such that

(52-3) $$P(x) = (x - r)Q(x) + P(r).$$

The polynomial $D(x)$ is a **factor** of the polynomial $P(x)$ if there exists a polynomial $Q(x)$ such that $P(x) = Q(x)D(x)$; that is, if after dividing $P(x)$ by $D(x)$ the remainder is zero. It is clear from Equation 52-3 that $x - r$ is a factor of $P(x)$ if, and only if, $P(r) = 0$. This result is known as the *Factor Theorem and its converse.*

Theorem 52-3   (The Factor Theorem). *If $P(x) = a_n x^n + \cdots + a_0$ is a polynomial of degree $n > 0$, and if r is a zero of $P(x)$, then $x - r$ is a factor of $P(x)$.*

Theorem 52-4   (Converse of the Factor Theorem). *If $P(x)$ is a polynomial and $x - r$ is a factor of $P(x)$, then r is a zero of $P(x)$.*

The Factor Theorem tells us that if we can find a zero of $P(x)$, then we can factor $P(x)$ into a product of a polynomial of lower degree than the degree of $P(x)$ and a linear factor. According to the Converse of the Factor Theorem, if we know a linear factor of $P(x)$, then we know a zero of $P(x)$.

EXAMPLE 52-3 Let $P(x) = x^3 - 8$. Factor $P(x)$.

Solution The number 2 is clearly a zero of $P(x)$. According to the Factor Theorem, $x - 2$ will therefore be a factor, and the other factor, found by dividing $P(x)$ by $x - 2$, will be a quadratic polynomial. Using long division, we get

$$x^3 - 8 = (x - 2)(x^2 + 2x + 4).$$

We can use the methods we covered in Section 51 to factor the quadratic polynomial $x^2 + 2x + 4$.

**PROBLEMS
52**

1. For the following choices of $P(x)$ and r, find $Q(x)$ and R such that $P(x) = Q(x)(x - r) + R$.
   (a) $P(x) = x^5 - x + 1, r = 2$ (b) $P(x) = x^6 + x^2 + 6, r = -2$
   (c) $P(x) = x^3 + 3, r = \sqrt{3}$ (d) $P(x) = x^3 - x, r = 2$
   (e) $P(x) = x^3 + ix - 2i, r = i$ (f) $P(x) = 2x^3 + ix^2 + 3 - i, r = -i$

2. Use the Remainder Theorem to find $P(r)$ for the following choices of $P(x)$ and r.
   (a) $P(x) = x^4 + 2x^3 + 3x^2 + 4x + 5, r = 1$
   (b) $P(x) = x^4 - 2x^3 + 3x^2 - 4x + 5, r = -1$
   (c) $P(x) = x^4 + 2ix^3 + 3x^2 + 4ix + 5, r = i$
   (d) $P(x) = x^4 - 2ix^3 + 3x^2 - 4ix + 5, r = -i$

3. Find the remainder if
   (a) $x^6 + 5$ is divided by $x - 10$.
   (b) $x^9 - \sqrt{2}$ is divided by $x - \sqrt{2}$.
   (c) $x^{26} + 2$ is divided by $x + i$.
   (d) $x^4 + x^3 + x^2 + x + 1$ is divided by $x - i$.

4. Determine k such that
   (a) $x - 1$ is a factor of $2x^3 + x^2 - 2kx + 1$
   (b) $x - 2$ is a factor of $x^3 + k^2x^2 - 2kx - 8$.
   (c) $x + 3$ is a factor of $2x^3 + kx^2 - 3kx + 90$.
   (d) $x - k$ is a factor of $3x^2 + 5x - 2$.

5. Use the Factor Theorem to show that
   (a) $x - a$ is a factor of $x^n - a^n$, where n is any positive integer.
   (b) $x + 2y^2$ is a factor of $x^5 + 32y^{10}$.
   (c) $x^2 - \pi$ is a factor of $x^8 - \pi^4$.
   (d) $x + y$ is a factor of $x^{1861} + y^{1861}$

6. Show that if $A(x)$ and $B(x)$ are two polynomials with a common zero r, then $x - r$ is a factor of the remainder after dividing $A(x)$ by $B(x)$.

7. If we divide a given polynomial $P(x)$ by the quadratic polynomial $x^2 + 1$, the remainder will be a linear polynomial $mx + b$ (why?). Show that $P(i) = b + mi$. What is $P(-i)$?

8. Show that we get the same remainder when we divide a given polynomial $P(x)$ by $x - \frac{1}{5}$ and by $5x - 1$.

9. (a) Show that if $P(x)$ is any polynomial and r is a non-zero number, then there exists a polynomial $Q(x)$ such that

$$P(x) = Q(x)(x^2 - r^2) + [(P(r) - P(-r))/2r]x + \tfrac{1}{2}(P(r) + P(-r)).$$

(b) What does this equation reduce to if $P(x)$ is an *even* polynomial (contains only even powers of x)?
(c) What does this equation reduce to if $P(x)$ is an *odd* polynomial (contains only odd powers of x)?

10. If $P(x)$ is a polynomial with real coefficients, then the graph of the equation $y = P(x)$ is a curve in the cartesian plane. For a given real number r, the Remainder Theorem says that this equation can be written as

$$y = (x - r)Q(x) + P(r).$$

Now replace $Q(x)$ with $Q(r)$ to obtain the linear equation $y = (x - r)Q(r) + P(r)$, whose graph, of course, is a line. Since we can expect that $Q(x)$ is close to $Q(r)$ when x is close to r, we can expect that this line approximates the graph of the equation $y = P(x)$ for x near r. In fact, this line is the *tangent line* to the graph at the point $(r, P(r))$. Sketch these graphs for the given $P(x)$ and r.

(a) $P(x) = x^2$, $r = 1$ (b) $P(x) = x^2$, $r = 0$
(c) $P(x) = 2x^2 + 3x - 1$, $r = -1$ (d) $P(x) = -x^2 + 2x + 3$, $r = 1$
(e) $P(x) = x^3$, $r = 1$ (f) $P(x) = x^3$, $r = 0$

53
FACTORING A GENERAL POLYNOMIAL

The Factor Theorem states that if r is a zero of the polynomial $P(x) = a_n x^n + \cdots + a_0$ (where $n > 0$), then there is a polynomial $Q(x) = a_n x^{n-1} + \cdots + q_0$ such that

(53-1) $$P(x) = (x - r)Q(x).$$

Obviously, we can't apply the theorem until we know something about the zeros of $P(x)$. We know how to find the zeros of first- and second-degree polynomials, but so far we have said nothing about polynomials of higher degree. In fact, we have not even said that polynomials of higher degree necessarily have any zeros. For example, it is not immediately evident that there is a number r that satisfies the equation

$$37x^7 - (5 + 2i)x^3 + 2x^2 + \pi = 0.$$

The following theorem does not help us to find the zeros of a polynomial, but it does tell us that they are there to be found. Usually, this theorem is

236

proved in a course in the "Theory of Complex Variables," which you may take several years hence. So we will skip the proof. We give this theorem its traditional name, but you should not infer that it is necessarily the most basic theorem in the entire field of algebra.

Theorem 53-1   (The Fundamental Theorem of Algebra). *If $P(x)$ is a polynomial of degree greater than zero, then there is at least one number r (not necessarily a real number) for which $P(r) = 0$.*

Now suppose that $P(x) = a_n x^n + \cdots + a_0$ is a polynomial of degree $n > 0$. According to the Fundamental Theorem of Algebra, there is a zero, call it r_1, of $P(x)$. Hence, according to Equation 53-1, there is a polynomial $Q_1(x) = a_n x^{n-1} + \cdots + q_1$ such that

(53-2)
$$P(x) = (x - r_1)Q_1(x).$$

If $n - 1 > 0$, we can apply the Fundamental Theorem to the polynomial $Q_1(x)$ to see that there is a number r_2 such that $Q_1(r_2) = 0$. Then according to the Factor Theorem, there is a polynomial $Q_2(x) = a_n x^{n-2} + \cdots + q_2$ such that

(53-3)
$$Q_1(x) = (x - r_2)Q_2(x).$$

From Equations 53-2 and 53-3, we have

(53-4)
$$P(x) = (x - r_1)(x - r_2)Q_2(x).$$

If the degree of $Q_2(x)$ (which is $n - 2$) is positive, then $Q_2(x)$ has a zero, r_3, so there is a polynomial $Q_3(x) = a_n x^{n-3} + \cdots + q_3$ such that $Q_2(x) = Q_3(x)(x - r_3)$, and hence

(53-5)
$$P(x) = (x - r_1)(x - r_2)(x - r_3)Q_3(x).$$

Clearly, we can continue this process until the degree of the quotient polynomial is zero—that is, for n steps. The last quotient polynomial is $Q_n(x) = a_n x^{n-n} = a_n$, and we see that

(53-6)
$$\boldsymbol{P(x) = a_n(x - r_1)(x - r_2) \cdots (x - r_n).}$$

We can state the result of our reasoning as a theorem.

Theorem 53-2   *If $P(x)$ is a polynomial of degree $n > 0$, then there are n numbers, $r_1, r_2, \ldots,$ and r_n (which need not all be different) such that Equation 53-6 is valid. In other words, any polynomial of degree n can be written as a product of n linear factors.*

We have already covered two special cases of this theorem (when $n = 1$ and $n = 2$) in Theorems 48-1 and 51-1, respectively.

From Equation 53-6 it is clear that the numbers $r_1, \ldots, r_n$ are precisely the zeros of $P(x)$. These numbers may not all be different, but there cannot be more than n different zeros of $P(x)$. Thus we have the following theorem.

Theorem 53-3 *A polynomial of degree n has at most n zeros.*

EXAMPLE 53-1 Suppose that there are 11 distinct zeros of the polynomial $P(x) = a_{10}x^{10} + a_9x^9 + \cdots + a_0$. What can we conclude about $P(x)$?

Solution Some of the numbers a_{10}, a_9, and so on, may be zero, so $P(x)$ is either a polynomial of degree $n \leq 10$, or $P(x)$ is the zero polynomial. But $P(x)$ cannot be a polynomial of degree $n \leq 10$ because it has 11 zeros, and this fact would contradict Theorem 53-3. Thus, $P(x)$ is the polynomial 0; that is, $a_0 = a_1 = \cdots = a_{10} = 0$.

The result of Example 53-1 is important, for on the basis of it we can conclude that if the set of zeros of our polynomial $P(x)$ contains 11 members, then it is the *entire set* of complex numbers. More generally, we have the following corollary of Theorem 53-3.

Corollary 53-1 *If the degrees of the polynomials $A(x)$ and $B(x)$ are not greater than n, and if the set $\{r \mid A(r) = B(r)\}$ contains at least $n + 1$ members, then $A(x) = B(x)$.*

Proof Let $P(x) = A(x) - B(x)$. Then $P(x)$ is either the polynomial 0 or a polynomial whose degree does not exceed n. But our hypothesis tells us that the set of zeros of $P(x)$ contains more than n members. Hence, according to Theorem 53-3, $P(x)$ cannot have a degree less than or equal to n; $P(x)$ must be the polynomial 0.

The zeros of a polynomial $P(x)$ form the set $\{r \mid P(r) = 0\}$. Let us speak of the set $\{r_1, r_2, \ldots, r_n\}$, where $r_1, r_2, \ldots, r_n$ are the numbers that appear in Equation 53-6, as the **extended set of zeros** of $P(x)$. The difference in the two sets is merely this: the members of the first set are all the different numbers that are zeros of $P(x)$, while the second set may contain repetitions. For example, $\{1, 3\}$ is the set of zeros of the polynomial

$$5(x - 1)^2(x - 3),$$

and $\{1, 1, 3\}$ is the extended set of zeros. Obviously, the concept of the extended set of zeros of a polynomial is not profound; it just gives us a convenient way to talk about the zeros of polynomials. If a number r appears twice in the extended set of zeros of a given polynomial $P(x)$, it is called a **double zero** of $P(x)$. A number appearing three times is a **triple zero.** In general, if a number appears m times in the extended set $\{r_1, r_2, \ldots, r_n\}$ of zeros of $P(x)$, then it is said to be a **zero of multiplicity m.** Thus, a number r is a zero of multiplicity m of the polynomial $P(x)$ if the factor $x - r$ appears m times in Equation 53-6. Our definition of the extended set of zeros of a polynomial allows us to replace Theorem 53-3 by the statement *the extended set of zeros of a polynomial of degree n has n members.*

EXAMPLE 53-2 Let $P(x) = 2x^3 - x^2 - 1$. If $r_1 = 1$ is a zero of $P(x)$, write $P(x)$ as a product of linear factors.

Solution Since r_1 is a zero of $P(x)$, we know from the Factor Theorem that $x - 1$ divides $P(x)$. Performing the division synthetically, we have

$$
\begin{array}{r|rrrr}
 & 2 & -1 & 0 & -1 \\
 &   & 2 & 1 & 1 \\
\hline
1 & 2 & 1 & 1 & 0.
\end{array}
$$

Thus, $P(x) = (2x^2 + x + 1)(x - 1)$. We then find the other zeros of $P(x)$ by solving the equation $2x^2 + x + 1 = 0$. If we use the Quadratic Formulas, we get

$$r_2 = \tfrac{1}{4}(-1 + i\sqrt{7}) \quad \text{and} \quad r_3 = -\tfrac{1}{4}(1 + i\sqrt{7}).$$

Thus,

$$P(x) = 2(x - 1)[x - \tfrac{1}{4}(-1 + i\sqrt{7})][x + \tfrac{1}{4}(1 + i\sqrt{7})].$$

EXAMPLE 53-3 Find the zeros of the polynomial $P(x) = x^3 - 2x^2 + 4x - 8$.

Solution By trial and error, we find that $P(2) = 0$. Then, by division, we find that $P(x) = (x - 2)(x^2 + 4)$. Hence, the remaining zeros of $P(x)$ are the solutions of the equation $x^2 + 4 = 0$, which are $2i$ and $-2i$. The zeros of $P(x)$ therefore form the set $\{2, 2i, -2i\}$. Theorem 53-3 tells us that these are all the zeros there are.

By performing the multiplication indicated on the right-hand side of Equation 53-6, we obtain the polynomial $a_n[x^n - (r_1 + \cdots + r_n)x^{n-1} + \cdots + (-1)^n r_1 \cdots r_n]$. Now we compare the coefficients of this polynomial with the coefficients of the polynomial on the left-hand side [our original polynomial $P(x)$], and we obtain a generalization of Equations 51-1:

(53-7) $r_1 + \cdots + r_n = -a_{n-1}/a_n \quad \text{and} \quad r_1 \cdots r_n = (-1)^n a_0/a_n.$

EXAMPLE 53-4 If we add all the members of the extended set of zeros of the polynomial $\pi x^{23} - 5x^{13} - ix^2 + \sqrt{83}$, what do we get?

Solution Of course, we have no idea of what the numbers in the extended set of zeros of this polynomial are. But since the coefficient of x^{22} is 0, the first of Equations 53-7 tells us that their sum is zero.

1. Find a polynomial of the form $x^n + a_{n-1}x^{n-1} + \cdots + a_1x + a_0$ that has the following set of zeros.

   (a) $\{1, i, -i\}$ (b) $\{1 + \sqrt{2}, 1 - \sqrt{2}, 1 + i, 1 - i\}$

   (c) $\{i, -1 + i, 1 + i\}$ (d) $\{\sqrt{3}, 2\sqrt{2}, \sqrt{5}\}$

2. The graph of a polynomial $P(x) = x^4 + a_3x^3 + a_2x^2 + a_1x + a_0$ intersects the X-axis in the points whose X-coordinates are $1, -1, 2$, and -2. Find $P(0)$.

3. Write $P(x)$ as a product of linear polynomials.

   (a) $P(x) = x^4 - 16$ (b) $P(x) = x^3 + x^2 - x - 1$

   (c) $P(x) = 12x^3 - 20x^2 + 11x - 2$ (d) $P(x) = 24x^3 + 10x^2 - 3x - 1$

       (one zero is $\frac{2}{3}$)   (one zero is $\frac{1}{3}$)

4. Let $P(x)$ be a polynomial of degree n with real coefficients.

   (a) Is it true that its graph must intersect the X-axis in n different points?

   (b) Can its graph intersect the X-axis in more than n points?

5. Let $A(x)$ and $B(x)$ be polynomials of degree 13 with real coefficients, and suppose that the graphs of the equations $y = A(x)$ and $y = B(x)$ intersect in (at least) 14 points. Show that the graphs actually coincide.

6. The Fundamental Theorem of Algebra guarantees that the function defined by the equation $f(x) = 3x^3 + 6x^2 - x + 2$ has at least one zero. Does the Theorem guarantee that the equation $g(x) = 3\sin^3 x + 6\sin^2 x - \sin x + 2$ defines a function that has at least one zero?

7. Find all the zeros of $P(x) = x^3 + 2x^2 - k^2x + 2k$ if -1 is one of them.

8. Let r_1, r_2, and r_3 be three given numbers, no two of which are equal, and let a_1, a_2, and a_3 be three other given numbers. These numbers determine the quadratic polynomial

   $$P(x) = \frac{a_1(x - r_2)(x - r_3)}{(r_1 - r_2)(r_1 - r_3)} + \frac{a_2(x - r_1)(x - r_3)}{(r_2 - r_1)(r_2 - r_3)} + \frac{a_3(x - r_1)(x - r_2)}{(r_3 - r_1)(r_3 - r_2)}.$$

   (a) Show that $P(x)$ is the only quadratic polynomial such that $P(r_1) = a_1$, $P(r_2) = a_2$, and $P(r_3) = a_3$.

   (b) Find the quadratic polynomial $P(x)$ such that $P(-1) = 3$, $P(0) = 1$, and $P(1) = -2$.

   (c) Find the quadratic polynomial whose graph contains the points $(0, 0)$, $(1, 1)$, and $(2, 8)$.

9. Prove that there is no polynomial $P(x)$ such that $P(x) = \sin x$.

10. Let r ($\neq 0$) be a zero of $P(x) = x^3 + bx^2 + cx + d$. Show that the two remaining members of the extended set of zeros of $P(x)$ are zeros of the polynomial $x^2 + (b + r)x - d/r$.

11. Show that $A(x)$ divides $B(x)$ if, and only if, the extended set of zeros of $A(x)$ is a subset of the extended set of zeros of $B(x)$.

12. Show that the set $\{z \mid 14z^9 - \pi z^8 + 2iz + \sqrt{13} = 0\}$ contains at least one member such that $|z| < 1$.

54
ZEROS OF POLYNOMIALS WITH REAL COEFFICIENTS

In this section we confine our attention to polynomials that have *real numbers* as coefficients. Since we must use some simple facts about the conjugate of a complex number, you might find it worthwhile to review Section 44 now.

Theorem 54-1 *Let $P(x) = a_n x^n + \cdots + a_0$ be a polynomial of degree $n > 0$ with real coefficients. Then if r is a zero of $P(x)$, so is its conjugate $\bar{r}$.*

Proof To say that r is a zero of $P(x)$ means that

(54-1)
$$P(r) = a_n r^n + \cdots + a_0 = 0.$$

If we take the conjugate of both sides of this equation, we obtain the equation

(54-2)
$$\overline{P(r)} = \overline{a_n r^n + \cdots + a_0} = \bar{0} = 0.$$

Theorem 44-1 states that the conjugate of the sum of two complex numbers is the same as the sum of the conjugates of the numbers. We can easily extend this statement to the sum of any number of terms, so Equation 54-2 can be written as

(54-3)
$$\overline{a_n r^n} + \overline{a_{n-1} r^{n-1}} + \cdots + \overline{a_0} = 0.$$

Theorem 44-1 also tells us that we can reverse the order of multiplication and conjugation, and hence

$$\overline{a_n r^n} = \bar{a}_n \bar{r}^n, \ \overline{a_{n-1} r^{n-1}} = \bar{a}_{n-1} \bar{r}^{n-1}.$$

and so on. According to Theorem 44-2, the conjugate of a real number is the number itself, and since the numbers a_k are real numbers in this case, Equation 54-3 can be written as

$$a_n \bar{r}^n + a_{n-1} \bar{r}^{n-1} + \cdots + a_0 = 0.$$

But this equation may be written as $P(\bar{r}) = 0$; that is, $\bar{r}$ is a zero of $P(x)$.

EXAMPLE 54-1 Illustrate Theorem 54-1 by letting $P(x) = x^3 - x^2 + 2$.

Solution You can easily verify that $\{-1, 1+i, 1-i\}$ is the set of zeros of $P(x)$. Since $\overline{-1} = -1$, $\overline{1+i} = 1-i$, and $\overline{1-i} = 1+i$, we see that the conjugate of every member of this set of zeros is also a member.

EXAMPLE 54-2 Two zeros of $P(x) = x^4 - 4x^3 + 14x^2 - 4x + 13$ are $r_1 = 2 + 3i$ and $r_2 = i$. Find the other two.

Solution According to Theorem 54-1, we can let $r_3 = \bar{r}_1 = 2 - 3i$ and $r_4 = \bar{r}_2 = -i$.

The equation

(54-4)
$$P(x) = a_n(x - r_1)(x - r_2) \cdots (x - r_n)$$

241

represents the "ultimate" factorization of a given polynomial $P(x) = a_n x^n + \cdots + a_0$, but it has one drawback. Even though the coefficients of $P(x)$ are real numbers, the coefficients of its linear factors needn't be; so now we raise the question, "How far can we go in factoring a polynomial $P(x)$ with real coefficients if we insist that its factors have real coefficients, too?"

The answer to this question is based on Theorem 54-1. That theorem tells us that if $r = a + bi$ (where a and b are real and $b \neq 0$) is a zero of $P(x)$, then so is $\bar{r} = a - bi$. Therefore, if $(x - r)$ appears in the linear factorization of $P(x)$ (Equation 54-4), then so must $(x - \bar{r})$. Now

$$(x - r)(x - \bar{r}) = x^2 - (r + \bar{r})x + r\bar{r} = x^2 - 2ax + a^2 + b^2,$$

so we see that the latter polynomial is a factor of $P(x)$. Further, since a and b are real numbers, this polynomial has real coefficients. This observation leads to the answer to our question about writing a polynomial with real coefficients as a product of "minimal" real factors; but first let us establish a fact that we will need in the course of the argument.

EXAMPLE 54-3   Suppose that $A(x)$ and $C(x)$, where $C(x) \neq 0$, are polynomials with real coefficients, and that $B(x)$ is a polynomial such that $A(x) = B(x)C(x)$. Show that the coefficients of $B(x)$ are real numbers. In other words, if the real polynomial $C(x)$ divides the real polynomial $A(x)$, then the quotient is a real polynomial.

Solution   Since each coefficient of $B(x)$ has the form $p + iq$, where p and q are real numbers, we can write $B(x) = P(x) + iQ(x)$, where $P(x)$ and $Q(x)$ are polynomials with real coefficients. Then the equation $A(x) = B(x)C(x)$ becomes

$$A(x) - P(x)C(x) = iQ(x)C(x).$$

This equality between polynomials implies equality between their coefficients. Since each coefficient of the polynomial on the left-hand side has the form $u + 0i$ and each coefficient of the polynomial on the right-hand side has the form $0 + iv$, we see that each polynomial must be 0. Therefore, $Q(x)C(x) = 0$. We are assuming that $C(x) \neq 0$, so $Q(x) = 0$, and hence $B(x) = P(x)$, a polynomial with real coefficients.

Now we are ready to prove the main theorem about factoring polynomials with real coefficients.

Theorem 54-2   *Suppose that $P(x) = a_n x^n + \cdots + a_0$ is a polynomial of degree $n > 0$ with real coefficients. Then it can be written as a_n times a product of polynomials with real coefficients, each of which has the form $x - r$ or $x^2 + px + q$.*

Proof   The proof is practically the same as the derivation of Equation 54-4. According to the Fundamental Theorem of Algebra, $P(x)$ has a zero, say r. If r is real, $x - r$ is a real factor of $P(x)$. If r is not real, we have just seen that $(x - r)(x - \bar{r})$, which is of the form $x^2 + px + q$, is a factor. Hence, we have either

$$P(x) = Q(x)(x - r) \quad \text{or} \quad P(x) = Q(x)(x^2 + px + q)$$

for some polynomial $Q(x)$. According to our result in Example 54-3, $Q(x)$ will have real coefficients, so we can apply the same reasoning to factor it, and so on.

EXAMPLE 54-4 Factor the polynomial $P(x) = x^3 - x^2 + 2$ into linear and quadratic factors with real coefficients.

Solution The zeros of this polynomial (see Example 54-1) form the set $\{-1, 1 + i, 1 - i\}$. Therefore,

$$\begin{aligned} P(x) &= (x + 1)[x - (1 + i)][x - (1 - i)] \\ &= (x + 1)(x^2 - 2x + 2). \end{aligned}$$

EXAMPLE 54-5 Write the polynomial of Example 54-2 as a product of real polynomials of the first or second degree.

Solution In Example 54-2 we found that $\{i, -i, 2 + 3i, 2 - 3i\}$ is the set of zeros of $P(x)$. Since

$$(x - i)(x + i) = x^2 + 1,$$

and

$$[x - (2 + 3i)][x - (2 - 3i)] = x^2 - 4x + 13,$$

we may write

$$x^4 - 4x^3 + 14x^2 - 4x + 13 = (x^2 + 1)(x^2 - 4x + 13).$$

EXAMPLE 54-6 Write the polynomial $x^4 + 1$ as a product of real polynomials of the first or second degree.

Solution First, we solve the equation $x^4 + 1 = 0$, that is, we find the fourth roots of -1. Since $-1 = \cos \pi + i \sin \pi$, the methods of Section 46 give us the four roots

$$\begin{aligned} \{\cos (\tfrac{1}{4}\pi + \tfrac{1}{2}k\pi) &+ i \sin (\tfrac{1}{4}\pi + \tfrac{1}{2}k\pi) \mid k = 0, 1, 2, 3\} \\ &= \{\tfrac{1}{2}\sqrt{2}(1 + i), -\tfrac{1}{2}\sqrt{2}(1 + i), \tfrac{1}{2}\sqrt{2}(1 - i), -\tfrac{1}{2}\sqrt{2}(1 - i)\} \end{aligned}$$

Hence,

$$\begin{aligned} x^4 + 1 = [x - \tfrac{1}{2}\sqrt{2}(1 + i)][x - \tfrac{1}{2}\sqrt{2}(1 - i)][x + \tfrac{1}{2}\sqrt{2}(1 - i)] \\ \times [x + \tfrac{1}{2}\sqrt{2}(1 + i)]. \end{aligned}$$

After some multiplication, this last expression becomes

$$x^4 + 1 = (x^2 - \sqrt{2}x + 1)(x^2 + \sqrt{2}x + 1).$$

1. Find the polynomial $P(x)$ of lowest degree, with real coefficients, and with 1 as the coefficient of the term of highest degree, if the following sets are subsets of the extended set of zeros of $P(x)$.

(a) $\{3, 2i\}$ (b) $\{2, 1 - i\}$ (c) $\{i, 2i\}$

(d) $\{1 - i, 1 + i\}$ (e) $\{1, 1, i, i\}$ (f) $\{2 + i, 2 + i, 2\}$

2. Can you find polynomials with integer coefficients whose sets of zeros have the following sets as subsets?

(a) $\{\sqrt{2}\}$ (b) $\{2 - \sqrt{2}\}$ (c) $\{-i\sqrt{2}\}$ (d) $\{2i - \sqrt{2}\}$

3. Find all the zeros of $P(x)$ if one zero is the given number r.

(a) $P(x) = 2x^3 + x^2 + 2x + 1$, $r = i$

(b) $P(x) = x^4 - 4x^3 + 7x^2 - 6x + 2$, $r = 1 + i$

(c) $P(x) = x^4 - 3x^3 + 6x^2 - 2x - 12$, $r = 1 - \sqrt{5}i$

(d) $P(x) = x^4 + x^3 + x^2 + x$, $r = i$

(e) $P(x) = x^3 - (1 + i)x^2 - (2 - i)x + 2i$, $r = i$

(f) $P(x) = x^4 + 4x^3 + 2x^2 - 4x - 3$, $r = \sqrt{i}$

4. Write $P(x)$ as a product of polynomials of the first and second degree with real coefficients.

(a) $x^3 + 8$ (b) $x^4 - 16$

(c) $x^3 + x^2 - 2$ (d) $x^4 + 16$

5. Let $a + bi$ be a zero of $P(x) = x^3 + px^2 + qx + r$, where p, q, and r are real numbers. Show by actual substitution that $a - bi$ is also a zero of $P(x)$.

6. Use the methods described in Section 46 to calculate the five 5th roots of 1, and hence solve the equation $x^5 - 1 = 0$. Verify that each of the roots is the complex conjugate of a root.

7. Prove that every cubic polynomial with real coefficients has at least one real zero.

8. Find a polynomial whose zeros are the negatives of the zeros of the polynomial

$$\sqrt{3}x^{17} + 1492x^9 - 3x^5 + \pi x^2 - 2^{-1/3}.$$

9. Suppose that $P(x)$ is a polynomial with complex coefficients and $\overline{P}(x)$ is the polynomial whose coefficients are the conjugates of $P(x)$. What is the relation between the sets of zeros of $P(x)$ and $\overline{P}(x)$? Can you convince yourself that the product $P(x)\overline{P}(x)$ is a polynomial with real coefficients? What about the sum $P(x) + \overline{P}(x)$? Discuss the set of zeros of the product $P(x)\overline{P}(x)$.

10. Suppose that $A(x)$ and $B(x)$ ($\neq 0$) are polynomials with real coefficients and that $Q(x)$ and $R(x)$ are the quotient and remainder that we obtain when we divide $A(x)$ by $B(x)$. Show that $Q(x)$ and $R(x)$ have real coefficients.

55

**THE
RATIONAL
ZEROS OF
POLYNOMIALS
WHOSE
COEFFICIENTS
ARE
INTEGERS**

In this section we will consider polynomials of the form $P(x) = a_n x^n + \cdots + a_0$ whose coefficients $a_0, a_1, \ldots, a_n$ are *integers*. Such polynomials, of course, have zeros, but they may or may not have zeros that are rational numbers. For example, $x^2 - 4$ has rational zeros, but $x^2 - 3$ doesn't. We will learn how to determine whether or not a polynomial with integral coefficients has rational zeros and, if so, how to find them.

We can illustrate the basic line of our argument by considering the polynomial $x^2 - 3$. If this polynomial had the rational number b/c as a zero, then it would be true that $b^2 = 3c^2$. This equation tells us that every factor of c must be a factor of b. Since we might as well assume at the start that b and c have no prime factors in common (b/c is in "lowest terms"), it follows that $c = 1$ or $c = -1$. The equation $b^2 = 3c^2$ also tells us that every factor of b is a factor of 3 or of c. Since we are ruling out the latter possibility, it follows that b divides 3. Hence, b is one of the members of the set $\{1, -1, 3, -3\}$, and so the *only possible* rational zeros of the polynomial $x^2 - 3$ are the numbers -1, 1, -3, and 3. Now we simply test these possibilities; the inequalities $(-1)^2 - 3 \neq 0$, $1^2 - 3 \neq 0$, $(-3)^2 - 3 \neq 0$, and $3^2 - 3 \neq 0$ show that they are *not* zeros. With the only possible rational zeros ruled out, we are left with the conclusion that the polynomial $x^2 - 3$ does not have *any* rational zeros.

Before taking up the problem for a general polynomial with integer coefficients, we should emphasize one point about integers. The fraction that represents a given rational number can always be reduced to lowest terms; that is, a rational number r can always be written as a quotient b/c, where b and c are integers with no common factors. We then say that b/c is the **reduced form** of r.

EXAMPLE 55-1   Suppose that the quotient b/c is the reduced form of a certain rational number. Further, suppose that there is an integer a and a positive integer n such that c is a factor of ab^n. What is the relation between the numbers c and a?

Solution   To get a clearer idea of what we have in mind here, let us start with a numerical example. Suppose that $b = 35$ and $c = 6$. Since $35 = 5 \cdot 7$, and $6 = 2 \cdot 3$, we see that the quotient $35/6$ is in reduced form. Now suppose that a is an integer and n is a positive integer such that 6 is a factor of $a \cdot 35^n$. Thus there is an integer q such that

$$a \cdot 5^n \cdot 7^n = 2 \cdot 3 \cdot q.$$

From this equation it is plain that 2 and 3 must be factors of a; that is, 6 is a factor of a.

The same result is true in general. Every prime factor of c is a factor of ab^n. Because it is not a factor of b (and hence not a factor of b^n), a prime factor of c must therefore be a factor of a. It follows that c itself is a factor of a.

We are now ready to prove the basic theorem for dealing with the rational zeros of a polynomial with integral coefficients.

Theorem 55-1 *If r is a rational zero of the polynomial $P(x) = a_n x^n + \cdots + a_0$, where the coefficients $a_0, \ldots, a_n$ are integers, and if b/c is the reduced form of r, then b is a factor of a_0 and c is a factor of a_n.*

Proof Since $r = b/c$ is a zero of $P(x)$,

(55-1)
$$a_n \left(\frac{b}{c}\right)^n + a_{n-1} \left(\frac{b}{c}\right)^{n-1} + \cdots + a_1 \left(\frac{b}{c}\right) + a_0 = 0.$$

We multiply both sides of this equation by c^n to obtain the equation

(55-2)
$$a_n b^n + a_{n-1} b^{n-1} c + \cdots + a_1 b c^{n-1} + a_0 c^n = 0,$$

which we write as

(55-3)
$$a_n b^n = (-a_{n-1} b^{n-1} - \cdots - a_0 c^{n-1})c.$$

From Equation 55-3 we see that c is a factor of $a_n b^n$. Since b/c is the reduced form of r, it follows from Example 55-1 that c is a factor of a_n.

We can also write Equation 55-2 as

(55-4)
$$a_0 c^n = (-a_n b^{n-1} - \cdots - a_1 c^{n-1})b.$$

This equation shows that b is a factor of $a_0 c^n$. We again turn to Example 55-1 (with the roles of b and c interchanged) to see that b is a factor of a_0.

It is important to note that Theorem 55-1 does *not* say that $P(x)$ has any rational zeros at all. It merely says that if it does, they must be of a certain form.

EXAMPLE 55-2 Find the rational zeros (if any) of the polynomial $P(x) = 2x^3 + 3x^2 + 2x + 3$.

Solution If $P(x)$ has a rational zero of reduced form b/c, then b must be a factor of 3 and c must be a factor of 2. The factors of 3 form the set $\{1, -1, 3, -3\}$, and $\{1, -1, 2, -2\}$ is the set of factors of 2. We therefore obtain the set of all *possible* rational zeros of $P(x)$ by forming all possible quotients of the members of the first set by members of the second. In this way we find that the set of rational zeros of $P(x)$ must be a subset of the set $\{1, -1, 3, -3, \frac{1}{2}, -\frac{1}{2}, \frac{3}{2}, -\frac{3}{2}\}$. Obviously, not all the eight members of this set can be zeros of $P(x)$ [since $P(x)$ has at most three zeros] and perhaps none of them is. To see which of them (if any) are zeros, we must calculate $P(1)$, $P(-1)$, and so on. We can calculate these numbers directly or we can use synthetic division, in which case the numbers will appear as remainders. Either method will reveal that $P(-\frac{3}{2}) = 0$, and that none of the other seven possible rational zeros is a zero.

EXAMPLE 55-3 Show that the polynomial $P(x) = x^3 + 3x - 5$ has no rational zeros.

Solution　If $r = b/c$ is a rational zero of $P(x)$, then b must be a factor of 5 and c must be a factor of 1. The possible choices for b are 1, -1, 5, and -5; for c, only 1 or -1. Therefore, the possible choices for r are the members of the set $\{1, -1, 5, -5\}$. But when we check these numbers, we find that none is a zero. Hence, $P(x)$ has no rational zeros. Of course, $P(x)$ has zeros, but they are not rational and may not be real. Incidentally, we can show that $P(x)$ must in fact have one irrational real zero (see Problem 54-7).

EXAMPLE 55-4　Find all the zeros of the polynomial of Example 55-2.

Solution　In Example 55-2 we saw that $-\frac{3}{2}$ is a zero of $P(x)$, so $x + \frac{3}{2}$ is a factor of $P(x)$. We find by dividing that $P(x) = 2(x^2 + 1)(x + \frac{3}{2})$. The other two zeros of $P(x)$ are the zeros of the quadratic polynomial $x^2 + 1$; namely, i and $-i$.

Although Theorem 55-1 pertains specifically to polynomials whose coefficients are integers, it may also be used to investigate the rational zeros of polynomials with rational coefficients. If we multiply a polynomial $P(x)$ by a non-zero number k, it is clear that $kP(x)$ is again a polynomial. In factored form,

(55-5)
$$kP(x) = ka_n(x - r_1) \cdots (x - r_n).$$

From Equation 55-5 it is obvious that the zeros of $kP(x)$ are the same as the zeros of $P(x)$. Now if the coefficients of $P(x)$ are rational, a proper choice of k (the product of all the denominators of the coefficients of $P(x)$, for example) will insure that $kP(x)$ has integral coefficients. We can therefore study the rational zeros of $P(x)$ by investigating the rational zeros of $kP(x)$.

EXAMPLE 55-5　Find all the rational zeros of $P(x) = \frac{2}{3}x^3 - \frac{1}{2}x^2 + \frac{2}{3}x - \frac{1}{2}$.

Solution　If we multiply $P(x)$ by 6, we get the equation

$$6P(x) = 4x^3 - 3x^2 + 4x - 3.$$

The rational zeros (if any) of this polynomial are members of the set $\{1, -1, 3, -3, \frac{1}{2}, -\frac{1}{2}, \frac{3}{2}, -\frac{3}{2}, \frac{1}{4}, -\frac{1}{4}, \frac{3}{4}, -\frac{3}{4}\}$. We find by checking that $\frac{3}{4}$ is the only rational zero of $6P(x)$; hence, $\frac{3}{4}$ is the only rational zero of $P(x)$.

PROBLEMS 55

1. Find all the rational zeros of $P(x)$ or show that none exist.
   (a) $P(x) = 2x^5 - x^4 - 2x + 1$　　(b) $P(x) = 2x^6 + 3x^3 + x$
   (c) $P(x) = 12x^4 + 4x^3 - 3x^2 - x$　(d) $P(x) = 3x^3 + 8x^2 + 19x + 10$
   (e) $P(x) = x^4 + x^2 + 2x + 6$　　　(f) $P(x) = 9x^4 - 9x^3 + 5x^2 + 4x - 4$

2. Find all the zeros of $P(x)$.
   (a) $P(x) = x^3 - 3x - 2$　　　　　(b) $P(x) = x^3 - 4x^2 - 5x + 14$
   (c) $P(x) = x^4 - 6x^3 - 3x^2 - 24x - 28$　(d) $P(x) = x^4 + 3x^3 + 4x^2 - 8$
   (e) $P(x) = 48x^4 - 52x^3 + 13x - 3$　　(f) $P(x) = 12x^4 + 7x^3 + 7x - 12$

3. Find all the rational zeros of $P(x)$ or show that none exist.
    (a) $P(x) = \frac{2}{3}x^3 + \frac{17}{6}x^2 - \frac{55}{6}x + 5$ (b) $P(x) = x^3 + \frac{3}{2}x^2 - \frac{2}{3}x + 1$
    (c) $P(x) = x^3 + \frac{1}{6}x^2 - 2x + \frac{5}{6}$ (d) $P(x) = x^3 + \frac{3}{2}x^2 + x + \frac{3}{2}$

4. To find the logarithms of the rational zeros of $P(x)$, you only need the logarithms of what sets of positive integers (at most)?
    (a) $P(x) = 6x^5 - 8x^2 - 12$ (b) $P(x) = 35x^{17} - 12x^3 + 24$
    (c) $P(x) = 35x^{17} - 12x^3 + 24x$ (d) $P(x) = 27x^{19} - 8x^3 + 1984$

5. If $P(x) = x^n + a_{n-1}x^{n-1} + \cdots + a_1x + a_0$ has *integer* coefficients, show that its only possible rational zeros are also integers.

6. Solve the equation $2 \sin^3 x + \sin^2 x - 13 \sin x + 6 = 0$.

7. Can the polynomial $\sqrt{2}x^{17} - 23x^8 + 3\sqrt{2}$ have any rational zeros?

8. A polynomial of the form $x^n + a_{n-1}x^{n-1} + \cdots + a_0$ can have rational coefficients but irrational zeros. If such a polynomial has only rational zeros, can it have irrational coefficients?

9. Find a set that certainly contains the rational zeros of the polynomial $15x^{31} - 36ix^{24} - 21x^5 + (2 - 3i)x + 42 + 30i$.

10. In Problem 54-7 we asked you to show that a cubic polynomial with real coefficients must have at least one real zero. Can you find an example of a cubic polynomial with integer coefficients that has no rational zero?

56
FINDING
REAL
ZEROS
OF
POLYNOMIALS
WITH REAL
COEFFICIENTS

Suppose that $P(x) = ax^n + \cdots + a_0$ is a polynomial of degree n with real coefficients. The Fundamental Theorem of Algebra tells us that $P(x)$ has at least one zero, and it can have as many as n different zeros. There are ingenious and complicated methods for finding the zeros of such a polynomial, but we have time and tools only to make a few simple observations about this problem. Furthermore, we will stick to the question of finding the *real* zeros of $P(x)$.

This problem can best be described graphically. We are looking for a number that is the X-coordinate of a point in which the graph of the polynomial intersects the X-axis. Let us consider the polynomial $P(x) = x^3 + x - 1$ as an example. From its graph in Fig- 56-1, it is obvious that $P(x)$ has a zero between 0 and 1. Suppose that we examine this "obvious" statement a little more closely. In drawing the graph of $P(x)$ we plotted certain points and joined them smoothly "by eye." Two of the plotted points are $(0, P(0)) = (0, -1)$ and $(1, P(1)) = (1, 1)$. These points lie on opposite sides of the X-axis, so it is only natural to suppose that the graph of $P(x)$ must intersect the X-axis in some point between 0 and 1. This statement is correct, but its proof belongs to higher mathematics. In a course in calculus you may see a proof of the following theorem.

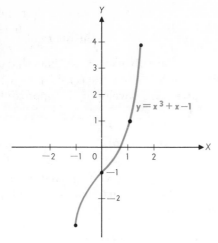

FIGURE 56-1

Theorem 56-1   *Let $P(x)$ be a polynomial with real coefficients, and suppose that a and b are two numbers such that $a < b$. If the two numbers $P(a)$ and $P(b)$ have opposite signs, then there is a number r between a and b such that $P(r) = 0$.*

It is this theorem that allows us to assert that since $P(0) = -1$ and $P(1) = 1$, the polynomial $P(x) = x^3 + x - 1$ has a zero between 0 and 1. Figure 56-1 suggests that this zero is slightly greater than $\frac{1}{2}$, and if we do a little calculating, we find that $P(.6) = -.184$ and $P(.7) = .043$. Since these two numbers have opposite signs, Theorem 56-1 tells us that our zero is a number between .6 and .7, and so on.

We can also use linear interpolation to approximate a zero of a polynomial. In Fig. 56-2 we have shown an arc of the graph of a polynomial $P(x)$. The numbers $P(a)$ and $P(b)$ have opposite signs, and r is a zero of $P(x)$ that lies between a and b. As an approximation to r we may take the number c, the X-coordinate of the point in which the line segment joining $(a, P(a))$ to

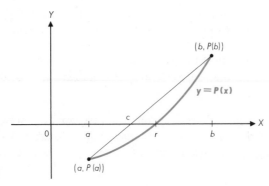

FIGURE 56-2

$(b, P(b))$ intersects the X-axis. To find c, we observe that the slope of this segment is the quotient $\dfrac{P(b) - P(a)}{b - a}$ and also the quotient $\dfrac{0 - P(a)}{c - a}$. When we equate these two numbers and solve for c, we find that $c = (aP(b) - bP(a))/(P(b) - P(a))$. Since we are going to consider this number as an approximation to the zero r, we have developed the approximation formula

$$(56\text{-}1) \qquad\qquad r \approx \frac{aP(b) - bP(a)}{P(b) - P(a)}.$$

For example, if $P(x) = x^3 + x - 1$, we have seen that $P(.6) = -.184$ and $P(.7) = .043$. Hence Formula 56-1 tells us that a zero of $P(x)$ that lies between .6 and .7 is approximately

$$\frac{.6(.043) - .7(-.184)}{.043 - (-.184)} = .68.$$

Now that we have the approximation .68 to a zero of our given polynomial, we could calculate the numbers $P(.681)$, $P(.682)$, $P(.679)$, and so on until we find two with opposite signs and then apply Formula 56-1 again. This technique is simple to describe, but is clearly tedious to carry out if our only calculating equipment is pencil and paper. Even if we have a desk calculator, it is slow work to get any kind of accuracy. But modern electronic computers work so fast that it took one of them less than 1 second to use this relatively inefficient method to find that (to 6 decimal places) the zero of $P(x)$ that we are talking about is .682328.

EXAMPLE 56-1 Discuss the real zeros of the polynomial $x^4 - 2x^3 - 3x^2 + 4x + 2$.

Solution A little calculating (we might use the Remainder Theorem and synthetic division) gives us the following table.

x	-2	-1	0	1	2	3
$P(x)$	14	-2	2	2	-2	14

From the table we see that $P(x)$ has zeros between -2 and -1, between -1 and 0, between 1 and 2, and between 2 and 3. Since $P(x)$ is a polynomial of degree 4, it has at most four zeros. Thus, there is exactly one zero between each of the pairs of integers listed. We can use Formula 56-1 to find that these zeros are approximated by the members of the set $\{-\frac{9}{8}, -\frac{1}{2}, \frac{3}{2}, \frac{17}{8}\}$.

Of course, we needn't restrict our use of graphical methods to solving polynomial equations. They (in particular, Approximation 56-1) are equally useful for solving an equation $f(x) = 0$, where f is not necessarily a polynomial function.

1. Bracket the real zeros of $P(x)$ between consecutive integers.
   (a) $P(x) = x^5 + 5x^4 - 3x^3 - 29x^2 + 2x + 23$
   (b) $P(x) = x^4 - 20x^2 - 21x - 22$
   (c) $P(x) = x^4 - x^2 - 4x + 4$
   (d) $P(x) = x^4 - 11x - 50$
   (e) $P(x) = x^4 + 5x^3 - 27x^2 - 21x - 14$
   (f) $P(x) = x^3 - 13x^2 + 7x - 1$

2. Find the real zero of $P(x)$ that lies between 1 and 2, correct to one decimal place.
   (a) $P(x) = 2x^3 - 5x^2 - x + 5$ (b) $P(x) = 3x^3 - x^2 - 8x - 2$
   (c) $P(x) = x^4 - 2x^2 - 1$ (d) $P(x) = x^5 + x^4 + x^3 - 2x^2 - 2x - 2$

3. Find the real zero of $P(x)$ that lies between 0 and 1, correct to two decimal places.
   (a) $P(x) = x^3 + 2x^2 - x - 1$ (b) $P(x) = x^3 + 2x^2 - 1$
   (c) $P(x) = x^4 - 5x^3 + 1$ (d) $P(x) = 2x^5 + 2x^3 - x^2 - 1$

4. Show that the polynomial $P(x) = x^3 - 3x^2 - 4x + 13$ has two real zeros between 2 and 3, even though $P(2)$ and $P(3)$ have the same algebraic sign. Can you convince yourself that if $P(a)P(b) > 0$, then the extended set of zeros of $P(x)$ has an even number of members between a and b, while the inequality $P(a)P(b) < 0$ implies that it has an odd number?

5. (a) How do you know that the polynomial $x^{12} + \pi x^9 + \sqrt{3}x^4 + 1492$ has no positive zeros?
   (b) How do you know the polynomial $x^{12} - \pi x^9 + \sqrt{3}x^4 + 1492$ has no negative zeros?

6. (a) Show that the real zeros (if there are any) of the polynomial $P(x) = x^{12} + 8x^2 - 8$ must lie in the interval $(-1, 1)$. [*Hint:* If $P(r) = 0$, then $r^2 - 1 < 0$.]
   (b) Can you convince yourself that $P(x)$ has at least two real zeros?
   (c) Can you convince yourself that if r and s satisfy the inequalities $0 < r < s$, then $P(r) < P(s)$? How does this inequality tell us that $P(x)$ can have at most one positive zero?
   (d) How do we know [from part (c)] that $P(x)$ has at most two real zeros?
   (e) How many non-real complex numbers are in the extended set of zeros of $P(x)$?
   (f) Can you convince yourself that at least two of these numbers have zero real part?
   (g) If you were told that $a + bi \in \{r \mid P(r) = 0\}$ and $ab \neq 0$, explain how you would automatically know three other members of the set of zeros of $P(x)$?

7. Show that the equation $\log x - x + 3 = 0$ has a solution that lies between 3 and 4. Does it have any other solutions?

8. There is a zero of the function that is defined by the equation $f(x) = \sin x - \log x$ that lies between 2 and 3. Find a two-place approximation to this zero.

9. The dimensions of a rectangular box are 1 foot, 2 feet, and 4 feet. In order to double the volume of the box, each edge is to be increased by the same amount. How much should each edge be increased?

10. An open box containing 20 cubic inches is to be made from a sheet of tin 6 inches by 10 inches by cutting equal squares out of the corners and turning up the edges. How large a square should be cut out?

11. Show that Formula 56-1 can be written as

$$r \approx \frac{a|P(b)| + b|P(a)|}{|P(b)| + |P(a)|}.$$

12. If the product of all the members of the extended set of zeros of the polynomial $P(x)$ with real coefficients is negative, the polynomial must have at least one real zero. See if you can concoct a simple graphical argument to convince yourself of this fact.

REVIEW PROBLEMS, CHAPTER SEVEN

You should be able to answer the following questions without referring back to the text.

1. If $A(x)$ and $B(x)$ are polynomials such that $A(x)B(x) = 1$, what can you conclude about $A(x)$ and $B(x)$?

2. Let k be a positive integer. Determine the values of k for which
   (a) $x + a$ is a factor of $x^k + a^k$.   (b) $x - ia$ is a factor of $x^k + a^k$.

3. For what value of c will the polynomials $2x^2 - x + c$ and $4x^2 + 4x + 1$ have a common factor?

4. Solve for x.
   (a) $\dfrac{1-c}{cx} = \dfrac{3cr+1}{1-x} + \dfrac{cr-1}{x-x^2}$ (b) $\dfrac{1}{kx^2+k} = \dfrac{k}{2k^2x+1}$
   (c) $\sqrt{x-2} = \sqrt{2x+5} + 3$ (d) $\sqrt{x-2} = \sqrt{2x+5} - 3$

5. If an airplane has enough fuel to fly for 16 hours at 400 miles per hour, how far can it travel from its base if it flies directly into a 50 mile per hour wind and returns with a 50 mile per hour tail wind?

6. Solve for x: $(1 - a^2)(x + a) - 2a(1 - x^2) = 0$.

7. Find the value of k such that k is the remainder after the polynomial $x^3 - kx^2 - 14x + 15k$ is divided by $x - 5$.

8. What is $P(x)$ if its extended set of zeros is the same as the extended set of zeros of the polynomial $2x^5 - 7x^2 + 3x + 4$ and $P(1) = 3$?

9. If 2 is a zero of the polynomial $x^3 + ax + b$, what quadratic equation must the other two zeros satisfy?

10. Solve for x: $(x + 2a)^3 + (x - a)^3 = (2x + a)^3$.

11. The zeros of the polynomial $x^2 + bx + c$ are the members of the set $\{-c + \sqrt{bc}, -c - \sqrt{bc}\}$. Find b and c if $c \neq 0$.

252

12. Let $P(x) = 2x^3 + x^2 + ax + b$. If a and b are real numbers and one zero of $P(x)$ is $1 + i$, find all the zeros of $P(x)$.

13. Show that
$$R^1 \cap \{(3 - 2i)x^{29} - (2 + i)x^8 + \pi + \sqrt{2}i = 0\}$$
$$= R^1 \cap \{3x^{29} - 2x^8 + \pi = 0\} \cap \{2x^{29} + x^8 - \sqrt{2} = 0\}.$$

14. The square of twice a number is larger than the square of the sum of the number and 1. Which numbers possess this property?

15. Let r_1 and r_2 be the zeros of the polynomial $ax^2 + bx + c$. Find the following numbers.

(a) $\dfrac{1}{r_1} + \dfrac{1}{r_2}$ (b) $r_1{}^2 + r_2{}^2$ (c) $r_1{}^3 + r_2{}^3$ (d) $|r_1 - r_2|$

MISCELLANEOUS PROBLEMS, CHAPTER SEVEN

The following exercises are designed to test your ability to apply your knowledge of the theory of equations to somewhat more difficult problems.

1. Solve the equation $4^{3x} - 2^{3x+1} + 1 = 0$.

2. Let $Q(x) = (x + 1)(x^2 + 1)(x^4 + 1)(x^8 + 1)(x^{16} + 1)$. Show that
$$(x - 1)Q(x) = x^{32} - 1.$$

3. Let $P(x)$ be a polynomial of degree n with real coefficients. Prove that no line can intersect the graph of $P(x)$ more than n times.

4. For what values of k will the difference between the zeros of $5x^2 + 4x + k$ be equal to the sum of the squares of the zeros?

5. Solve the inequality $\dfrac{6x + 7}{x} > x$.

6. For what real values of k are the zeros of the polynomial $x^2 - 2kx - 4k + 21$ real numbers?

7. A rectangular strip of carpet 3 feet wide is laid diagonally across the floor of a room 9 feet by 12 feet so that each of the four corners of the strip touches a wall. How long is the strip?

8. Let $P(x) = x^n + \cdots + a_1 x + a_0$ be a polynomial with integral coefficients such that $P(0)$ and $P(1)$ are odd integers. Show that $P(x)$ does not have any rational zeros.

9. If $P(x)$ is a polynomial and a is a number, then according to Theorems 52-1 and 52-2 we can write $P(x) = Q(x)(x - a) + P(a)$. If r is a zero of $P(x)$, we therefore have $0 = Q(r)(r - a) + P(a)$, and hence $r = a - P(a)/Q(r)$. If a is fairly close to r, we might expect that $Q(r)$ is fairly close to $Q(a)$, and so we have the approximation formula
$$r \approx a - \frac{P(a)}{Q(a)}.$$

(a) Apply this formula with $P(x) = x^3 + x - 1$ and $a = .7$ to obtain the approximation $.683$, and compare this result with the approximation we obtained in Section 56.

(b) If b is a positive number, its square root is the positive zero of the polynomial $P(x) = x^2 - b$. If a is an approximation to this square root, show that the approximation formula of this question gives the further approximation

$$\sqrt{b} \approx \frac{1}{2}\left(a + \frac{b}{a}\right).$$

Use this formula twice, starting with $a = 3$, to approximate $\sqrt{10}$, and use logarithms to see how accurate your result is.

10. If $y = ax^2 + bx + c$, then by completing the square we may write

$$y = a\left(x + \frac{b}{2a}\right)^2 + c - \frac{b^2}{4a}.$$

(a) From this equation deduce that if a, b, and c are real numbers, then

$$y \le c - \frac{b^2}{4a} \text{ if } a < 0, \text{ and } y \ge c - \frac{b^2}{4a} \text{ if } a > 0.$$

(b) A man has 100 ft of wire and wishes to fence in a rectangular area by driving 2 stakes in a line parallel to a stone wall and putting his wire along 3 sides of the enclosure so that the wall forms the fourth side. Where should he drive his stakes in order to enclose the maximum area?

11. (a) If $\{r_1, r_2, r_3\}$ is the extended set of zeros of the polynomial $\pi x^3 + 2ix^2 - \sqrt{19}x + 17$, what is the number $r_1r_2 + r_1r_3 + r_2r_3$?

(b) If $\{r_1, \ldots, r_n\}$ is the extended set of zeros of the polynomial $P(x) = a_nx^{n-1} + \cdots + a_0$, can you express the number

$$r_1r_2 + \cdots + r_1r_n + r_2r_3 + \cdots + r_{n-1}r_n$$

(the sum of all possible products r_ir_j, where $i < j$) in terms of the coefficients of $P(x)$?

12. In certain applications of mathematics we meet **stable** polynomials, polynomials with real coefficients all of whose zeros have negative real parts. How are the (real) numbers a, b, and c related if $ax^2 + bx + c$ is a stable polynomial?

13. Suppose that $P(x) = a_nx^n + a_{n-1}x^{n-1} + \cdots + a_0$.

(a) Find a polynomial whose extended set of zeros consists of the reciprocals of the members of the extended set of zeros of $P(x)$.

(b) Find a polynomial whose extended set of zeros consists of the negatives of the members of the extended set of zeros of $P(x)$.

(c) Find a polynomial whose extended set of zeros is obtained from the extended set of zeros of $P(x)$ by adding r to each member.

(d) Find a polynomial whose extended set of zeros is obtained from the extended set of zeros of $P(x)$ by multiplying each member by r.

Systems
of
Equations

eight

Suppose we know that the area of a rectangular file card is 24 square inches and its perimeter is 20 inches. How can we determine its dimensions? In this problem we are given *two* pieces of information, and we wish to find *two* quantities, length and width. One way to solve it is to denote the length of one side of the card by x and the length of the other side by y and use the information we have to write two equations in x and y:

$$xy = 24$$
$$2x + 2y = 20.$$

The solutions of this *system of equations* are the pairs $(x, y) = (6, 4)$ and $(x, y) = (4, 6)$, so the dimensions of our card are 4 inches by 6 inches.

This type of problem, in which we have a number of unknown quantities and a given set of relations among these unknowns, occurs frequently in the applications of mathematics. In this chapter we will discuss methods by which we can find the unknowns, concentrating on cases in which the number of unknowns is the same as the number of equations.

A **solution** of a system of equations consists of an array of numbers $(x_1, y_1, \ldots)$ that satisfy all the equations of the system. For example, you can readily verify by substitution that the number pair $(x, y) = (1, -1)$ is a solution of the system

(57-1)
$$x - 2y = 3$$
$$2x^2 + 3y^2 = 5.$$

How do we find this solution, and does this system have other solutions?

The most direct way to treat our system is to solve one of the equations for one unknown in terms of the other and substitute the result in the remaining equation, thus reducing the problem to one equation in one unknown. We now solve this equation in one unknown and work our way back to a solution of the original system. Thus, the equation $x - 2y = 3$ yields $x = 2y + 3$, and when we substitute this expression for x in the second equation, we get the equation $2(2y + 3)^2 + 3y^2 = 5$. This quadratic equation simplifies to $11y^2 + 24y + 13 = 0$, or, in factored form, $(y + 1)(11y + 13) = 0$. Its solution set is $\{-1, -\frac{13}{11}\}$, and we find x from the equation $x = 2y + 3$. Thus, if $y = -1$, we have $x = 2(-1) + 3 = 1$, and if $y = -\frac{13}{11}$, we have $x = \frac{7}{11}$. The pairs $(x, y) = (1, -1)$ and $(x, y) = (\frac{7}{11}, -\frac{13}{11})$ are therefore the solutions of our given System 57-1.

EXAMPLE 57-1   Solve the system of equations

(57-2)
$$4y + 2\cos x = -3$$
$$2y + 6\cos x = 1.$$

Solution   The second equation yields $y = \frac{1}{2} - 3\cos x$, which we substitute in the first:

$$4(\tfrac{1}{2} - 3\cos x) + 2\cos x = -3.$$

This equation in one unknown immediately simplifies to

$$\cos x = \tfrac{1}{2},$$

and hence $x \in \{\frac{1}{3}\pi + 2k\pi\} \cup \{-\frac{1}{3}\pi + 2k\pi\}$. Since $y = \frac{1}{2} - 3\cos x$, and $\cos x = \frac{1}{2}$, we have $y = \frac{1}{2} - \frac{3}{2} = -1$. Finally, then, the solution set of System 57-2 is the set of pairs of numbers $\{(\frac{1}{3}\pi + 2k\pi, -1)\} \cup \{(-\frac{1}{3}\pi + 2k\pi, -1)\}$. Some number pairs belonging to this set are $(\frac{1}{3}\pi, -1)$, $(-\frac{1}{3}\pi, -1)$, $(\frac{5}{3}\pi, -1)$, and so on.

In theory, we may apply the procedure we have just outlined to a system of n equations in n unknowns, where n is any positive integer (for example, to a system of five equations in five unknowns). The first equation (or any one for that matter) is used to express one of the unknowns in terms of the other unknowns. This expression is then substituted in the remaining equations, leaving us with a new system that has one less equation and one less unknown. We repeat the process until just one equation in one unknown remains. Then we solve this last equation and work our way back to find the other unknowns.

As with equations in one unknown, we have a natural definition of equivalence. Two systems of equations are **equivalent** if every solution of either one is also a solution of the other.

The systems

$$7x + 3y = 4$$
$$x + y = 0$$

and

$$x = 1$$
$$x + y = 0,$$

for example, are equivalent, since the only solution of each system is the number pair $(1, -1)$.

When faced with a system of equations, we naturally try to replace it with an equivalent system that is easier to solve. *The following operations transform a system of equations into an equivalent system.*

Operation 57-1. *Interchanging the position of two equations.*

Operation 57-2. *Replacing an equation with a non-zero multiple of itself.* (A *multiple of an equation* is the equation that results when we multiply both sides of the equation by the same number.)

Operation 57-3. *Replacing an equation with the equation that results when that equation is added to another equation of the system.* (*Adding two equations* means, of course, adding the corresponding sides of the equations.) For example, we may replace the second equation of a system of four equations with the sum of the second and third equations.

Operations 57-2 and 57-3 are frequently combined as follows.

Operation 57-4. *Replacing an equation with the sum of a non-zero multiple of itself and a multiple of another equation of the system.* For example, we may replace the third equation of a system of three equations with twice the third equation plus -5 times the first.

EXAMPLE 57-2 Replace the following system with an equivalent system:

(57-3)
$$x^2 + xy + y^2 = 2$$
$$x^2 + y^2 = 4.$$

Solution 1 We can replace the first equation with -1 times the second plus the first (Operation 57-4). This operation yields the equivalent system

(57-4)
$$xy = -2$$
$$x^2 + y^2 = 4.$$

Of course, what we have done is to "subtract the second equation from the first."

Solution 2 If we replace the first equation with twice the first equation minus the second, we get another system equivalent to System 57-3:

(57-5)
$$x^2 + 2xy + y^2 = 0$$
$$x^2 + y^2 = 4.$$

257

There are any number of systems that are equivalent to System 57-3. We are naturally interested in finding an equivalent system that is easy to solve. System 57-4, for example, *is* easy to solve. The first equation yields $y = -2/x$, which we substitute in the second to obtain $x^2 + 4/x^2 = 4$, an equation in one unknown. Now we multiply by x^2 and factor:

$$x^4 - 4x^2 + 4 = 0,$$
$$(x^2 - 2)^2 = 0,$$
$$x^2 = 2.$$

What we have shown so far is that if the pair (x, y) is a solution of System 57-4, then $x \in \{\sqrt{2}, -\sqrt{2}\}$. The equation $y = -2/x$ gives us y. If $x = \sqrt{2}$, then $y = -2/\sqrt{2} = -\sqrt{2}$. If $x = -\sqrt{2}$, then $y = -2/(-\sqrt{2}) = \sqrt{2}$. Thus, Systems 57-3, 57-4, and 57-5 each have the solution set

$$\{(\sqrt{2}, -\sqrt{2}), (-\sqrt{2}, \sqrt{2})\}.$$

PROBLEMS
57

1. Is the given number pair (x, y) a solution of the accompanying system?

(a) $2x + y = 1$ (b) $\sqrt{x} + y = 0$ (c) $\sqrt{x} - y^2 = 1$
$x^3 - y^2 = -1$ $x - y^2 = 0$ $\log x^y = 6$
$(2, -3)$ $(4, 2)$ $(100, 3)$

(d) $2 \sin x \cos y = -1$ (e) $8^x + 8^y = \frac{5}{2}$
$\cos^2 x + \sin^2 y = 1$ $xy^{-2} = 3$
$(\frac{3}{2}\pi, \frac{1}{3}\pi)$ $(\frac{1}{3}, -\frac{1}{3})$

(f) $\sin (\text{Cos}^{-1} x) + \cos (\text{Sin}^{-1} y) = \sqrt{3}$
$\cos (\text{Sin}^{-1} x) + \sin (\text{Cos}^{-1} y) = \sqrt{3}$
$(\frac{1}{2}, -\frac{1}{2})$

2. Solve one equation for one unknown and substitute in the other equation to find the solutions of the following systems.

(a) $xy = -2$ (b) $3y + ix = -5$
$2x + y = 0$ $iy - 3x = -7i$

(c) $y^2 - \log x = 2$ (d) $x - \sin y = 1$
$y - \log x = 0$ $x \sin y = 2$

(e) $x - y = 1$ (f) $x^3 - y^2 = 4x$
$x^2 + xy + y^2 = 7$ $y^2 = x(x - 4)$

3. Which of the following systems are equivalent to the system

$$x^2 + y^2 = 7$$
$$xy = 1?$$

(a) $2xy = 2$ (b) $(x + y)^2 = 9$
$x^2 + y^2 = 7$ $xy = 1$

(c) $(x - y)^2 = 5$ (d) $(x + y)^2 = 9$
$xy = 1$ $(x - y)^2 = 5$

(e) $x + y = 3$ (f) $|x + y| = 3$
$xy = 1$ $xy = 1$

4. Replace each of the following systems with an equivalent system you consider simpler.

(a) $x^2 - 2y^2 = -1$
$x^2 - 3y^2 = -2$

(b) $x^2 + xy - y^2 = 1$
$(x + iy)^2 = 1$

(c) $x^2 - xy + y^2 = 3$
$x^3 + y^3 = 6$

(d) $|x + y| = 3$
$|x - y| = 1$

5. The sum of the circumferences of two circles is 16π feet, and their combined area is 34π square feet. Find their radii.

6. Find values of a and b so that $x - 1$ and $x + 2$ are factors of the polynomial

$$x^4 - x^3 - x^2 + ax + b.$$

7. If we add 3 times the first equation of the system

$$x^2 y + xy^2 = -1$$
$$x^3 + y^3 = 3$$

to the second, we get an equivalent system. Solve either system.

8. Explain why the system of equations

$$\sin (x + y) = \log y$$
$$\cos 2^{xy} = 3x - 2y$$

is equivalent to the single equation

$$|\sin (x + y) - \log y| + |\cos 2^{xy} - 3x + 2y| = 0.$$

Is it always possible to replace a system of equations with a single equation?

9. Solve each of the following systems.

(a) $y = \sin x$
$x = \sin y$

(b) $y = \log x$
$x = \log y$

(c) $y = x^2$
$x = y^2$

(d) $y = \text{Tan}^{-1} x$
$x = \text{Tan}^{-1} y$

58
SYSTEMS OF LINEAR EQUATIONS. TRIANGULAR FORM

A **linear equation** in n unknowns $x_1, x_2, \ldots, x_n$ has the form

$$a_1 x_1 + a_2 x_2 + \cdots + a_n x_n = b,$$

where $a_1, a_2, \ldots, a_n$, and b are given numbers. For example, $3x - 4y = 5$ and $6x - 7y + iz - 3 = 0$ are linear equations. We shall devote the next few sections to a study of systems of linear equations.

The simplest systems of linear equations are those in **triangular form.** A system of linear equations in n unknowns $x_1, x_2, \ldots, x_n$ is in triangular form if the first equation involves only x_1; the second equation involves x_2 and perhaps x_1, but not $x_3, x_4, \ldots$, and x_n; the third equation involves x_3 and perhaps x_1 and x_2, but not $x_4, \ldots$, and x_n; and so on. The following system, for example, is in triangular form:

(58-1)
$$2x = 4$$
$$3x - 2y = 8$$
$$x + y - 4z = 1.$$

The first equation tells us that $x = 2$. Once we know that $x = 2$, it is clear from the second equation that $y = -1$, and then we can solve the third equation to find $z = 0$. Thus, the solution of System 58–1 is $(x, y, z) = (2, -1, 0)$.

Because it is easy to solve a system of equations in triangular form, we adopt the following method of solving a general system of linear equations. *By using Operations 57-1, 57-2, 57-3, and 57-4, we try to replace a given system of linear equations with an equivalent system in triangular form.* "Most" systems can be reduced to triangular form in this way, and, as we shall see later, our efforts will not be wasted if we attempt the reduction on a system that is not equivalent to a system in triangular form. We can describe the procedure by applying it to some typical examples.

EXAMPLE 58-1   Solve the system of equations

(58-2)
$$2x + 3y = 4$$
$$6x - 2y = -10.$$

Solution   We first multiply the second equation by $\frac{1}{2}$ to get the equivalent system

$$2x + 3y = 4$$
$$3x - y = -5.$$

Now we replace the first equation with the sum of the first equation and 3 times the second, and we have the equivalent triangular system

(58-3)
$$11x \quad\ \ = -11$$
$$3x - y = -5.$$

It is easy to see that the solution of this system (and hence of our original system) is $(x, y) = (-1, 2)$.

We can describe our reduction to a triangular system less formally. First, for simplicity, we "divide out" the common factor 2 in the second equation. Then we "combine" the two equations in such a way as to produce a new first equation that does not contain y.

EXAMPLE 58-2   Solve the system

$$3x - 2y = 7$$
$$2x - iy = 6i.$$

Solution   By replacing the first equation with the sum of -2 times the second and i times the first, we get the equivalent triangular system

$$(-4 + 3i)x \quad\quad = -5i$$
$$2x - iy = 6i.$$

Therefore,

$$x = \frac{-5i}{-4 + 3i} = \frac{5i(4 + 3i)}{25} = -\frac{3}{5} + \frac{4}{5}i,$$

and

$$y = \frac{2x - 6i}{i} = -\frac{22}{5} + \frac{6}{5}i.$$

EXAMPLE 58-3   Solve the system of equations

(58-4)
$$\begin{aligned} \tfrac{1}{2}x - \tfrac{2}{3}y + z &= \tfrac{8}{3} \\ 2x - y + z &= 6 \\ 3x + 2y &= 4. \end{aligned}$$

Solution   (*Note.* It takes longer to describe the solution process than it does to carry it out. You are warned, however, against trying to take too many steps at once—it is all too easy to make little errors when solving linear systems.) To simplify things, we multiply the first equation by 6 and change the order of the equations, and we obtain the system

(58-5)
$$\begin{aligned} 3x + 2y &= 4 \\ 3x - 4y + 6z &= 16 \\ 2x - y + z &= 6. \end{aligned}$$

Now we multiply the third equation by 6 and subtract it from the second to obtain a system in which z is missing from the second equation:

(58-6)
$$\begin{aligned} 3x + 2y &= 4 \\ -9x + 2y &= -20 \\ 2x - y + z &= 6. \end{aligned}$$

Next we get an equivalent triangular system by subtracting the second of Equations 58-6 from the first:

(58-7)
$$\begin{aligned} 12x &= 24 \\ -9x + 2y &= -20 \\ 2x - y + z &= 6. \end{aligned}$$

System 58-7 easily yields the solution $(x, y, z) = (2, -1, 1)$. Needless to say, it is wise to check the final results in the original system.

EXAMPLE 58-4   Solve the system

(58-8)
$$\begin{aligned} x - 6y + 2z &= 5 \\ 2x - 3y + z &= 4 \\ 3x + 4y - z &= -2. \end{aligned}$$

Solution   When we add the third equation to the second, and twice the third equation to the first, we get the equivalent system

(58-9)
$$\begin{aligned} 7x + 2y &= 1 \\ 5x + y &= 2 \\ 3x + 4y - z &= -2. \end{aligned}$$

261

Next we subtract twice the second of Equations 58–9 from the first to get

$$
\begin{aligned}
-3x && &= -3 \\
5x + y && &= 2 \\
3x + 4y - z &= -2.
\end{aligned}
$$

The solution then comes easily: $(x, y, z) = (1, -3, -7)$. Notice that the last equation was used to eliminate one of the unknowns from all the other equations, then the next to the last equation was used to eliminate another unknown from the equation above it. Although this procedure stops after two steps in our example, it is clear that it is a direct method that can be used to reduce a system consisting of more linear equations in more unknowns to triangular form.

All the operations involved in solving a system of linear equations are the simplest possible—multiplication, division, addition, and subtraction. But there are a lot of operations involved. Because these systems have great practical importance, many schemes have been, and are still being, devised to solve them in the most efficient way possible. Some of these methods are especially suited for the modern electronic computers. Each method has its advantages and its drawbacks. All involve large amounts of computation. As general methods go, the practice of reducing the system to triangular form is probably as efficient as any, and in the next section we introduce some notation that simplifies this method.

PROBLEMS
58

1. Find (x, y) by first finding an equivalent triangular system.
   (a) $2x + 4y = 0$
   $3x - 2y = -8$
   (b) $x - 5y = 3$
   $2x + y = 6$
   (c) $5x - 4y = 1$
   $6x - 3y = 3$
   (d) $y - x = -5$
   $2x - y = 7$
   (e) $\frac{1}{3}x + y = -3$
   $x - \frac{1}{4}y = 4$
   (f) $\frac{1}{2}x - \frac{1}{3}y = -5$
   $x - .1y = 24$

2. Find (x, y) by first finding an equivalent triangular system.
   (a) $ix - 2y = -5$
   $3x + 2iy = 7i$
   (b) $(1 + i)x - 3y = -1 - 3i$
   $x + (1 - i)y = 3 - i$
   (c) $x + 2y = 3 + i(x + 1)$
   $x - 2y = 3 - i(3 - x)$
   (d) $3x - iy = ix - 3y$
   $5x + 2iy = 2ix + 5y$

3. Solve the following systems by first finding an equivalent triangular system.
   (a) $x + 2y = 0$
   $x - 2y = 4$
   $2x + 4y - 2z = -6$
   (b) $x + 3y - 4z = 9$
   $2x - y + 2z = -2$
   $4x - 6y + z = 2$
   (c) $3x - 4y + 2w = 1$
   $2x - y + w = 2$
   $x + 5y - z = 5$
   $y - 3z + w = -1$
   (d) $x_1 - x_4 = 1$
   $x_1 + x_3 + x_4 = 3$
   $x_1 + x_2 + x_4 = 2$
   $2x_1 + 3x_3 + 5x_4 = 6$

262

4. Find (x, y) by first making a substitution that yields a linear system; for example, let $u = 1/x$ and $v = 1/y$ in part (a).

   (a) $3/x + 4/y = -6$

       $1/x - 5/y = 17$

   (b) $2x^2 + 3y^2 = 5$

       $x^2 + 2y^2 = 3$

   (c) $5 \log x - \log y = 11$

       $\log x + 3 \log y = -1$

   (d) $4 \sin x - \sin y = 2$

       $2 \sin x + 3 \sin y = 10$

5. Use a system of two equations in two unknowns to find what quantities of coffee worth 75¢ a pound and \$1.15 a pound are needed to produce a blend worth 91¢ a pound.

6. The sum of the digits of a two-digit integer is 13. The number formed by reversing the digits is 27 greater than the original number. Find the number.

7. (a) Show that both $(1, 2, -1)$ and $(6, 5, -8)$ are solutions of the system

$$2x - y + z = -1$$
$$x + 3y + 2z = 5$$
$$x - 4y - z = -6.$$

(b) In fact, show that $(1 + 5t, 2 + 3t, -1 - 7t)$ is a solution of this system for each number t.

(c) Show that if (x, y, z) and (u, v, w) are solutions of our system, then $(x - u, y - v, z - w)$ is a solution of the system we get when we replace each of the numbers on the right-hand sides of our equations with zero.

8. The graphs of the equations $x - 2y - 3 = 0$, $2x + y + 1 = 0$, and $3x + 4y + 5 = 0$ are lines. Show that they intersect in a point.

9. Solve for (x, y).

   (a) $3[\![x]\!] - 2[\![y]\!] = 8$

       $2[\![x]\!] + 5[\![y]\!] = -1$

   (b) $3|x| - 2|y| = 8$

       $2|x| + 5|y| = -1$

   (c) $\log \dfrac{x^3}{y^2} = 8$

       $\log x^2 y^5 = -1$

   (d) $3 \cdot 10^x - 2 \cdot 10^{-y} = 8$

       $2 \cdot 10^x + 5 \cdot 10^{-y} = -1$

10. To what extent can you use Operations 57-1, 57-2, and 57-3 to solve the system of inequalities

$$3x - 2y < 8$$
$$2x + 5y < -1?$$

59
MATRICES

Reducing a system of linear equations to triangular form requires us to write a number of equivalent systems. We can save a lot of the work involved by using a notation in which the symbols for the unknowns need not be copied down every time we write out a new equivalent system.

    To illustrate this notation, consider the system

(59-1)
$$\tfrac{1}{2}x - 3y + z = 5$$
$$x + 4y - 3z = 2$$
$$2x \qquad + \tfrac{1}{4}z = -1.$$

The coefficients of the unknowns can be exhibited as an array of numbers:

(59-2)

$$\begin{bmatrix} \frac{1}{2} & -3 & 1 \\ 1 & 4 & -3 \\ 2 & 0 & \frac{1}{4} \end{bmatrix}.$$

Such an array is called a **matrix.** In particular, if there are the same number of rows and columns in the array, it is a **square matrix.** The square Matrix 59-2 is the **coefficient matrix** of System 59-1. The numbers on the right-hand sides of the equations in System 59-1 can be displayed as a matrix with three rows and one column:

(59-3)

$$\begin{bmatrix} 5 \\ 2 \\ -1 \end{bmatrix}.$$

We combine this matrix with the coefficient matrix to get the **augmented matrix** of a system. Thus, the matrix

$$\begin{bmatrix} \frac{1}{2} & -3 & 1 & 5 \\ 1 & 4 & -3 & 2 \\ 2 & 0 & \frac{1}{4} & -1 \end{bmatrix}$$

is the augmented matrix of System 59-1. (The vertical bar between the third and fourth columns is not really a part of matrix notation; we use it here to remind us that part of our matrix represents one side of a system of equations and part the other.)

It is easy to construct the augmented matrix of a given system of linear equations. We must arrange the system so that each column of the matrix represents the coefficients of the same unknown, of course, and we must decide which column corresponds to which unknown. But once we have disposed of these points, we have a unique matrix that represents the system. Conversely, it is easy to find the system of equations of a given augmented matrix. Thus, the augmented matrix and the system of equations say the same thing, but the matrix says it more concisely.

For us a matrix is nothing more than a device to save the trouble of writing out the x's, y's, z's, and the like, of a system of linear equations. This simple application does not begin to suggest how useful matrices are in all sorts of mathematical activities these days. If you continue the study of mathematics, you can expect to become something of an expert in matrix theory.

The operations we use to obtain equivalent systems of equations from a given system can easily be expressed as operations to be performed on the rows of numbers in the augmented matrix of the system. These operations always produce a matrix of an equivalent system of equations and can be summarized as follows.

> **Operation 59-1.** *Interchanging two rows in a matrix.*

> **Operation 59-2.** *Multiplying all the elements of some row by the same non-zero number.* (We say that we *multiply the row* by the number.)

> **Operation 59-3.** *Replacing a row with the sum of a non-zero multiple of itself and a multiple of another row.* (*Adding two rows* means, of course, adding corresponding elements.)

We will solve systems of linear equations by using these operations to produce a triangular coefficient matrix, that is, a matrix of a system of equations in triangular form.

EXAMPLE 59-1 Use matrix notation to solve System 58-2.

Solution The system is

$$2x + 3y = 4$$
$$6x - 2y = -10,$$

so the corresponding matrix is

$$\left[\begin{array}{cc|c} 2 & 3 & 4 \\ 6 & -2 & -10 \end{array}\right].$$

We multiply the second row by $\frac{1}{2}$ to obtain the matrix

$$\left[\begin{array}{cc|c} 2 & 3 & 4 \\ 3 & -1 & -5 \end{array}\right].$$

We now add 3 times the second row to the first to obtain the matrix

$$\left[\begin{array}{cc|c} 11 & 0 & -11 \\ 3 & -1 & -5 \end{array}\right].$$

Since the coefficient matrix is now triangular, our matrix represents the triangular system

$$11x \qquad = -11$$
$$3x - y = -5,$$

which readily yields $(x, y) = (-1, 2)$.

EXAMPLE 59-2 Use matrix operations to solve System 58-8.

Solution The matrix that corresponds to System 58-8 is

$$\left[\begin{array}{ccc|c} 1 & -6 & 2 & 5 \\ 2 & -3 & 1 & 4 \\ 3 & 4 & -1 & -2 \end{array}\right].$$

If we add the last row to the second row, and add twice the last row to the first, we obtain the matrix

$$\begin{bmatrix} 7 & 2 & 0 & 1 \\ 5 & 1 & 0 & 2 \\ 3 & 4 & -1 & -2 \end{bmatrix}.$$

Now subtracting twice the second row from the first produces the matrix

$$\begin{bmatrix} -3 & 0 & 0 & -3 \\ 5 & 1 & 0 & 2 \\ 3 & 4 & -1 & -2 \end{bmatrix}.$$

Our coefficient matrix is triangular. In fact, we have the matrix of the triangular System 58–10, from which we easily obtained the solution of the original system.

EXAMPLE 59-3 Solve the system of equations

$$\begin{aligned} 2x - y + z - w &= -1 \\ x + 3y - 2z &= -5 \\ 3x - 2y + 4w &= 1 \\ -x + y - 3z - w &= -6. \end{aligned}$$

(59-4)

Solution The matrix of this system is

(59-5)

$$\begin{bmatrix} 2 & -1 & 1 & -1 & -1 \\ 1 & 3 & -2 & 0 & -5 \\ 3 & -2 & 0 & 4 & 1 \\ -1 & 1 & -3 & -1 & -6 \end{bmatrix}.$$

Our goal is to reduce the coefficient matrix to triangular form, and we will start by making the first three elements of the fourth column 0 (replacing the colored numbers with 0's). To do so, we multiply the last row by 4 and add the resulting row to the third; then we subtract the last row from the first to produce the matrix

(59-6)

$$\begin{bmatrix} 3 & -2 & 4 & 0 & 5 \\ 1 & 3 & -2 & 0 & -5 \\ -1 & 2 & -12 & 0 & -23 \\ -1 & 1 & -3 & -1 & -6 \end{bmatrix}.$$

The fourth column is now in satisfactory form, so we concentrate on the first three rows and work to make the first two colored elements in the third column 0. We multiply the second row by 6. Then we subtract the third row from this new second row. Now we multiply the first row by 3 and add the third row. These operations yield the matrix

(59-7)

$$\begin{bmatrix} 8 & -4 & 0 & 0 & -8 \\ 7 & 16 & 0 & 0 & -7 \\ -1 & 2 & -12 & 0 & -23 \\ -1 & 1 & -3 & -1 & -6 \end{bmatrix}.$$

Finally, we add the second row to 4 times the first to cancel the first colored element of the second column, and we obtain the matrix

(59-8)

$$\begin{bmatrix} 39 & 0 & 0 & 0 & | & -39 \\ 7 & 16 & 0 & 0 & | & -7 \\ -1 & 2 & -12 & 0 & | & -23 \\ -1 & 1 & -3 & -1 & | & -6 \end{bmatrix}.$$

This matrix is the matrix of the system of equations in triangular form

(59-9)

$$\begin{aligned} 39x & & & & = -39 \\ 7x + 16y & & & & = -7 \\ -x + 2y - 12z & & & = -23 \\ -x + y - 3z - w & = -6. \end{aligned}$$

From System 59-9, we readily find the solution to be $(x, y, z, w) = (-1, 0, 2, 1)$.

A little practice with matrix operations will enable you to pick up a fair degree of speed in solving systems of linear equations. There is nothing fixed about the sequence of operations. Our systems could have been reduced to triangular form in a slightly different manner without affecting the solution. The exact procedure to follow is a matter of individual choice.

PROBLEMS 59

1. Form an augmented matrix for each of the following systems.

    (a) $x - 2y = 2$
    $y - 3z = 3$
    $x + 3 = 4z$

    (b) $x_1 - x_5 = 3$
    $x_2 + x_3 + x_4 = 7$
    $3x_3 - x_4 + x_5 = 6$
    $x_2 + x_5 = 3$
    $x_1 + x_3 + x_4 + 6 = x_5$

    (c) $x - z = 2$
    $y - w = -1$
    $w - x = 3$
    $z - y = 0$

    (d) $a - 4d = 5c$
    $b + 2c = 2d$
    $c - b = a$
    $d + 3a = -b$

2. Use matrix notation to find (x, y).

    (a) $2x + 3y = -1$
    $3x - y = -7$

    (b) $2x + 3y = 8$
    $3x - y = 1$

    (c) $2x + 3y = 0$
    $x + 2y = -1$

    (d) $2x + 3y = 12$
    $x + 2y = 8$

    (e) $2y = -2$
    $x - y = 2$

    (f) $2y = 4$
    $x - y = 0$

3. Use matrix notation to solve the following systems.

    (a) $2x + 3y - z = -3$
    $x - y + 2z = 6$
    $x + 2y + z = 1$

    (b) $2x + 3y - z = 0$
    $x - y + 2z = 0$
    $x + 2y + z = 0$

    (c) $x + 4y - 2z = 0$
    $-2x + y = 0$
    $x - y + z = 0$

    (d) $x + 4y - 2z = 0$
    $-2x + y = -4$
    $x - y + z = 3$

267

(e) $$\begin{aligned} x + y &= 0 \\ y + z &= 0 \\ z + w &= 0 \\ x - y + z - w &= 4 \end{aligned}$$ (f) $$\begin{aligned} x + y &= 3 \\ y + z &= 0 \\ z + w &= -1 \\ x - y + z - w &= 0 \end{aligned}$$

4. Use matrix notation to solve the following "homogeneous" systems.

(a) $$\begin{aligned} 2x - y + z &= 0 \\ x + 3y + 2z &= 0 \\ 2x - 4y - z &= 0 \end{aligned}$$ (b) $$\begin{aligned} 2x - y + z &= 0 \\ -x + 3y + 2z &= 0 \\ 3x + 2y + 3z &= 0 \end{aligned}$$

5. Solve the system of equations

$$\begin{aligned} 2x + y + 2z &= u \\ x - 4y + 2z &= v \\ x + y + z &= w. \end{aligned}$$

6. Find the angles of a triangle if one angle is 20° more than the sum of the other two, and 80° more than the positive difference of the other two.

7. Let $P(x) = x^3 + ax^2 + bx + c$. If we know that $P(-1) = -1$, $P(1) = 5$, and $P(2) = 11$, find the values of a, b, and c.

8. Three pipe lines supply an oil reservoir. The reservoir can be filled by pipes A and B running for 10 hours, by pipes B and C running for 15 hours, or by pipes A and C running for 20 hours. How long does it take to fill the tank if (a) all three pipes run? (b) pipe A is used alone?

9. You are told that a bag of 30 coins contains nickels, dimes, and quarters amounting to $3, and that there are twice as many nickels as there are dimes. Should you believe it?

10. Solve the system of equations:

$$\begin{aligned} x \cos \alpha + y \sin \alpha &= u \\ -x \sin \alpha + y \cos \alpha &= v. \end{aligned}$$

11. (a) Explain how Operation 59-3 "includes" Operation 59-2.
    (b) Can you show that Operation 59-3 also "includes" Operation 59-1?

60
DETERMINATIVE, DEPENDENT, AND INCONSISTENT SYSTEMS

Most of the systems of linear equations that we have seen thus far have had exactly one solution. We will call such a system of linear equations a **determinative** system. Not all systems are determinative. A system may have no solution, in which case we say that the system is **inconsistent.** At the other extreme, a system may have infinitely many solutions, in which case the system is **dependent.** The method of solving systems of linear equations by trying to reduce them to triangular form is still applicable to inconsistent and dependent systems. In fact, this method will enable us to tell when we are dealing with an inconsistent or a dependent system. The following examples illustrate how we can detect these situations.

EXAMPLE 60-1   Solve the system of equations

$$2x - y + z = 1$$
(60-1)
$$x + 2y - z = 3$$
$$x + 7y - 4z = 2.$$

Solution   In matrix notation, the reduction to triangular form appears as

$$\begin{bmatrix} 2 & -1 & 1 & | & 1 \\ 1 & 2 & -1 & | & 3 \\ 1 & 7 & -4 & | & 2 \end{bmatrix} \rightarrow \begin{bmatrix} 9 & 3 & 0 & | & 6 \\ 3 & 1 & 0 & | & 10 \\ 1 & 7 & -4 & | & 2 \end{bmatrix} \rightarrow \begin{bmatrix} 0 & 0 & 0 & | & -24 \\ 3 & 1 & 0 & | & 10 \\ 1 & 7 & -4 & | & 2 \end{bmatrix}.$$

To go from the second matrix to the third, we subtract 3 times the second row from the first. The purpose of this operation, of course, is to remove the 3 in the first row, but in the process we also remove the 9. The final matrix can be considered as the matrix of the system

$$0 \qquad\qquad = 24$$
$$3x + y \qquad = 10$$
$$x + 7y - 4z = 2.$$

Since this system contains the false statement $0 = 24$, some explanation is in order. Logically, we did nothing wrong. We are simply asserting that *if* there is a solution (x, y, z) of System 60-1, *then* $0 = 24$ But $0 \neq 24$, so we must conclude that *the system does not have a solution*. System 60-1 is inconsistent.

EXAMPLE 60-2   Solve the system of equations

$$2x - y + z = 1$$
(60-2)
$$x + 2y - z = 3$$
$$x + 7y - 4z = 8.$$

Solution   (Notice that this system is very similar to the system in Example 60-1.) In this case, the matrix reduction is

$$\begin{bmatrix} 2 & -1 & 1 & | & 1 \\ 1 & 2 & -1 & | & 3 \\ 1 & 7 & -4 & | & 8 \end{bmatrix} \rightarrow \begin{bmatrix} 9 & 3 & 0 & | & 12 \\ 3 & 1 & 0 & | & 4 \\ 1 & 7 & -4 & | & 8 \end{bmatrix} \rightarrow \begin{bmatrix} 0 & 0 & 0 & | & 0 \\ 3 & 1 & 0 & | & 4 \\ 1 & 7 & -4 & | & 8 \end{bmatrix}.$$

The last matrix is the matrix of the system

$$0 \qquad\qquad = 0$$
(60-3)
$$3x + y \qquad = 4$$
$$x + 7y - 4z = 8.$$

Unlike Example 60-1 (which yielded $0 = 24$), the first equation here is of absolutely no help. There is nothing false about the statement that $0 = 0$, but it doesn't tell us anything we didn't already know. There are many solutions of System 60-3. For instance, we could choose $x = 1$; then $y = 1$ and $z = 0$. Or if $x = 0$, then $y = 4$ and $z = 5$. Indeed, if t is any number, then a solution of System 60-3 (and hence of System 60-2) is

$(x, y, z) = (t, 4 - 3t, 5 - 5t)$. Therefore, we will write the solution set of this system as

$$\{(x, y, z) = (t, 4 - 3t, 5 - 5t)\}.$$

System 60-2 is dependent. Observe that if we first choose $y = s$, we obtain the solution set

$$\{(x, y, z) = (\tfrac{1}{3}(4 - s), s, \tfrac{5}{3}(s - 1))\}.$$

This set looks different from our other solution set, but of course it isn't. Can you show that these sets are equal?

If a system consists of two linear equations in two unknowns, and if all the given numbers are real, there is a simple way of interpreting geometrically the meaning of inconsistency and dependence. Recall (Section 48) that the graph of a linear equation $ax + by = c$, where a, b, and c are real numbers, is a line. Each equation of the pair

(60-4)
$$ax + by = c$$
$$dx + ey = f$$

therefore represents a line. There are three possibilities concerning the intersection of these lines.

Case I. The lines represented by Equations 60-4 may intersect in a single point (x_0, y_0).

In this case the point (x_0, y_0) is the one and only point that belongs to both lines. Hence, the coordinates of this point are the only numbers that satisfy both equations, and so there is only one solution of System 60-4. In geometric terms, finding the solution of a determinative system of two linear equations means finding the coordinates of the point of intersection of the two lines represented by the equations.

Case II. The lines of System 60-4 may be parallel and non-intersecting.

In this case there are no points that belong to both lines. Therefore, there are no points whose coordinates satisfy both equations. Thus, there is no solution of System 60-4; the system is inconsistent.

Case III. Both equations of System 60-4 may represent the same line.

In this case every point that belongs to one line belongs to the other. In other words, every pair of numbers (x, y) that satisfies one equation, satisfies the other, and System 60-4 is dependent.

The results of this analysis may be put in tabular form.

TYPE OF SYSTEM	DESCRIPTION OF GRAPH
Determinative	Two intersecting lines
Inconsistent	Two parallel lines
Dependent	One line

1. Find a solution (x, y, z) of System 60-2 such that
   (a) $x = 2$ (b) $y = -5$ (c) $x = 1 + i$ (d) $y = 1 + 3i$

2. The augmented matrix of a given system of equations is

$$\begin{bmatrix} 24 & 0 & 0 & 0 \\ 1 & 2 & 0 & 2 \\ 3 & 4 & 1 & 5 \end{bmatrix}.$$

   Is the system determinative?

3. Determine whether the following systems are determinative, inconsistent, or dependent.

   (a) $2x - 4y = 5$
   $12y - 6x = -15$

   (b) $z - 4y = 4$
   $3x - 2y - z = 5$
   $2x - z = 2$

   (c) $2x - 3y + z = 1$
   $x - 4y - z = 1$
   $x + 2y + 3z = 4$

   (d) $5x + 6y = z$
   $7x + 6y - 2z + 8 = 0$
   $3x + 2y - z = -5$

   (e) $2a + c + 3 = b$
   $a + b = 3 + c$
   $a + 2c + 4 = b$

   (f) $2a + c + 3 = b$
   $a + b = 1 + c$
   $a + 2c + 4 = b$

4. Determine whether the following systems are determinative, inconsistent, or dependent.

   (a) $2x - 6y + 2z + 10w = -7$
   $2x + 4y - z - 5w = 6$
   $x - 3y + z + 5w = -3$
   $2x + 2y - z + 3w = 4$

   (b) $x + w = 3$
   $y + z = -1$
   $x + z = 1$
   $y + w = 1$

5. In each case find the point of intersection of the pair of given lines, or show that the lines do not intersect.

   (a) $2x - 3y = -1$
   $x + 3y = 4$

   (b) $x - y = 5$
   $x + \frac{1}{2} = \frac{1}{2}x + \frac{3}{2}$

   (c) $y = \frac{3}{4}x + 1$
   $3x - 4y = 2$

   (d) $2x - 3y = 0$
   $x - 4y = 0$

6. Solve, if possible, the following systems.

   (a) $4x + 9y - z = 0$
   $x + 4y - z = -1$
   $x - 3y + 2z = 3$

   (b) $4x + 9y - z = 0$
   $x + 4y - z = 1$
   $x - 3y + 2z = -3$

   (c) $4x + 9y - z = 1$
   $x + 4y - z = 2$
   $x - 3y + 2z = 3$

   (d) $4x + 9y - z = 0$
   $x + 4y - z = 0$
   $x - 3y + 2z = 0$

7. Find the value of k such that the following system is *not* determinative.

$$kx + 2y - 2z = 2$$
$$x - 2y + z = 4$$
$$3x - 3y + z = 1$$

8. The sum of the digits of a three-digit number is 14 and the middle digit is the sum of the other two digits. If the last two digits are interchanged, we get a number that is 45 less than the original number. Find the original number. Can you solve the problem if all the information is the same, except that the number obtained when the last two digits are interchanged is 36 less than the original number?

9. Four high schools, South, East, North, and West, have a total enrollment of 1000 students. South reports that 10% of the students make A's, 25% B's, and 50% C's; East reports 15% A's, 25% B's, and 55% C's; North reports 25% A's, 15% B's, and 35% C's; West reports 15% A's, 20% B's and 40% C's. Is this information consistent with the fact that of the total student enrollment in all four schools 15% receive A's, 25% B's, and 50% C's?

10. Discuss the problem of finding the coefficients a, b, and c so that the graph of the polynomial equation $y = ax^2 + bx + c$ contains three given points (x_1, y_1), (x_2, y_2) and (x_3, y_3).

61
SYSTEMS INVOLVING QUADRATIC EQUATIONS

Sometimes we have to solve a system of equations in which products of the unknowns appear in one or more of the equations. Systems 57-1 and 57-3 are examples. Solving the most general system of two quadratic equations in two unknowns is a complicated operation, and we will spare you the details. We will only consider some isolated types of systems involving quadratic equations that occur most often in practice.

If the system consists of one linear and one quadratic equation, we can solve it by the method of substitution suggested in Section 57.

EXAMPLE 61-1 Solve the system

(61-1)
$$2x + y = 1$$
$$3x^2 - xy - y^2 = -2.$$

Solution The first equation yields $y = 1 - 2x$, and when we substitute this result in the second equation we get

$$3x^2 - x(1 - 2x) - (1 - 2x)^2 = -2,$$

so that $x^2 + 3x + 1 = 0$. The solution set of this equation is $\{-\frac{3}{2} + \frac{1}{2}\sqrt{5}, -\frac{3}{2} - \frac{1}{2}\sqrt{5}\}$, and we obtain corresponding values of y from the relation $y = 1 - 2x$. In this way we see that the set of solutions of System 61-1 is $\{(-\frac{3}{2} + \frac{1}{2}\sqrt{5}, 4 - \sqrt{5}), (-\frac{3}{2} - \frac{1}{2}\sqrt{5}, 4 + \sqrt{5})\}$.

The substitution method may also be used when neither of the equations is linear, as illustrated by the next example.

EXAMPLE 61-2 A publisher wishes to print a book with 1-inch margins at the top, bottom, inside, and outside of each page. Each page is to have an area of 42 square inches, and the printed area is to be 20 square inches. What must be the dimensions of the page?

Solution   Suppose that one edge of the page is x inches long and the other edge is y inches long. Then, since the area of the page is to be 42 square inches,

(61-2)
$$xy = 42.$$

The dimensions of the printed area are $(x - 2)$ and $(y - 2)$, so we also have

(61-3)
$$(x - 2)(y - 2) = 20.$$

Equations 61-2 and 61-3 taken together yield a system of two quadratic equations:

(61-4)
$$xy = 42$$
$$xy - 2x - 2y = 16.$$

If we solve the first of these equations for y and substitute the result in the second, we get the equation

$$42 - 2x - \frac{84}{x} = 16.$$

This equation reduces to

$$x^2 - 13x + 42 = 0 \quad \text{or} \quad (x - 6)(x - 7) = 0,$$

so it has the two solutions 6 and 7. Corresponding values of y are $\frac{42}{6} = 7$ and $\frac{42}{7} = 6$. Therefore, the original system of equations has two solutions, $(x, y) = (6, 7)$ and $(x, y) = (7, 6)$. These solutions both say the same thing; the page should measure 6 inches by 7 inches.

Another non-linear system that is easy to solve consists of two equations of the form $ax^2 + by^2 = c$. The equations of such a system become linear if we substitute u for x^2 and v for y^2.

EXAMPLE 61-3   Solve the system

(61-5)
$$3x^2 + 4y^2 = 8$$
$$x^2 - y^2 = 5.$$

Solution   Substituting u for x^2 and v for y^2 leads to the system of linear equations

$$3u + 4v = 8$$
$$u - v = 5.$$

The solution of this linear system is $(u, v) = (4, -1)$. It follows that the solution set of System 61-5 is

$$\{(2, i), (2, -i), (-2, i), -2, -i)\}.$$

Other systems that involve quadratic equations may be solved by replacing the given system with an equivalent system that is easier to solve.

EXAMPLE 61-4   Solve the system

(61-6)
$$3x^2 - xy - y^2 = 9$$
$$8x^2 - 5xy - 4y^2 = 18.$$

Solution If we subtract twice the first equation from the second, we obtain the equivalent system

$$3x^2 - xy - y^2 = 9$$
$$2x^2 - 3xy - 2y^2 = 0.$$

After the second equation is factored, this system becomes

$$3x^2 - xy - y^2 = 9$$
$$(2x + y)(x - 2y) = 0.$$

Now we see that our system is equivalent to *two* systems:

(61-7)
$$3x^2 - xy - y^2 = 9$$
$$2x + y = 0$$

and

(61-8)
$$3x^2 - xy - y^2 = 9$$
$$x - 2y = 0.$$

Thus, the solution set of our original System 61-6 is the union of the solution sets of Systems 61-7 and 61-8. We can solve both of these systems by substitution. Substituting $y = -2x$ in the first equation of System 61-7 gives us the simple equation $x^2 = 9$. Hence, $x = -3$ or $x = 3$. Since $y = -2x$, we therefore see that the solution set of System 61-7 is $\{(-3, 6),$ $(3, -6)\}$. Similarly, we find that $\{(-2, -1), (2, 1)\}$ is the solution set of System 61-8. Thus, the solution set of System 61-6 is

$$\{(-3, 6), (3, -6), (-2, -1), (2, 1)\}.$$

The methods we have discussed in this section are by no means exhaustive. It can be shown that any system of equations of the form

(61-9)
$$ax^2 + bxy + cy^2 + dx + ey = f$$
$$gx^2 + hxy + ky^2 + lx + my = n$$

may be solved algebraically. (That is, there are algebraic methods for determining whether or not such a system has a solution, and, if so, finding all the solutions.)

PROBLEMS
61

1. Solve the following systems.

(a)
$$x - 2y = 3$$
$$x^2 + xy - y^2 = -1$$

(b)
$$xy + 6 = 0$$
$$x + y - 1 = 0$$

(c)
$$y^2 = 4x$$
$$x + y = 3$$

(d)
$$y^2 = 2x^2$$
$$x^2 = 2y^2$$

(e)
$$x - y = 0$$
$$x^2 - y^2 = 0$$

(f)
$$y = mx + c$$
$$\frac{x^2}{a^2} - \frac{y^2}{b^2} = 1$$

274

2. Solve the following systems.

(a) $2x^2 - y^2 = -2$
$x^2 - 3y^2 = -11$

(b) $4x^2 + 9y^2 = -40$
$3x^2 - 5y^2 = 17$

(c) $2x^2 - y^2 = -3$
$3ix^2 + 2y^2 = 8i$

(d) $x^2 + 4y^2 = 0$
$2x^2 - 3y^2 = 11$

(e) $x^2 + y^2 = 0$
$x^2 - 3xy - 2y^2 = 0$

(f) $x^2 + y^2 = 0$
$x^2 - 3xy - 2y^2 = 6 - 6i$

3. Solve the following systems.

(a) $x^2 - 2xy - y^2 = -4$
$x^2 - 3xy + y^2 = -4$

(b) $2x^2 - 3xy - 2y^2 = 0$
$x^2 + xy - y^2 = 45$

(c) $6x^2 + 3xy - 2y^2 = 4$
$4x^2 + 2xy - y^2 = 4$

(d) $xy + 4y^2 = 0$
$x^2 + 2xy - 4y^2 = 0$

4. How long are the sides of a right triangle if the hypotenuse is 34 inches long and the area is 240 square inches?

5. A box with a volume of 80 cubic inches is to be constructed by cutting 2-inch squares from the corners of a rectangular piece of cardboard, and then bending up the sides and tying a piece of string around them to hold them up. If the piece of string has to be 27 inches long (including 1 inch for making the knot), what should be the dimensions of the original piece of cardboard?

6. Solve the system

$$x^3 - y^3 = 7$$
$$x - y = 1.$$

7. The area of the base of a rectangular box is a square feet, the area of one side is b square feet, and the area of the end is c square feet. What is the volume of the box?

8. Solve the system of equations

$$(2x - y + 3)(x + y - 3) = 0$$
$$(x - 2y + 3)(x - 5y - 3) = 0.$$

9. Show that if you know how to find the zeros of polynomials of degrees up to 4, then you can always solve a system of equations of the form of System 61-9.

10. Show that there are no real numbers x and y such that

$$(x^2 - y^2 - 2x)^2 + (y - 3x + 1)^2 = 0.$$

62
THE DETERMINANT OF A MATRIX OF ORDER 2

For us, a matrix is simply an array of numbers written in rows and columns. If the number of rows of a matrix is the same as the number of columns, then the matrix is a square matrix. The **order** of a square matrix is the number of its rows (or columns). The matrix

$$\begin{bmatrix} 1 & 2 & 3 \\ 4 & 5 & 6 \\ 7 & 8 & 9 \end{bmatrix},$$

for example, is a square matrix of order 3. In particular, the square matrix

$$\begin{bmatrix} a & b \\ c & d \end{bmatrix}$$

is the coefficient matrix of the system of linear equations

(62-1)
$$\begin{aligned} ax + by &= r \\ cx + dy &= s. \end{aligned}$$

With each square matrix we associate a number, called the *determinant* of the matrix. The rules for finding the determinant of a matrix are quite complicated, especially for matrices of high orders. Therefore, in this section we will restrict ourselves to matrices of order 2, for which it is relatively easy to do the necessary calculations. We will develop some of the theorems of determinants for matrices of order 2, and in the next section we will discuss (without proofs) the situation for matrices of higher orders.

Definition 62-1   *The **determinant** of the square matrix* $\begin{bmatrix} a & b \\ c & d \end{bmatrix}$ *is the number* $ad - bc$, *and we denote this number by* $\begin{vmatrix} a & b \\ c & d \end{vmatrix}$.

Thus,

$$\begin{vmatrix} \boldsymbol{a} & \boldsymbol{b} \\ \boldsymbol{c} & \boldsymbol{d} \end{vmatrix} = \boldsymbol{ad} - \boldsymbol{bc}.$$

EXAMPLE 62-1   Find the determinant of the matrix $\begin{bmatrix} 1 & 2 \\ 2 & 3 \end{bmatrix}$.

Solution   According to Definition 62-1,

$$\begin{vmatrix} 1 & 2 \\ 2 & 3 \end{vmatrix} = 1 \cdot 3 - 2 \cdot 2 = -1.$$

Now let us look at System 62–1. It would be ridiculous to consider a system in which all the coefficients a, b, c, and d are zero, so we will assume that at least one of them is not zero. Then, by interchanging the equations or by interchanging x and y and relettering if necessary, we can always arrange the system so that $d \neq 0$. So let us assume that $d \neq 0$ and solve System 62-1 by matrix reduction:

$$\left[\begin{array}{cc|c} a & b & r \\ c & d & s \end{array} \right] \rightarrow \left[\begin{array}{cc|c} ad - bc & 0 & rd - bs \\ c & d & s \end{array} \right].$$

(We multiplied the first row of the first matrix by d and subtracted b times the second row to get the second matrix.)

System 62-1 is therefore equivalent to the triangular system

(62-2)
$$(ad - bc)x \qquad = rd - bs$$
$$cx \qquad + dy = s.$$

It is easy to give a complete discussion of this system. If the number $ad - bc$ is not 0, we can solve the first of Equations 62-2 for x and then the second for y, so the system is determinative. If $ad - bc = 0$, however, there are two possibilities. Either $rd - bs = 0$, in which case any choice of x will satisfy the first of Equations 62-2, or else $rd - bs \neq 0$, in which case no number will satisfy the first equation. In the first case, the system is dependent, and in the second case it is inconsistent. Thus, from the triangular System 62-2 we read the following facts about the equivalent System 62-1:

(i) If $ad - bc = 0$ and $rd - bs \neq 0$, the system is inconsistent.

(ii) If $ad - bc = 0$ and $rd - bs = 0$, the system is dependent.

(iii) If $ad - bc \neq 0$, the system is determinative.

(We have said that the case $a = b = c = d = 0$ is too trivial to study, but for the sake of completeness, we might mention that System 62-1 obviously cannot be determinative in this case.) These statements tell us that the determinant $ad - bc$ of the coefficient matrix of our system of equations determines (hence its name) whether or not the system is determinative. We state this important result as a theorem.

Theorem 62-1   *A system of two linear equations in two unknowns is determinative if, and only if, the determinant of the coefficient matrix is not zero.*

If the numbers r and s are both 0, then System 62-1 is said to be **homogeneous,** and it reduces to the system

(62-3)
$$ax + by = 0$$
$$cx + dy = 0.$$

Obviously, one solution of System 62-3 is the pair $(0, 0)$. A question that arises very often in the application of mathematics is, "Under what circumstances does a homogeneous system possess solutions other than the 'trivial' solution $(0, 0)$?" Theorem 62-1 gives us a partial answer. If the determinant $\begin{vmatrix} a & b \\ c & d \end{vmatrix}$ is not 0, System 62-3 is determinative; that is, it has exactly one solution. And since the pair $(0, 0)$ is a solution, we see that there can be no others. Thus, System 62-3 can have a "non-trivial" solution only if the determinant of the coefficient matrix is zero. The complete answer to our question is contained in the next theorem.

The homogeneous System 62-3 has a solution in addition to the trivial solution $(0, 0)$ *if, and only if, the determinant of the coefficient matrix is* 0.

Proof We have just proved half of this theorem; we still must show that if $ad - bc = 0$, then System 62-3 has a solution (x, y) in addition to $(0, 0)$. If all the coefficients $a, b, c,$ and d are 0, then any number pair (x, y) satisfies System 62-3. So let us suppose that at least one of the coefficients, say d, is not 0. Then the pair $(d, -c)$ is not the pair $(0, 0)$, and $(d, -c)$ satisfies the second equation of System 62–3, as we can see by substitution. If we substitute d for x and $-c$ for y in the left-hand side of the first of Equations 62-3, we obtain $ad - bc$, the determinant of the coefficient matrix. This number is 0 by hypothesis, so the pair $(d, -c)$ satisfies both equations of System 62–3 and is not the pair $(0, 0)$.

Now let us suppose that System 62-1 is determinative. From Equations 62-2 we see that

$$x = \frac{rd - bs}{ad - bc}.$$

The numerator and denominator of this fraction can be written in determinant notation,

$$rd - bs = \begin{vmatrix} r & b \\ s & d \end{vmatrix} \quad \text{and} \quad ad - bc = \begin{vmatrix} a & b \\ c & d \end{vmatrix},$$

so we have

(62-4)
$$x = \frac{\begin{vmatrix} r & b \\ s & d \end{vmatrix}}{\begin{vmatrix} a & b \\ c & d \end{vmatrix}}.$$

Notice that we can obtain the determinant in the numerator from the determinant of the coefficient matrix by replacing the coefficients of the x terms with the numbers on the right-hand sides of Equations 62-1. Similarly, it is easy to verify that y can be written in determinant notation as

(62-5)
$$y = \frac{\begin{vmatrix} a & r \\ c & s \end{vmatrix}}{\begin{vmatrix} a & b \\ c & d \end{vmatrix}}.$$

Here we obtain the determinant in the numerator from the determinant of the coefficient matrix by replacing the coefficients of the y terms with the numbers on the right-hand sides of Equations 62-1.

EXAMPLE 62-2   Write the solution of the system

$$2x + \pi y = \sqrt{17}$$
$$ix - 23y = 89$$

in terms of determinants.

Solution   The determinant of the coefficient matrix of this system is $-46 - \pi i$, which is not 0. Hence, we may use Equations 62-4 and 62-5 to write

$$x = \begin{vmatrix} \sqrt{17} & \pi \\ 89 & -23 \end{vmatrix} \bigg/ \begin{vmatrix} 2 & \pi \\ i & -23 \end{vmatrix} \quad \text{and} \quad y = \begin{vmatrix} 2 & \sqrt{17} \\ i & 89 \end{vmatrix} \bigg/ \begin{vmatrix} 2 & \pi \\ i & -23 \end{vmatrix}.$$

The following theorem lists a number of useful facts concerning the determinant of a matrix of order 2.

Theorem 62-3   *For a matrix of order 2, it is true that:*

(i) *If the rows of one matrix are the columns (in the same order) of another, then the two matrices have the same determinant; that is,*

$$\begin{vmatrix} a & b \\ c & d \end{vmatrix} = \begin{vmatrix} a & c \\ b & d \end{vmatrix}.$$

(ii) *Interchanging two rows (or columns) of a matrix changes the algebraic sign of the determinant; that is,*

$$\begin{vmatrix} c & d \\ a & b \end{vmatrix} = - \begin{vmatrix} a & b \\ c & d \end{vmatrix} \quad \text{and} \quad \begin{vmatrix} b & a \\ d & c \end{vmatrix} = - \begin{vmatrix} a & b \\ c & d \end{vmatrix}.$$

(iii) *If any row (or column) of a matrix is multiplied by a number k, the determinant also is multiplied by the number k; for example,*

$$\begin{vmatrix} ka & kb \\ c & d \end{vmatrix} = k \begin{vmatrix} a & b \\ c & d \end{vmatrix} \quad \text{and} \quad \begin{vmatrix} ka & b \\ kc & d \end{vmatrix} = k \begin{vmatrix} a & b \\ c & d \end{vmatrix}.$$

(iv) *If a multiple of any row (or column) is added to any other row (or column) the value of the determinant is unchanged; for example,*

$$\begin{vmatrix} a + kc & b + kd \\ c & d \end{vmatrix} = \begin{vmatrix} a & b \\ c & d \end{vmatrix} \quad \text{and} \quad \begin{vmatrix} a + kb & b \\ c + kd & d \end{vmatrix} = \begin{vmatrix} a & b \\ c & d \end{vmatrix}.$$

Because the proof of this theorem consists of a number of straightforward applications of Definition 62-1, we leave it to you. For example, the following chain of equations verifies statement (iii):

$$\begin{vmatrix} ka & kb \\ c & d \end{vmatrix} = kad - kbc = k(ad - bc) = k \begin{vmatrix} a & d \\ c & d \end{vmatrix}.$$

279

1. Find the determinants of the following matrices.

(a) $\begin{bmatrix} 2 & 1 \\ -3 & 4 \end{bmatrix}$

(b) $\begin{bmatrix} 4 & 3i \\ i & 1 \end{bmatrix}$

(c) $\begin{bmatrix} 2i & i \\ -3i & 4i \end{bmatrix}$

(d) $\begin{bmatrix} 2-i & 3+2i \\ 3-2i & 2+i \end{bmatrix}$

(e) $\begin{bmatrix} \cos 4 & \sin 4 \\ -\sin 4 & \cos 4 \end{bmatrix}$

(f) $\begin{bmatrix} 3^x & 2^y \\ 2^{-y} & 3^{-x} \end{bmatrix}$

2. Solve the following equations.

(a) $\begin{vmatrix} 2 & x \\ 3 & 12 \end{vmatrix} = 0$

(b) $\begin{vmatrix} x-3 & 0 \\ x^2+9x-2 & x+2 \end{vmatrix} = 0$

(c) $\begin{vmatrix} 1 & x \\ 2x-1 & 3x^2-3x-3 \end{vmatrix} = 0$

(d) $\begin{vmatrix} x^2-2x-5 & x^2+3x-2 \\ 2x-1 & 3x+2 \end{vmatrix} = 0$

3. Solve the following inequalities.

(a) $\begin{vmatrix} 2 & x \\ 3 & 12 \end{vmatrix} < 0$

(b) $\begin{vmatrix} 1 & x \\ 2x-1 & 3x^2-3x-3 \end{vmatrix} > 0$

4. Find three solutions of the system

$$\tfrac{1}{2}x - 2y = 0$$
$$-4x + 16y = 0.$$

5. Write the solution of each system in terms of determinants.

(a) $2x - 3y = 4$
    $x + y = 7$

(b) $ix - \pi y = 2$
    $3y = 5$

(c) $\quad x \cos t + y \sin t = u$
    $-x \sin t + y \cos t = v$

(d) $x \log 6 - 2y = 3$
    $x \log 2 + y = 1$

6. Prove that if the determinant of the coefficient matrix of System 62-3 is zero, then for any number k, the pair $(kd, -kc)$ is a solution of the system.

7. Prove that the determinant of a matrix of order 2 that has two identical columns (or two identical rows) is zero.

8. Verify the following identities.

(a) $\begin{vmatrix} a+u & b \\ c+v & d \end{vmatrix} = \begin{vmatrix} a & b \\ c & d \end{vmatrix} + \begin{vmatrix} u & b \\ v & d \end{vmatrix}$

(b) $\begin{vmatrix} a+u & b+v \\ c & d \end{vmatrix} = \begin{vmatrix} a & b \\ c & d \end{vmatrix} + \begin{vmatrix} u & v \\ c & d \end{vmatrix}$

9. If you know that the determinants of two second-order matrices are equal are the matrices identical? What about the converse?

10. Compute the following determinant.

$$\begin{vmatrix} \begin{vmatrix} 1 & -2 \\ 5 & -3 \end{vmatrix} & \begin{vmatrix} 4 & 6 \\ -1 & 1 \end{vmatrix} \\ \begin{vmatrix} 7 & 8 \\ 1 & 1 \end{vmatrix} & \begin{vmatrix} 3 & -1 \\ 4 & -3 \end{vmatrix} \end{vmatrix}$$

11. Show that if $\begin{vmatrix} a & b \\ c & d \end{vmatrix} = 0$ and $d \neq 0$, then there is a number r such that $a = rb$ and $c = rd$.

12. Give a complete proof of Theorem 62-3.

The determinant of the matrix $\begin{bmatrix} a & b \\ c & d \end{bmatrix}$ is the number $ad - bc$. If $b = 0$, our matrix is in triangular form, and its determinant is simply the number ad, the product of the **diagonal numbers** a and d of the matrix. By applying the operations of interchanging the position of two rows, multiplying the elements of a row by a given number, and adding to each element of one row the same multiple of the corresponding element of another row, any square matrix can be reduced to "triangular form," in which all the numbers above the diagonal numbers are 0. Theorem 62-3 tells us how these operations affect the determinant of a matrix of order two, so one method of calculating the determint of such a matrix consists of the following steps:

(i) Reduce the matrix to triangular form, taking into account the effect of our matrix operations on the value of the determinant.

(ii) Multiply the diagonal numbers of the resulting matrix

Table 63-1 shows how operations on matrices affect the corresponding determinants. The equations in the third column of the table come from Theorem 62-3, but the middle column needs some explaining. When we solved systems of equations in Section 59 by reducing their coefficient matrices to triangular form, it was not necessary to keep track of our operations as we went from one matrix to the next. Here it is, so we will augment our notation a little. Instead of writing just an arrow ($\rightarrow$) between successive matrices, we will (as illustrated in the table) write an arrow and a number. The number

Table 63-1

ITEM	MATRIX OPERATIONS	DETERMINANTS OF CORRESPONDING MATRICES
Operation 59-1 *Interchanging rows*	$\begin{bmatrix} a & b \\ c & d \end{bmatrix} \xrightarrow{-1} \begin{bmatrix} c & d \\ a & b \end{bmatrix}$	$\begin{vmatrix} a & b \\ c & d \end{vmatrix} = - \begin{vmatrix} c & d \\ a & b \end{vmatrix}$
Operation 59-2 *Multiplying a row by* *a non-0 number k*	$\begin{bmatrix} a & b \\ c & d \end{bmatrix} \xrightarrow{\frac{1}{k}} \begin{bmatrix} ka & kb \\ c & d \end{bmatrix}$	$\begin{vmatrix} a & b \\ c & d \end{vmatrix} = \frac{1}{k} \begin{vmatrix} ka & kb \\ c & d \end{vmatrix}$
Operation 59-3 *Adding a multiple of* *one row to another row*	$\begin{bmatrix} a & b \\ c & d \end{bmatrix} \xrightarrow{1} \begin{bmatrix} a+kc & b+kd \\ c & d \end{bmatrix}$	$\begin{vmatrix} a & b \\ c & d \end{vmatrix} = \begin{vmatrix} a+kc & b+kd \\ c & d \end{vmatrix}$

is the number by which we multiply the determinant of the matrix at the head of the arrow to obtain the determinant of the matrix at the tail. If we have a whole chain of matrices, the determinant of the first is obtained by multiplying the determinant of the last by the product of all the numbers above the arrows in the chain.

EXAMPLE 63-1 Find the number $\begin{vmatrix} 1 & 2 \\ 3 & 4 \end{vmatrix}$ by following steps (i) and (ii).

Solution We first reduce the matrix $\begin{bmatrix} 1 & 2 \\ 3 & 4 \end{bmatrix}$ to triangular form:

$$\begin{bmatrix} 1 & 2 \\ 3 & 4 \end{bmatrix} \xrightarrow{-1} \begin{bmatrix} 3 & 4 \\ 1 & 2 \end{bmatrix} \xrightarrow{1} \begin{bmatrix} 1 & 0 \\ 1 & 2 \end{bmatrix}.$$

As you can see, we first interchanged the rows of the given matrix, and then we subtracted twice the second row from the first. We use the resulting chain to calculate the required determinant:

$$\begin{vmatrix} 1 & 2 \\ 3 & 4 \end{vmatrix} = (-1)(1) \begin{vmatrix} 1 & 0 \\ 1 & 2 \end{vmatrix} = -2.$$

Of course, it is easy to calculate determinants of matrices of order two directly, instead of going through the rather complicated procedure we have just outlined. But this more complicated procedure can also be used to find the determinants of matrices of higher order. Although there are various shortcuts that one learns through experience, there is no really simple way to calculate such determinants. The following theorem, which we will not attempt to prove, tells us something about determinants of matrices of higher order.

Theorem 63-1 *With a square matrix of any order we can associate a number, called the* ***determinant*** *of the matrix, such that:*

(a) *Statements* (i), (ii), (iii), *and* (iv) *in Theorem 62-3 are valid, and*
(b) *The determinant of a triangular matrix in which all the numbers above the diagonal numbers are 0 is the product of the diagonal numbers.*

Now let us use steps (i) and (ii) to find the determinant of a matrix of order 3.

EXAMPLE 63-2 Evaluate the determinant $\begin{vmatrix} 1 & 3 & 2 \\ 1 & 1 & -1 \\ 6 & 4 & 2 \end{vmatrix}$.

Solution We first reduce the matrix $\begin{bmatrix} 1 & 3 & 2 \\ 1 & 1 & -1 \\ 6 & 4 & 2 \end{bmatrix}$ to triangular form:

$$\begin{bmatrix} 1 & 3 & 2 \\ 1 & 1 & -1 \\ 6 & 4 & 2 \end{bmatrix} \xrightarrow{2} \begin{bmatrix} 1 & 3 & 2 \\ 1 & 1 & -1 \\ 3 & 2 & 1 \end{bmatrix} \xrightarrow{1} \begin{bmatrix} 1 & 3 & 2 \\ 4 & 3 & 0 \\ 3 & 2 & 1 \end{bmatrix}$$

$$\xrightarrow{1} \begin{bmatrix} -5 & -1 & 0 \\ 4 & 3 & 0 \\ 3 & 2 & 1 \end{bmatrix} \xrightarrow{-1} \begin{bmatrix} 4 & 3 & 0 \\ -5 & -1 & 0 \\ 3 & 2 & 1 \end{bmatrix} \xrightarrow{1} \begin{bmatrix} -11 & 0 & 0 \\ -5 & -1 & 0 \\ 3 & 2 & 1 \end{bmatrix}.$$

You can easily see what operations we used to go from one matrix to the next in this chain. For example, the 2 above the first arrow indicates that the second matrix is obtained from the first by multiplying a row by $\frac{1}{2}$.

282

The determinant of the first matrix is the product of the determinant of the last matrix and all the numbers above the arrows in between. Since the determinant of our final triangular matrix is the product of its diagonal numbers, we therefore have

$$\begin{vmatrix} 1 & 3 & 2 \\ 1 & 1 & -1 \\ 6 & 4 & 2 \end{vmatrix} = 2 \cdot 1 \cdot 1 \cdot (-1) \cdot 1 \cdot \begin{vmatrix} -11 & 0 & 0 \\ -5 & -1 & 0 \\ 3 & 2 & 1 \end{vmatrix}$$

$$= (-2)(-11)(-1)(1) = -22.$$

Theorems that correspond to Theorems 62-1 and 62-2 are also valid for systems of linear equations in more than two unknowns. We state them here, but we will not prove them.

Theorem 63-2 *A system of n linear equations in n unknowns is determinative if, and only if, the determinant of the coefficient matrix is not zero.*

Theorem 63-3 *A system of n linear and homogeneous equations in n unknowns has solutions in addition to the trivial solution $(0, 0, \ldots, 0)$ if, and only if, the determinant of the coefficient matrix is 0.*

You might verify that the system

$$2x - y + z = 0$$
$$x - 2y + 3z = 0$$
$$3x \qquad - z = 0$$

has the solution $(x, y, z) = (1, 5, 3)$ in addition to the trivial solution $(0, 0, 0)$, and that the determinant of its coefficient matrix is 0. Can you find all the solutions of this system?

It is a simple process, but a long one, to show that if the system

$$a_1x + b_1y + c_1z = d_1$$
$$a_2x + b_2y + c_2z = d_2$$
$$a_3x + b_3y + c_3z = d_3$$

is determinative, then its solution is given by the equations

$$x = \frac{\begin{vmatrix} d_1 & b_1 & c_1 \\ d_2 & b_2 & c_2 \\ d_3 & b_3 & c_3 \end{vmatrix}}{\begin{vmatrix} a_1 & b_1 & c_1 \\ a_2 & b_2 & c_2 \\ a_3 & b_3 & c_3 \end{vmatrix}}, \quad y = \frac{\begin{vmatrix} a_1 & d_1 & c_1 \\ a_2 & d_2 & c_2 \\ a_3 & d_3 & c_3 \end{vmatrix}}{\begin{vmatrix} a_1 & b_1 & c_1 \\ a_2 & b_2 & c_2 \\ a_3 & b_3 & c_3 \end{vmatrix}}, \quad z = \frac{\begin{vmatrix} a_1 & b_1 & d_1 \\ a_2 & b_2 & d_2 \\ a_3 & b_3 & d_3 \end{vmatrix}}{\begin{vmatrix} a_1 & b_1 & c_1 \\ a_2 & b_2 & c_2 \\ a_3 & b_3 & c_3 \end{vmatrix}}.$$

This statement can be generalized to apply to systems of n linear equations in n unknowns, where n is any positive integer.

Since calculating a determinant by our steps (i) and (ii) is a tedious business, most people who work with determinants have memorized schemes for evaluating the determinant of a square matrix of order 3. It turns out that

(63-1)
$$\begin{vmatrix} a_1 & a_2 & a_3 \\ b_1 & b_2 & b_3 \\ c_1 & c_2 & c_3 \end{vmatrix} = a_1b_2c_3 + a_2b_3c_1 + a_3b_1c_2 - a_3b_2c_1 - a_1b_3c_2 - a_2b_1c_3.$$

This formula can be remembered by the devices illustrated in Fig. 63-1 (*which cannot be used for finding determinants of matrices of higher orders*).

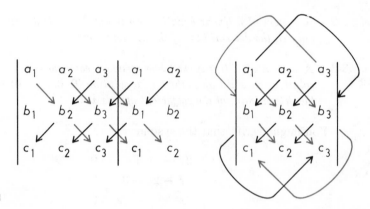

FIGURE 63-1

In the scheme on the left-hand side of the figure, we copy down two extra columns of numbers (the first two columns of our given matrix) and from the resulting array obtain the terms in the sum in Equation 63-1 by multiplying as indicated by the arrows. To the terms that arise from multiplying along the colored arrows we affix the positive sign, and to the terms that arise from multiplying along the black arrows we affix the negative sign. On the right-hand side of the figure we have tried to show the same thing without copying the extra columns of numbers. It really isn't such a bad scheme, once you get used to it.

PROBLEMS
63

1. Compute the determinant of each matrix by the method illustrated in Example 63-1.

(a) $\begin{bmatrix} 1 & -1 \\ 3 & 1 \end{bmatrix}$

(b) $\begin{bmatrix} 6 & 4 \\ 0 & -2 \end{bmatrix}$

(c) $\begin{bmatrix} 1 & \tan x \\ \tan x & -1 \end{bmatrix}$

(d) $\begin{bmatrix} \cos t & -\sin t \\ \sin t & \cos t \end{bmatrix}$

2. Evaluate the following determinants.

(a) $\begin{vmatrix} 2 & 0 & 0 \\ 3 & -1 & 0 \\ 4 & -5 & 3 \end{vmatrix}$

(b) $\begin{vmatrix} 2 & 3 & 0 \\ -1 & 0 & 0 \\ 5 & 4 & 3 \end{vmatrix}$

(c) $\begin{vmatrix} 1 & 1 & 1 \\ 1 & 1 & 0 \\ 1 & 0 & 0 \end{vmatrix}$

(d) $\begin{vmatrix} 1 & 2 & 3 & 4 \\ 5 & 6 & 7 & 0 \\ 8 & 9 & 0 & 0 \\ 10 & 0 & 0 & 0 \end{vmatrix}$

3. Evaluate the following determinants, either by "steps (i) and (ii)" or by the scheme illustrated in Fig. 63-1.

(a) $\begin{vmatrix} 2 & -1 & 0 \\ -1 & -1 & 2 \\ 3 & 0 & -2 \end{vmatrix}$

(b) $\begin{vmatrix} 2 & -3 & 0 \\ 1 & 0 & 1 \\ 2 & 3 & 1 \end{vmatrix}$

(c) $\begin{vmatrix} 0 & 1 & 0 \\ 2 & 0 & 1 \\ 1 & 1 & 1 \end{vmatrix}$

(d) $\begin{vmatrix} 1 & 2 & 3 \\ 4 & 0 & 5 \\ 0 & 6 & 0 \end{vmatrix}$

4. Evaluate the following determinants.

(a) $\begin{vmatrix} 1 & 2 & 1 & 2 \\ -1 & 2 & 1 & -2 \\ -1 & -2 & 1 & 2 \\ -1 & -2 & 1 & -2 \end{vmatrix}$

(b) $\begin{vmatrix} 2 & 0 & 1 & 3 \\ 0 & 0 & 5 & 1 \\ -1 & 0 & -1 & 0 \\ 0 & 4 & 0 & 6 \end{vmatrix}$

(c) $\begin{vmatrix} 0 & 1 & 0 & 0 \\ 0 & 0 & 1 & 0 \\ 0 & -1 & 0 & 1 \\ 0 & 0 & -1 & 0 \end{vmatrix}$

(d) $\begin{vmatrix} 10 & 6 & 3 & 2 \\ 9 & 6 & 3 & 2 \\ 7 & 5 & 3 & 2 \\ 4 & 3 & 2 & 1 \end{vmatrix}$

5. Convince yourself that the determinant of a square matrix that has two identical columns is 0.

6. Write the solution of each of the following systems in determinant form. How much work would it be to complete the solution, that is, to evaluate the determinants? Try it, and compare with our earlier methods.

(a)
$$3x - 2y = -3$$
$$x - y + 2z = -4$$
$$-x + y - z = 3$$

(b)
$$x_1 + x_3 + x_4 = 2$$
$$2x_1 + x_2 + x_3 = 1$$
$$x_1 + x_2 + x_4 = 3$$
$$x_2 - 2x_3 - x_4 = 0$$

7. Solve the equation $\begin{vmatrix} x & 0 & 0 & 1 \\ 0 & x & 0 & 0 \\ 0 & 0 & x & 0 \\ 1 & 0 & 0 & x \end{vmatrix} = 0.$

8. For what values of k is the following system determinative?

$$kx + y = 1$$
$$y + z = 1$$
$$z + x = 1$$

285

9. For what values of k does the following system have a solution other than the trivial solution $(0, 0, 0)$?

$$-2x + 2y + 2z = kx$$
$$6x - 2y - 6z = ky$$
$$-3x + 2y + 3z = kz$$

10. Show that if the lines $a_1x + b_1y = c_1$, $a_2x + b_2y = c_2$ and $a_3x + b_3y = c_3$ have a common point, then

$$\begin{vmatrix} a_1 & b_1 & c_1 \\ a_2 & b_2 & c_2 \\ a_3 & b_3 & c_3 \end{vmatrix} = 0.$$

(*Hint:* Use the fact that a certain system of three linear homogeneous equations in three unknowns has a non-trivial solution.)

11. Show that $\begin{vmatrix} a & b & 0 \\ c & d & 0 \\ e & f & g \end{vmatrix} = g \begin{vmatrix} a & b \\ c & d \end{vmatrix}$.

REVIEW PROBLEMS, CHAPTER EIGHT

You should be able to answer the following questions without referring back to the text.

1. Solve the following system of equations.

$$\log (x^2/y^3) = -8$$
$$\log (x^3y^2) = 1$$

2. Determine the numbers a, b, and c so that the graph of the equation $y = ax^2 + bx + c$ contains the points $(-1, 6)$, $(0, 3)$, and $(1, 4)$.

3. Let $u = 2 + 3i$, $v = 4 - i$, and $w = 7i$. Find real numbers s and t such that $su + tv = w$.

4. Solve the system

$$x + 2y + 3z = 0$$
$$3x + y + 2z = 0$$
$$-2x + y + z = 0.$$

5. Show that the three lines $x - y + 1 = 0$, $2x + y - 2 = 0$, and $x + y - 3 = 0$ do not intersect in a point.

6. The number pair $(0, 0)$ is a solution of the system

$$(2 - k)x + y = 0$$
$$3x - ky = 0.$$

For what values of k does the system have a solution other than $(0, 0)$?

7. Show that the following system is inconsistent unless $2r - s - 3t = 0$.
$$x + 3y = r$$
$$2x - 3z = s$$
$$2y + z = t$$

8. Let a, b, c, d, r, and s be given numbers. Suppose x and y are numbers such that
$$ax + by = r$$
$$cx + dy = s,$$
and suppose (u, v) is a solution of the system
$$au + cv = 0$$
$$bu + dv = 0.$$
Show that $ru + sv = 0$.

9. An automobile factory makes both six- and eight-cylinder cars that are priced at \$2800 and \$3100, respectively. During one week the plant used 2560 pistons and produced cars worth \$1,049,000. How many cars of each type were made?

10. Suppose four horses—A, B, C, and D—are entered in a race and the odds on them, respectively, are 6 to 1, 5 to 1, 4 to 1, and 3 to 1. If you bet \$1 on A, then you receive \$6 if A wins; thus you realize a net gain of \$5. You lose your dollar if A loses. How should you bet your money to guarantee that you win \$12 no matter how the race comes out?

11. Describe the set of points $\{(x, y) \mid y = 3x - 1 \text{ and } y \geq 2x + 1\}$.

12. For simplicity, we have confined our attention to systems of equations with the same number of unknowns as equations. If we have a different number of equations and unknowns, our reduction technique leads to "echelon form," rather than triangular form, but there is no real conceptual difference. Solve the following systems of equations.

(a) $3x - 2y + z = 0$
$x + 3y - 2z = 2$

(b) $3x + y = 4$
$2x - 3y = -1$
$x - 2y = -1$

(c) $3x + y = -1$
$2x - 3y = 4$
$x - 2y = -1$

(d) $x_1 + x_2 + x_3 + x_4 + x_5 + x_6 + x_7 + x_8 = 2$
$x_1 - x_2 + x_3 - x_4 + x_5 - x_6 + x_7 - x_8 = 0$

MISCELLANEOUS PROBLEMS, CHAPTER EIGHT

The exercises that follow are designed to test your ability to apply your knowledge of systems of equations and the theory of matrices to more difficult problems.

1. Solve the following system of equations.

(a) $xy = 4$
$y = 2x^2 - 4x + 2$

(b) $2^x = 32^y$
$x^2 + 25y^2 + 7x + 5y - 10 = 0$

2. Solve the following system.

$$
\begin{aligned}
x_1 + x_2 &= 10 \\
x_2 + x_3 &= 9 \\
x_3 + x_4 &= 8 \\
x_4 + x_5 &= 7 \\
x_5 + x_6 &= 6 \\
x_6 + x_7 &= 5 \\
x_7 + x_8 &= 4 \\
x_8 + x_9 &= 3 \\
x_1 + x_9 &= 2
\end{aligned}
$$

3. Solve the following equations.

(a) $\begin{vmatrix} x & 1 & -2 \\ -2 & x+1 & 1 \\ -1 & 3 & x-6 \end{vmatrix} = 0$ (b) $\begin{vmatrix} x+9 & -3-2i & -12 \\ -8i & x+2i & 8i \\ 9+2i & -3-2i & x-12-2i \end{vmatrix} = 0$

4. Suppose you have a combination of ten coins consisting of nickels, dimes, and quarters whose total value is \$1.25. How many of each coin do you have?

5. Suppose you have a rectangular sheet of paper. Can you trim one edge of the paper to obtain a rectangle whose area and perimeter are half the area and perimeter of the original sheet? Can you do it by trimming two edges?

6. Show that

$$
\begin{vmatrix} x & y & 1 \\ x_1 & y_1 & 1 \\ x_2 & y_2 & 1 \end{vmatrix} = 0
$$

is the equation of the line that contains the points (x_1, y_1) and (x_2, y_2).

7. Show that $\begin{vmatrix} 1 & 1 & 1 \\ x_1 & x_2 & x_3 \\ x_1^2 & x_2^2 & x_3^2 \end{vmatrix} = (x_3 - x_1)(x_2 - x_1)(x_3 - x_2)$.

8. Suppose that we are given three complex numbers, $u = a + bi$, $v = c + di$, and $w = e + fi$.
   (a) Under what conditions can you be sure that there are *real* numbers s and t such that $su + tv = w$?
   (b) What about finding *complex* numbers s and t so that the equation holds?

9. Show that Equation 63-1 can be written as

$$
\begin{vmatrix} a_1 & a_2 & a_3 \\ b_1 & b_2 & b_3 \\ c_1 & c_2 & c_3 \end{vmatrix} = a_1 \begin{vmatrix} b_2 & b_3 \\ c_2 & c_3 \end{vmatrix} - a_2 \begin{vmatrix} b_1 & b_3 \\ c_1 & c_3 \end{vmatrix} + a_3 \begin{vmatrix} b_1 & b_2 \\ c_1 & c_2 \end{vmatrix}.
$$

10. Show that

$$
\overline{\begin{vmatrix} a & b \\ c & d \end{vmatrix}} = \begin{vmatrix} \bar{a} & \bar{b} \\ \bar{c} & \bar{d} \end{vmatrix}.
$$

Do you think this equation is true for determinants of matrices of higher order?

Enumeration, the Binomial Theorem, and Probability

nine

One of the significant trends in the modern-day applications of mathematics is the increasing use of probability and statistics to describe the world in which we live. To obtain a working knowledge of these topics would require much more space than we have available. There are, however, many basic concepts that are interesting in themselves and which do not require extensive treatment. In this chapter we shall consider some of these elementary notions that are fundamental to an understanding of probability and statistics.

64
THE FUNDAMENTAL PRINCIPLE OF ENUMERATION AND FACTORIAL NOTATION

A typical question that we will meet later in this chapter is, "How many members are there in the set of all possible bridge hands?" This number is huge; we could never hope to find it by listing all the hands and simply counting them. The way to answer our question is to apply the *Fundamental Principle of Enumeration*, which we introduce in this section.

Let us start with an example. A freshman engineering student has a choice of six engineering departments in which to enroll (electrical, mechanical, and so on), and after graduation a choice of three services (Army, Navy, and Air Force) in which to serve. How many different academic-service careers can he choose (Electrical-Army, Aeronautical-Air Force, and so forth)? We can use a diagram to help us count the members of this set of academic-service careers. In Fig. 64-1 we have shown the "paths" that our student may

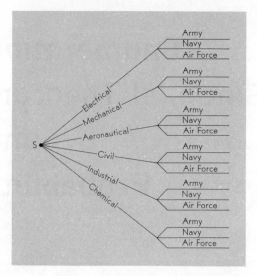

Army
Navy
Air Force

Army
Navy
Air Force

Army
Navy
Air Force

Army
Navy
Air Force

Army
Navy
Air Force

Army
Navy
Air Force

Electrical
Mechanical
Aeronautical
Civil
Industrial
Chemical

S

FIGURE 64-1

follow. Such a diagram is called a **tree.** The student starting at the point S may choose any one of the six engineering branches of the tree, and then a service branch. We find the number of possible choices he might make simply by counting the number of service branches shown. There are 18. Notice that each of the 6 engineering branches is the base of 3 service branches. Therefore, the total number of possibilities facing our student can be expressed as the product $6 \cdot 3 = 18$.

In general, when we wish to enumerate a set we will think of its members as the possible **outcomes** of certain **operations.** Thus our student's set of possible academic-service careers is the collection of possible outcomes of the operation of selecting a course of study and then a branch of the armed services. Clearly, this operation is a "composite" operation; it is composed of the operation of selecting a course of study, followed by the operation of selecting a service branch. To express the number of possible outcomes of a composite operation in terms of the number of possible outcomes of the operations of which it is composed, we make use of the following **Fundamental Principle of Enumeration.**

Fundamental Principle of Enumeration. Suppose we have two operations—Operation (i) and Operation (ii)—such that there are m possible outcomes of Operation (i) and, after Operation (i) has been performed, n possible outcomes of Operation (ii). Then the set of possible outcomes of the composite operation that consists of performing Operation (i) and then Operation (ii) has mn members.

Sec. 64
The Fundamental
Principle of
Enumeration
and Factorial
Notation

Thus, our student was faced with 6 possible outcomes of his operation of selecting an academic major subject and then 3 possible outcomes of the operation of selecting a service career. We have already observed that the composite operation has $6 \cdot 3 = 18$ possible outcomes, as the Fundamental Principle says it should. The Fundamental Principle may be extended to any number of operations.

EXAMPLE 64-1   How many four letter code words can be formed by using only the letters A, B, C, and D?

Solution   We will think of the operation of forming a four-letter word as being composed of the four operations of selecting a first letter, selecting a second letter, selecting a third letter, and selecting a fourth letter. Each of these operations has 4 possible outcomes (we could select any one of the 4 letters A, B, C, and D), and hence the Fundamental Principle says that the composite operation has $4 \cdot 4 \cdot 4 \cdot 4 = 4^4$ possible outcomes.

Notice carefully that the Fundamental Principle asks us to count the possible outcomes of Operation (ii) *after* Operation (i) has been performed. The following example may seem to be the same as the one we just completed, but it isn't.

EXAMPLE 64-2   How many four-letter code words can be formed by using 4 alphabet blocks on which are printed the letters A, B, C, and D?

Solution   Again we think of our operation of constructing a code word as being composed of the operations of selecting a first letter, selecting a second letter, selecting a third letter, and selecting a fourth letter. As before, the first operation has 4 possible outcomes; we can select any one of the 4 blocks. But *after the first operation has been performed*, one block is missing, so there are only 3 possible outcomes of the second operation. After the second operation has been performed, there are only two blocks left, so there are only 2 possible outcomes of the third operation. After the third operation has been performed, there is only one block left, so there is just 1 possible outcome of the fourth operation. Therefore, according to the Fundamental Principle, there are

$$4 \cdot 3 \cdot 2 \cdot 1 = 24$$

possible code words that we can form from our four blocks. The difference between this example and Example 64–1, of course, is that in the earlier example we allowed such words as AABA, and here we don't.

EXAMPLE 64-3   A six-man committee wishes to choose a chairman, vice-chairman, and a secretary. How many lists of officers are possible?

Solution   There are 6 choices for chairman. After the chairman has been selected, there remain 5 possibilities for vice-chairman, and then 4 possibilities for secretary. There are therefore $6 \cdot 5 \cdot 4 = 120$ possible lists of officers.

The answer to the question asked in Example 64-2 was the product $1 \cdot 2 \cdot 3 \cdot 4$. Such products of successive integers will appear so many times in the following pages that it is convenient to have a special notation for them.

Definition 64-1 *Let n be a positive integer. The symbol $n!$ is read "n factorial," and it represents the product of all the positive integers from 1 to n; that is,*

$$n! = 1 \cdot 2 \cdot 3 \cdots n.$$

The symbol 0! is defined as $0! = 1$. (Notice the similarity between this meaning and the special definition for zero exponents.)

For example, $4! = 1 \cdot 2 \cdot 3 \cdot 4 = 24$, and $6! = 1 \cdot 2 \cdot 3 \cdot 4 \cdot 5 \cdot 6 = 720$.

EXAMPLE 64-4 If n and r are positive integers, and $r < n$, show that

(64-1) $$n! = r!(r+1)(r+2) \cdots n.$$

Solution Since $r! = 1 \cdot 2 \cdots r$, we see that

$$r!(r+1)(r+2) \cdots n = 1 \cdot 2 \cdots r(r+1)(r+2) \cdots n,$$

which is precisely what Equation 64-1 says.

We can use Equation 64-1 to simplify expressions that involve factorials.

EXAMPLE 64-5 Simplify the expression $\dfrac{5! - 4!}{6!}$.

Solution According to Equation 64-1, $5! = 4! \cdot 5$, and $6! = 4! \cdot 5 \cdot 6$. Hence,

$$\frac{5! - 4!}{6!} = \frac{4!(5-1)}{4! \cdot 5 \cdot 6} = \frac{4}{30} = \frac{2}{15}.$$

EXAMPLE 64-6 Solve the equation $\dfrac{(n+2)!}{n!} = 56$.

Solution According to Equation 64-1,

$$\frac{(n+2)!}{n!} = \frac{n!(n+1)(n+2)}{n!} = (n+1)(n+2).$$

Thus,

$$(n+1)(n+2) = 56,$$
$$n^2 + 3n - 54 = 0,$$
$$(n-6)(n+9) = 0,$$

and hence $n \in \{6, -9\}$. Since we defined factorials for non-negative numbers only, the symbol $(-9)!$ is meaningless. The solution to our problem is therefore $n = 6$.

Sec. 64

The Fundamental
Principle of
Enumeration
and Factorial
Notation

If n is a large integer, then $n!$ will be an extremely large number. For example, $20! = 2{,}432{,}902{,}008{,}176{,}640{,}000$. Computations with such large numbers are much easier when we use logarithms. Table IV lists the common logarithms of $n!$ for values of n between 1 and 100.

EXAMPLE 64-7 Calculate $N = \dfrac{33! \, 37!}{51!}$.

Solution From Table IV,

$$\log N = \log 33! + \log 37! - \log 51!$$
$$= 36.9387 + 43.1387 - 66.1906$$
$$= 13.8868.$$

From Table I we therefore find that N is approximately 7.53×10^{13}.

**PROBLEMS
64**

1. Simplify the following numbers.

   (a) $\dfrac{11!}{8!}$

   (b) $\dfrac{4! - 7!}{5! - 8!}$

   (c) $\dfrac{(n!)^2}{(n+1)!(n-1)!}$

   (d) $\dfrac{(2n)!}{2n!}$

   (e) $\dfrac{(n+1)! + (n-1)!}{n!}$

   (f) $\dfrac{n! + (n-1)!}{n! - (n-1)!}$

2. Use Tables I and IV to calculate the following numbers.

   (a) $\dfrac{35!}{7!}$

   (b) $35! \, 7!$

   (c) $35! - 7!$

   (d) $35!^{7!}$

3. What is the smallest positive integer n such that $n! + 1$ is not a prime number?

4. (a) How many four-digit numbers are there? (The first digit cannot be 0.)
   (b) How many four-digit even numbers are there?

5. A nickel, a dime, and a quarter are thrown into the air. In how many different heads-tails arrangements can they land?

6. Hamlet and Laertes engage in a "best two out of three" fencing match. Draw a tree showing the possible outcomes.

7. How many three-letter code words are possible if we insist that the letter q must *always* be followed by u?

8. How many seating arrangements are possible for a class of 25 students in a classroom with 60 chairs? Suppose the front row contains 10 chairs and that there are 5 girls in the class. How many seating arrangements are possible if the front row is reserved for the 5 girls and the other rows are reserved for the boys?

9. Suppose that mathematics is offered at 8, 9, 10, 1, and 2 o'clock; English at 9 and 2 o'clock; and chemistry at 8, 10, and 1 o'clock. How many different schedules are possible for a student who is taking these 3 courses?

10. How many possible four-letter code words can be written in which (a) the first letter does not appear again in the word? (b) the same letter is not permitted to appear side by side?

11. (a) Find the smallest positive integer n such that $\sin n!\,\pi r = 0$ for each number r in the set $\{\frac{5}{7}, \frac{2}{3}, \frac{8}{5}, \frac{1}{2}, \frac{3}{4}\}$.
    (b) With this choice of n, find $\cos n!\,\pi r$ for each r in the set.

12. Is $(nm)!$ equal to $n!\,m!$? If not, which is the larger number?

65
PARTITIONING A SET

Suppose an instructor decides to assign seats to a class of 12 students as follows: 6 are to sit in the first row, 3 are to sit in the second row, and 3 are to sit in the third row. When he assigns these seats, he is **partitioning** his set of students into 3 subsets, the set of students who are to sit in the first row, the set of students who are to sit in the second row, and the set of students who are to sit in the third row. In this section, we will learn how to enumerate the possible outcomes of this operation of partitioning a set. We can build our formula for solving a general partitioning problem on the formula for a special case which we will take up first.

Suppose that our instructor simply wants to list his 12 students in his roll book. We can think of this listing operation as a partitioning process whereby he divides his class into 12 subsets, each of which contains one member. These subsets are the subset that consists of the first student on the list, the subset that consists of the second student on the list, and so on. Such a partition of a set gives us a **permutation** of the elements of the set. Thus a permutation of a set of objects is simply an ordering of them. It is easy to count the number of possible outcomes of the operation of ordering a class of 12 students. This operation is composed of the 12 operations of picking the first student to be listed, picking the second student, picking the third student, and so on. Obviously, there are 12 possible outcomes of the first operation. After this operation has been performed, there are 11 possible outcomes of the second operation, then 10 possible outcomes of the third, and so on. There are therefore $12 \cdot 11 \cdot 10 \cdots 2 \cdot 1 = 12!$ permutations of our 12 students. The same argument gives us a general permutation formula.

Theorem 65-1   *There are $n!$ permutations of n objects (n is a positive integer).*

For example, at the end of the season, the Big Ten football teams might be listed in any one of $10! = 3,628,800$ different orders.

To see how this special result enables us to solve a general partitioning problem, we return to our teacher and his class of 12 students, which he wants to partition into subsets made up of 6 students sitting in the first row, 3 students in the second row, and 3 students in the third row (he is not concerned with the order of the students within the rows). We will use the symbol $\begin{pmatrix} 12 \\ 6, 3, 3 \end{pmatrix}$ to denote the number of possible outcomes of the operation of forming such a partition. To find a formula for $\begin{pmatrix} 12 \\ 6, 3, 3 \end{pmatrix}$, we will write an equation that contains this number and solve it. According to Theorem 65-1, the operation of ordering our class has 12! possible outcomes. We can think of this operation as being composed of the following four operations:

> Operation (i)—assigning the students to their proper rows,
> Operation (ii)—ordering the 6 students of the first row,
> Operation (iii)—ordering the 3 students of the second row,
> Operation (iv)—ordering the 3 students of the third row.

We have said that Operation (i) has $\binom{12}{6,\,3,\,3}$ possible outcomes. The other three operations result in permutations of the students in the various rows, so we already know (Theorem 65-1) that they have 6!, 3!, and 3! possible outcomes. Hence, there are $\binom{12}{6,\,3,\,3}$ 6! 3! 3! possible outcomes of the composite operation of ordering the students in the class. Since this number is 12!, we have

$$12! = \binom{12}{6,\,3,\,3} 6!\,3!\,3!, \quad \text{and thus} \quad \binom{12}{6,\,3,\,3} = \frac{12!}{6!\,3!\,3!}.$$

The general partitioning formula is obtained by exactly the same sort of reasoning, and we state the result as a theorem.

Theorem 65-2 *Suppose we wish to partition a set of n elements into p subsets, the first subset containing r_1 elements, the second subset containing r_2 elements, ..., and the pth subset containing r_p elements, where*

$$r_1 + r_2 + \cdots + r_p = n.$$

Then the number of possible outcomes of this operation is given by the formula

(65-1)
$$\binom{n}{r_1,\,r_2,\,\ldots,\,r_p} = \frac{n!}{r_1!\,r_2! \cdots r_p!}.$$

(The formula also applies when some of the r's are 0.)

EXAMPLE 65-1 Suppose our instructor decides to assign grades to his 12 students as follows: 0 A's, 2 B's, 2 C's, 3 D's, and 5 E's. How many outcomes of this grading operation are possible?

Solution Our instructor is partitioning his class into subsets containing 0, 2, 2, 3, and 5 members. According to Equation 65-1, there are

$$\binom{12}{0,\,2,\,2,\,3,\,5} = \frac{12!}{0!\,2!\,2!\,3!\,5!} = 166{,}320$$

possible outcomes of such a partitioning.

295

A permutation of n objects is an ordering of them; and Theorem 65-1 tells us that there are $n!$ permutations of n objects, assuming that we can tell one arrangement of the objects from another. For example, our theorem tells us that there are 7! permutations of 7 alphabet blocks that spell the word DIVIDED. But some of these orderings are indistinguishable from others. If one arrangement can be obtained from another by interchanging the I's, for example, we can't tell them apart. Thus, there are fewer than 7! *distinguishable* permutations of these blocks. We can use Equation 65-1 to count the distinguishable permutations as follows. The blocks fill 7 spaces in a row. Our problem is to partition these 7 spaces into 4 subsets: a subset of 3 spaces that will contain the D's, a subset of 2 spaces that will contain the I's, a subset of 1 space that will contain the V, and a subset of 1 space that will contain the E. There are

$$\binom{7}{3, 2, 1, 1} = \frac{7!}{3!\,2!\,1!\,1!} = 420$$

possible outcomes of this partitioning problem and hence 420 distinguishable permutations of the letters of the word DIVIDED.

EXAMPLE 65-2   Suppose we have 6 signal flags—3 red, 2 yellow, and 1 blue—and we wish to make signals by arranging them vertically on a staff. How many different signals are possible?

Solution   Our problem is obviously that of counting the distinguishable permutations of the flags, and we have just seen that we can use Equation 65-1 for this purpose. Thus, we can fly

$$\binom{6}{3, 2, 1} = \frac{6!}{3!\,2!\,1!} = 60$$

different signals.

Before we bring this section to a close, we will introduce still another use of the word "permutation." A permutation of a set of objects is an ordering of them. If we have a set of n objects and if we select r (where $r \leq n$) of them and order the selected objects, then we have what is known as a *permutation of n objects taken r at a time*. For example, when a coach selects a 9-man baseball team from a squad of 23, he has produced a permutation of 23 objects taken 9 at a time. We use Equation 65-1 to count the number of permutations of n objects taken r at a time. For we can think of our problem as that of partitioning the given set into $r + 1$ subsets. The first subset consists of the first element of our arrangement of the r selected elements, the second subset consists of the second element, . . . , the rth subset consists of the rth element, and the $(r + 1)$st subset consists of the $n - r$ elements that we don't select. There are

$$\binom{n}{1, 1, \ldots, 1, n - r} = \frac{n!}{1!\,1!\cdots 1!\,(n - r)!} = \frac{n!}{(n - r)!}$$

possible such partitions, and so we have proved the following theorem.

Theorem 65-3   *There are* $\dfrac{n!}{(n-r)!}$ *permutations of n objects taken r at a time.*

EXAMPLE 65-3   A football coach has 20 men, each capable of playing any one of the 7 line positions, and 16 men, each capable of playing any one of the 4 backfield positions. How many teams can he field?

Solution   We will think of the operation of selecting a team as being composed of Operation (i)—selecting the linemen, and Operation (ii)—selecting the backs. When we fill our 7 line positions from the 20 qualified men, we are forming a permutation of 20 objects taken 7 at a time. According to Theorem 65-3, there are $\dfrac{20!}{13!}$ such permutations. Similarly, a backfield is one of the permutations of 16 objects taken 4 at a time, and there are $\dfrac{16!}{12!}$ of them. Thus, there are $\dfrac{20!}{13!}$ possible outcomes of Operation (i) and $\dfrac{16!}{12!}$ possible outcomes of Operation (ii), and hence there are $\dfrac{20!}{13!}\dfrac{16!}{12!}$ possible outcomes of the composite operation of selecting a full team. How large is this number?

PROBLEMS 65

1. Use Tables I and IV to evaluate the following numbers.

   (a) $\dbinom{25}{5,\,10,\,10}$ (b) $\dbinom{15}{5,\,5,\,5}$ (c) $\dbinom{52}{26,\,26}$

   (d) $\dbinom{6}{0,\,1,\,2,\,3}$ (e) $\dbinom{52}{52}$ (f) $\dbinom{100}{25,\,25,\,25,\,25}$

2. Show that Equation 65-1 can be written as
   $$\binom{r_1+r_2+\cdots+r_p}{r_1,\,r_2,\,\ldots,\,r_p} = \frac{(r_1+r_2+\cdots+r_p)!}{r_1!\,r_2!\cdots r_p!}$$

3. In the game of bridge, four players (called N, E, S, and W) are each dealt 13 cards. How many deals are possible?

4. A group of 100 freshmen is to be assigned to dormitories as follows: 39 to Short Hall, 28 to Long Hall, and the rest to Yew Hall.

   (a) How many different dormitory assignments are possible?
   (b) If the Dean of Students considers each possibility for 1 second, how long will it take him (working around the clock) to get the rooms assigned?

5. At the beginning of class, an instructor asks for 4 volunteers from his class of 30, each to go to the board and do one of the day's 4 homework problems. How many different outcomes are possible? What really happens?

6. A committee of 9 people is to be split into 3 subcommittees that contain 4, 3, and 2 members.

   (a) How many different subcommittee assignments are possible?
   (b) Suppose that we originally started with 18 people from whom we choose the 3 subcommittees of 4, 3, and 2 members?

7. A row of alphabet blocks spells OSHKOSH.

   (a) How many distinguishable rearrangements are possible?
   (b) What if we allow blocks to be turned upside down as well as moved around?

8. Four men, 3 women, and 2 children stand in a row to pose for pictures. A photo is taken of each possible lineup. How many photos are taken? When the pictures are developed they are so blurred that you can identify an individual only as a man, woman, or child. How many "different" pictures are there?

9. From a squad of 15 men, the basketball coach must choose a team consisting of 2 guards, 1 center, and 2 forwards.

   (a) If every man can play every position, how many teams can he field?
   (b) If 5 of the 15 men can play only guard, and each of the other 10 men can play either forward or center (but not guard), how many teams can he field?
   (c) If 5 of the 15 men can play only guard, and each of the other 10 men can play any position, how many teams can he field?

10. Three men and 3 women are to be seated around a round table. Two seating arrangements are considered the same if one arrangement can be "rotated" into the other. How many ways may the men and women be seated if

    (a) there are no restrictions on where they sit?
    (b) men and women must be seated alternately?

    The type of permutations involved in part (a) of this problem are termed **circular permutations.** What is the number of circular permutations of n objects?

66
COMBINATIONS

When we ask the question, "How many different 13-card bridge hands are possible?," we are really asking for the number of 13-member subsets that we can form from a 52-member deck of cards. Forming such a subset (that is, dealing out a 13-card hand) partitions the deck into two subsets, the 13-card subset that constitutes the bridge hand and the 39-card subset that constitutes the remainder of the deck. So we can use our partitioning formula (Equation 65-1) to count the number of possible 13-card bridge hands. There are $\begin{pmatrix} 52 \\ 13, 39 \end{pmatrix} = \dfrac{52!}{13!\,39!}$ of them.

More generally, suppose we have a set of n objects, and we wish to select a subset of them that contains r members (where, of course, $0 \leq r \leq n$). We call such a subset a **combination** of n objects taken r at a time. Thus a bridge hand is one of the combinations of 52 objects taken 13 at a time. As with the bridge hands, when we form a combination of n things taken r at a time, we are really partitioning our given set into two subsets, the desired set of r objects and the remaining set of $n - r$ objects. Therefore, our partitioning Formula 65-1 tells us that a set of n elements has $\begin{pmatrix} n \\ r, n - r \end{pmatrix} = \dfrac{n!}{r!\,(n - r)!}$ different subsets that contain r elements. It is customary to use an abbreviated notation and to write $\dbinom{n}{r}$ in place of $\dbinom{n}{r,\, n-r}$. Therefore,

(66-1) $$\binom{n}{r} = \frac{n!}{r!(n-r)!} = \frac{n(n-1)\cdots(n-r+1)}{r!},$$

and we have the following theorem.

Theorem 66-1 *There are $\dbinom{n}{r}$ combinations of n objects taken r at a time.*

Other symbols, such as C_r^n, are sometimes used to denote the number of combinations of n objects taken r at a time. The symbols P_r^n then denote the number of permutations of n objects taken r at a time.

EXAMPLE 66-1 Suppose the U.S. Senate consists of 57 Democrats and 43 Republicans. How many different 11-member committees can be formed? (Two committees are considered to be different if one of them has at least one member who is not also a member of the other.) How many committees consisting of 7 Democrats and 4 Republicans are possible?

Solution To answer the first question, we simply need to count the number of possible 11-member subsets of the 100-member U.S. Senate. According to Theorem 66-1, there are $\dbinom{100}{11} = \dfrac{100!}{11!\,89!}$ of them.

We consider the operation of selecting a committee of the type described in the second question to be composed of the operation of selecting the Democratic members and the operation of selecting the Republican members. The first operation chooses one of the combinations of 57 objects taken 7 at a time, and the second operation chooses one of the combinations of 43 objects taken 4 at a time. Therefore, the composite operation has

$$\binom{57}{7}\binom{43}{4} = \frac{57!}{7!\,50!} \cdot \frac{43!}{4!\,39!}$$

possible outcomes. Use logarithms to compare the answers to the two questions of this example.

Students are frequently confused by the distinction between a *combination* of n objects taken r at a time and a *permutation* of n objects taken r at a time. The difference is this. To form a combination of n objects taken r at a time, we simply select a subset of r objects from a given set of n objects. To form a permutation of n objects taken r at a time, we again select a subset of the original set of n objects, *and we also order the subset.* Thus, the ordered set of cards that consists of the 2, 3, 4, 5, and 6 of spades is one of the permutations of the 52 cards of a standard bridge deck taken 5 at a time. Another such permutation is the ordered set that consists of the 3, 2, 4, 5, and 6 of spades. These permutations are different. But to a poker player, these cards represent the same hand. He is only interested in the *set* that consists of these cards, not in their order. This set is one of the combinations of 52 cards taken 5 at a time.

EXAMPLE 66-2   A grandfather has one penny, one nickel, one dime, one quarter, and one half-dollar in his pocket. He draws out one coin and gives it to his eldest grandchild, then another coin goes to the middle grandchild, and finally a third coin is given to the youngest grandchild. Discuss the situation from the point of view of the grandfather and from the point of view of the children.

Solution   As far as the grandfather is concerned, at the end of the operation he will have 3 fewer coins than he started with. He is not interested in the order in which he distributes the coins; only their sum concerns him. Therefore, he uses the combination formula to see that there are $\binom{5}{3} = 10$ different sets of 3 coins that he can select from among the 5 coins in his pocket. Hence, there are 10 different sums that he can dispense in this way.

    To the children, on the other hand, the order in which the coins are drawn *is* important, because it determines who gets how much. Therefore, they use the permutation formula (Theorem 65-3) to see that there are $\frac{5!}{2!} = 60$ possible outcomes of their grandfather's generous act.

It is sometimes necessary to consider situations in which the objects from which combinations are to be selected are not all distinct. For example, suppose that the grandfather of Example 66-2 had one penny, one nickel, one dime, one quarter, and two half-dollars in his pocket. From his point of view, every sum (that is, every total payout) that occurs in Example 66-2 could arise in this case, and in addition there are those sums that occur when two of the three coins drawn are half-dollars. In fact, there are exactly four additional sums, since the two half-dollars may be paired with any one of the remaining four coins. Thus, the total number of sums in this case is $\binom{5}{3} + 4 = 14$. We won't attempt to write a formula covering all such examples; each case can be treated individually. If the numbers involved are not too large, it may help to draw a tree to illustrate the problem.

EXAMPLE 66-3   An urn contains two red, two green, one blue, and one white ball. How many different color combinations of two balls may be selected?

Solution   The number of combinations in which there are two different colors is clearly the number of combinations of four colors taken two at a time, that is, $\binom{4}{2}$. In addition, there are those combinations that consist of two red or two green balls. Thus, the total number of color combinations is

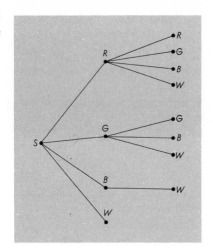

FIGURE 66-1

$\binom{4}{2} + 2 = 8$. The tree in Fig. 66-1 illustrates this example. Notice that we were careful not to repeat branches that would give us the same combination but a different permutation.

PROBLEMS
66

1. Calculate the following numbers.

(a) $\binom{6}{3}$ (b) $\binom{15}{2}$ (c) $\binom{n}{1}$

(d) $\binom{n}{n-1}$ (e) $\binom{n}{n}$ (f) $\binom{n-1}{n-2}$

2. Solve the following equations for n.

(a) $\binom{n}{1} = 10$ (b) $\binom{n}{2} = 45$

(c) $\binom{n+1}{n-1} = 21$ (d) $\binom{n+1}{3} = 2\binom{n}{2}$

3. Verify the following identities.

(a) $\binom{n}{n-r} = \binom{n}{r}$ (b) $\binom{n}{r-1} + \binom{n}{r} = \binom{n+1}{r}$

4. Show that $\binom{n}{r} r!$ is the number of permutations of n things taken r at a time.

5. We modified Example 66-2 and discussed the modification from the point of view of the grandfather. Discuss it from the point of view of the grand-children.

6. How many basketball games are played in the Big Ten if each team plays every other team twice?

7. How many different bridge hands of 13 cards can be dealt from a deck of 52 cards? Use logarithms to compute your answer.

8. An instructor has two algebra classes, one containing 30 students, the other containing 23. The Dean says he must flunk 6 students in the first class and 4 in the second. How many different groups of 10 sad students are possible? Would there be more or fewer possibilities if the Dean's edict had simply been to flunk 10 of the 53 students?

9. Consider 7 points, no three of which are collinear.
   (a) How many lines are determined?
   (b) How many triangles whose vertices are the given points are determined?

10. Ten couples are planning a picnic. They agree that 3 men will build the fire and that 7 women will be in charge of the food.
    (a) How many groups of workers are possible?
    (b) How many groups of workers are possible if only 1 member of each couple is given a job?

11. A group of 23 students decides to tar and feather the Dean. Five students are assigned to the tar detail and 3 are to collect feathers. How many different groups of workers are possible?

12. Explain why

$$\binom{p+q+r}{p,\,q,\,r} = \binom{p+q+r}{p}\binom{q+r}{r}.$$

67
THE BINOMIAL THEOREM

We call the expression $(a+b)$ a **binomial** to indicate that it is the sum of two terms. Let us calculate a few powers of this binomial by direct multiplication:

$$
\begin{aligned}
(a+b)^1 &= a+b \\
(a+b)^2 &= (a+b)(a+b) \\
&= a^2 + ba + ab + b^2 \\
(67\text{-}1) \qquad &= a^2 + 2ab + b^2 \\
(a+b)^3 &= (a+b)(a+b)(a+b) \\
&= a^3 + a^2b + aba + ba^2 + b^2a + bab + ab^2 + b^3 \\
&= a^3 + 3a^2b + 3ab^2 + b^3.
\end{aligned}
$$

The longer we continue this process, the more tedious it gets. Our goal in this section is to develop a formula for expanding a binomial power without using direct multiplication.

From the definition of the power of a number,

$$(67\text{-}2) \qquad (a+b)^n = (a+b)(a+b) \cdots (a+b),$$

302

where there are n factors on the right-hand side of Equation 67-2. Our problem is to express this product as a sum of terms such as those found in Equations 67-1. We therefore must first determine precisely what kind of terms will appear in the expansion.

Each term in the expansion of $(a + b)^n$ arises from the multiplication of a's and b's, where we choose either the number a or the number b from each of the n parentheses on the right-hand side of Equation 67-2. We might, for example, say that we got the term aba in the expansion of $(a + b)^3$ shown in Equations 67-1 by choosing a number a from the first and third parentheses, and b from the second. If we choose the number b from r parentheses, then we must choose the number a from $n - r$ parentheses, since there are n parentheses altogether. When we multiply these a's and b's, we get the product $a^{n-r}b^r$, where r is one of the numbers $0, 1, 2, \ldots, n$. Thus, if $n = 6$, the set of terms of the expansion is $\{a^6, a^5b, a^4b^2, a^3b^3, a^2b^4, ab^5, b^6\}$. Now we know *what* terms appear in the expansion, and our next step is to find out *how many* of each there are.

For a given number r, the product $a^{n-r}b^r$ may arise in several different ways. For example, the product $a^{n-2}b^2$ can occur when we choose b's from the first two parentheses (and a's from the rest), again when we choose b's from the first and third parentheses, and so on. In other words, we obtain the term $a^{n-2}b^2$ by choosing b's from 2 members of the set of n parentheses on the right-hand side of Equation 67-2. Thus, each subset of 2 parentheses leads to a term $a^{n-2}b^2$ in the expansion of $(a + b)^n$, so the number of such terms is the number of subsets of 2 parentheses. Therefore, since our set of n parentheses contains $\binom{n}{2}$ subsets of 2 parentheses, we see that there are $\binom{n}{2}$ such terms in the expansion. By reasoning in exactly the same way for any integer r, we see that the number of terms of the form $a^{n-r}b^r$ that occur in the expansion of $(a + b)^n$ is $\binom{n}{r}$. This result is stated in the following theorem.

Theorem 67-1   (The Binomial Theorem).   *If n is a positive integer and a and b are any two numbers, then*

$$(a + b)^n = \binom{n}{0} a^n b^0 + \binom{n}{1} a^{n-1} b^1 + \cdots$$

$$(67\text{-}3) \qquad + \binom{n}{r} a^{n-r} b^r + \cdots + \binom{n}{n-1} ab^{n-1} + \binom{n}{n} a^0 b^n$$

$$= a^n + na^{n-1}b + \cdots + \binom{n}{r} a^{n-r} b^r + \cdots + nab^{n-1} + b^n.$$

Since the numbers $\binom{n}{r}$ appear as coefficients in the expansion of a power of a binomial, they are frequently called the **binomial coefficients.**

EXAMPLE 67-1   Calculate $(a+b)^5$.

Solution   According to Equation 67-3,

$$(a+b)^5 = a^5 + 5a^4 + \binom{5}{2} a^3b^2 + \binom{5}{3} a^2b^3 + 5ab^4 + b^5.$$

When the binomial coefficients that appear in this expansion are evaluated, we find that

$$(a+b)^5 = a^5 + 5a^4b + 10a^3b^2 + 10a^2b^3 + 5ab^4 + b^5.$$

EXAMPLE 67-2   Show that

(67-4) $$\binom{n}{0} + \binom{n}{1} + \binom{n}{2} + \cdots + \binom{n}{n-1} + \binom{n}{n} = 2^n$$

for any positive integer n.

Solution   The Binomial Theorem is true when a and b are any two numbers; in particular, it is true when the two numbers are both 1. Thus,

$$(1+1)^n = \binom{n}{0} 1^n + \binom{n}{1} 1^{n-1}1^1 + \cdots + \binom{n}{n} 1^n,$$

$$2^n = \binom{n}{0} + \binom{n}{1} + \cdots + \binom{n}{n}.$$

Let us look a little more closely at Equation 67-4. The symbol $\binom{n}{0}$ denotes the number of 0-member subsets of a set of n elements, $\binom{n}{1}$ is the number of 1-member subsets, . . . , and $\binom{n}{n}$ is the number of n-member subsets. When we add these numbers together, we obviously get *the total number of subsets* of a set of n elements. Equation 67-4 says that this number is 2^n. Another line of reasoning also shows that a set of n elements has 2^n subsets. We can think of the operation of choosing a subset as being composed of n operations. The first operation consists of deciding whether to select or not to select some particular element of the original set as a member of the subset. Then we move on to another member and decide whether or not to select *it* as a member of the subset, and so on, until we have checked all n members. Each of these n selection operations has 2 possible outcomes—selection or rejection—so the composite operation of selecting a subset has 2^n possible outcomes.

EXAMPLE 67-3   Find the binomial expansion of $(x^2 - 1/y)^4$.

Solution   We can use the Binomial Theorem with the substitution $a = x^2$ and

304

$b = -1/y$ to obtain

$$\left(x^2 - \frac{1}{y}\right)^4 = (x^2)^4 + 4(x^2)^3\left(-\frac{1}{y}\right) + \binom{4}{2}(x^2)^2\left(-\frac{1}{y}\right)^2$$

$$+ 4(x^2)\left(-\frac{1}{y}\right)^3 + \left(-\frac{1}{y}\right)^4$$

$$= x^8 - \frac{4x^6}{y} + \frac{6x^4}{y^2} - \frac{4x^2}{y^3} + \frac{1}{y^4}.$$

EXAMPLE 67-4   Compute an approximate value of $(1.01)^7$.

Solution   Using the Binomial Theorem, we find that

$$(1.01)^7 = (1 + .01)^7$$
$$= 1^7 + 7(1)^6(.01) + 21(1)^5(.01)^2 + \cdots + (.01)^7$$
$$= 1 + 7(.01) + 21(.01)^2 + 35(.01)^3 + \cdots + (.01)^7$$
$$= 1 + .07 + .0021 + .000035 + \cdots + .00000000000001.$$

It appears that, to three decimal places, $(1.01)^7 = 1.072$.

An argument that is very similar to our discussion of the Binomial Theorem leads to the **Multinomial Theorem.** Suppose that p and n are positive integers. Then the power $(x_1 + x_2 + \cdots + x_p)^n$ can be written as the product

$$(x_1 + \cdots + x_p)^n = (x_1 + \cdots + x_p)(x_1 + \cdots + x_p)\cdots(x_1 + \cdots + x_p).$$

This product will be a sum of terms, which we obtain by choosing an x from each of the n parentheses on the right-hand side of the equation and multiplying the resulting n numbers. For example, we could choose x_1 from the first parentheses and x_2 from all the others and thus obtain the term $x_1 x_2^{n-1}$. In general, a term will have the form $x_1^{r_1} x_2^{r_2} \cdots x_p^{r_p}$, where we have chosen x_1 from r_1 of the parentheses, x_2 from r_2 of the parentheses, ..., x_p from r_p of the parentheses. Since there are n parentheses altogether, we see that $r_1 + \cdots + r_p = n$. To find the number of such terms, we observe that our choices of the x's partition the parentheses into p subsets—the subset consisting of the r_1 parentheses from which we chose x_1, the subset consisting of the r_2 parentheses from which we chose x_2, and so on. Hence, we can use our partitioning formula to see that there are $\binom{n}{r_1, \ldots, r_p}$ terms of the form $x_1^{r_1} \cdots x_p^{r_p}$ in our product. Therefore, the multinomial $(x_1 + \cdots + x_p)^n$ can be expressed as the sum of terms of the form

$$\binom{n}{r_1, r_2, \ldots, r_p} x_1^{r_1} x_2^{r_2} \cdots x_p^{r_p},$$

where we choose all possible sets $\{r_1, r_2, \ldots, r_p\}$ of p non-negative integers whose sum is n.

EXAMPLE 67-5 Find $(a+b+c)^2$.

Solution Here $n = 2$ and $p = 3$. Our given product can be written as a sum of terms of the form $\begin{pmatrix} 2 \\ r_1, r_2, r_3 \end{pmatrix} a^{r_1} b^{r_2} c^{r_3}$, where r_1, r_2, and r_3 are non-negative numbers whose sum is 2. Thus we have

$$(a+b+c)^2 = \begin{pmatrix} 2 \\ 2, 0, 0 \end{pmatrix} a^2 b^0 c^0 + \begin{pmatrix} 2 \\ 0, 2, 0 \end{pmatrix} a^0 b^2 c^0 + \begin{pmatrix} 2 \\ 0, 0, 2 \end{pmatrix} a^0 b^0 c^2$$

$$+ \begin{pmatrix} 2 \\ 1, 1, 0 \end{pmatrix} a^1 b^1 c^0 + \begin{pmatrix} 2 \\ 1, 0, 1 \end{pmatrix} a^1 b^0 c^1 + \begin{pmatrix} 2 \\ 0, 1, 1 \end{pmatrix} a^0 b^1 c^1$$

$$= a^2 + b^2 + c^2 + 2ab + 2ac + 2bc.$$

PROBLEMS 67

1. Find and simplify the fifth term in the expansions of the following binomials.

   (a) $(x+y)^8$ (b) $(2x+y)^{11}$ (c) $\left(2x^2 + \dfrac{iy}{2}\right)^9$ (d) $\left(2x^2 + \dfrac{i}{2y}\right)^9$

2. Find and simplify the "middle term" (or middle two terms) of the expansions of the following binomials.

   (a) $(x - 2y^{-1})^6$ (b) $(x^{-1} + 2y)^6$ (c) $(\sqrt{a} - i\sqrt{b})^7$ (d) $\left(\sqrt{a} + \dfrac{i}{\sqrt{b}}\right)^7$

3. Write the complete expansions of the following binomials.

   (a) $(1+i)^6$ (b) $(1-i)^5$ (c) $(a+bi)^4$ (d) $(u-iv)^5$

4. Find and simplify the fourth term in the expansions of the following binomials.

   (a) $\left(2u - \dfrac{1}{2u}\right)^9$ (b) $\left(2x^2 + \dfrac{1}{2x}\right)^{10}$ (c) $(\tan x - 2 \cot x)^7$

   (d) $(3 \cos t - \tfrac{1}{9} \sec t)^8$ (e) $(2^x - 3^x)^6$ (f) $(2^x - 4^x)^7$

5. Use the Binomial Theorem to find approximations to the following numbers.

   (a) $(1.02)^{10}$ (b) $(.98)^4$ (c) $(102)^5$ (d) 99^5

6. Find the term that involves x^2 in the expansions of the following binomials.

   (a) $(a - x^{1/2})^6$ (b) $(1/x - x)^4$ (c) $(3x^{-3} + 2x^2)^6$ (d) $(\sqrt{x} + 1/\sqrt{x})^{12}$

7. What is the binomial expansion of $(a - b)^n$?

8. Show that $\sin^{10} x + \cos^{10} x = 1 - 5 \cos^2 x + 10 \cos^4 x - 10 \cos^6 x + 5 \cos^8 x$.

9. (a) Show that $\begin{pmatrix} n \\ 0 \end{pmatrix} - \begin{pmatrix} n \\ 1 \end{pmatrix} + \begin{pmatrix} n \\ 2 \end{pmatrix} - \begin{pmatrix} n \\ 3 \end{pmatrix} + \cdots + (-1)^n \begin{pmatrix} n \\ n \end{pmatrix} = 0.$

   (b) Use the result of part (a) and Example 67-2 to show that if n is even, then

$$\begin{pmatrix} n \\ 0 \end{pmatrix} + \begin{pmatrix} n \\ 2 \end{pmatrix} + \begin{pmatrix} n \\ 4 \end{pmatrix} + \cdots + \begin{pmatrix} n \\ n \end{pmatrix} = 2^{n-1}.$$

10. (a) If n is a positive integer, show that

$$(2i)^n = (1+i)^{2n} = 1 + \binom{2n}{1} i - \binom{2n}{2} - \binom{2n}{3} i + \cdots + (-1)^n.$$

(b) Use this formula to conclude that

$$1 - \binom{14}{2} + \binom{14}{4} - \binom{14}{6} + \binom{14}{8} - \binom{14}{10} + \binom{14}{12} - 1 = 0.$$

11. According to DeMoivre's Theorem (Section 46),

$$(\cos \theta + i \sin \theta)^2 = \cos 2\theta + i \sin 2\theta.$$

Now if the left side is multiplied out, we obtain the equation

$$\cos^2 \theta - \sin^2 \theta + 2i \sin \theta \cos \theta = \cos 2\theta + i \sin 2\theta.$$

Hence, $\cos 2\theta = \cos^2 \theta - \sin^2 \theta$ and $\sin 2\theta = 2 \sin \theta \cos \theta$. Use DeMoivre's Theorem and the Binomial Theorem to develop a formula for $\sin 4\theta$.

12. Expand the following multinomials and simplify.
(a) $(a+b+c)^3$ (b) $(a+b+c+d)^2$
(c) $(u-v+w)^2$ (d) $(1+x+x^2)^2$
(e) $(\cos t + \sin t + i)^2$ (f) $(2^x + 3^x + 4^x)^2$

68
THE SIGMA NOTATION

Each term of the sum in Equation 67-3 can be obtained from the expression $\binom{n}{r} a^{n-r} b^r$ by replacing the letter r with one of the numbers $0, 1, 2, \ldots, n$. The Binomial Theorem, therefore, gives the following "recipe" for expanding the expression $(a+b)^n$: "Replace the letter r in the expression $\binom{n}{r} a^{n-r} b^r$ with each of the numbers $0, 1, 2, \ldots, n$ in turn, and add up all the terms that result." Directions like these occur so often in mathematics that we have a special symbol for them. If $A(r)$ is some mathematical expression and n is a positive integer, then the symbol $\sum_{r=0}^{n} A(r)$ means, "Successively replace the letter r in the expression $A(r)$ with the numbers $0, 1, 2, \ldots, n$ and add up the resulting terms." The symbol Σ is the Greek letter sigma, and it is used to suggest "sum," since we are to add a number of terms. This notation is merely a symbolism by which we avoid having to write out the sum $A(0) + A(1) + A(2) + \cdots + A(n)$. We needn't start counting at $r = 0$, but could start at any number less than n. Thus,

$$\sum_{r=1}^{n} A(r) = A(1) + A(2) + \cdots + A(n)$$

and

$$\sum_{r=6}^{8} A(r) = A(6) + A(7) + A(8),$$

for example. The letter r is called the **index of summation.** We may use other letters as the index of summation and we still have the same sum. For example,

$$\sum_{k=7}^{10} A(k) = A(7) + A(8) + A(9) + A(10) = \sum_{r=7}^{10} A(r).$$

EXAMPLE 68-1   Evaluate $\sum_{r=0}^{5} r^2$.

Solution $\sum_{r=0}^{5} r^2 = 0^2 + 1^2 + 2^2 + 3^2 + 4^2 + 5^2 = 55.$

You will often meet the sigma notation in later mathematics courses. *In terms of this symbolism the Binomial Theorem reads*

(68-1) $$(a + b)^n = \sum_{r=0}^{n} \binom{n}{r} a^{n-r} b^r.$$

Let us examine the coefficients of the terms in the expansion of $(a + b)^n$. The term involving $a^{n-r}b^r$ is

$$\binom{n}{r} a^{n-r} b^r = \frac{n(n-1) \cdots (n-r+1)}{r!} a^{n-r} b^r.$$

Thus, we may express the coefficient of $a^{n-r}b^r$ as a quotient. To get the dividend we calculate a product of integers—the first integer is n, each succeeding integer is obtained by subtracting 1 from the factor that precedes it, and the last integer in the product is 1 more than the exponent of a. The divisor is the factorial of the exponent of b. Once we note this fact, we can readily calculate, for example, that the coefficient of the term a^7b^3 in the expansion of $(a + b)^{10}$ is $\dfrac{10 \cdot 9 \cdot 8}{3!} = 120.$

Next we observe that there is one term (the number 1) in the expansion of $(a + b)^0$, two terms in the expansion of $(a + b)^1$, three terms in the expansion of $(a + b)^2$, and in general, $n + 1$ terms in the expansion of $(a + b)^n$. An ingenious scheme for calculating the binomial coefficients, known as **Pascal's Triangle,** is illustrated below. The first row lists the coefficients

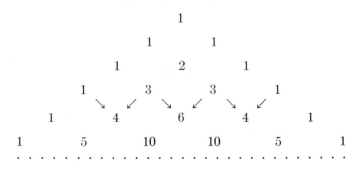

in the expansion of $(a + b)^0$, the second row lists the coefficients in the expansion of $(a + b)^1$, the third row lists the coefficients for $(a + b)^2$, and so on. Each number in the array, except the 1's on the edges, is obtained by adding the two numbers nearest it in the row above. As the arrows indicate, in the fifth row we have $4 = 1 + 3$, $6 = 3 + 3$, and $4 = 3 + 1$, for example. This method yields the binomial coefficients because (as we asked you to show in Problem 66-3b)

$$(68\text{-}2) \qquad \binom{n+1}{r} = \binom{n}{r-1} + \binom{n}{r}.$$

Thus, the number 6 in the fifth row of Pascal's Triangle is $\binom{4}{2}$, and Equation 68-2 states that $\binom{4}{2} = \binom{3}{1} + \binom{3}{2} = 3 + 3$. You should also note the symmetry of Pascal's Triangle; it is an expression of the identity $\binom{n}{n-r} = \binom{n}{r}$ (Problem 66-3a).

Another way to evaluate the binomial coefficients is based on a relationship between $\binom{n}{r+1}$ and $\binom{n}{r}$ that we will now proceed to find. According to Equation 66-1,

$$\binom{n}{r+1} = \frac{n(n-1)\cdots(n-r)}{(r+1)!}$$

$$= \left[\frac{n(n-1)\cdots(n-r+1)}{r!}\right]\frac{(n-r)}{(r+1)}.$$

The fraction inside the brackets is $\binom{n}{r}$, so we have the formula we are looking for:

$$(68\text{-}3) \qquad \binom{n}{r+1} = \frac{(n-r)}{(r+1)}\binom{n}{r}.$$

Equation 68-3 is an example of a *recursion formula*. If we write the binomial coefficients for a given number n in a sequence

$$\binom{n}{0}, \ \binom{n}{1}, \ \ldots, \ \binom{n}{r}, \ \binom{n}{r+1}, \ \ldots, \ \binom{n}{n},$$

the recursion formula enables us to find, one by one, all the binomial coefficients that follow one that we know. But we know that $\binom{n}{0} = 1$ for any n, so we can use Equation 68-3 to find, one by one, the values of all the binomial coefficients for any positive integer n.

EXAMPLE 68-2 Find the first four terms of the expansion of $(x+y)^{21}$.

Solution We use our recursion formula with $n = 21$, together with the fact that $\binom{21}{0} = 1$, to write

$$\binom{21}{0} = 1$$

$$\binom{21}{1} = \frac{21 - 0}{1} \cdot 1 = 21 \quad (r = 0 \text{ in Equation 68-3})$$

$$\binom{21}{2} = \frac{21 - 1}{2} \cdot 21 = 210$$

$$\binom{21}{3} = \frac{21 - 2}{3} \cdot 210 = 1330.$$

Thus, $(x+y)^{21} = x^{21} + 21x^{20}y + 210x^{19}y^2 + 1330x^{18}y^3 + \cdots$.

By modifying our summation notation slightly, we can write the Multinomial Theorem in a fairly simple form. We stated this theorem in words in the last section, and when these words are translated into symbols, we have the equation

(68-4)
$$(x_1 + x_2 + \cdots + x_p)^n = \sum_{r_1 + \cdots + r_p = n} \binom{n}{r_1, r_2, \ldots, r_p} x_1{}^{r_1} x_2{}^{r_2} \cdots x_n{}^{r_n}.$$

In the context of this theorem, the symbol $\displaystyle\sum_{r_1 + \cdots + r_p = n}$ means that we are to add together all possible terms in which the indices $r_1, r_2, \ldots, r_p$ are non-negative integers whose sum is n. Using this notation, the sum in Example 67-5 is written

$$\sum_{r_1 + r_2 + r_3 = 2} \binom{2}{r_1, r_2, r_3} a^{r_1} b^{r_2} c^{r_3}.$$

PROBLEMS 68

1. Find the fifth term in the expansions of the following binomials.
   (a) $(a - b)^{12}$ (b) $(2x^3 + 3x^{-2})^7$ (c) $(1 + ix)^8$ (d) $(\tan x - \cot x)^9$

2. Find the first four terms in the expansions of the following binomials.
   (a) $(x + 1)^{100}$ (b) $(1 + x)^{100}$ (c) $\left(x + \dfrac{1}{x}\right)^{19}$ (d) $(u + iv)^{30}$

3. Calculate $\left(x - \dfrac{1}{x}\right)^6$.

4. Write out the indicated sum, but *do not* perform the addition.

   (a) $\displaystyle\sum_{r=1}^{4} \cos r$ (b) $\displaystyle\sum_{r=0}^{5} \frac{x^r}{r!}$ (c) $\displaystyle\sum_{r=0}^{5} (-1)^r x^{5-r}$

(d) $\displaystyle\sum_{r=0}^{3} \frac{(2r)!}{2r!}$ 　　　　(e) $\displaystyle\sum_{r+s=3} \binom{3}{r,\,s} x^r$ 　　　　(f) $\displaystyle\sum_{r+s+t=3} \binom{3}{r,\,s,\,t} x^{rst}$

5. Show that $(a-b)^n = \displaystyle\sum_{r=0}^{n} (-1)^r \binom{n}{r} a^{n-r} b^r$.

6. Expand $(x+y+z)^4$ by first writing it as $[x+(y+z)]^4$. Also use the Multinomial Theorem.

7. Let n be a positive integer and x and h be any numbers. Expand and simplify the expression $\dfrac{(x+h)^n - x^n}{h}$.

8. Let $A(n) = (1+1/n)^n$, and show that $A(n) = \displaystyle\sum_{r=0}^{n} \binom{n}{r} (1/n)^r$. Calculate $A(1)$, $A(2)$, and $A(5)$.

9. Evaluate the quotient $\displaystyle\sum_{r=0}^{10} \binom{10}{r} 3^{10-r} 4^r \bigg/ \sum_{r=0}^{8} \binom{8}{r} 4^{8-r} 3^r$

10. Solve for x.

(a) $\displaystyle\sum_{r=0}^{7} \binom{7}{r} x^{7-r} 5^r = 0$ 　　　　　　(b) $\displaystyle\sum_{r=0}^{5} (-1)^r \binom{5}{r} x^{5-r} 7^r = 0$

(c) $\displaystyle\sum_{r=0}^{3} \binom{3}{r} x^r + 8 = 0$ 　　　　　　(d) $\displaystyle\sum_{r=1}^{3} \binom{3}{r} 2^{3-r} x^r = 0$

11. Show that if $p + q = 1$, then $\displaystyle\sum_{r=0}^{n} \binom{n}{r} p^{n-r} q^r = 1$ for each positive integer n.

12. Show that for each pair of positive integers p and n we have

$$\sum_{r_1 + \cdots + r_p = n} \binom{n}{r_1,\, r_2,\, \ldots,\, r_p} = p^n.$$

69
PROBABILITY

If a person draws a card from a standard bridge deck of 52 cards, what are his chances of turning up an ace? What are his chances of drawing the ace of spades? What are your chances of getting a C or better in mathematics? Questions of this kind involve the theory of probability, which we will discuss briefly in the next few sections.

　　We analyze these questions by considering the drawing of a card from a deck or the assigning of a grade in mathematics as an **operation.** For each operation there is a set of possible outcomes, the **outcome set** U. For card drawing, the outcome set U is the set of cards in a bridge deck, and for assigning grades, $U = \{A, B, C, D, F\}$. To each outcome $u \in U$ we assign a non-negative number $p(u)$, the **probability** of the occurrence of the outcome, in such a way that the sum of these numbers is equal to 1. For example, the probabilities that you will receive the various possible grades in mathematics might be given by the following table:

A	B	C	D	F
.1	.2	.3	.2	.2

Thus we have $p(A) = .1$, $p(B) = .2$, and so on. Obviously, it is no easy matter to assign such probabilities so that they correspond to "reality," but let us suppose that the probabilities have been assigned as shown. Observe that the sum of these numbers is 1, as we said it should be. Then, by definition, the probability that you will get a C or better is simply the sum of the probabilities that you will get an A, B, or C; that is, $.1 + .2 + .3 = .6$.

We do the same sort of thing for the card drawing operations. Here it seems reasonable to assume that the 52 possible outcomes of the operation of drawing a card from a bridge deck are *equally likely*. Therefore, we assign the same probability to each. Since the 52 probabilities are to add up to 1, we assign the probability of $\frac{1}{52}$ to each possible outcome. The probability of drawing the ace of spades is therefore $\frac{1}{52}$. To find the probability of drawing an ace, we add the probabilities of drawing the ace of spades, the ace of hearts, the ace of diamonds, and the ace of clubs, and we obtain the number $\frac{4}{52} = \frac{1}{13}$.

In our brief look at the theory of probability, we shall usually consider operations, such as rolling dice or drawing cards from a deck, that have outcomes which most people would call **equally likely.** The distinguishing feature of a set of equally likely outcomes is that each of its members is assigned the same probability. Therefore, if the outcome set has $n(U)$ members, each member u must be assigned the probability $p(u) = 1/n(U)$. Suppose that we consider certain of our outcomes to be successes (for example, we considered it a "success" to draw an ace in our first card-drawing operation), and let us denote by S the subset of the outcome set U that contains these successful outcomes. Then we define the probability $p(S)$ that our operation will result in a successful outcome to be the sum of the probabilities that are assigned to the individual members of S. If the outcomes are equally likely, each member has the probability $1/n(U)$, and so [assuming that S has $n(S)$ members] this sum is $n(S)$ times $1/n(U)$. We therefore have the following theorem.

Theorem 69-1 *If the outcomes of a certain operation are equally likely, then the probability that the operation will result in one of the members of a subset S of the outcome set U is given by the equation*

$$p(S) = \frac{n(S)}{n(U)}.$$

In our first card-drawing operation, for example, U is the set of cards in a bridge deck and S is the set of aces. Therefore, $n(U) = 52$ and $n(S) = 4$. The probability of drawing an ace is thus $p(S) = \frac{4}{52} = \frac{1}{13}$.

EXAMPLE 69-1   What is the probability of making a 3 in one throw of a die?

Solution   Throwing a die can result in six possible numbers, all of which, we suppose, are equally likely. So we have $U = \{1, 2, 3, 4, 5, 6\}$, and $S = \{3\}$. Thus, the probability of making a 3 is $p(S) = \dfrac{n(S)}{n(U)} = \frac{1}{6}$.

EXAMPLE 69-2   What is the probability of making a 7 in one throw of a pair of dice?

Solution   We could say that the possible outcomes of a throw of a pair of dice are the 11 numbers in the set $\{2, 3, \ldots, 12\}$. *But these outcomes are not equally likely.* For example, the only way for a 2 to occur is for a 1 to appear on each die, but there are two ways to make a 3, so we would not expect these outcomes to be assigned the same probability. Therefore, we consider our outcome set to be the set of pairs $U = \{(1, 1), (1, 2), (2, 1), \ldots, (6, 6)\}$, where the first member of the pair is the number of spots that turn up on one die, and the second member is the number of spots that turn up on the other. We assume that these outcomes are equally likely, and now we will count them. Each number that can turn up on one die can be paired with the six possible numbers that can turn up on the other, so $n(U) = 6 \cdot 6 = 36$. The successful outcomes of our operation are those pairs of numbers that add up to 7, so $S = \{(1, 6), (2, 5), (3, 4), (4, 3), (5, 2), (6, 1)\}$, and hence $n(S) = 6$. The probability of making a 7 is therefore $p(S) = \frac{6}{36} = \frac{1}{6}$.

Whenever we use Theorem 69-1 to calculate probabilities, you are to assume that the phrase "number of outcomes" means "number of equally likely outcomes." We will find that our previous study of combinations is useful in counting outcomes.

EXAMPLE 69-3   What is the probability of being dealt a five-card poker hand that contains only spades?

Solution   There are $\dbinom{52}{5}$ possible five-card poker hands, so $n(U) = \dbinom{52}{5}$. The "successful" hands are those that can be formed from the 13 spades, and therefore $n(S) = \dbinom{13}{5}$. Hence, the probability of being dealt a hand containing five spades is

$$p(S) = \binom{13}{5} \Big/ \binom{52}{5} = \frac{13!\,5!\,47!}{8!\,5!\,52!} = .0025.$$

We have denoted by S the subset of the outcome set U that contains the "successful" outcomes of an operation. Let us call all of the other outcomes "failures" and denote the set of failures by F. The probability $p(S)$ that our operation will result in a successful outcome is the sum of the probabilities of the individual members of S, and the probability $p(F)$ that our operation will result in a failure is the sum of the probabilities of the individual members

313

of F. Since the sum of the probabilities of all the members of U is 1, and since each member of U belongs to exactly one of the sets S or F, it is clear that

$$p(S) + p(F) = 1.$$

Thus, we have shown that the probability of failure is 1 minus the probability of success and the probability of success is 1 minus the probability of failure. Sometimes it is easier to calculate $p(F)$ and subtract it from 1 than it is to compute $p(S)$ directly.

EXAMPLE 69-4   What is the probability of being dealt a five-card poker hand that contains at least one spade?

Solution   In this case, the set F of failures consists of the five-card hands that contain no spades at all. There are 39 non-spades in a bridge deck, from which we can form $\binom{39}{5}$ five-card hands. There are $\binom{52}{5}$ possible five-card hands altogether, so the probability of being dealt a spade-free hand is

$$p(F) = \binom{39}{5} \Big/ \binom{52}{5} = \frac{39!\,5!\,47!}{5!\,34!\,52!} = .22.$$

The probability of receiving at least one spade is therefore $1 - p(F) = .78$.

The expectation of winning a prize is defined as the value of the prize times the probability of winning it. If a game has n outcomes $u_1, u_2, \ldots, u_n$ with probabilities $p(u_1), p(u_2), \ldots, p(u_n)$ (where $\sum_{r=1}^{n} p(u_r) = 1$), and if the amounts to be paid on the corresponding outcomes are $A_1, A_2, \ldots, A_n$, then the **average expectation,** or simply the **expectation,** of the game is

$$E = \sum_{r=1}^{n} p(u_r)A_r = p(u_1)A_1 + p(u_2)A_2 + \cdots + p(u_n)A_n.$$

EXAMPLE 69-5   There are three one-dollar bills and two five-dollar bills in a box. You are allowed to draw one bill and keep it. What is the expectation of this game? (This amount might be considered the price you should pay for the privilege of playing.)

Solution   The probability of drawing a one-dollar bill is $\frac{3}{5}$, and the probability of drawing a five-dollar bill is $\frac{2}{5}$. Thus the expectation is

$$E = (\tfrac{3}{5})(1.00) + (\tfrac{2}{5})(5.00) = 2.60.$$

PROBLEMS
69

1.   What is the probability that the sum 11 will appear in a single throw of 2 dice?

2.   A card is drawn from a deck. What is the probability that it will be a black jack?

3. A box contains 5 red, 7 white, and 4 black balls.
   (a) What is the probability of drawing at random one red ball?
   (b) One black ball?
   (c) Two white balls?

4. What is the probability of your being dealt a bridge hand (13 cards) that consists of all honor cards (A, K, Q, J, or 10's)?

5. One card is drawn from a deck.
   (a) What is the probability that it is either a king or a queen?
   (b) Neither a king nor a queen?

6. Four shoes are selected at random from a pile consisting of eight left shoes and twelve right shoes. What is the probability that we end up with two pairs of shoes?

7. A coin is tossed 3 times.
   (a) What is the probability of its coming up heads all 3 times?
   (b) What is the probability of tossing at least 1 tail?

8. (a) What is the probability that a sum less than 7 will appear in a single throw of 2 dice?
   (b) What is the probability that a sum of 13 will appear?

9. In a single throw of a pair of dice what is the probability that
   (a) the same number will turn up on both dice?
   (b) a six will turn up on at least one die?
   (c) the number on each die will be less than 5?
   (d) the number on one die will be 3 more than the number on the other?

10. Two jokers are added to a standard bridge deck and a hand containing n cards is to be dealt from this deck. The hand will be called a success if it contains at least one joker.
    (a) Find $p(S)$ if $n = 5$.
    (b) Find $p(F)$ if $n = 10$.
    (c) Find the smallest number n for which $p(S) > \frac{1}{2}$.

11. (a) What is the probability of your throwing a sum less than 5 in a single roll of a pair of dice?
    (b) What is the expectation of the game if a person wins $2 for throwing a sum less than 5 and $1 for any other sum?

12. A bag contains ten slips that are numbered from 1 to 10. If you draw a number divisible by 3, you win $3. If the number you pick is divisible by 5, then you win $5. Otherwise you win nothing. What is the expectation of the game?

13. In poker, what is the probability of being dealt
    (a) a royal flush, (ace, king, queen, jack, ten of the same suit)?
    (b) a straight flush (five cards in numerical sequence of the same suit, but not a royal flush)?
    (c) a flush (five cards in the same suit but not a straight or royal flush)?

315

We know (Example 69-2) that the probability of making a 7 in a single throw of a pair of dice is $\frac{1}{6}$. Similarly, we can calculate that the probability of making an 11 is $\frac{1}{18}$. In this section, we will learn how to use this information to answer such questions as, "What is the probability of making a 7 *or* an 11 in a single throw of a pair of dice?" and "What is the probability of making a 7 followed by an 11 in two throws?"

The first question is easier to answer than the second. To find the probability that the operation of throwing a pair of dice should be a success, we have to figure out what the successful outcomes are, attach a probability to each, and add up these probabilities of individual successful outcomes. In our example, successful outcomes are of two types, those pairs of numbers whose sum is 7 and those pairs whose sum is 11. Suppose that we let S_1 and S_2 denote these sets of pairs of numbers. Then their union $S_1 \cup S_2$ is the set of all successful outcomes, and the probability of success (throwing a 7 or an 11) is just the number $p(S_1 \cup S_2)$ that we obtain by adding together the probabilities of the individual members of this set. To find this number, we may first add the probabilities of the individual members of S_1, and to this sum add the probabilities of the individual members of S_2. The sum of the probabilities of the members of S_1 is $p(S_1)$, and when we add the sum of the probabilities of the members of S_2, our total is $p(S_1) + p(S_2)$. This number is therefore the sum of the probabilities of the members of $S_1 \cup S_2$; that is,

$$(70\text{-}1) \qquad\qquad \boldsymbol{p(S_1 \cup S_2) = p(S_1) + p(S_2)}.$$

Since $p(S_1) = \frac{1}{6}$ and $p(S_2) = \frac{1}{18}$, we see that the probability of making a 7 *or* an 11 in a single throw of a pair of dice is $\frac{1}{6} + \frac{1}{18} = \frac{2}{9}$.

The argument we just used is simple, but it is incomplete. Here is an example that illustrates the pitfall you must avoid. Suppose that we draw a single card from a standard bridge deck. What is the probability that it will be either an ace or a spade? Superficially, the problem sounds like the problem of throwing a 7 or an 11. But there is a difference. As before, we can think of the successful outcomes as being of two types—drawing a spade and drawing an ace—and we can form two subsets S_1 and S_2 of the outcome set that are made up of these two types of success. The successful outcomes of our operation are the members of the union $S_1 \cup S_2$, and we find the probability of success by adding the probabilities of the individual members of this set. In this case, however, one possible successful outcome (drawing a spade ace) belongs to both S_1 and S_2, so if we tried to obtain the probability $p(S_1 \cup S_2)$ by adding the numbers $p(S_1)$ and $p(S_2)$, we would be counting the probability of this outcome twice.

In order to use Equation 70-1 to calculate a probability, we must be sure that the sets S_1 and S_2 have no members in common; that is, that $S_1 \cap S_2 = \varnothing$. In this case, we say that S_1 and S_2 are **disjoint** subsets of U. More generally, if $S_1, S_2, \ldots, S_k$ is a collection of subsets of U, such that the intersection of every pair of them is empty, we say that the collection is **pairwise disjoint.** The result expressed by Equation 70-1 can easily be extended to more than two subsets, and we have the following theorem.

Theorem 70-1   *If $S_1, S_2, \ldots, S_k$ are pairwise disjoint subsets of the outcome set of an operation, then the probability that the operation will result in one of the members of their union is given by the equation*

$$p(S_1 \cup S_2 \cup \cdots \cup S_k) = p(S_1) + p(S_2) + \cdots + p(S_k).$$

Our second type of combined probability problem involves a study of composite operations. Thus, when we ask for the probability of throwing a 7 followed by an 11, our operation (two throws of a pair of dice) is composed of two single throws. Let us consider another simple example of a composite operation. Suppose we have two boxes, box a_1 that contains a red ball, a yellow ball, and a green ball, and box a_2 that contains a blue ball and a white ball. We are to select a box and draw a ball from it. What is the probability of drawing a red ball? Here we are dealing with a *two stage* composite operation, composed of the basic operations of selecting a box and selecting a ball. We will assume that the outcomes of the first basic operation are equally likely, so the probability of selecting box a_1 is $\frac{1}{2}$. Assuming that we have selected that box, the probability of drawing a red ball is $\frac{1}{3}$. The tree in Fig. 70-1

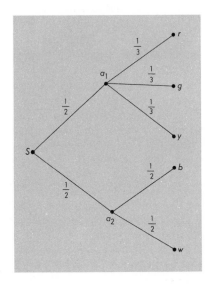

FIGURE 70-1

illustrates this two stage operation; the probability of each outcome is shown on the branch leading to the outcome. It seems reasonable to assume that if we performed this two stage operation a large number of times, say m times, then we would select box a_1 about $\frac{1}{2}m$ times, and in about $\frac{1}{3}$ of these $\frac{1}{2}m$ times we would draw a red ball. In other words, it seems reasonable to suppose that we would draw a red ball about $\frac{1}{3} \cdot \frac{1}{2}m = \frac{1}{6}m$ times out of the m times we perform the two stage operation. Therefore, we take the probability of drawing a red ball to be $\frac{1}{6}$, which is just the product of the probabilities written on the branches of the tree in Fig. 70-1 that lead to the red ball outcome.

317

On the basis of the example in the preceding paragraph, we can formulate a principle for assigning probabilities to the outcomes of a two stage operation, but first we will introduce some notation. If the outcomes of the first and second operations of a two stage composite operation are partitioned into pairwise disjoint subsets $A_1, A_2, \ldots, A_r$ and $B_1, B_2, \ldots, B_s$, respectively, then we use the symbol $p(B_j \mid A_i)$ to denote the **conditional probability** that the second outcome should be a member of B_j on the assumption that the first outcome belongs to A_i. We will use the symbol $p(A_iB_j)$ to denote the probability that successive outcomes are in A_i and B_j. Thus, in the example illustrated by the tree in Fig. 70-1, we have $A_1 = \{a_1\}$, $A_2 = \{a_2\}$, $B_1 = \{r\}$, $B_2 = \{y\}$, $B_3 = \{g\}$, $B_4 = \{b\}$, and $B_5 = \{w\}$. Here $p(A_1)$, the probability of picking box a_1, is $\frac{1}{2}$. The conditional probability of drawing a red ball, on the assumption that we have first chosen box a_1, is the number $p(B_1 \mid A_1) = \frac{1}{3}$. And we have decided to take the product, $\frac{1}{6}$, of these numbers to be the probability $p(A_1B_1)$ of successively choosing box a_1 and a red ball. We therefore have the equation $p(A_1B_1) = p(A_1)p(B_1 \mid A_1)$, which gives us our rule for assigning probabilities to the outcomes of two stage composite operations. For each pair of indices (i, j), we will take

(70-2)
$$p(A_iB_j) = p(A_i)p(B_j \mid A_i).$$

EXAMPLE 70-1 Show that Equation 70-2 gives the correct probability of making a 7 followed by an 11 in two throws of a pair of dice.

Solution We will first calculate this probability by using Theorem 69-1. There are 36 equally likely outcomes for both the first and second throws of the dice, so the Fundamental Principle of Enumeration tells us that there are $36 \cdot 36$ equally likely outcomes for our two stage operation. There are 6 successful outcomes for the first throw and 2 for the second, so there are $6 \cdot 2$ outcomes of the two stage operation that are successful. Hence, the probability of making a 7 followed by an 11 is $\dfrac{6 \cdot 2}{36 \cdot 36} = \dfrac{1}{108}$. Now we will use Equation 70-2 to obtain this same result. Let S_1 and S_2 be the sets of successful outcomes for the two throws. Then $p(S_1) = \frac{1}{6}$. Since the probability of throwing an 11 on the second throw does not depend on the outcome of the first throw, $p(S_2 \mid S_1) = p(S_2) = \frac{1}{18}$. Thus Equation 70-2 gives us $p(S_1S_2) = \frac{1}{6} \cdot \frac{1}{18} = \frac{1}{108}$, as we were to show. We will have more to say about the equation $p(S_2 \mid S_1) = p(S_2)$ later in this section.

The technique we used in Example 70-1 can be used to show that Equation 70-2 gives correct results in coin flipping operations, dice throwing operations, and so on, but we won't prove that Equation 70-2 is the only "reasonable" way to assign probabilities to the outcomes of a general two stage operation. In Problem 70-13 we give you a hint as to how you can show that our assignment does have a key property of a probability measure; that is, the sum of the probabilities that we assign to all of the individual outcomes of the two stage operation is equal to 1.

In the example illustrated by the tree in Fig. 70-1, the probability of drawing a red ball in the second operation cannot be given until the outcome of the first operation is known. In one case it is $\frac{1}{3}$, and in the other case it is 0; it depends on which box you select. But in the two stage operation of throwing a pair of dice twice, the probability of a particular sum occurring on the second throw does not depend on the outcome of the first throw. This fact showed up in our solution to Example 70-1 when we wrote the equation $p(S_2 \mid S_1) = p(S_2)$. When the probabilities of the outcomes of the second operation of a two stage operation do not depend on the outcome of the first operation, we say that the two operations are **independent.** In the case of independent operations, the conditional probabilities reduce to simple probabilities; that is, $p(B_j \mid A_i) = p(B_j)$. Thus, *in the case of independent operations*, Equation 70-2 reduces to

(70-3) $$p(A_iB_j) = p(A_i)p(B_j).$$

EXAMPLE 70-2 A box contains 5 red, 7 white, and 8 black balls. What is the probability of drawing a red ball and then a white ball if (i) the first ball drawn is replaced, or (ii) the first ball drawn is not replaced in the box?

Solution The probability of drawing a red ball is $p(R) = \frac{5}{20} = \frac{1}{4}$ in either case. In the first case, the two operations are independent, and so $p(W \mid R) = p(W) = \frac{7}{20}$. We use Equation 70-3 to see that $p(RW) = p(R)p(W) = \frac{1}{4} \cdot \frac{7}{20} = \frac{7}{80}$. In the second case, the two operations are not independent, and $p(W \mid R) = \frac{7}{19} \neq p(W)$. We use Equation 70-2 to find that, in this case, $p(RW) = \frac{1}{4} \cdot \frac{7}{19} = \frac{7}{76}$.

Our method of calculating probabilities of outcomes in a two stage operation can be extended to operations that have three or more stages in a natural way, as our next example shows.

EXAMPLE 70-3 What is the probability of making a 7 at least once in three throws of a pair of dice?

Solution Here is another case where it is easier to calculate the probability of failure and subtract it from 1 than it is to calculate the probability of success directly. Thus, if F_1, F_2, and F_3 are the sets of failures in the three throws, we know that $p(F_1) = p(F_2) = p(F_3) = \frac{5}{6}$. Since the three throws of a pair of dice are independent operations, we use Equation 70-3 to find that the probability of failure for the three stage operation is $p(F_1F_2F_3) = \frac{5}{6} \cdot \frac{5}{6} \cdot \frac{5}{6} = \frac{125}{216}$. The probability of throwing at least one seven in three throws is therefore $1 - p(F_1F_2F_3) = \frac{91}{216} = .42$.

We can combine our methods of computing probabilities to treat even more complicated cases. Here is an example.

EXAMPLE 70-4 Find the probability that in two throws of a pair of dice we make either two 7's or a 7 and an 11.

319

Solution   If we let $S = \{7\}$ and $E = \{11\}$, then the successful outcomes of our two stage operation are $SS = \{(7, 7)\}$, $SE = \{(7, 11)\}$, and $ES = \{(11, 7)\}$. These outcome subsets are pairwise disjoint, so $p(SS \cup SE \cup ES) = p(SS) + p(SE) + p(ES)$. Since the two stage operation consists of independent operations, we can use Equation 70-3 to compute the probabilities in this sum. Thus we have

$$p(SS \cup SE \cup ES) = p(S)p(S) + p(S)p(E) + p(E)p(S)$$
$$= \tfrac{1}{6} \cdot \tfrac{1}{6} + \tfrac{1}{6} \cdot \tfrac{1}{18} + \tfrac{1}{18} \cdot \tfrac{1}{6} = \tfrac{5}{108} = .046.$$

PROBLEMS
70

1. A number is chosen at random from the set $\{1, 2, \ldots, 50\}$.
   (a) What is the probability that it will be divisible by 3 or by 17?
   (b) What is the probability that it will be divisible by 3 or by 16?
   (c) What is the probability that it will be divisible by 3 and by 17?
   (d) What is the probability that it will be divisible by 3 and by 16?

2. Use Theorem 70-1 to calculate the probability of making a number less than 7 with one roll of a pair of dice.

3. A coin is tossed twice. What is the probability that
   (a) a head will come up on the first toss and a tail on the second?
   (b) a tail will come up on the first toss and a head on the second?
   (c) one head and one tail will come up on the two tosses?

4. A pair of dice is thrown twice. What is the probability that
   (a) two 8's will fall?
   (b) two 10's will fall?
   (c) two 8's or two 10's will fall?
   (d) either an 8 or a 10 will fall on both throws?

5. A box contains 5 red, 6 white, and 9 black balls. Two balls are selected at random—that is, successively without being replaced in the box. What is the probability that
   (a) both balls will be red?
   (b) both balls will be of the same color?
   (c) at least one black ball will be drawn?
   (d) the balls will be of different color?

6. What is the probability of throwing either a 7 or an 11 on 5 successive throws of a pair of dice?

7. Two boxes each contain 6 red poker chips and 4 white poker chips. One chip is selected at random from the first box and placed in the second What is the probability that
   (a) a chip selected at random from the second box will be white?
   (b) two chips selected at random from the second box will both be red?

8. A coin is tossed. If a head appears you draw 2 cards from a bridge deck, and if a tail turns up you draw 1 card. The drawing of a spade is called a success. Find $p(S \mid H)$ and $p(HS)$. What is the probability that the composite operation will be a success?

320

9. Two boxes each contain a certain number of bills. You can select a box at random and select a bill at random. What is the mathematical expectation of the game if

(a) one box contains three $10 bills and one $5 bill, and the second box contains three $5 bills and one $1 bill?

(b) one box contains three $10 bills and one $1 bill, and the second box contains four $5 bills?

(c) one box contains three $10 bills and three $5 bills, and the second box contains one $5 bill and one $1 bill?

10. A box contains 4 red and 6 white balls. One ball is drawn from the box and replaced by a white ball. Then two more balls are drawn. What is the probability that the second drawing will produce two balls of the same color?

11. How many pairs of dice must be thrown simultaneously in order that the probability of making a 2 on at least one of the pairs will be greater than $\frac{1}{2}$? How many times must you throw one pair of dice in order that the probability of making a 2 on the next throw will be greater than $\frac{1}{2}$?

12. Four recruits whose shoe sizes are 7, 8, 9, and 10 report to the supply clerk to be issued shoes. The supply clerk selects one pair of shoes of each of the four required sizes and hands them at random to the men. What is the probability that

(a) no man will receive the correct size?

(b) exactly 2 men will receive the correct size?

(c) exactly 3 men will receive the correct size?

(d) each man will receive the correct size?

13. Suppose that the outcomes of the first and second operations of a two stage composite operation are partitioned into pairwise disjoint subsets $A_1, A_2, \ldots,$ A_r and $B_1, B_2, \ldots, B_s$, respectively. Then for any fixed A_i, $\sum\limits_{j=1}^{s} p(B_j \mid A_i) = 1$. Use this fact and Equation 70-2 to show that $\sum\limits_{j=1}^{s} \sum\limits_{i=1}^{r} p(A_i B_j) = 1$.

14. In order for Equation 70-1 to be valid, we have to assume that $S_1 \cap S_2 = \varnothing$. Can you convince yourself that the equation $p(S_1 \cup S_2) = p(S_1) + p(S_2) - p(S_1 \cap S_2)$ is valid even when $S_1 \cap S_2 \neq \varnothing$? Show that this formula gives you the correct probability of drawing at random a black card or a face card from a standard bridge deck.

71
THE BINOMIAL DISTRIBUTION

Suppose we flip a coin ten times and ask for the probability of obtaining exactly 6 heads. Here we have a ten stage composite operation that consists of independent repetitions of one basic operation. We will begin our study of such operations by analyzing a particular example.

EXAMPLE 71-1 What is the probability of making exactly three 7's in five throws of a pair of dice?

321

Solution We are dealing with a five stage operation; its outcomes are sequences of five numbers, each number in the sequence being an outcome of the basic operation of throwing a pair of dice. An example of a successful sequence of throws is three consecutive 7's followed by two non-7's. We will label this sequence $(7, 7, 7, \text{—}, \text{—})$. Other successful sequences are $(7, \text{—}, 7, \text{—}, 7)$, $(\text{—}, 7, \text{—}, 7, 7)$, and so on. To find the probability of success of our five stage operation, we must find the probabilities of these individual successful outcomes and add them. So our first task is to find the probability of these successful sequences; suppose we start with the sequence $(7, \text{—}, 7, 7, \text{—})$. The probability of throwing a 7 is $\frac{1}{6}$, and the probability of throwing a non-7 is $1 - \frac{1}{6} = \frac{5}{6}$. Since our five throws of the dice are independent, we may use Equation 70-3 (extended to five factors) to see that the probability of the successful sequence $(7, \text{—}, 7, 7, \text{—})$ is $\frac{1}{6} \cdot \frac{5}{6} \cdot \frac{1}{6} \cdot \frac{1}{6} \cdot \frac{5}{6}$. Exactly the same sort of reasoning shows that the probability of the successful sequence $(7, 7, \text{—}, \text{—}, 7)$ is $\frac{1}{6} \cdot \frac{1}{6} \cdot \frac{5}{6} \cdot \frac{5}{6} \cdot \frac{1}{6}$, and so on Now we observe that each of these probabilities is equal to $(\frac{1}{6})^3(\frac{5}{6})^2$, so to obtain the sum of the probabilities of the individual successful sequences, we may simply count them and multiply their number by $(\frac{1}{6})^3(\frac{5}{6})^2$. To count the successful sequences, we observe that each can be constructed by replacing three blanks in the array $(\text{—}, \text{—}, \text{—}, \text{—}, \text{—})$ with 7's; that is, by choosing three places out of five to be labeled "7." Since there are $\binom{5}{3}$ ways of choosing three things from among five, it follows that there are $\binom{5}{3}$ successful sequences. Therefore, the probability of obtaining a successful sequence (making exactly three 7's) is

$$p_3 = \binom{5}{3}(\tfrac{1}{6})^3(\tfrac{5}{6})^2 = .0322.$$

Exactly the same sort of argument works in the general case of n independent repetitions of a given basic operation. We state the result as a theorem.

Theorem 71-1 (The Binomial Law). *If the probability of success for a single operation is p, and if q = 1 − p, then the probability p_r of obtaining exactly r successes when the operation is independently repeated n times is given by the formula*

(71-1)
$$p_r = \binom{n}{r} p^r q^{n-r}.$$

EXAMPLE 71-2 What is the probability of making at least two 7's in three rows of a pair of dice?

Solution Let p_r denote the probability of making exactly r 7's in three throws. Since the probability p of success in a single throw is $\frac{1}{6}$, we have (Theorem 71-1)

$$p_2 = \binom{3}{2}(\tfrac{1}{6})^2(\tfrac{5}{6})^1 = \frac{15}{216} \quad \text{and} \quad p_3 = \binom{3}{3}(\tfrac{1}{6})^3(\tfrac{5}{6})^0 = \frac{1}{216}.$$

We want to find the probability of making *at least* two 7's in three throws—that is, the probability of making exactly two 7's or exactly three 7's. Thus, the required probability is

$$p_2 + p_3 = \tfrac{16}{216} = \tfrac{2}{27} = .0741.$$

Suppose that we roll a die and consider it a successful throw if a 1 or a 2 turns up. The probability of success is clearly $p = \tfrac{1}{3}$. Now suppose we throw 25 dice and consider the number of 1's and 2's that appear. If p_r denotes the probability that exactly r dice will turn up a 1 or a 2, the Binomial Law tells us that $p_r = \binom{25}{r} (\tfrac{1}{3})^r (\tfrac{2}{3})^{25-r}$. Table 71-1 gives p_r for $0 \leq r \leq 25$.

Table 71-1

r	p_r	r	p_r	r	p_r
0	.00004	9	.15802	18	.00007
1	.00050	10	.12642	19	.00001
2	.00297	11	.08619	20	.00000
3	.01139	12	.05028	21	.00000
4	.03131	13	.02514	22	.00000
5	.06575	14	.01077	23	.00000
6	.10959	15	.00395	24	.00000
7	.14872	16	.00123	25	.00000
8	.16732	17	.00033		Total 1.00000

The last six entries are 0 to five decimal places

As r increases from 0 to 8, the numbers p_r in the table also increase. As r increases from 8 to 25, the numbers p_r decrease. The **most likely number** of successes is therefore 8. Now we want to see how to find this number without going to the trouble of constructing a table.

Suppose that p_r is the maximal probability in a given binomial distribution. Then, of course, it is at least as large as p_{r+1} and p_{r-1}; that is,

$$p_r \geq p_{r+1} \quad \text{and} \quad p_r \geq p_{r-1}.$$

According to Theorem 71-1,

$$\frac{p_r}{p_{r+1}} = \binom{n}{r} p^r q^{n-r} \Big/ \binom{n}{r+1} p^{r+1} q^{n-r-1} = \binom{n}{r} q \Big/ \binom{n}{r+1} p.$$

We can use Equation 68-3 to simplify the right-hand side of this last equation, and we get

$$\frac{p_r}{p_{r+1}} = \frac{(r+1)q}{(n-r)p}.$$

Therefore,

$$\frac{p_r}{p_{r+1}} \geq 1 \qquad \text{if, and only if,} \qquad \frac{(r+1)q}{(n-r)p} \geq 1.$$

In other words,

$$p_r \geq p_{r+1} \qquad \text{if, and only if,} \qquad (r+1)q \geq (n-r)p.$$

When we use the equation $p + q = 1$ to simplify the last inequality, we find that

(71-2) $\qquad p_r \geq p_{r+1} \qquad$ if, and only if, $\qquad r + 1 \geq (n+1)p.$

Similarly,

(71-3) $\qquad p_r \geq p_{r-1} \qquad$ if, and only if, $\qquad r \leq (n+1)p.$

Inequalities 71-2 and 71-3 tell us that if p_r is the *maximal probability*, then the index r must satisfy the inequalities

(71-4) $$(n+1)p - 1 \leq r \leq (n+1)p.$$

Observe now that r is an *integer*. If $(n+1)p$ is not an integer, then there is precisely one integer that satisfies Inequalities 71-4, namely, $r = [\![(n+1)p]\!]$. We have the following theorem for finding the most likely number of successes in a given binomial distribution.

Theorem 71-2   *Let p be the probability of a successful outcome of a given operation. If $(n+1)p$ is not an integer, then the most likely number of successful outcomes of the n stage composite operation that consists of n independent repetitions of the given operation is $[\![(n+1)p]\!]$.*

Let us apply Theorem 71–2 to the data in Table 71-1. We see that $(n+1)p = \frac{26}{3}$, and so $[\![(n+1)p]\!] = [\![\frac{26}{3}]\!] = 8$. Thus, the most likely number of successes as calculated by the formula in Theorem 71-2 is the same as the number that we found by inspecting the table.

If the number $(n+1)p$ is an integer, Theorem 71-2 does not quite apply. There are *two* possible integers that satisfy Inequalities 71-4, namely, $(n+1)p$ and $(n+1)p - 1$. It is not hard to see that, in this case, $p_{(n+1)p} = p_{(n+1)p-1}$, and this number is the maximal probability. Thus, *if $(n+1)p$ is an integer, then the numbers $(n+1)p$ and $(n+1)p - 1$ are the two most likely numbers of successes.*

You should not get the impression that if r is the most likely number of successes, then the probability p_r is large. It is fairly obvious that even though p_r is the largest probability of a binomial distribution, p_r is very small if the number of trials n is very large.

EXAMPLE 71-3 Suppose a box contains 4 red balls and 1 white ball. An operation consists of drawing a ball and then replacing it in the box, and it is a success if a white ball is drawn. If the operation is repeated 100 times, what is the most likely number of successes, and what is the probability of obtaining the most likely number of successes?

Solution The number of operations is $n = 100$; the probability of success in each is $p = \frac{1}{5}$. Thus, $(n + 1)p = \frac{101}{5}$ and according to Theorem 71-2, the most likely number of successes is $[\![\frac{101}{5}]\!] = 20$. According to Equation 71-1,

$$p_{20} = \binom{100}{20} (\tfrac{1}{5})^{20}(\tfrac{4}{5})^{80}.$$

If we use logarithms and Tables I and IV, we find that $p_{20} = .0994$. Thus, even though the most likely number of successes is 20, the chances are less than 1 in 10 of having exactly 20 successes.

PROBLEMS
71

1. An operation consists of drawing a card at random from a standard bridge deck, noting its value, and replacing it.
   (a) What is the probability that exactly 1 ace is noted in 2 such draws?
   (b) What if the first card is not replaced?

2. The data in Table 71-1 tell us that the most likely number of successes is 8. What is the probability of 7, 8, or 9 successes?

3. What is the probability of passing a 10-question true-false examination by guessing, if a passing score is (a) 60%? (b) 70%?

4. A coin is tossed 10 times.
   (a) What is the probability that exactly 5 tosses will come up heads?
   (b) What is the probability that at least 5 heads will turn up?
   (c) How many heads are most likely to turn up?

5. If you throw 12 dice, how many would most likely turn up 1? What is the probability that exactly this many would be 1?

6. A box contains 10 white and 3 red balls. An operation consists of drawing two balls and then replacing them in the box. What is the most likely number of successes in 10 operations if an operation is called a success when
   (a) both balls are white?
   (b) the balls are of different color?
   (c) both balls are of the same color?

7. A man offers to bet you even money that when throwing a pair of dice he will throw a 7 before the sixth roll. Should you take the bet?

8. Three cards are drawn from a deck and then replaced. What is the most likely number of times that 3 cards of the same suit will be drawn if 100 drawings are made? What is the probability of obtaining this most likely number?

9. A pair of dice is thrown. You win if a 7 or an 11 appears. You lose if a 2 or a 12 appears, and if any other number appears, the throw is a draw. If the dice are thrown 20 times, what is the most likely number of times that

(a) you will win?

(b) you will either win or lose?

10. Three possible answers are provided for each of the 25 questions of a multiple choice examination. If you get 23 or more right you receive an A; 20 to 22, a B; 16 to 19, a C; and 13 to 15, a D. Find the probability that:

(a) You can get a B or better by guessing.

(b) You can get a C by guessing.

(c) You can pass the exam by guessing.

(d) You can get a B or better if you know the answer to 13 questions and guess on the rest of them.

(e) You can get a C if you proceed as in (d).

(f) You can pass the exam if you proceed as in (d).

11. A die is thrown 6 times. Define the function f by the statement: $f(x)$ is the probability that the number 1 will occur x times. What is the domain of f? Draw a graph of f. Calculate $\sum_{r=0}^{2} f(r)$.

12. A coin is tossed 15 times. Define the function p by the statement: $p(x)$ is the probability that x of the tosses will come up heads. For what values of x is $p(x)$ a maximum? What is this maximum value? Compute $\sum_{r=6}^{9} p(r)$.

13. Suppose the probability that a certain operation has a successful outcome is p, and we repeat it (independently) n times. Show that if $(n+1)p$ is an integer, then $p_{(n+1)p} = p_{(n+1)p-1}$, and this number is the largest member of the set $\{p_0, p_1, \ldots, p_n\}$. (As in the text, p_r denotes the probability that the n repetitions should result in exactly r successes.)

72
EMPIRICAL PROBABILITY

As we said earlier, we have confined our examples of probability to situations in which most people would agree that certain outcomes are "equally likely." It seems clear that the probability of throwing a 1 in a single throw of a die is $\frac{1}{6}$, that the probability of drawing the ace of spades from a deck of cards is $\frac{1}{52}$, and so on. These are examples of what statisticians call *a priori* **probability.** But there are situations in which it is not possible to determine the probability of success by finding the ratio of the number of possible successful outcomes to the number of equally likely outcomes. Sometimes the determination of the probability of an outcome is found experimentally. A probability that is found experimentally is called an **empirical probability.**

In order to determine an empirical probability of a certain outcome, we make a frequency count. Let us suppose that we are given a die so obviously imperfect that we doubt that the probability of throwing a 3, for example, is $\frac{1}{6}$.

In order to determine the probability that a 3 will occur, we make a number of throws, say N of them. If S denotes the number of times that a 3 appears, then we take the empirical probability of a 3 appearing to be S/N. For example, if a 3 appeared 30 times in 100 throws, then the empirical probability would be $\frac{3}{10}$. Of course, the confidence we place in this probability depends a great deal on the number N. If N were 20, it is doubtful that the result would be at all reliable, but if N were 100,000, then we would expect the ratio S/N to represent rather accurately the probability of a 3 occurring. Essentially, the process of finding the empirical probability of a particular outcome by making frequency counts is an attempt to predict future behavior by past performance. The more information we possess about the history of an operation, the surer we feel about our predictions concerning the future outcome of this operation.

EXAMPLE 72-1 In May, Bill Clout was batting .333 for the local baseball team. What is the probability that he will get at least one hit in four official appearances at the plate?

Solution We take the probability of Clout's getting a hit in one official appearance as $\frac{1}{3}$. The probability of his failing to hit is therefore $\frac{2}{3}$, and hence the probability of his failing to hit in four consecutive appearances is $(\frac{2}{3})^4 = \frac{16}{81}$. The probability of his getting at least one hit is thus $1 - \frac{16}{81} = \frac{65}{81} = .802$.

There are two things we should note in Example 72-1. First, even though Clout is averaging one hit in three times at the plate, we do not conclude that he will surely get at least one hit in four appearances. We only say that he probably will, and we are about .8 sure. Second, we would feel more confident about this figure if our observations had been made near the end of the season when Clout had been at bat many times, rather than during the opening weeks of the season when he had batted only a few times.

To illustrate further the necessity for being careful in making predictions about empirical probability, let us consider the following situation. Suppose we have a die that we suspect, but do not know, is weighted so as to make a 1 or a 2 appear more often than these numbers would on a true die. To test this hypothesis, we throw the die 25 times and note the number of times that a 1 or a 2 appears. Suppose there are 13 throws that produce a 1 or a 2. What conclusion can we draw? Since the *a priori* probability of making a 1 or a 2 is $\frac{1}{3}$, we expect only about 8 such throws. Therefore, we ask whether the occurrence of 13 throws that result in 1's or 2's is a sufficiently rare event to justify a conclusion that the die is loaded in favor of 1's and 2's. Table 71-1 gives the probability of making a 1 or a 2 exactly r times in 25 throws of a "perfect" die. It tells us that the probability of 13 such throws is very small, only about .025. But the most likely number of successful throws, 8, also has a small probability, only about .167. The question of whether or not our actual result deviates enough from the "expected" result to enable us to draw any conclusion about the "fairness" of the die is one that is extensively treated in statistics, and we will not be able to pursue the matter much farther here. Obviously, such questions are important ones. In this case, we might observe from our table that for a fair die the probability of as many as 13 throws out of 25 turning up 1 or 2 is about .04 (this number is the sum of the probabilities of 13 successes, 14 successes, and so on). Because this number is so small, we are tempted to conclude that our actual result was not due to chance but rather that the die was biased in favor of 1's and 2's. On the other hand, we might decide that the number .04 is too large to allow us to conclude from our experiment that such a bias exists. Determining what is conclusive evidence in such instances is really a matter of agreement concerning the limits of variation allowed.

PROBLEMS
72

1. A coin is tossed 10 times and comes up heads each time. What is the empirical probability that the coin will come up heads on the 11th toss? What is the *a priori* probability that the coin will come up heads on the 11th toss? What is the *a priori* probability that a coin will come up heads 10 consecutive times?

2. Suppose that birth records indicate that the number of boys and the number of girls born each year are equal. What is the probability that a family of 7 children selected at random will consist of
   (a) all boys?
   (b) exactly 4 boys?
   (c) more boys than girls?

3. A drug that was used to treat 1000 patients with a certain illness was found to be of benefit in 200 cases. A doctor uses the drug to treat 10 of his patients. What is the probability that
   (a) none of the patients will benefit from the use of the drug?
   (b) all the patients will benefit from the use of the drug?
   (c) more than half the patients will benefit from the use of the drug?

4. In a primary election in a certain city, 4000 people register as Democrats and 3000 people register as Republicans. The day before the general election you choose 10 names from the phone book and make calls to remind them to vote. Assuming there are no Independents, what is the probability that you contact an equal number of Democrats and Republicans?

5. Suppose that the weather bureau records show that the 300 days of the month of April for the past 10 years could be grouped into 120 sunny days and 180 cloudy days or grouped into 150 warm days and 150 cold days. What is the probability that there will be no more than 1 week of cloudy weather in the month of April? What is the probability that exactly 15 days in the month of April will be warm and sunny?

6. If past records indicate that 25% of the students who enroll for a certain mathematics course do not make a passing grade, what is the probability that no students will fail in a class of 30? How many students are most likely to fail?

7. The Red Sox have beaten the White Sox in 9 of the 15 games played so far this season, and now they are preparing to play a 3-game series.
   (a) What is the probability that the Red Sox will win exactly 1 game?
   (b) What is the most likely number of games that the White Sox will win?

8. The mortality rate of male smokers aged 40–69 is about twice that of non-smokers, and we will assume that about half the men in this age bracket smoke. Of 25 corpses in this age group, what is the most likely number of ex-smokers? According to Table 71-1, what is the likelihood that there are 20 or more non-smokers among the 25 deceased?

9. Suppose that one-third of the men in a certain community are college graduates. The district attorney finds 10 old grads in a group of 25 tax dodgers. Is he justified in concluding that college men are more likely to cheat on their taxes than non-college men? A doctor finds only 3 college grads among 25 men with callouses on their hands. Is he justified in concluding that college men do not work as hard as other men?

10. A man mixes 2000 white balls and 1000 red balls in a large box and offers you the following bet. You are to pick 3 numbers and then choose 25 balls from the box. If the number of red balls you draw is one of your 3 chosen numbers, you win. Otherwise, you lose. What numbers should you pick? Should you play the game at even money? What if you are allowed to pick 4 numbers?

REVIEW PROBLEMS, CHAPTER NINE

You should be able to answer the following questions without referring back to the text.

1. How many even two-digit numbers can be formed from the set $\{1, 2, 3, 4, 5\}$,
   (a) allowing repetitions?　　　　　(b) not allowing repetitions?

2. Simplify the following expressions.
   (a) $\sum_{r=1}^{3} \cos 3! \, \pi/r$　　　　　(b) $\sum_{r=0}^{3} \cos \binom{3}{r} \pi$

3. **A diagonal** of a polygon is a line that joins two non-adjacent vertices. How many diagonals does a pentagon have? How many diagonals does an n-sided polygon have?

4. How many 7-letter code words can be formed from a set of 26 alphabet blocks? How many of these words contain the letter A?

5. Into Box 1 your instructor puts alphabet blocks A, B, and C; into Box 2 he puts D and F. You roll a die. If the resulting number is divisible by 3, you draw your grade from Box 1; otherwise, you draw from Box 2. What are your chances of failing the course? What are your chances of passing?

6. Explain why $(a+b)^n = \sum_{r+s=n} \binom{n}{r, \, s} a^r b^s$.

329

7. An instructor with a class of 40 students calls on 5 students chosen at random every day for recitation. What is the probability that an unprepared student will escape detection for 5 consecutive days?

8. Compute the number $\displaystyle\sum_{r=0}^{10} \binom{10}{r} \cos^{20-2r} t \sin^{2r} t$.

9. What is the probability of failing to throw either a 7 or an 11 in three consecutive rolls of a pair of dice? Is this number less than $\frac{1}{2}$ or greater?

10. A baby has 10 letter blocks, 1 A, 2 B's, 3 E's, 1 L, and 3 R's. He places them all in a row and all right side up. What is the probability that they spell BEER BARREL? If he selects 4 blocks and places them right side up, what is the probability that they will spell BEER?

11. An examination of medical records reveals that the mortality rate of a certain disease is .2. A doctor is treating 6 patients with this disease. How many patients are most likely to survive? What is the probability that less than half of his patients will survive?

12. Jack has 3 coins and Jill has 4 coins. They both toss all their coins and whoever gets the greatest number of heads wins. If each gets the same number of heads, Jack will win. What is the probability that Jill will win?

MISCELLANEOUS PROBLEMS, CHAPTER NINE

These exercises are designed to test your ability to apply your knowledge of enumeration, the Binomial Theorem, and probability to somewhat more difficult problems.

1. Solve the equation $k \dbinom{n}{2} = \dbinom{2n}{n}$ for positive integers k and n.

2. Prove that $\displaystyle\sum_{r=0}^{n} \binom{n}{r}^2 = \binom{2n}{n}$.

3. Evaluate $\log \left(\displaystyle\sum_{r=0}^{10} \binom{10}{r} \cos^{20} x \tan^{2r} x \right)$.

4. A deck is made up of cards that are numbered from 1 to 10 inclusive. You draw two cards from the deck at random. If the sum of the numbers on the cards you draw is even, you win $1, and if the sum is odd, you lose $1. Is the game fair?

5. Three boxes each contain 5 white, 3 red, and 2 blue poker chips. One chip is selected at random from the first box and placed in the second. Two chips are then selected at random from the second box and placed in the third box. Finally, 3 chips are selected at random from the third box. What is the probability that all three chips will be of different colors?

6. Suppose that you draw balls successively without replacing them from a box that contains 3 red and 7 white balls until you pick a red ball. How many drawings will you most likely make?

7. In an average hour, five buses and two taxis come down Smith's street. What is the probability that he can catch a bus in a given 5-minute period? What is the probability that he can catch some kind of a ride?

8. (a) Show how DeMoivre's Theorem and the Binomial Theorem combine to tell us that, for a given positive integer n,

$$\sum_{r=0}^{n} \binom{n}{r} i^r \cos^{n-r} x \sin^r x = \cos nx + i \sin nx.$$

(b) Infer from this equation that

$$\cos nx = \cos^n x - \binom{n}{2} \cos^{n-2} x \sin^2 x + \cdots + (-1)^t \binom{n}{2t} \cos^{n-2t} x \sin^{2t} x,$$

where $t = [\![n/2]\!]$.

9. Let p_r be the probability of making r heads in five tosses of a coin. Define the function f by the equations

$$f(x) = p_r \qquad \text{if } r - \tfrac{1}{2} < x \leq r + \tfrac{1}{2} \quad (r = 0, 1, 2, 3, 4, 5).$$

Draw the graph of f. Define the function F by the statement: $F(x)$ is the probability that the number of heads made in five tosses is less than or equal to x. Draw the graph of F.

10. From a box containing 10 five-dollar bills and 10 one-dollar bills you make 3 successive draws of 1 bill each. If the bill is a one, you return it, but if it is a five, you keep it. What is the expectation of this game? What is your most likely winning?

11. Box A contains 3 red balls and 1 white ball. Box B contains 3 white balls and 1 red ball. A ball is selected at random from box A and placed in box B; then a ball is selected at random from box B and returned to box A. This procedure is continued until either (a) 1 box contains 4 red balls and the other box contains 4 white balls, or (b) each box contains 2 red and 2 white balls. If the game ends in situation (a), you receive \$1, and if it ends in situation (b), you receive nothing. How much should you pay to play the game?

12. Suppose that $\{r_1, r_2, \ldots, r_p\}$ is a set of positive integers. Show that

$$\binom{r_1 + r_2 + \cdots + r_p}{r_1, r_2, \ldots, r_p}$$
$$= \binom{r_1 + \cdots + r_p}{r_1} \binom{r_2 + \cdots + r_p}{r_2} \cdots \binom{r_{p-1} + r_p}{r_{p-1}} \binom{r_p}{r_p}.$$

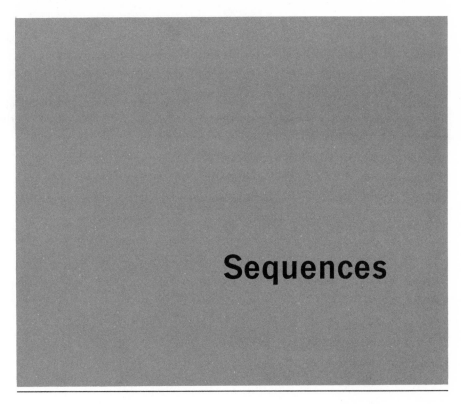

Sequences

ten

You will recall from Chapter Two that a function is a set of pairs. The first elements of these pairs make up the domain of the function, and the second elements constitute its range. As far as the general definition of a function goes, the elements of the domain and range can be any kinds of objects whatever. For us, they are mathematical entities such as angles, points, and the like. Most of the time, of course, they are numbers. In this chapter we will discuss a very special class of functions, the **sequences.** Sequences are simply functions whose domains are sets of integers. Perhaps such functions seem too specialized to be worth a whole chapter, but traditionally sequences have played a central role in mathematical analysis. They have been extensively studied, and our chapter is only a brief introduction to a few of the topics in the theory of sequences.

73
BASIC
CONCEPTS

Sequences have acquired a special notation of their own, and our first task is to explain this notation. If an integer n is in the domain of a function a, then the corresponding element in the range of a is denoted by $a(n)$. In sequence notation, we denote this element by a_n. Thus a sequence is a set of pairs, $\{(n, a_n)\}$, whose first members are integers. It is customary to abbreviate this notation and to speak of "the sequence $\{a_n\}$." Usually, we don't specifically mention the domain of a sequence. Most of our sequences will have the entire set of positive integers as their domains, and if a sequence has some other domain, it will be clear from the context what it is. The numbers $a_1, a_2, a_3, \ldots$ are called the **terms** of the sequence $\{a_n\}$, and because they are paired with the integers $1, 2, 3, \ldots$, we speak of the *first term* a_1, the *second term* a_2, and so on.

EXAMPLE 73-1   Find the second, fifth, and fourteenth terms of the sequence $\left\{\dfrac{(-1)^n}{(2n-1)!}\right\}$.

Solution   If we denote this sequence by $\{a_n\}$, where $a_n = \dfrac{(-1)^n}{(2n-1)!}$, we see that

$$a_2 = (-1)^2/3! = \tfrac{1}{6},\ a_5 = (-1)^5/9! = -1/9!\quad \text{and}\quad a_{14} = 1/27!.$$

The terms of a sequence need not be given by a simple arithmetic formula, as in the last example. Any rule that assigns a number a_n to each integer n will define a sequence. Furthermore, a sequence is independent of the choice of the letter we use as the index. Thus, $\{n!\}$ and $\{k!\}$ are the same sequence.

EXAMPLE 73-2   Let $\{a_n\}$ be the sequence of digits in the decimal representation of $\sin \tfrac{1}{3}\pi$. Find the first four terms of the sequence.

Solution   We know that $\sin \tfrac{1}{3}\pi = \tfrac{1}{2}\sqrt{3} = .866025 \ldots$. Thus $a_1 = 8$, $a_2 = 6$, $a_3 = 6$, and $a_4 = 0$. We have no simple algebraic formula to calculate a_n for each n, but the number a_{34}, for example, is completely determined, and we could find it if we cared to.

A sequence may also be given by specifying the first term and then stating a *recursion formula* that tells how to find each remaining term from the term that precedes it.

EXAMPLE 73-3   Let $\{a_n\}$ be the sequence in which $a_1 = 1$, and $a_n = 3a_{n-1} - 1$ for $n > 1$. Find a_2, a_3, and a_4.

Solution   When we set $n = 2$ in the recursion formula $a_n = 3a_{n-1} - 1$, we see that $a_2 = 3a_1 - 1$. Since $a_1 = 1$, this equation tells us that $a_2 = 3 \cdot 1 - 1 = 2$. Now we set $n = 3$ in the recursion formula to find that $a_3 = 3a_2 - 1$. We have already found that $a_2 = 2$, so we see that $a_3 = 3 \cdot 2 - 1 = 5$. Similarly, $a_4 = 3a_3 - 1 = 14$. It is clear that we can find any term, for example, a_{1776}, simply by plodding along a step at a time. You can easily check to see that the formula

(73-1)
$$a_n = \tfrac{1}{2}(3^{n-1} + 1)$$

gives the correct value of the four terms we have found so far. You should also be able to show that the equation $a_n = 3a_{n-1} - 1$ is satisfied. "Mathematical induction" (which we shall discuss in the next section) is required to show that Formula 73-1 is the *only* formula for a_n such that both $a_1 = 1$ and $a_n = 3a_{n-1} - 1$.

Recursion formulas are especially suitable for use with digital computers, the high-speed machines that do most of our calculating today. We program the machine to go from one term of a sequence to the next, punch in the first term, and turn it on. A modern machine can turn out terms of a sequence at an amazing rate.

Another common way to specify a sequence is to list the first few terms and let the reader use them to infer the general pattern. There is really no logical basis for saying that the fifth term of the sequence $\{1, \frac{1}{2}, \frac{1}{3}, \frac{1}{4}, \ldots\}$ is $\frac{1}{5}$, but that is the number that would come to most people's minds, so a sequence is often specified in this way.

With each sequence $\{a_n\}$ we can associate a sequence $\{S_n\}$ whose terms are defined by the equation

(73-2)
$$S_n = \sum_{k=1}^{n} a_k = a_1 + a_2 + \cdots + a_n.$$

This sequence $\{S_n\}$ is the sequence of **partial sums** of $\{a_n\}$. We have $S_1 = a_1, S_2 = a_1 + a_2 = S_1 + a_2, S_3 = a_1 + a_2 + a_3 = S_2 + a_3$, and so on. Clearly, for each $n > 1$,

(73-3)
$$S_n = S_{n-1} + a_n.$$

EXAMPLE 73-4   Find the first five terms of the sequence of partial sums of the sequence $\{a_n\} = \{1/n(n+1)\}$, and also find a formula that gives S_n for each positive integer n.

Solution   We have

$$S_1 = a_1 = \frac{1}{1 \cdot 2} = \frac{1}{2},$$

$$S_2 = S_1 + a_2 = \frac{1}{2} + \frac{1}{2 \cdot 3} = \frac{2}{3},$$

$$S_3 = S_2 + a_3 = \frac{2}{3} + \frac{1}{3 \cdot 4} = \frac{3}{4},$$

$$S_4 = S_3 + a_4 = \frac{3}{4} + \frac{1}{4 \cdot 5} = \frac{4}{5},$$

$$S_5 = S_4 + a_5 = \frac{4}{5} + \frac{1}{5 \cdot 6} = \frac{5}{6}.$$

When we look at these terms, we are strongly tempted to conclude that $S_6 = \frac{6}{7}$, $S_7 = \frac{7}{8}$, and in general that $S_n = \dfrac{n}{n+1}$. This conclusion is correct, for

$$a_k = \frac{1}{k(k+1)} = \frac{1}{k} - \frac{1}{k+1},$$

and therefore

$$S_n = a_1 + a_2 + a_3 + \cdots + a_n$$

$$= \left(1 - \frac{1}{2}\right) + \left(\frac{1}{2} - \frac{1}{3}\right) + \left(\frac{1}{3} - \frac{1}{4}\right) + \cdots + \left(\frac{1}{n} - \frac{1}{n+1}\right).$$

When we remove the parentheses, this sum "collapses" to

$$S_n = 1 - \frac{1}{n+1} = \frac{n}{n+1}.$$

PROBLEMS
73

1. Find the first five terms of each of the following sequences.
   (a) $\{1 + (-1)^n\}$ (b) $\{3i^n\}$
   (c) $\left\{\dfrac{2 \cdot 4 \cdot 6 \cdots 2n}{(2n)!}\right\}$ (d) $\{\operatorname{Sin}^{-1}(-1)^n\}$
   (e) $\{\cos \frac{1}{2} n\pi\}$ (f) $\left\{\dfrac{1}{n}\log 100^n\right\}$

2. Find the first five terms of the sequence $\{a_n\}$ if you are given the following information. Can you find a formula for a_n?
   (a) $a_1 = 3$, $a_n = 2a_{n-1}$ for $n > 1$
   (b) $a_1 = 2$, $a_n = na_{n-1}$ for $n > 1$
   (c) $a_1 = 2$, $a_n = 1/a_{n-1}$ for $n > 1$
   (d) $a_1 = 3$, $a_n = a_{n-1}^2$ for $n > 1$
   (e) $a_1 = 1$, $a_n = \cos(na_{n-1}\pi)$ for $n > 1$
   (f) $a_1 = 0$, $a_2 = 2$, and $a_n = \frac{1}{2}(a_{n-1} + a_{n-2})$ for $n > 2$
   (g) $a_1 = 0$, $a_2 = 2$, and $a_n = [\![\frac{1}{2}(a_{n-1} + a_{n-2})]\!]$ for $n > 2$
   (h) $a_1 = 1$, $a_2 = 4$, and $a_n = \sqrt{a_{n-1}a_{n-2}}$ for $n > 2$

3. Find a simple formula for a_n such that the first few terms of $\{a_n\}$ are as follows.
   (a) $\frac{5}{2}, \frac{5}{4}, \frac{5}{8}, \frac{5}{16}, \frac{5}{32}, \ldots$ (b) $4, 7, 10, 13, 16, 19, \ldots$
   (c) $2, 5, 10, 17, 26, 37, \ldots$ (d) $1, 0, -1, 0, 1, 0, -1, 0, \ldots$
   (e) $5, 5^2, (5^2)^2, ((5^2)^2)^2, \ldots$ (f) $1, 1, -1, -1, 1, 1, -1, -1, \ldots$

4. Find the fifth partial sum S_5 for each of the following sequences. Can you find a formula for S_n?
   (a) $\{1\}$ (b) $\{\log n\}$ (c) $\{(-1)^n\}$
   (d) $\{1 + (-1)^n\}$ (e) $\{(n+1)! - n!\}$ (f) $\{\log(a_{n+1}/a_n)\}$

5. If we have a given sequence $\{a_k\}$, we can find its sequence of partial sums. Conversely, suppose that we have a given sequence $\{S_n\}$. Show that we can

336

find a sequence $\{a_k\}$ of which the given sequence is the sequence of partial sums. Find a formula for a_k for the following choices of S_n.

(a) $\{2/(n+1)\}$ (b) $\{\log n\}$ (c) $\{n\}$

(d) $\{n^2\}$ (e) $\{n(n+1)/2\}$ (f) $\{3 - 3^{-n}\}$

6. Find the first five terms of the sequence $\{x_n\}$, where x_n is given by the following formula.

(a) $x_n = \displaystyle\sum_{r=0}^{n} \cos r\pi$ (b) $x_n = \displaystyle\sum_{r=0}^{n} |\cos r\pi|$

(c) $x_n = \displaystyle\sum_{r=0}^{n} \binom{n}{r}$ (d) $x_n = \displaystyle\sum_{r=0}^{n-1} 2^r$

7. A certain yeast plant matures in one hour; each hour thereafter it buds off one new plant, which also matures in one hour and starts to produce new plants, and so on. Thus, starting with 1 plant, there will be 1 plant at the end of the first hour, 2 plants at the end of the second hour, 3 plants at end of three hours, and 5 plants at end of four hours. How many plants are there at the end of seven hours? (The sequence defined in this way is an example of what is called a *Fibonacci Sequence*.)

8. Let a_n be the digit in the nth decimal place of the decimal expression for $\frac{1}{13}$. Write the first eight terms of the sequence $\{a_n\}$ and find a_{1984}.

9. Show that a formula for the nth term of the sequence of Example 73-2 is

$$a_n = [\![10^n \sin \tfrac{1}{3}\pi]\!] - 10[\![10^{n-1} \sin \tfrac{1}{3}\pi]\!].$$

10. Let a_n be the area of the regular n-sided polygon inscribed in the unit circle. Find a formula for $a_n (n \geq 3)$. From the geometry of the situation, guess an approximation to a_{1066}.

The sequence $\{a_n\}$ of Example 73-3 was specified by (i) giving the first term and (ii) giving a *recursion formula*, that is, a rule for calculating the successor of any term. Since a_1 is known, we can use the rule to calculate its successor a_2. Then we are in a position to calculate a_3, and so on. If we want the term a_{123}, we can keep going until we get to it. But this process requires 123 steps—steps that can be avoided by developing a formula that expresses the nth term a_n directly in terms of n. We usually arrive at such a formula by guesswork, but we establish its validity by the process of **mathematical induction**, which is based on the following axiom.

    The Induction Axiom. The only sequence of numbers $\{a_n\}$ that has the properties

    (i) $a_1 = 0$, and

    (ii) for each index k such that $a_k = 0$, we have $a_{k+1} = 0$,

is the sequence in which every term is zero; that is, $a_n = 0$ for every index n.

Our statement of the Induction Axiom is pretty formal; it can be worded more simply: "If the first term of a sequence is 0, and if every term that is 0 is followed by a term that is 0, then every term of the sequence is 0." Notice how Properties (i) and (ii) tell us that $a_3 = 0$. We are given that $a_1 = 0$ [Property (i)]. Hence, Property (ii) tells us that this 0 term is followed by a 0 term; that is, $a_2 = 0$. But now we can use Property (ii) again to see that the 0 second term is followed by a 0 term, so $a_3 = 0$. The axiom asserts that if n is any given positive integer, we can carry out this process n times and arrive at the conclusion that $a_n = 0$.

Although we have worded the Induction Axiom in a form that is most useful for our present study of sequences, the axiom is really one of the axioms of the system of positive integers. Let us look at it from a point of view that brings out this fact more clearly. Suppose that we denote by Z the set of indices of terms of the sequence $\{a_n\}$ that are 0. In symbols, $Z = \{n \mid a_n = 0\}$. For example, we can word Property (i) (which says that $a_1 = 0$) as $1 \in Z$. Property (ii) says that if $k \in Z$, then $k + 1 \in Z$. The Induction Axiom states that when Properties (i) and (ii) hold, then *every* term of the sequence is 0; that is, Z is the *entire* set of positive integers. The integer $k + 1$ is called the "successor" of the integer k, so we can word the induction axiom as follows: *The only set of positive integers that contains the integer 1 and the successor of each of its members is the entire set P of positive integers.* There are a number of different-sounding, but logically equivalent, useful ways of stating the Induction Axiom.

Beginners usually find the Induction Axiom a little subtle to apply. It is easy to check whether or not Property (i) holds, but how do we verify Property (ii)? We must show that if k is an index such that $a_k = 0$, then $a_{k+1} = 0$. To this end we *assume* that $a_k = 0$ and show that the equality $a_{k+1} = 0$ stems logically from this assumption. We do not have to *prove* that $a_k = 0$ for any k except $k = 1$. Perhaps a simple example will clear up this point.

EXAMPLE 74-1 Let $\{a_n\}$ be a sequence such that $a_1 = t$, where t is a given number, and $a_{n+1} = a_n^2$ for $n \geq 1$ (that is, each term after the first is the square of the preceding term). Show that this sequence possesses Property (ii) of the Induction Axiom.

Solution Each term of our sequence is the square of its predecessor, so if that predecessor is 0, the term is 0. Thus, it is obvious that 0 terms are followed by 0 terms; the sequence has Property (ii) of the Induction Axiom. By itself, Property (ii) is not enough to guarantee that we are dealing with a sequence consisting only of 0's. We still need to know whether or not $a_1 = 0$ [Property (i) holds]. By hypothesis, $a_1 = t$, so if $t = 0$, our sequence has Property (i) and hence consists only of 0's. If $t \neq 0$, then the sequence does not have Property (i) [although it still has Property (ii)], and the Induction Axiom does not apply.

The next theorem simply puts the Induction Axiom into words that apply more directly to the problems of this chapter.

Theorem 74-1   *If* $\{a_n\}$ *and* $\{b_n\}$ *are two sequences that have the properties*

(i) $a_1 = b_1$ *and*

(ii) *for each index* k *such that* $a_k = b_k$, *we have* $a_{k+1} = b_{k+1}$,

then $a_n = b_n$ *for every index* n.

Proof   Let $\{c_n\}$ be the sequence that is defined by the equation $c_n = a_n - b_n$. We are trying to show that $a_n = b_n$, which amounts to showing that $c_n = 0$. To verify this latter equation, we shall show that the terms of the sequence $\{c_n\}$ satisfy both properties of the Induction Axiom.

According to the first condition of this theorem,

$$c_1 = a_1 - b_1 = 0.$$

Further, if $c_k = 0$, then $a_k = b_k$. Thus, according to the second condition of this theorem, $a_{k+1} = b_{k+1}$, and hence

$$c_{k+1} = a_{k+1} - b_{k+1} = 0.$$

Therefore, for each index k such that $c_k = 0$, we have $c_{k+1} = 0$. Thus we see that the sequence $\{c_n\}$ satisfies both conditions of the Induction Axiom, and hence $c_n = 0$ for every index n. It follows that $a_n = b_n$ for every index n.

We can put Theorem 74-1 in other words as follows. *If the first terms of two sequences coincide, and if the successors of coincident terms are also coincident, then the two sequences are coincident.*

When a sequence is defined in terms of a recursion formula, we may be able to guess the formula for the nth term after writing out a number of the early terms. Theorem 74-1 provides a way to determine whether or not our guess is correct.

EXAMPLE 74-2   Let $\{a_n\}$ be the sequence for which $a_1 = 3$, and $a_n = na_{n-1}$ for $n > 1$. Find a formula for a_n.

Solution   The first few terms of the sequence $\{a_n\}$ are 3, $2 \cdot 3$, $3 \cdot 2 \cdot 3$, $4 \cdot 3 \cdot 2 \cdot 3$, and $5 \cdot 4 \cdot 3 \cdot 2 \cdot 3$. These terms suggest that the formula we seek is $a_n = 3n!$. To prove that this formula is correct, we let $\{b_n\}$ be the sequence in which $b_n = 3n!$ and apply Theorem 74-1. Clearly, $a_1 = b_1$, so Condition (i) of Theorem 74-1 is satisfied. It follows from the recursion formula for the original sequence that

$$a_{k+1} = (k+1)a_{(k+1)-1} = (k+1)a_k$$

for any positive integer k. If, for some index k, $a_k = b_k = 3k!$, we see that

$$a_{k+1} = (k+1) \cdot 3k! = 3(k+1)! = b_{k+1}.$$

Hence, the sequences $\{a_n\}$ and $\{b_n\}$ also satisfy Condition (ii) of Theorem 74-1, and they are therefore the same sequence; that is, $a_n = b_n = 3n!$ for every positive integer n.

EXAMPLE 74-3 Show that for every positive integer n,

$$\sum_{r=1}^{n} r = 1+2+\cdots+n = \frac{n(n+1)}{2}.$$

Solution Consider the sequences $\{a_n\}$ and $\{b_n\}$, where

$$a_n = \sum_{r=1}^{n} r \quad \text{and} \quad b_n = \frac{n(n+1)}{2}.$$

We have $a_1 = 1$, and $b_1 = \frac{1}{2} \cdot 1 \cdot 2 = 1$, so $a_1 = b_1$. Suppose that for some index k, $a_k = b_k$. By definition,

$$a_{k+1} = \sum_{r=1}^{k+1} r = \sum_{r=1}^{k} r + (k+1) = a_k + (k+1).$$

Since we have assumed that $a_k = b_k$, we may write

$$a_{k+1} = b_k + (k+1) = \frac{k(k+1)}{2} + (k+1)$$

$$= \frac{(k+1)(k+2)}{2} = b_{k+1}.$$

The two conditions of Theorem 74-1 are therefore satisfied, and hence we can conclude that $a_n = b_n$ for every n.

EXAMPLE 74-4 Use Mathematical Induction to prove DeMoivre's Theorem:

$$(\cos \theta + i \sin \theta)^n = \cos n\theta + i \sin n\theta.$$

Solution The first terms of the sequences $\{(\cos \theta + i \sin \theta)^n\}$ and $\{\cos n\theta + i \sin n\theta\}$ obviously coincide. Now suppose that for some index k we have

$$(\cos \theta + i \sin \theta)^k = \cos k\theta + i \sin k\theta.$$

Therefore,

$$(\cos \theta + i \sin \theta)^{k+1} = (\cos k\theta + i \sin k\theta)(\cos \theta + i \sin \theta)$$

$$= \cos (k+1)\theta + i \sin (k+1)\theta \quad \text{(Equation 45-2)}.$$

Thus, if the kth terms of our sequences coincide, so do the $(k+1)$st terms. Conditions (i) and (ii) of Theorem 74-1 are both satisfied, and hence the desired equation is established for every positive integer n.

PROBLEMS 74

1. Let $\{a_n\}$ be a sequence with first term $a_1 = 3$.
   (a) If $a_n = -a_{n-1}$, show that $a_n = 3(-1)^{n-1}$.
   (b) If $a_n = a_{n-1} + 5$, show that $a_n = 5n - 2$.
   (c) If $a_n = a_{n-1}^2$, show that $a_n = 3^{2^{n-1}}$.
   (d) If $a_n = a_{n-1}/n$, show that $a_n = 3/n!$.
   (e) If $a_n = \sin \frac{1}{2}a_{n-1}\pi$, show that $a_n = -1$ for $n > 1$.
   (f) If $a_n = [\![\text{Tan}^{-1} a_{n-1}]\!]$, show that $a_n = 0$ for $n > 2$.

2. Which of the sequences of the preceding problem possess Property (ii) of the Induction Axiom?

3. Let $\{a_n\}$ be a sequence with the first term $a_1 = 2$. Find a formula for a_n in terms of n if a_n and a_{n-1} satisfy the following relation.

(a) $a_n = \frac{1}{2}a_{n-1}$ (b) $a_n = \frac{1}{2} + a_{n-1}$

(c) $a_n = -\dfrac{n-1}{n}a_{n-1}$ (d) $a_n = \dfrac{\sin(n-1)t}{\sin nt}a_{n-1}$

(e) $a_n = 3na_{n-1}$ (f) $3^{a_n} = 2^{a_{n-1}}$

4. If $\{a_n\}$ is the sequence with $a_1 = 3$ and $a_n = 1/a_{n-1}$ for $n > 1$, show that

$$a_n = \frac{2 + (-1)^{n-1}}{2 + (-1)^n}.$$

5. Use mathematical induction to show that the formula $a_n = \frac{1}{2}(3^{n-1} + 1)$ is the only formula that yields the sequence of Example 73-3.

6. Let $\{a_n\}$ be a sequence with a given first term a_1. Suppose that $a_n = a_{n-1} + d$ (where d is a given number) for $n > 1$. Show that $a_n = a_1 + (n-1)d$.

7. Let $\{a_n\}$ be a sequence with a given first term a_1. Suppose that $a_n = ra_{n-1}$ (where r is a given number) for $n > 1$. Show that $a_n = ra^{n-1}$.

8. Use mathematical induction to verify the following equations.

(a) $2 + 4 + 6 + \cdots + 2n = n(n+1)$
(b) $1 + 3 + 5 + \cdots + (2n-1) = n^2$
(c) $1^2 + 2^2 + \cdots + n^2 = n(n+1)(2n-1)/6$

(d) $\dfrac{1}{1 \cdot 2} + \dfrac{1}{2 \cdot 3} + \cdots + \dfrac{1}{n(n+1)} = \dfrac{n}{n+1}$

(e) $\sum_{r=1}^{n} (\tfrac{1}{2})^r = 1 - (\tfrac{1}{2})^n$

(f) $\sum_{r=1}^{n} 2(-1)^r = (-1)^n - 1$

9. If $a_n = 3n^2 - 2n + 1$, find two possible recursion formulas that relate a_n and a_{n-1}.

10. Another axiom that is equivalent to the Induction Axiom is the statement "every non-empty set of positive integers has a smallest member." If we accept this statement, then it is easy to show that a sequence $\{a_n\}$ with our Properties (i) and (ii) must be a sequence of 0's. See if you can fill in the details of the following outline of a proof. We let N be the set $\{n \mid a_n \neq 0\}$. Then the statement that each term of the sequence $\{a_n\}$ is the number 0 becomes the equation $N = \varnothing$ (the empty set). If $N \neq \varnothing$, then it contains a smallest member. This smallest member is not 1 (why?), so it can be written as $k + 1$. Hence $k \notin N$, and therefore $a_k = 0$. But this equation implies that $a_{k+1} = 0$; that is, $k+1 \in N$. This contradiction proves that it *cannot* be true that $N \neq \varnothing$, so N must be $\varnothing$.

An **arithmetic progression** (abbreviated as *A.P.*) is a sequence $\{a_n\}$ that is determined by the equation

(75-1) $$a_n = a_1 + (n-1)d,$$

where a_1 and d are given numbers.

EXAMPLE 75-1   Find the first five terms of the A.P. $\{a_n\}$ if $a_1 = 5$, and $d = -2$.

Solution   According to Formula 75-1, $a_n = 5 - 2(n - 1) = 7 - 2n$. Hence, $a_1 = 5$, $a_2 = 3$, $a_3 = 1$, $a_4 = -1$, and $a_5 = -3$.

If a_{n-1} and a_n are any two consecutive terms of the A.P. that is determined by Equation 75-1, then

$$a_n - a_{n-1} = a_1 + (n - 1)d - a_1 - (n - 2)d = d.$$

The number d is therefore the difference between two consecutive terms of the progression. Conversely, suppose that $\{a_n\}$ is a sequence such that $a_n - a_{n-1} = d$ for every integer $n > 1$; that is, $a_n = a_{n-1} + d$. Therefore, $a_2 = a_1 + d$. Now it follows that $a_3 = a_2 + d = a_1 + 2d$. Similarly, $a_4 = a_3 + d = a_1 + 3d$, and so on. From these few steps, it is already quite evident that $a_n = a_1 + (n - 1)d$; that is, that $\{a_n\}$ is an A.P. To *prove* this statement, we must use mathematical induction (we asked you to do so in Problem 74-6).

EXAMPLE 75-2   The numbers 5, 8, 11, 14, and 17 are the first five terms of an A.P. $\{a_n\}$. Find the formula for a_n, and find the 12th term.

Solution   We can compute d from any two consecutive terms; for instance, $d = 11 - 8 = 3$, or $d = 14 - 11 = 3$. Since $a_1 = 5$, we see that $a_n = 5 + (n - 1)3 = 3n + 2$, and therefore $a_{12} = 38$.

EXAMPLE 75-3   Suppose $\{a_n\}$ is an A.P. with $a_4 = 24$ and $a_8 = 36$. Find the formula for a_n.

Solution   The required formula is $a_n = a_1 + (n - 1)d$, and we are to determine the numbers a_1 and d. Now $a_4 = a_1 + 3d$, and $a_8 = a_1 + 7d$, and since we know the values of a_4 and a_8, we have a system of two linear equations:

$$a_1 + 3d = 24$$
$$a_1 + 7d = 36.$$

These equations can be easily solved and yield $d = 3$ and $a_1 = 15$. It follows that $a_n = 15 + (n - 1)3 = 12 + 3n$.

There is a simple way to find the sum of a given number of terms of an A.P. To illustrate the method of finding the formula for the sum of n terms of an A.P., let us find the sum of the first seven terms of the A.P. $\{3n - 1\}$. We first write the sum in the order that the terms appear in the A.P. Then we reverse the order and add the two equations:

$$S_7 = 2 + 5 + 8 + 11 + 14 + 17 + 20$$
$$S_7 = 20 + 17 + 14 + 11 + 8 + 5 + 2$$

$$2S_7 = 22 + 22 + 22 + 22 + 22 + 22 + 22$$

$$2S_7 = 7 \cdot 22$$

$$S_7 = \tfrac{1}{2} \cdot 7 \cdot 22 = 77.$$

Using this calculation as a model, we will derive the formula for the sum of the first n terms of an A.P. $\{a_n\}$. First, observe that we obtain each term of our A.P. by adding the number d to its predecessor or by subtracting d from its successor. Thus, for example, $a_2 = a_1 + d$, $a_3 = a_2 + d = a_1 + 2d$, and so on; and $a_{n-1} = a_n - d$, $a_{n-2} = a_{n-1} - d = a_n - 2d$, and so on. As in our numerical example, we now write the sum $S_n = \sum_{r=1}^{n} a_r$ in two orders and add the two equations:

$$S_n = a_1 + (a_1 + d) + (a_1 + 2d) + \cdots + (a_n - 2d) + (a_n - d) + a_n$$
$$S_n = a_n + (a_n - d) + (a_n - 2d) + \cdots + (a_1 + 2d) + (a_1 + d) + a_1$$

$$2S_n = (a_1 + a_n) + (a_1 + a_n) + (a_1 + a_n) + \cdots + (a_1 + a_n) + (a_1 + a_n)$$
$$= n(a_1 + a_n).$$

We solve this last equation for S_n, and we find that the formula for the sum of the first n terms of an A.P. is

(75-2)
$$\sum_{r=1}^{n} a_r = S_n = n\left(\frac{a_1 + a_n}{2}\right).$$

Formula 75-2 states that *a sum of consecutive terms of an A.P. is obtained by multiplying the average of the first and last terms considered by the number of terms being added.*

EXAMPLE 75-4 Find the sum of the first 100 positive odd integers.

Solution The sequence of positive odd integers is the A.P. $\{2n - 1\}$. The first term is 1, the difference d is 2, and $a_{100} = 199$. According to Formula 75-2,

$$S_{100} = 100(1 + 199)/2 = 100^2 = 10,000.$$

EXAMPLE 75-5 A carpenter was hired to construct 192 window frames. The first day he made five frames, and each day thereafter he made two more frames than he had the day before. How long did it take him to finish the job?

Solution The numbers of frames completed each day form an A.P. whose first few terms are 5, 7, 9, ... Thus, $a_1 = 5$, and $d = 2$. Let n be the number of days required to complete the job. Then $S_n = 192$, and $a_n = 5 + 2(n - 1) = 2n + 3$. Hence,

$$S_n = n(5 + 2n + 3)/2 = n(n + 4),$$
$$192 = n(n + 4),$$
$$n^2 + 4n - 192 = 0,$$
$$(n - 12)(n + 16) = 0.$$

It follows that $n = 12$ or $n = -16$. Since n must be positive, we see that the solution is $n = 12$.

EXAMPLE 75-6   If $\{a_n\}$ is an A.P. with $d = -2$ and $S_8 = 72$, find the terms a_1 and a_8.

Solution   The two fundamental Formulas 75-1 and 75-2 lead to the following system of equations for a_1 and a_8:

$$a_8 = a_1 - 14$$
$$72 = 4(a_1 + a_8).$$

We solve this system and find that $a_1 = 16$ and $a_8 = 2$.

There is a close relationship between A.P.'s and linear functions. You will recall that a linear function f is defined by an equation of the form $f(x) = mx + b$. The domain of f is understood to be the set of all real numbers, unless otherwise specified. If we form the sequence $\{a_n\}$, where $a_n = f(n)$, we see that

$$a_n = m \cdot n + b = m + b + (n - 1)m$$
$$= f(1) + (n - 1)m$$
$$= a_1 + (n - 1)m.$$

Thus, every linear function f determines an A.P. $\{f(n)\}$. Conversely, if we are given an A.P. $\{a_n\}$, with a common difference d, then you can easily verify that the linear function defined by the equation $f(x) = dx + (a_1 - d)$ has the property that $f(n) = a_n$ for each positive integer n. In other words, *we can consider an A.P. as a linear function whose domain consists of the positive integers*. The relationship between an A.P. and a linear function is illustrated graphically in Fig. 75-1. Observe that, geometrically speaking, the common

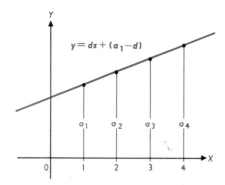

FIGURE 75-1

difference d is the slope of the graph of the linear function that generates the A.P.

PROBLEMS 75

1.   Suppose that $\{a_n\}$ is an A.P.
(a) If $a_1 = 2$ and $d = -3$, find a_7 and S_7.
(b) If $a_1 = 2$ and $a_7 = 20$, find S_7 and d.

344

(c) If $a_1 = 2$ and $S_7 = -70$, find a_7 and d.

(d) If $a_3 = 2$ and $a_8 = -8$, find a_1, d, and S_8.

(e) If $d = 2$ and $S_{12} = 204$, find a_1 and a_{12}.

(f) If $a_1 = -3$, $d = 3$, and $S_n = 27$, find a_n.

2. How many positive integral multiples of 13 are less than 300? What is their sum?

3. Two hundred logs are stacked in the following manner: 20 logs in the bottom row, 19 in the next row, and so on. How many rows are there, and how many logs are in the top row?

4. Each of the following equations defines a linear function. Find the sum of the first five terms of the A.P. $\{f(n)\}$.

(a) $f(x) = 3x + 2$ (b) $f(x) = -3x + 2$

(c) $f(x) = ix + 5$ (d) $f(x) = 1 - 3x$

5. Find the following sums.

(a) $\sum_{r=1}^{10} 2r$ (b) $\sum_{r=1}^{10} (2r + 5)$ (c) $\sum_{r=1}^{10} (5 - 2r)$ (d) $\sum_{r=3}^{13} (5 - 2r)$

6. A ladder has 12 equally spaced rungs, the bottom rung being 18 inches long and the top rung 12 inches long. How many feet of material went into making the rungs of this ladder?

7. (a) Show that the sum of the first n odd integers is n^2.

(b) Find the sum of the first n even integers.

8. The acceleration of a body falling in a vacuum near the surface of the earth is 32 feet per second per second. Each second that the body falls, it falls 32 feet farther than it did the second before. If a body falls from rest, it falls 16 feet in the first second. How far does it fall in 10 seconds? How far does it fall in t seconds? (We are assuming that t is a positive integer here, but the formula is valid for non-integral positive numbers, too.)

9. On January 1, 1960, Mr. Smith deposited $1000 in the bank at 5% interest. Each January 1 thereafter he gave the interest from the account to the Red Cross and deposited another $1000. How much did the Red Cross receive on January 1, 1972? To what total did that bring Mr. Smith's gifts to the Red Cross? How much did he have in the bank on January 2, 1972?

10. Show that the sum S_n of n terms of our A.P. is given by the equation $S_n = na_1 + \frac{1}{2}n(n-1)d$. Solve this equation for n, taking into consideration the fact that n must be positive.

11. Suppose that $\{a_n\}$ and $\{b_n\}$ are sequences such that $a_1 = 2$, $b_1 = 1$,

$$a_n = b_{n-1} + 3, \quad \text{and} \quad b_n = a_{n-1} + 2.$$

Find formulas that express a_n and b_n in terms of n. (*Hint:* You may want to consider separately the cases of odd and even indices.)

12. See if you can work through the steps of the following alternative derivation of Equation 75-2; it will be a good test of your understanding of the sigma notation:

$$S_n = \sum_{r=1}^{n} a_r = \sum_{r=1}^{n} [a_1 + (r-1)d]$$

$$= \sum_{r=1}^{n} a_1 + d \sum_{r=1}^{n} (r-1) = na_1 + d \sum_{r=1}^{n-1} r.$$

From the result of Example 74-3, we see that $\sum_{r=1}^{n-1} r = \frac{1}{2}n(n-1)$, so we have

$$S_n = na_1 + \frac{1}{2}n(n-1)d = \frac{1}{2}n[a_1 + a_1 + (n-1)d] = \frac{1}{2}n(a_1 + a_n).$$

76
GEOMETRIC PROGRESSIONS

Another type of sequence that frequently appears in mathematical applications is the **geometric progression** (**G.P.**). A sequence $\{a_n\}$ is a G.P. if each term is obtained from the formula

(76-1) $$a_n = a_1 r^{n-1},$$

where a_1 is a given first term and r is a given number. From this equation we see that $a_{n+1} = a_1 r^n$, and hence

$$\frac{a_{n+1}}{a_n} = \frac{a_1 r^n}{a_1 r^{n-1}} = r.$$

Thus, the *quotient* of two consecutive terms of a G.P. is constant, whereas in an A.P., the *difference* between two consecutive terms is constant. Conversely, if $\{a_n\}$ is a sequence which satisfies the recursion formula $a_{n+1} = ra_n$, then $a_2 = ra_1$, $a_3 = ra_2 = r^2 a_1$, $a_4 = ra_3 = r^3 a_1$, and so on. It is plain that $a_n = r^{n-1}a_1$, so our sequence is a G.P. To *prove* this statement, we must use mathematical induction (as we asked you to in Problem 74-7).

EXAMPLE 76-1 The numbers 1, 2, 4, 8, and 16 are the first five terms of a G.P. $\{a_n\}$. Find the formula for a_n and use it to calculate a_7.

Solution Clearly, $a_1 = 1$ and $r = \frac{2}{1} = 2$. Hence, $a_n = 1 \cdot 2^{n-1} = 2^{n-1}$, and therefore $a_7 = 2^6 = 64$.

EXAMPLE 76-2 The fifth term of a G.P. $\{a_n\}$ is 4, and the ratio of two consecutive terms if $\frac{2}{3}$. Find the formula for a_n.

Solution We know that $r = \frac{2}{3}$, and hence $a_5 = a_1(\frac{2}{3})^4$. But we also know that $a_5 = 4$, so we have $4 = a_1(\frac{2}{3})^4$. Therefore, $a_1 = 4(\frac{2}{3})^{-4}$, and

$$a_n = 4(\tfrac{2}{3})^{-4}(\tfrac{2}{3})^{n-1} = 4(\tfrac{2}{3})^{n-5}.$$

It is easy to find a formula for the nth partial sum S_n of a G.P. We first write

(76-2) $$S_n = a_1 + a_1 r + a_1 r^2 + \cdots + a_1 r^{n-1}.$$

Then we multiply both sides of Equation 76-2 by the number r to obtain the equation

(76-3)
$$rS_n = a_1 r + a_1 r^2 + \cdots + a_1 r^{n-1} + a_1 r^n.$$

Now we subtract each side of the Equation 76-3 from the corresponding side of Equation 76-2 to obtain the equation

(76-4)
$$(1 - r)S_n = a_1 - a_1 r^n.$$

If $r \neq 1$, we may solve Equation 76-4 for S_n:

(76-5)
$$S_n = \sum_{k=1}^{n} a_1 r^{k-1} = a_1 \frac{1 - r^n}{1 - r}.$$

EXAMPLE 76-3   If $a_1 = 3$, $r = 2$, and $a_N = 768$ in a G.P., find N and S_N.

Solution   The two fundamental formulas associated with geometric progressions are Equations 76-1 and 76-5. After we substitute the given values of a_1, r, and a_N in these equations, we have the system of equations

$$768 = 3 \cdot 2^{N-1}$$

$$S_N = 3 \frac{1 - 2^N}{1 - 2} = 3(2^N - 1).$$

From the first of these equations we obtain

$$2^{N-1} = 256 = 2^8,$$

so $N - 1 = 8$ and $N = 9$. Since $2^{N-1} = 256$, we have $2^N = 256 \cdot 2 = 512$, and the second equation yields

$$S_9 = 3(512 - 1) = 1533.$$

EXAMPLE 76-4   When a certain golf ball is dropped on a piece of pavement, it bounces to a height of three-fifths the distance from which it fell. If the ball is dropped from a height of 100 inches, how far has it traveled when it hits the ground for the tenth time? How high will it rise on the next bounce?

Solution   The first time the ball hits the ground, it has fallen from a height of 100 inches. It then bounces to a height of $100(\frac{3}{5})$ inches, from which height it hits the ground for the second time. It then bounces to a height of $100(\frac{3}{5})(\frac{3}{5}) = 100(\frac{3}{5})^2$ inches, and so on. The tenth time the ball hits the ground it will have fallen from a height of $100(\frac{3}{5})^9$ inches and will then bounce to a height of $100(\frac{3}{5})^{10}$ inches. This figure answers the second question. To answer the first question, we observe that, except for the original drop, the ball must bounce from the ground to a "peak" height and fall back to the ground on each bounce. Hence, the total distance D the ball travels before it hits the ground for the tenth time is

$$D = 100 + 2 \cdot 100(\tfrac{3}{5}) + 2 \cdot 100(\tfrac{3}{5})^2 + \cdots + 2 \cdot 100(\tfrac{3}{5})^9.$$

Starting with the second term, the sum on the right is the ninth partial sum of the G.P. with $a_1 = 200(\frac{3}{5})$ and $r = \frac{3}{5}$. From Formula 76-5 we see that

$$D = 100 + 200(\tfrac{3}{5}) \frac{1 - (\tfrac{3}{5})^9}{1 - \tfrac{3}{5}} = 100 + 300[1 - (\tfrac{3}{5})^9]$$

$$= 400 - 300(\tfrac{3}{5})^9.$$

We use logarithms to show that $(\frac{3}{5})^9$ is about .01, and so $D = 397$.

In the same way, we can show that when the ball hits the ground for the hundredth time, it will have traveled $400 - 300(\frac{3}{5})^{99}$ inches; indeed, when the ball hits the ground for the nth time (where n is any positive integer), it will have traveled $400 - 300(\frac{3}{5})^{n-1}$ inches. Thus, no matter how often the ball bounces, the distance it will travel will be less than 400 inches.

In Section 18 we said that an exponential function is defined by an equation of the form $f(x) = b^x (b > 0)$. Let us extend this definition slightly and speak of functions defined by equations of the form $f(x) = cb^x$ as exponential functions, too. A G.P. is related to such exponential functions in the same way that an A.P. is related to a linear function. Suppose the exponential function f is defined by the equation $f(x) = cb^x (b > 0)$, and let $a_n = f(n)$ for each positive integer n. Then we see that

$$a_n = f(n) = cb^n = (cb)b^{n-1}$$
$$= f(1)b^{n-1} = a_1 b^{n-1}.$$

Thus our exponential function f defines a G.P. in which $r = b$ and $a_1 = cb$. Conversely, a G.P. $\{a_n\}$ (with $r > 0$) determines an exponential function f, defined by the equation $f(x) = a_1 r^{x-1}$, that possess the property that $a_n = f(n)$. The relationship between a G.P. and an exponential function is illustrated graphically in Fig. 76-1.

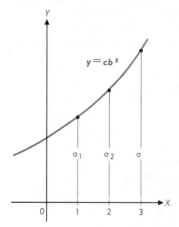

FIGURE 76-1

1. Suppose that $\{a_n\}$ is a G.P.
   (a) If $a_1 = 3$ and $r = 5$, find a_5 and S_5.
   (b) If $a_1 = 2$ and $a_2 = 2i$, find a_8 and S_8.
   (c) If $a_1 = 3$, $r = 2$, and $S_n = 765$, find n.
   (d) If $a_1 = 5$ and $a_5 = 80$, find all possible values of r and S_5.
   (e) If $a_1 = 1$ and $S_{12} = 42$, how many possible values of r are there?

2. Formula 76-5 is meaningless if $r = 1$. What is the formula for S_n if $r = 1$?

3. Find the values of each of the following sums.

   (a) $\displaystyle\sum_{k=1}^{5} 3 \cdot 2^{k-1}$ (b) $\displaystyle\sum_{k=0}^{4} 3 \cdot 2^{k}$ (c) $\displaystyle\sum_{k=1}^{8} 2^{-k}$

   (d) $\displaystyle\sum_{k=0}^{100} i^{k}$ (e) $\displaystyle\sum_{k=1}^{5} (2^{k} + 2k)$ (f) $\displaystyle\sum_{k=1}^{8} (3^{k} + \tfrac{1}{3}^{k})$

4. In 1970, the population of Tunisia was about 5.1 million, and it was growing at the rate of 3% per year. At that rate, what would the population be in 1984? What would be the first year in which the population is double the 1970 figure?

5. You have 2 parents, 4 grandparents, 8 great-grandparents, and so on. How many people does this total if we go back 10 generations?

6. If we know that 3 numbers, a, b, and c, are consecutive terms of both an A.P. and a G.P., find all possible values of a, b, and c.

7. (a) If $\{a_n\}$ is a G.P. of positive terms, show that $\{\log a_n\}$ is an A.P.
   (b) If $\{a_n\}$ is an A.P., show that $\{2^{a_n}\}$ is a G.P.

8. Find the sum of all the positive integral powers of 3 that are less than 2000.

9. Convince yourself that

$$\sum_{k=1}^{n} a_1 r^{k-1} = \sum_{k=0}^{n-1} a_1 r^{k}.$$

10. According to the previous problem, Equation 76-5 can also be written as

$$\sum_{k=0}^{n-1} a_1 r^{k} = a_1 \frac{1 - r^{n}}{1 - r}.$$

Let a and b be two different numbers and set $a_1 = 1$ and $r = a/b$ in this equation. Show that the result can be simplified to

$$\frac{b^{n} - a^{n}}{b - a} = \sum_{k=0}^{n-1} b^{n-k-1} a^{k}.$$

11. On January 1, 1960, Mr. Jones (a neighbor of Mr. Smith of Problem 75-9) deposited $1000 in the bank at 5% interest, and on each January 1 thereafter he deposited another $1000, letting his interest remain in the account. On January 2, 1972, he subtracted the total amount of money he had deposited over the years from the amount in the account and gave the difference to the Red Cross. How much did the Red Cross get?

12. If u is a complex number such that $|u| = 1$ and n is a positive integer, the equation $w^n = u$ has n solutions, $w_1, w_2, \ldots, w_n$. Show that these numbers can be arranged as n consecutive terms of a G.P. What is their sum?

13. By referring to Fig. 76-1, convince yourself that the area of the region that is bounded by the graph of the equation $y = cb^x$, the X-axis, the Y-axis, and the vertical line whose equation is $x = n$ (n is a given positive integer) is a number between

$$c\,\frac{1-b^n}{1-b} \quad \text{and} \quad cb\,\frac{1-b^n}{1-b}.$$

77

ARITHMETIC AND GEOMETRIC MEANS. THE SUM OF A G.P.

The **mean** (more precisely, the **arithmetic mean**) of two numbers a and b is their average, $(a+b)/2$. Thus, the mean of the numbers 3 and 7 is 5. If we observe that the three numbers 3, 5, 7 are the consecutive terms of an A.P., we are led to a generalization of the concept of the arithmetic mean of two numbers. Instead of inserting just one number between two given numbers, we can insert 5, 10, or, in general, n numbers.

Definition 77-1   *If we choose n numbers, each between two given numbers a and b, in such a way that the set of numbers that consists of the chosen numbers and the given numbers a and b forms the first $n+2$ terms of an A.P., then we call the chosen numbers the **nth arithmetic means** of the numbers a and b.*

EXAMPLE 77-1   Find the third arithmetic means of the numbers -3 and 9.

Solution   According to Definition 77-1, we must find three numbers, a_2, a_3, and a_4, such that the numbers $a_1 = -3$, a_2, a_3, a_4, and $a_5 = 9$ form the first five terms of an A.P. To find the common difference d of successive terms of this A.P., we simply substitute the given values of a_1 and a_5 in the equation $a_5 = a_1 + 4d$:

$$9 = -3 + 4d.$$

Thus, $d = 3$, and so $a_2 = 0$, $a_3 = 3$, and $a_4 = 6$.

The **geometric mean** of two positive numbers a and b is the number $\sqrt{ab}$. For example, the geometric mean of 2 and 8 is 4. If we observe that the three numbers 2, 4, 8 are consecutive terms of a G.P., we are led to a generalization of the concept of the geometric mean of two numbers.

Definition 77-2   *If we choose n positive numbers, each between two given positive numbers a and b, in such a way that the set of numbers that consists of the chosen numbers and the given numbers a and b forms the first $n+2$ terms of a G.P., then we call the chosen numbers the **nth geometric means** of the numbers a and b.*

EXAMPLE 77-2   Find the fourth geometric means of the numbers $\frac{1}{2}$ and 16.

Solution   According to Definition 77-2, we must find four positive numbers, a_2, a_3, a_4, and a_5, such that the numbers $a_1 = \frac{1}{2}$, a_2, a_3, a_4, a_5, and $a_6 = 16$ are the first six terms of a G.P., and we find the (positive) ratio r of successive terms of this G.P. by substituting the given values of a_1 and a_6 in the equation $a_6 = a_1 r^5$:

$$16 = \tfrac{1}{2}r^5.$$

Thus, $r = 2$, and the required means are therefore the numbers $a_2 = 1$, $a_3 = 2$, $a_4 = 4$, and $a_5 = 8$.

EXAMPLE 77-3   If a and b are two positive numbers, show that their first arithmetic mean is not less than their first geometric mean.

Solution   Since the arithmetic mean of a and b is $\frac{1}{2}(a+b)$ and the geometric mean is $\sqrt{ab}$, we must show that

$$\sqrt{ab} \leq \tfrac{1}{2}(a+b).$$

In order to demonstrate the correctness of this inequality, we first rewrite it:

$$2\sqrt{ab} \leq (a+b),$$
$$4ab \leq (a+b)^2,$$
$$4ab \leq a^2 + 2ab + b^2.$$

But this last inequality can be rearranged as

$$0 \leq a^2 - 2ab + b^2,$$
$$0 \leq (a-b)^2.$$

The final inequality is obviously correct, and we can reverse the steps in our reasoning to solve the problem.

In Example 76-4, we found that the distance (in inches) traveled by a certain golf ball in n bounces was

$$D_n = 400 - 300(\tfrac{3}{5})^{n-1}.$$

We therefore concluded that no matter how long the ball bounces, it can never travel more than 400 inches. But if n is a very large integer, for example, 1,000,000, then the distance D_n is almost 400. In fact, since we can make the number $(\tfrac{3}{5})^{n-1}$ as small as we please by choosing a large enough value for n, we can make the difference between D_n and 400 as close to zero as we please. We say that the **limit** of D_n is 400.

Let $\{a_n\}$ be the G.P. that is defined by Equation 76-1. Then according to Equation 76-5, we can write the formula for the nth partial sum of the G.P. as

(77-1) $$S_n = \frac{a_1}{1-r} - \left(\frac{a_1}{1-r}\right)r^n.$$

If $|r| < 1$, we can make the second term of the right-hand side of Equation 77-1 as close to zero as we wish by making n large enough. (You can probably convince yourself of this fact by considering the value of an expression like $(\frac{3}{4})^n$, when n is a large number.) It follows that if $|r| < 1$, then $a_1/(1 - r)$ closely approximates the sum S_n when n is a large number; in fact, the larger the number n, the closer the approximation. To express the fact that we can make the difference between S_n and $a_1/(1 - r)$ as small as we please by making n large enough, we write

$$(77\text{-}2) \qquad \lim_{n \uparrow \infty} S_n = \sum_{k=1}^{\infty} a_k = \sum_{k=1}^{\infty} a_1 r^{k-1} = \frac{a_1}{1 - r} \qquad (\text{if } |r| < 1).$$

The symbol $\lim_{n \uparrow \infty} S_n$ is read "the limit of S_n as n increases without bound." The symbol $\sum_{k=1}^{\infty} a_k$ is read "the sum of the sequence $\{a_k\}$."

EXAMPLE 77-4 Find the limit of the partial sums of the G.P. $\{(\frac{1}{2})^n\}$.

Solution Since $a_1 = \frac{1}{2}$ and $r = \frac{1}{2}$, we have, according to Equation 77-2,

$$\lim_{n \uparrow \infty} S_n = \frac{\frac{1}{2}}{1 - \frac{1}{2}} = 1.$$

In Fig. 77-1 the numbers S_1, S_2, S_3, and S_4 are shown plotted on the num-

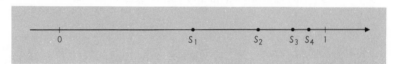

FIGURE 77-1

ber scale. We see that each time a point S_{k+1} is plotted, it bisects the segment between the points S_k and 1. Thus, even though the point S_n is *never* the point 1, it is clear that the point S_n is *very close to* the point 1 when n is a large number.

EXAMPLE 77-5 Find $\sum_{k=0}^{\infty} (\frac{1}{3})^k$.

Solution We are to find the sum of a geometric progression whose first term is 1 and for which $r = \frac{1}{3}$. In Problem 76-9, we pointed out that

$$S_n = \sum_{k=0}^{n-1} a_1 r^k,$$

so Equation 77-2 can also be written as

$$\sum_{k=0}^{\infty} a_1 r^k = \frac{a_1}{1 - r} \qquad (\text{if } |r| < 1).$$

In our present example, $a_1 = 1$ and $r = \frac{1}{3}$, so this equation becomes

$$\sum_{k=0}^{\infty} (\tfrac{1}{3})^k = \frac{1}{1 - \frac{1}{3}} = \tfrac{3}{2}.$$

1. Find the second arithmetic means of the following numbers.
   (a) 6 and 12   (b) −3 and 9   (c) 9 − 5i and 3 + 10i
   (d) 0 and log 125   (e) 3^x and 3^y (f) a and b

2. Find the third geometric means of the following numbers.
   (a) 16 and 81   (b) π^4 and 10^4
   (c) 3 and 15   (d) $\sin^4 t$ and $\cos^4 t$

3. Find the value of each of the following sums.

   (a) $\sum_{k=0}^{\infty} (-\frac{1}{3})^k$ (b) $\sum_{k=1}^{\infty} (-\frac{1}{3})^k$ (c) $\sum_{k=0}^{\infty} 6(.58)^k$

   (d) $\sum_{k=7}^{\infty} (\frac{3}{5})^k$ (e) $\sum_{k=0}^{\infty} \sin^{2k} .6$ (f) $\sum_{k=0}^{\infty} (\log 5)^k$

4. The number .111 ... can be considered as the sum of the G.P. $\{(\frac{1}{10})^n\}$. In other words, .111 ... $= \sum_{k=1}^{\infty} (\frac{1}{10})^k = \frac{9}{10}$. Use this idea to find the rational numbers expressed by the following infinite decimals.
   (a) .222 ...   (b) .999 ...   (c) .1010101 ...   (d) 3.123123123 ...

5. Find 3 numbers a, b, and c that are between 2 and 18 such that their sum is 25, the numbers 2, a, and b are consecutive terms of an A.P., and the numbers b, c, and 18 are consecutive terms of a G.P.

6. For some choices of x, the "sum" $\sum_{k=0}^{\infty} \cos^{2k} x$ makes sense; for some it doesn't. Which are which? Find the sum when it exists.

7. Inside a square that has sides 1 foot long a second square is constructed whose vertices are the midpoints of the sides of the first square. Similarly, a third square is constructed whose vertices are the midpoints of the second square. If this sequence of constructions is continued,
   (a) find the area of the 10th square,
   (b) find the area of the nth square,
   (c) find the sum of the areas of the first 10 squares,
   (d) find the limit of the sum of the areas of the squares.

8. A function f whose domain is the set of all positive numbers is defined by the equation

$$f(x) = \sum_{k=0}^{\infty} (1+x)^{-k}.$$

Sketch the graph of f.

9. Suppose we are told that the zeros of the polynomial $P(x) = ax^3 + bx^2 + cx + d$ form an A.P. Explain why $x + \frac{1}{3}b$ is a factor of $P(x)$.

10. Suppose $\{a_n\}$ is an A.P. What can you say about the sequences $\{-a_n\}$, $\{1/a_n\}$, $\{3a_n\}$, $\{a_n^3\}$, $\{3 + a_n\}$, $\{10^{a_n}\}$, $\{\log |a_n|\}$, and $\{|a_n|\}$? What if our given sequence $\{a_n\}$ is a G.P.?

The terms of the sequences we have talked about up to now have been *numbers*. In this section we are going to discuss sequences whose terms are *statements*; that is, with each positive integer n we will associate some mathematical statement P_n. Here are some examples. What is P_3 in each illustration? Is it a true statement?

Illustration 78-1 P_n is the statement "$1 + 2 + \cdots + n = \frac{1}{2}n(n+1)$."

Illustration 78-2 P_n is the statement "$1 + \dfrac{1}{1!} + \cdots + \dfrac{1}{n!} \leq 3 - \dfrac{1}{n}$."

Illustration 78-3 P_n is the statement "$(n+1)^n < n^{n+1}$."

Illustration 78-4 P_n is the statement "$2^n - 1$ is a perfect square."

We shall explain one way to determine whether or not a sequence $\{P_n\}$ of statements contains only true statements. For any integer n, the corresponding statement P_n of Illustration 78-1 is true, as we showed in Example 74-3. We shall show in Example 78-1 that all the statements of the sequence of Illustration 78-2 are also true. In Illustration 78-3, statements P_1 and P_2 are false, but we shall show (in Example 78-3) that all the other statements in that sequence are true. The only true statement of the sequence of Illustration 78-4 is P_1, but the methods we will discuss in this section are not the way to show that a statement such as P_3 is false.

Our basic tool for dealing with sequences of statements is the following theorem, whose proof is based on the Induction Axiom.

Theorem 78-1   *If $\{P_n\}$ is a sequence of statements that has the properties*

(i) *P_1 is true and*
(ii) *for each index k such that P_k is true, the statement P_{k+1} is also true,*
then P_n is a true statement for every positive integer n.

Proof   Since the Induction Axiom deals with sequences of numbers, and we are working with sequences of statements, our first task is to replace the sequence $\{P_n\}$ of statements with a sequence $\{a_n\}$ of numbers. To this end, we simply let $a_n = 0$ if P_n is a true statement, and $a_n = 1$ if P_n is false. Thus, Property (i) tells us that $a_1 = 0$, and Property (ii) says that if $a_k = 0$, then $a_{k+1} = 0$. But these properties are exactly the ones (according to the Induction Axiom) that guarantee that $\{a_n\}$ is a sequence consisting entirely of 0 terms. In other words, all the terms of the sequence $\{P_n\}$ are true statements.

EXAMPLE 78-1   Use Theorem 78-1 to show that

$$1 + \frac{1}{1!} + \frac{1}{2!} + \cdots + \frac{1}{n!} \leq 3 - \frac{1}{n}$$

for every positive integer n.

Solution We are to verify that the sequence $\{P_n\}$ of Illustration 78-2 consists only of true statements. Since P_1 is the statement "$1 + \dfrac{1}{1!} \leq 3 - \dfrac{1}{1}$", it is true, and so we see that our sequence possesses Property (i) of Theorem 78-1. Now suppose that k is an index such that P_k is a true statement, that is, that

$$1 + \frac{1}{1!} + \cdots + \frac{1}{k!} \leq 3 - \frac{1}{k}.$$

It therefore follows that

$$1 + \frac{1}{1!} + \cdots + \frac{1}{k!} + \frac{1}{(k+1)!} \leq 3 - \frac{1}{k} + \frac{1}{(k+1)!}$$

$$\leq 3 - \frac{1}{k} + \frac{1}{k(k+1)}$$

$$= 3 - \frac{1}{k+1}.$$

This final inequality says that P_{k+1} is true, and so we have shown that the sequence $\{P_n\}$ also possesses Property (ii) of Theorem 78-1. It therefore follows that all the statements in the sequence are true, as we were to show.

EXAMPLE 78-2 Let x be any real number. Show that $|\sin nx| \leq n|\sin x|$ for every positive integer n.

Solution Let $\{P_n\}$ be the sequence in which P_n is the statement "$|\sin nx| \leq n |\sin x|$." Then P_1 is the statement "$|\sin x| \leq |\sin x|$," which is surely true. Now we must show that the truth of the statement P_k implies that statement P_{k+1} also is true. For any k, we have

$$|\sin (k+1)x| = |\sin (kx + x)|$$
$$= |\sin kx \cos x + \cos kx \sin x|$$
$$\leq |\sin kx \cos x| + |\cos kx \sin x|$$
$$\leq |\sin kx| + |\sin x|.$$

Hence, if P_k is true (that is, if $|\sin kx| \leq k |\sin x|$), we see that

$$|\sin (k+1)x| \leq |\sin kx| + |\sin x| \leq k |\sin x| + |\sin x|$$
$$= (k+1)|\sin x|.$$

Thus, the statement P_{k+1},

$$|\sin (k+1)x| \leq (k+1)|\sin x|,$$

is true whenever P_k is true. The two conditions of Theorem 78-1 are satisfied, so we conclude that statement P_n is true for any positive integer n.

EXAMPLE 78-3 Show that, for $n \geq 3$, $(n+1)^n < n^{n+1}$.

Solution With a little algebra, we can show that the above inequality is equivalent to the statement

$$\left(\frac{n+1}{n}\right)^n < n.$$

We will call this last inequality the statement P_n. Since $\left(\dfrac{1+1}{1}\right)^1 > 1$, we see that P_1 is false, and so the sequence $\{P_n\}$ does not possess Property (i) of Theorem 78-1. We will now show, however, that it does possess Property (ii). For any positive integer k is it clearly true that

$$\frac{k+2}{k+1} < \frac{k+1}{k}.$$

Hence,

$$\left(\frac{k+2}{k+1}\right)^{k+1} < \left(\frac{k+1}{k}\right)^{k+1} = \left(\frac{k+1}{k}\right)^{k} \cdot \left(\frac{k+1}{k}\right).$$

If P_k is true (that is, if $\left(\dfrac{k+1}{k}\right)^{k} < k$) then

$$\left(\frac{k+2}{k+1}\right)^{k+1} < k\left(\frac{k+1}{k}\right) = k+1,$$

which simply states that P_{k+1} is true. Hence, $\{P_n\}$ possesses Property (ii) —that is, every true statement of the sequence is followed by a true statement. If we replace our original sequence $\{P_n\}$ with a sequence $\{Q_n\}$ that is obtained from $\{P_n\}$ by neglecting the first two terms ($Q_1 = P_3, Q_2 = P_4$, and so on), it is clear that $\{Q_n\}$ will also possess Property (ii). The sequence $\{Q_n\}$ is the sequence of statements

$$\text{``}(n+3)^{n+2} < (n+2)^{n+3}.\text{''}$$

The sequence $\{Q_n\}$ also possesses Property (i), for Q_1 is the statement

$$\text{``}(1+3)^3 < (1+2)^4\text{''}$$

or

$$\text{``}64 < 81.\text{''}$$

According to Theorem 78-1, it follows that statement Q_n is true for each n, and therefore statement P_n is true for each $n \geq 3$.

PROBLEMS
78

1. Use mathematical induction to show that for each positive integer n, $\log n < n$.

2. Use mathematical induction to show that if $0 < a < b$, then $a^n < b^n$ for each positive integer n.

3. Suppose that $\{a_n\}$ is a sequence of positive numbers and that there is a number r and an index N such that $a_{n+1}/a_n < r$ for each $n \geq N$. Show that $a_{N+p} < a_N r^p$ for each positive integer p.

4. There is a positive integer N such that $\dfrac{2^n n!}{n^n} \leq 1$ for $n \geq N$. Find N, and verify this inequality.

5. Use mathematical induction to prove that the following statements are true for every positive integer n.
   (a) $n^2 - 3n + 4$ is an even number.
   (b) $2n^3 - 3n^2 + n$ is divisible by 6.

6. Suppose P_n is false for every index n. Does the sequence $\{P_n\}$ possess Property (ii) of Theorem 78-1? Suppose a sequence $\{P_n\}$ possesses Property (ii) and there are exactly two indices i and j such that P_i and P_j are false statements, but P_n is true for each index n that is different from i and j. What are the numbers i and j?

7. Suppose that the domain of a certain function f is the set of positive integers and that $f(m + n) = f(m) + f(n)$ for each pair m and n of positive integers. Show that $f(n) = nf(1)$.

8. Suppose we know (by some means or other) that $5 + 1 = 1 + 5$ and that the associative law of addition is valid. Prove that $5 + n = n + 5$ for each positive integer n. (To establish the fundamental laws of arithmetic—Equations 1-1 to 1-5—for the positive integers, we have to resort to mathematical induction. As we said, the Induction Axiom is one of the basic axioms that determine the positive integers, so naturally we have to use it to develop their basic properties.)

9. Let $\{a_n\}$ be the Fibonacci Sequence that is defined by the equations
$$a_1 = 1, \; a_2 = 2, \; a_{n+2} = a_n + a_{n+1}.$$
   Prove that $\sum_{r=1}^{n} a_r = a_{n+2} - 2$.

10. Use mathematical induction to show that for each positive integer n, $[\![n + x]\!] = n + [\![x]\!]$. How can you show that this equation is also true if $n \leq 0$?

11. Suppose that $a \geq -1$. Show that for every positive integer n
$$(1 + a)^n \geq 1 + na.$$
   Where does your proof break down if $a < -1$?

REVIEW PROBLEMS, CHAPTER TEN

You should be able to answer the following questions without referring back to the text.

1. (a) Let $\{a_n\}$ and $\{b_n\}$ be A.P.'s. Show that $\{a_n + b_n\}$ is an A.P.
   (b) Let $\{a_n\}$ and $\{b_n\}$ be G.P.'s. Show that $\{a_n b_n\}$ is a G.P.

2. (a) Let $\{a_n\}$ be an A.P. If b_n is the arithmetic mean of a_n and a_{n+1}, is the sequence $\{b_n\}$ an A.P.? If c_n is the geometric mean of a_n and a_{n+1}, is $\{c_n\}$ an A.P.? Is $\{c_n\}$ a G.P.?
   (b) Answer the questions in part (a) if $\{a_n\}$ is a G.P.

357

3. Calculate the following sums.

(a) $\displaystyle\sum_{k=1}^{30} \log k$ (b) $\displaystyle\sum_{k=1}^{50} \log k^2$ (c) $\displaystyle\sum_{k=10}^{60} \log k$ (d) $\displaystyle\sum_{k=20}^{30} \log \sqrt{k}$

4. If $\{a_n\}$ is a G.P., what is $\sin\left(\dfrac{\log |a_{13}| - \log |a_3|}{\log |a_{21}| - \log |a_{17}|}\right)\pi$?

5. Evaluate the following.

(a) $i^{\sum\limits_{r=1}^{100} (r-3)}$

(b) $(-1)^{\sum\limits_{r=0}^{n} 2r}$

(c) $\sin\left(\displaystyle\sum_{k=3}^{\infty} 2^{-k}\pi\right)$

(d) $\mathrm{Cos}^{-1}\left(\displaystyle\sum_{k=2}^{\infty} (\tfrac{1}{2})^k\right)$

6. Let a, b, and c be given numbers. Show that the sequence formed by taking differences of consecutive terms of the sequence $\{an^2 + bn + c\}$ is an A.P.

7. Let $P(n)$ be the statement "$n^5 < 5^n$." Find a positive integer N such that $P(n)$ is true for each $n \geq N$ and prove it.

8. Is it true that $|\cos nx| \leq n |\cos x|$ for each positive integer n and real number x?

9. There is an old story about a worker who asked to be paid according to an innocent-sounding scheme. His employer was to put one grain of wheat on the first square of a chessboard, two grains on the second square, four on the third, and so on. A chessboard contains 64 squares. How many grains of wheat did the worker collect? (You will need logarithms to help you estimate this huge number.)

10. Mr. Jones started smoking 4 cigarettes a day when he was 17. Each year thereafter, he increased his consumption by 2 cigarettes a day. When he finally died (of lung cancer), he was smoking two packs a day (40 cigarettes). How old was he? How many cigarettes had he smoked during his life?

11. The sequence $\{a_{2n}\}$ is the *subsequence* of $\{a_n\}$ that consists of terms with even indices. If $\{a_n\}$ is an A.P., show that $\{a_{2n}\}$ is an A.P., and if $\{a_n\}$ is a G.P., show that $\{a_{2n}\}$ is a G.P.

MISCELLANEOUS PROBLEMS, CHAPTER TEN

These problems are designed to test your ability to apply your knowledge of sequences and mathematical induction to somewhat more difficult problems.

1. For what positive integers is the inequality $\log n! > n$ true? Use mathematical induction to prove your answer.

2. The product $P_n = a_1 \cdot a_2 \cdots a_n$ is called the **nth partial product** of the sequence $\{a_n\}$. Find a formula for the nth partial product of a G.P.

3. Billiard balls are placed in a rack in the form of an equilateral triangle. There are 5 balls along each side of the rack. A second layer of balls is placed on top

of those in the rack so that there are 4 balls on a side; a third layer is placed on the second layer, and so on, so that a pyramid of balls is formed with one ball on top. How many balls does the pyramid contain? Find a formula for the number of balls in the pyramid if the bottom layer has n balls along each side of the rack.

4. A sequence $\{y_n\}$ is called an **harmonic progression** if the sequence $\left\{\dfrac{1}{y_n}\right\}$ is an A.P.

(a) Write the first 4 terms of an H.P. if $y_1 = 1$, $y_2 = \frac{1}{2}$.

(b) Write the first 4 terms of an H.P. if $y_1 = 2$, $y_2 = \frac{3}{2}$.

(c) Make a logical definition of the harmonic mean of two numbers and find a formula for the harmonic mean of two numbers a and b.

5. Suppose that $a_n = a_1 + (n-1)d$, where d is an integer. Show that $\{[\![a_n]\!]\}$ is an A.P.

6. Suppose that $\{a_n\}$ is an A.P. whose successive terms differ by d. Show that for given positive integers n and p,

$$\sum_{r=1}^{n} a_{r+p} = npd + \sum_{r=1}^{n} a_r.$$

7. A family buys a \$110 item, agreeing to pay \$10 per month for 12 months. How do you figure the monthly interest rate?

8. Geometrically, the set of numbers $\{x \mid 0 \leq x \leq 1\}$ is the interval of the number scale between the points 0 and 1. Suppose that we color the points of the "middle third" of this interval ($\{x \mid \frac{1}{3} < x < \frac{2}{3}\}$) blue, thus leaving us with the uncolored intervals $\{x \mid 0 \leq x \leq \frac{1}{3}\}$ and $\{x \mid \frac{2}{3} \leq x \leq 1\}$. Now we color the middle thirds of each of these intervals blue, and so on. If we continue this process indefinitely, will we color all the points of the original interval? How "long" will the eventual colored set be?

9. Suppose that $\{a_n\}$ and $\{b_n\}$ are sequences such that $a_1 = 5$, $b_1 = -1$, $a_n = 2b_{n-1}$, and $b_n = 3a_{n-1}$ for $n > 1$. What is a_{1776}? b_{1493}?

10. (a) Show that, for each positive integer n, $\log n < \frac{1}{2}n$.

(b) Show that the inequality of part (a) is equivalent to the inequality

$$(n^2)^{1/n^2} < 10^{1/n}.$$

(c) Now convince yourself that if n is a very large number, $(n^2)^{1/n^2}$ is close to 1.

(d) Use the result of Example 78-3 to show that if $m \geq n^2$, then

$$m^{1/m} \leq (n^2)^{1/n^2}.$$

(e) Hence, convince yourself that if m is a very large number, then $m^{1/m}$ is close to 1.

Topics
in
Analytic Geometry

eleven

The number scale is a device for geometrically displaying the set R^1 of real numbers. Similarly, we can introduce a coordinate system into the plane and thereby associate geometric points with *pairs* of numbers. This association allows us to express geometric problems analytically and analytic problems geometrically. Therefore, the study of geometry by coordinate methods is called **analytic geometry.** We have had numerous occasions in this book to introduce ideas from analytic geometry—the distance formula, equations of lines, and so on. In this chapter, we will take up a few other topics. Analytic geometry and calculus go hand-in-hand. The more you know about one topic, the easier it is to understand the other. The brief introduction to analytic geometry that you find in this book will be enough to get you started in calculus.

79
SUBSETS
OF
THE PLANE

Plane geometric figures, such as circles, lines, and so on, are simply sets of points, subsets of the plane. To specify a figure in geometry, therefore, we describe a collection of points. For example, a circle of radius 5 is the set of points that are 5 units from a given point, the center of the circle. Now suppose we introduce a coordinate system into our plane. Then each point is represented by a pair of numbers, and we can use numerical relations to specify plane point sets. For example, the set $\{(x, y) \mid xy < 0\}$ is the set of pairs of numbers that have opposite signs. Graphically, such a pair of numbers is plotted as a point in Quadrant II or in Quadrant IV of the coordinate plane, and so we can picture the set $\{(x, y) \mid xy < 0\}$ as the plane set Quadrant II ∪ Quadrant IV. This set is called the graph of the relation $xy < 0$. In general, the graph of a relation is the set of points in a coordinate plane whose coordinates satisfy the given relation. Most of the relations whose graphs we will study will be *equations* in x and y. Two relations are said to be **equivalent** if they have the same graphs in a coordinate plane. These ideas are, of course, familiar to you, for we have been using them constantly since Chapter 2.

EXAMPLE 79-1   Give a geometric description of, and sketch, the graph of the equation

$$(x - 2)^2 + (y - 6)^2 = 25.$$

Solution   We can find any number of points of the graph of this equation simply by replacing x with a "suitable" number and solving the resulting equation for y. Thus, if we replace x with 2, we obtain the equation $(y - 6)^2 = 25$, whose solutions are 1 and 11. Therefore, the points $(2, 1)$ and $(2, 11)$ are points of our graph. Similarly, we find the points $(-1, 2)$ and $(-1, 10)$ by letting $x = -1$ and solving for y, and so on. We could continue to calculate more points indefinitely and plot them. The more points we plot, the more our figure will "take shape," but this method will never yield the complete picture of the graph of our equation. Let us therefore turn to the equivalent equation

$$\sqrt{(x - 2)^2 + (y - 6)^2} = 5$$

to find a geometric description of our graph. We recognize the left-hand side of this equation as the expression for the distance between the points (x, y) and $(2, 6)$. Thus, in words, the equation says, "The distance between the points (x, y) and $(2, 6)$ is 5." Therefore, a point (x, y) belongs to the graph of the equation $(x - 2)^2 + (y - 6)^2 = 25$ if, and only if, it is a point of the circle of radius 5 whose center is the point $(2, 6)$. We have sketched this circle in Fig. 79-1.

In Example 79-1, we started with an equation and found a geometric description of its graph. Now let us look at an example in which we proceed in the opposite direction. We will start with a point set (described geometrically) in a coordinate plane and find an equation of which the set is the graph. We speak of such an equation as an *equation of the set*. To find an equation of a given set, we must translate the geometric language that tells us that a point (x, y) belongs to the set into a numerical relation between x and y.

EXAMPLE 79-2   Find an equation of the perpendicular bisector of the segment that joins the points $(-5, 4)$ and $(1, -2)$.

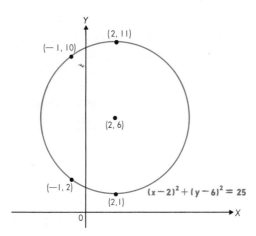

FIGURE 79-1

Solution   From the geometric description of our set, it is easy to make the sketch shown in Fig. 79-2. The given conditions tell us that a point (x, y) of our

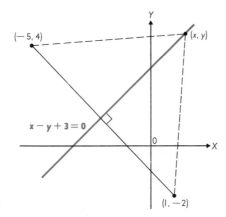

FIGURE 79-2

set is equidistant from the points $(-5, 4)$ and $(1, -2)$. We use the distance formula to express this statement as the equation

$$\sqrt{(x+5)^2 + (y-4)^2} = \sqrt{(x-1)^2 + (y+2)^2}.$$

Conversely, each point whose coordinates satisfy this equation is equidistant from the points $(-5, 4)$ and $(1, -2)$, and therefore is a point of the perpendicular bisector of the segment that joins these points. This equation is perfectly correct, but we can write it more simply. If we square both sides of our equation, multiply out the expressions in the parentheses, and collect terms, we obtain the equation $x - y + 3 = 0$, which is also an equation of the perpendicular bisector.

In the preceding example, we found two equations of the same set of points in a coordinate plane. Instead of speaking of such equations as equivalent equations, we will often speak of them as *two forms of the same equation.*

363

Furthermore, in the same way that we identify a pair of coordinates (x, y) with the point it represents, we will talk about the equation of a point set in a coordinate plane as if the equation *were* the set. Thus, in the preceding example we might speak of the perpendicular bisector $x - y + 3 = 0$.

The intersection of the graphs of two relations is the intersection of two subsets of the plane and consists of those points whose coordinates satisfy both relations.

EXAMPLE 79-3   Find the intersection of the circle that we plotted in Example 79-1 and the line that we plotted in Example 79-2.

Solution   We must find all pairs of numbers that satisfy the two equations

$$(x - 2)^2 + (y - 6)^2 = 25 \quad \text{and} \quad x - y + 3 = 0,$$

so our problem simply amounts to solving this system of equations. From the second equation, we have $y = x + 3$, and when we replace y with $x + 3$ in the first equation we have

$$(x - 2)^2 + (x - 3)^2 = 25,$$
$$x^2 - 5x - 6 = 0,$$
$$(x + 1)(x - 6) = 0.$$

The numbers -1 and 6 are solutions of this equation, and since $y = x + 3$, we see that our intersection consists of the points $(-1, 2)$ and $(6, 9)$.

**PROBLEMS
79**

1.   Because the relations that define them involve only x or only y, the following subsets of a coordinate plane consist of "strips" parallel to the X- or the Y-axis. Describe these sets geometrically.

   (a) $\{(x, y) \mid y > 0\}$　　　　　　(b) $\{(x, y) \mid |x| = 1\}$
   (c) $\{(x, y) \mid -2 < y < 1\}$　　　(d) $\{(x, y) \mid \sin \pi x = 0\}$

2.   Describe the following sets of pairs of numbers geometrically.

   (a) $\{(x, y) \mid xy = 0\}$　　　　　(b) $\{(x, y) \mid x/y = 0\}$
   (c) $\{(x, y) \mid x + y = 0\}$　　　(d) $\{(x, y) \mid x/y < 0\}$
   (e) $\{(x, y) \mid x \text{ and } y \text{ are integers}\}$　(f) $\{(x, y) \mid x \text{ or } y \text{ is an integer}\}$

3.   Find relations between x and y that specify that a point (x, y) belongs to the following sets.

   (a) The set of points to the right of the Y-axis.
   (b) The set of points below the line $y = -4$.
   (c) The circle of radius 3 whose center is the origin.
   (d) The line that bisects Quadrant II and Quadrant IV.
   (e) (the X-axis) $\cap$ (the Y-axis).
   (f) (the X-axis) $\cup$ (the Y-axis).

4.   Sketch the graphs of the following relations.

   (a) $(x - 2)^2 + (y + 1)^2 < 9$　　(b) $(x - 2)^2 + (y + 1)^2 > 9$
   (c) $y < 2x - 1$　　　　　　　　(d) $3x + 2y > 6$
   (e) $y > \sin x$　　　　　　　　(f) $x < \text{Tan}^{-1} y$

5. Sketch the graphs of the following equations.

(a) $y = |\sin \pi x|$ (b) $y = |\log x|$

(c) $y = |2x - 4|$ (d) $y = |\tan x|$

6. Find equations of the following sets of points.

(a) The circle of radius 1 whose center is the point $(2, -3)$.

(b) The perpendicular bisector of the segment joining the points $(-3, 1)$ and $(1, 7)$.

(c) The points that are twice as far from the point $(1, -2)$ as they are from the point $(2, -1)$.

(d) The points that are equidistant from the X- and Y-axes.

(e) The line that is parallel to the X-axis and contains the point $(2, -3)$.

(f) The points whose Y-coordinates are 3 units less than their X-coordinates.

7. Find the intersection of the graphs of the given pairs of equations.

(a) $y = x^2$ and $x = y^2$

(b) $y - 3x = 1$ and $x + 3y = 1$

(c) $y = \sin x$ and $y = \tan x$

(d) $(x - 2)^2 + (y + 1)^2 = 4$ and $y = x - 5$

(e) $y = x^2$ and $2y - x = 1$

(f) $x^2 - y^2 = 3$ and $x^2 + y^2 = 5$

8. Sketch the graphs of the following equations.

(a) $\sin \pi y = 0$ (b) $\log |x| = 0$

(c) $\log (y - x) = 0$ (d) $\operatorname{Sin}^{-1} (y - 2x) = \frac{1}{2}$

(e) $|\sin x| = \sin x$ (f) $[\![y]\!] = y$

9. (a) Explain why the point (x, y) is $|x|$ units from the Y-axis and $|y|$ units from the X-axis.

(b) What equation expresses the fact that a point (x, y) is twice as far from the X-axis as it is from the Y-axis?

(c) Sketch the graph of this equation.

(d) Indicate graphically those points that are *more* than twice as far from the X-axis as they are from the Y-axis.

10. A set of points is **convex** if for each pair of points that belong to the set, the line segment joining them also belongs to the set. Which of the sets of Problem 79-1 are convex?

80
LINES

As we have gone along, we have learned a lot about lines, but these important results are scattered throughout the book, so we will now bring them together and add some new ones. As we saw in Section 48, the general equation of a line has the form

(80-1) $$Ax + By + C = 0,$$

where we must assume that $(A, B) \neq (0, 0)$. If B is not 0, our line is not parallel to the Y-axis, and we can write its equation in the form

(80-2) $$y = mx + b.$$

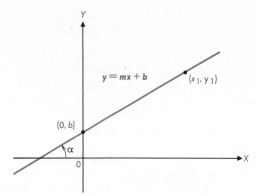

FIGURE 80-1

The number m is the slope of the line, and b is its Y-intercept. Therefore, we speak of Equation 80-2 as the **slope-intercept** form of the equation of a line. Figure 80-1 illustrates these concepts. In Section 40 we found that the slope of a line is the tangent of its angle of inclination, the angle α that the line makes with the positive X-axis. If our line contains the point (x_1, y_1), its equation can be written (explain why) as

(80-3)
$$y - y_1 = m(x - x_1).$$

We speak of this equation as the **point-slope** form of the equation of a line.

EXAMPLE 80-1   Find the equation of the line that makes an angle of 60° with the positive X-axis and contains the point $(-1, 2)$.

Solution   The angle of inclination of our given line is 60°, so its slope is $m = \tan 60° = \sqrt{3}$. Now we know the slope of our line and a point that it contains, so Equation 80-3 becomes $y - 2 = \sqrt{3}(x + 1)$; that is,

$$y = \sqrt{3}x + \sqrt{3} + 2.$$

Notice that the Y-intercept of the line is the number $\sqrt{3} + 2 \approx 3.7$. Make a sketch of this line.

In Section 16 we showed that parallel lines have the same slopes. Now let us find the condition that two lines be perpendicular. Figure 80-2 shows two lines $y = m_1 x + b_1$ and $y = m_2 x + b_2$. The condition that the lines be perpendicular is that their angles of inclination should differ by 90°. Thus, if α_2 is the larger angle, we must have $\alpha_2 = \alpha_1 + 90°$, and hence $\tan \alpha_2 = \tan(\alpha_1 + 90°) = -\cot \alpha_1$. Conversely, if α_1 and α_2 are two angles between 0° and 180° and such that $\tan \alpha_2 = -\cot \alpha_1$, then they differ by 90°. But $\tan \alpha_2 = m_2$ and $\cot \alpha_1 = 1/m_1$, so we see that we can write the condition that the lines be perpendicular as $m_2 = -1/m_1$; that is, $m_1 m_2 = -1$. We state this result as a theorem.

Theorem 80-1   *Two lines $y = m_1 x + b_1$ and $y = m_2 x + b_2$ are perpendicular if, and only if, $m_1 m_2 = -1$.*

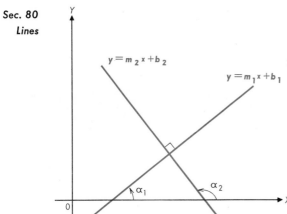

FIGURE 80-2

EXAMPLE 80-2   Find the equation of the line that is perpendicular to the line $3x - y - 7 = 0$ at the point $(2, -1)$.

Solution   We can substitute the coordinates $(2, -1)$ into the point-slope Equation 80-3, and we see that our desired line has the equation

$$y + 1 = m(x - 2),$$

where the slope m is a number to be determined. When we write the equation of our given line in the form $y = 3x - 7$, we see that its slope is 3. Therefore, the slope m of the desired perpendicular line satisfies the equation $3m = -1$; that is, $m = -\frac{1}{3}$. Hence, the equation of this line becomes $y + 1 = -\frac{1}{3}(x - 2)$, which simplifies to $x + 3y + 1 = 0$.

EXAMPLE 80-3   Find the perpendicular distance between the point $(7, 2)$ and the line $3y = 4x + 3$.

Solution   (The discussion that follows will be much clearer if you draw a figure illustrating the problem and add to it as you read along.) Our usual distance formula gives us the distance between two *points;* here we want the distance between a *point* and a *line.* So to use the distance formula, we will have to find the point of our given line in which the perpendicular from the point $(7, 2)$ intersects it. The slope of the given line is easily seen to be $\frac{4}{3}$, so Theorem 80-1 tells us that the slope of a perpendicular line is $-\frac{3}{4}$. The equation of the perpendicular that contains the point $(7, 2)$ is therefore $y - 2 = -\frac{3}{4}(x - 7)$; that is, $3x + 4y = 29$. We solve this equation simultaneously with the equation of the given line, and we find that they intersect in the point $(3, 5)$. The distance between this point and the point $(7, 2)$ is

$$\sqrt{(7 - 3)^2 + (2 - 5)^2} = \sqrt{16 + 9} = 5.$$

which is the desired number. There are formulas for the distance between a point and a line, but we won't take the time to develop them. The method we used in this example can always be used to find such a distance.

1. Find the slope and Y-intercept of the following lines.
   (a) $y = 2x - 3$ (b) $y = 3 - 2x$
   (c) $2x + 3y = 4$ (d) $x - 2y = 5$

2. Find the equations of the following lines.
   (a) The line containing the points $(2, 3)$ and $(6, 5)$.
   (b) The line containing the points $(-1, 8)$ and $(1, 2)$.
   (c) The line containing the point $(2, 4)$ and having slope -3.
   (d) The line containing the point $(-1, 3)$ and having slope 2.
   (e) The line containing the point $(1, -1)$ and making an angle of $120°$ with the positive X-axis.
   (f) The line containing the point $(-2, 2)$ and making an angle of $-45°$ with the positive X-axis.

3. A certain line contains the point $(3, -4)$. Find its equation in the following cases.
   (a) It is parallel to the line $2x + 3y = 5$.
   (b) It is perpendicular to the line $2x + 3y = 5$.
   (c) It is tangent to the circle whose center is the origin and whose radius is 5.
   (d) It contains the midpoint of the line segment joining the points $(2, -1)$ and $(4, 5)$.
   (e) It contains the point of intersection of the lines $2x - y + 4 = 0$ and $x - 3y + 7 = 0$.
   (f) It intersects the X-axis in a point that is 5 units from the origin.

4. The slope (or the angle of inclination) and one point of a line are given. Is the Y-intercept positive or negative?
   (a) $m = \frac{22}{7}$, $(1, \pi)$ (b) $m = \sqrt{2}$, $(1, 1.414)$
   (c) $\alpha = 1$ radian, $(1, \sqrt{3})$ (d) $\alpha = 2$ radians, $(-2, 3)$

5. (a) How far is the point $(1, 8)$ from the line $4x - 3y = 5$?
   (b) How far apart are the lines $4x - 3y = 5$ and $4x - 3y = 30$?

6. Explain why Theorem 80-1 does not apply to the case of a horizontal line and a vertical line.

7. Show that the equation of the line that contains the point (x_1, y_1) and has an angle of inclination α is $(y - y_1) \cos \alpha = (x - x_1) \sin \alpha$.

8. Show that an equation of the line whose Y-intercept is $b \neq 0$ and whose X-intercept is $a \neq 0$ is

$$\frac{x}{a} + \frac{y}{b} = 1.$$

This equation is called the **intercept form** of the equation of the line. Use it to write an equation of the line that contains the points $(11, 0)$ and $(0, 7)$.

9. A circle is drawn so that it is tangent to the line $3x - 4y + 24 = 0$ at the point $(0, 6)$ and also tangent to the line $y = -3$. There are two possible choices for the center of such a circle. Find them.

10. Explain why $\begin{vmatrix} x & y & 1 \\ x_1 & y_1 & 1 \\ x_2 & y_2 & 1 \end{vmatrix} = 0$ is the equation of the line containing the points (x_1, y_1) and (x_2, y_2).

11. Sketch the graphs of the following relations.
    (a) $|3x - 4y| = 5$ (b) $[\![3x - 4y]\!] = 5$
    (c) $|3x - 4y| \leq 5$ (d) $[\![3x - 4y]\!] \leq 5$

12. Sketch the graph of the equation $y = [\![x]\!]x - \frac{1}{2}[\![x]\!][\![x+1]\!]$. (This formula looks a lot more fearsome than it really is.)

81
THE CIRCLE

Geometrically speaking, a circle is a point set in the plane, each of whose members is the same distance from a given point called the center of the circle. We wish to translate this geometric statement into analytic language, so let us suppose we are talking about a circle whose radius is a given positive number r and whose center is a point (h, k) of the coordinate plane. If (x, y) is a point of our circle, then its distance from (h, k) is the number

$$\sqrt{(x - h)^2 + (y - k)^2},$$

and hence the statement that this distance is r can be expressed as the equation $\sqrt{(x - h)^2 + (y - k)^2} = r$. This equation will look a little nicer if we square both sides, and when we do we get the following result: *The equation of the circle with a radius of r and with the point (h, k) as center is*

(81-1) $$(x - h)^2 + (y - k)^2 = r^2.$$

For example, the equation of the circle with a radius of 3 and $(0, -1)$ as center is $x^2 + (y + 1)^2 = 9$; that is, $x^2 + y^2 + 2y - 8 = 0$.

EXAMPLE 81-1   Find the center and radius of the circle
$$x^2 - 4x + y^2 + 8y - 5 = 0.$$

Solution   We can find the radius and the coordinates of the center by inspection (and, incidentally, verify the fact that the graph of the given equation *is* a circle) if we write the equation in the form of Equation 81-1. In order to do so we complete the square (Section 49) as follows:

$$(x^2 - 4x \quad) + (y^2 + 8y \quad) = 5,$$
$$(x^2 - 4x + 4) + (y^2 + 8y + 16) = 5 + 4 + 16,$$
$$(x - 2)^2 + (y + 4)^2 = 25.$$

From our final form of the given equation, we see that its graph is the circle with a radius of 5 and with the point $(2, -4)$ as its center.

EXAMPLE 81-2 Find the center of the circle that contains the three points $(0, 0)$, $(3, 3)$, and $(2, 4)$.

Solution If we denote the coordinates of the center of our circle by (h, k) and its radius by r, then its equation will be Equation 81-1; we are to find the numbers h and k. Since the three given points are to be points of the circle, the coordinates of each must satisfy this equation. So we substitute the coordinates of our three points in Equation 81-1 to obtain the system of equations

$$(0 - h)^2 + (0 - k)^2 = r^2$$
$$(3 - h)^2 + (3 - k)^2 = r^2$$
$$(2 - h)^2 + (4 - k)^2 = r^2.$$

After some simplification, this system takes the form

$$h^2 + k^2 = r^2$$
$$18 - 6h - 6k + h^2 + k^2 = r^2$$
$$20 - 4h - 8k + h^2 + k^2 = r^2.$$

Now we subtract the first equation from the second and the third and obtain a system of two equations in the two unknowns h and k:

$$18 - 6h - 6k = 0$$
$$20 - 4h - 8k = 0.$$

These equations immediately yield $h = 1$ and $k = 2$, so the center of our circle is the point $(1, 2)$.

For the circle with a radius of r and with the origin as center, Equation 81-1 reduces to

(81-2) $$x^2 + y^2 = r^2.$$

This equation is "simpler" than Equation 81-1. When dealing with geometric problems in terms of coordinates, it is wise to choose the coordinate axes so as to make your equations as simple as possible. Sometimes it is even convenient to change coordinate axes in order to simplify things. One method of changing the coordinate axes is to perform a *translation of axes*, a technique we will discuss in the next section.

EXAMPLE 81-3 Jones' head has the equation $x^2 + y^2 = 1$, and an axe is swung along the line $3x + 2y = 10$. Is Jones singular or plural?

Solution The condition that our given curves intersect is that the system of equations

$$x^2 + y^2 = 1$$
$$3x + 2y = 10$$

has a solution pair (x, y) consisting of real numbers. Let us write the second equation as $2y = 10 - 3x$ and square both sides:

$$4y^2 = 100 - 60x + 9x^2.$$

370

In this equation, we replace y^2 with $1 - x^2$ (from the first equation), and the resulting equation simplifies to the quadratic equation

$$13x^2 - 60x + 96 = 0.$$

We are really not interested in *solving* this equation; we only need to know *whether* or not it has (real) solutions. So we calculate its discriminant:

$$60^2 - 4 \cdot 13 \cdot 96 = 3600 - 4992.$$

Because this number is negative, we know that the solutions of our quadratic equation are not real numbers. Hence, the solutions of our original system of equations are not pairs of real numbers. They therefore do not represent points of the plane, and so we conclude that the given curves do not intersect.

PROBLEMS
81

1. Find the equations of the circles with the given points as centers and the given numbers as radii.
   (a) $(2, -3)$, 5
   (b) $(-2, -2)$, 1
   (c) $(\cos 3, \sin 3)$, 1
   (d) $(\log \frac{1}{3}, \log 3)$, $\log 100$

2. Complete the square to find the center and radius of each of the following circles.
   (a) $x^2 + y^2 + 4x - 2y - 4 = 0$
   (b) $x^2 + y^2 - 4y = 0$
   (c) $x^2 + y^2 - x - 2y = 0$
   (d) $36x^2 + 36y^2 - 24x + 36y - 131 = 0$

3. Find the equation of the circle C in the following cases.
   (a) $(2, -3)$ is the center of C, and $(1, 1)$ is one of its points.
   (b) $(0, 0)$, $(0, 2)$, and $(2, 0)$ are points of C.
   (c) $(0, -1)$, $(2, 1)$, and $(-2, 5)$ are points of C.
   (d) C is tangent to the X- and Y-axes, lies in Quadrant II, and has a radius of 5.
   (e) The center of C is the point $(8, 14)$ and C just touches the circle
   $$x^2 + y^2 + 2x - 4y - 20 = 0.$$
   (f) The center of C is the point $(2, 6)$ and C just touches the circle
   $$(x + 1)^2 + (y - 2)^2 = 225.$$
   (g) The center of C is a point of the Y-axis, its radius is 5, and it contains the point $(4, 2)$.
   (h) C lies in Quadrant I, has a radius of 1, and is tangent to the X-axis and the line $y = x$.

4. Describe the following sets geometrically.
   (a) $\{(x, y) \mid x^2 + y^2 \geq 9\}$
   (b) $\{(x, y) \mid x^2 + y^2 - 4x + 2y \leq 4\}$
   (c) $\{(x, y) \mid 4 \leq x^2 + y^2 \leq 25\}$
   (d) $\{(x, y) \mid |x^2 + y^2 - 4x + 2y| < 4\}$

5. For each real number c, the equation $(x - c)^2 + y^2 = c^2$ represents a circle. We call the collection of such circles (for all possible choices of c) a *family* of circles. Describe the members of this family.

6. A basketball goal has the equation $(x+2)^2+(y-1)^2 = 5$, and a ball with a radius of 1 comes down with its center at the point $(-1, 2)$. Does the ball touch the rim?

7. For each choice of t, the pair of equations $x = 5 \cos t$ and $y = 5 \sin t$ gives us a pair of numbers (x, y). For example, if we choose $t = 0$, we get the point $(5, 0)$. For $t = \frac{1}{2}\pi$, we have $(0, 5)$, and so on. Show that the set of all such points is a circle. Describe geometrically the set

$$\{(x, y) \mid x = -3 \sin t, y = 3 \cos t, t \in R^1\}.$$

8. A point moves so that it is always three times as far from the point $(4, -3)$ as it is from the point $(0, 1)$. Describe the path of the moving point.

9. The equation $x^2 + y^2 + ax + by + c = 0$ does not represent a circle for every choice of a, b, and c (for example, not when $a = b = 0$ and $c = 1$). What relation among the coefficients a, b, and c is necessary in order that the graph should be a circle?

10. Explain why the following relations are analytic representations of the circular disk of radius r whose center is the origin. What would the relations be if the center of the disk were the point (h, k)?
   (a) $x^2 + y^2 < r^2$ (b) $[[(x^2 + y^2)r^{-2}]] = 0$

11. Sketch the graphs of the following equations.
   (a) $x^2 + 2xy + y^2 = 36$ (b) $(x - |x|)^2 + 4y^2 = 4$
   (c) $(|x| + x)^2(|y| + y)^2 = 36$ (d) $x^2 + y^2 + 2|x| - 4y = 0$
   (e) $[[x]]^2 + [[y]]^2 = 13$ (f) $[[x^2 + y^2]] = 13$

82
TRANSLATION
OF
AXES.
SYMMETRY

A change of coordinate axes can frequently simplify a problem in analytic geometry, and among the most useful transformations of coordinates are the *translations*. Suppose that we have a set of coordinate axes, shown in Fig. 82-1 as solid lines. We have labeled these axes the X-axis and the Y-axis. We now select another set of coordinate axes, shown in Fig. 82-1 as dashed lines, and label them the $\overline{X}$-axis and the $\overline{Y}$-axis. The $\overline{X}$-axis is

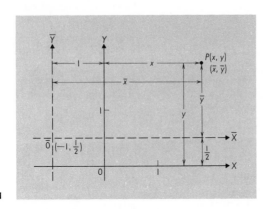

FIGURE 82-1

parallel to the X-axis and $\frac{1}{2}$ unit above it. The $\overline{Y}$-axis is parallel to the Y-axis and 1 unit to the left of it. We say that we have obtained the $\overline{X}\overline{Y}$-axes by a *translation*. Any point P in the plane can be described by either its XY-coordinates (x, y) or its $\overline{X}\overline{Y}$-coordinates $(\overline{x}, \overline{y})$. From Fig. 82-1 we see that the relation between the numbers x and $\overline{x}$ is $\overline{x} = x + 1$ and between the numbers y and $\overline{y}$ is $\overline{y} = y - \frac{1}{2}$. Let us write these equations as

$$\overline{x} = x - (-1)$$
$$\overline{y} = y - \frac{1}{2}.$$

(82-1)

Now observe that the origin of the $\overline{X}\overline{Y}$-coordinate system has $\overline{X}\overline{Y}$-coordinates $(0, 0)$ and XY-coordinates $(-1, \frac{1}{2})$. Thus, Equations 82-1 tell us that we obtain the $\overline{X}\overline{Y}$-coordinates of a point P by subtracting the XY-coordinates of the origin of the $\overline{X}\overline{Y}$-system from the XY-coordinates of P.

Let us generalize the problem we have just been looking at and consider a coordinate change in which we obtain new coordinates $(\overline{x}, \overline{y})$ for a point with original coordinates (x, y) by using the formulas

$$\overline{x} = x - h$$
$$\overline{y} = y - k,$$

(82-2)

where h and k are given numbers. It is clear from Equations 82-2 that the origin of the $\overline{X}\overline{Y}$-coordinate system—that is, the point whose $\overline{X}\overline{Y}$-coordinates are $(0, 0)$—is the point whose XY-coordinates are (h, k). It is just as easy to locate the new coordinate axes. The $\overline{Y}$-axis is the set of points for which $\overline{x} = 0$. According to Equations 82-2, this equation says that $x = h$. In other words, *the line $x = h$ is the $\overline{Y}$-axis*. Similarly, *the line $y = k$ is the $\overline{X}$-axis*. Thus, we see that the coordinate change described by Equations 82-2 produces $\overline{X}$- and $\overline{Y}$-axes that are parallel to the X- and Y-axes, have the same directions, and intersect in the point whose XY-coordinates are (h, k). Equations 82-2 are called the **transformation of coordinate equations** for a **translation** of axes.

EXAMPLE 82-1   Find a translation of axes that simplifies the equation $x^2 + y^2 - 4x + 8y + 11 = 0$.

Solution   We recognize this equation as the equation of a circle, and by completing the square, we can write it as

$$(x - 2)^2 + (y + 4)^2 = 9.$$

Now we set

(82-3) $\overline{x} = x - 2$ and $\overline{y} = y + 4,$

and the equation of our circle takes the simple form

$$\overline{x}^2 + \overline{y}^2 = 9.$$

Equations 82-3 represent a translation; the origin of our new axis system being the point $(2, -4)$ of the old system. This point, of course, is the center of our circle.

The next example illustrates another technique for finding a new
coordinate system, with respect to which a given equation takes a simpler
form. We could have used this method in our preceding example, too.

EXAMPLE 82-2 Find a translation of axes that simplifies the equation

$$xy + 3x - 4y = 5.$$

Solution We will replace x with $\bar{x} + h$ and y with $\bar{y} + k$ in the given equation, and
then choose h and k so that the resulting equation is simpler than the
original. When we make the replacement, our equation becomes

$$(\bar{x} + h)(\bar{y} + k) + 3(\bar{x} + h) - 4(\bar{y} + k) = 5,$$
$$\bar{x}\bar{y} + (k + 3)\bar{x} + (h - 4)\bar{y} + hk + 3h - 4k = 5.$$

Now if we set $h = 4$ and $k = -3$, this equation takes the simple form
$\bar{x}\bar{y} - 12 + 12 + 12 = 5$, or $\bar{y} = -7/\bar{x}$. An easy way to sketch the graph
of the original equation, therefore, is to draw an $\bar{X}\bar{Y}$-coordinate system
whose origin is the point $(4, -3)$ of the XY-system. Then, relative to this
new system, sketch the graph of the equation $\bar{y} = -7/\bar{x}$. This example is
illustrated in Fig. 82-2.

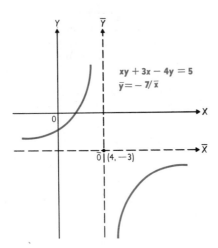

xy + 3x − 4y = 5
ȳ = − 7/ x̄

FIGURE 82-2

A circle is a highly symmetrical figure, and if its center is the origin, we
can use this symmetry to good effect when we draw its graph. Thus, we need
only plot the part of the graph that lies in the first quadrant and "reflect" that
part across the coordinate axes to get the remainder of the figure. Many of the
curves we are going to discuss in the remainder of this chapter are also symmetrical, so before we turn to them, let us say a few words of review about the
symmetry of graphs in general. We originally met these ideas in Section 30,
where we discussed the symmetry of the graphs of the trigonometric functions.
You might also look back to Section 35 and interpret the idea of a "phase
shift" in terms of a translation of axes.

A plane point set S is **symmetric with respect to a line** L if for each point $P \in S$ there is a point $Q \in S$ such that the line L is the perpendicular bisector of the segment PQ. In particular, a set S of the coordinate plane is symmetric with respect to the Y-axis if whenever $(x, y) \in S$, we also have $(-x\ y) \in S$, because, as you can see, the Y-axis is the perpendicular bisector of the segment that joins the points (x, y) and $(-x, y)$. Thus, *the graph of a relation is symmetric with respect to the Y-axis if whenever the pair (x, y) satisfies the relation, the pair $(-x, y)$ also satisfies it.* A similar statement guarantees symmetry with respect to the X-axis.

A set S is **symmetric with respect to a point** Q if for each point $P \in S$ there is a point $R \in S$ such that the point Q is the midpoint of the segment PR. In particular, a set S of the coordinate plane is symmetric with respect to the origin if whenever $(x, y) \in S$, we also have $(-x, -y) \in S$, because the origin is clearly the midpoint of the segment that joins the points (x, y) and $(-x, -y)$. Thus, *the graph of a relation is symmetric with respect to the origin if whenever the pair (x, y) satisfies the relation, the pair $(-x, -y)$ also satisfies it.*

We see immediately that if the pair (x, y) satisfies the equation $x^2 + y^2 = 9$, then the pairs $(-x, y)$, $(x, -y)$, and $(-x, -y)$ do, too. These statements say that the circle $x^2 + y^2 = 9$ is symmetric with respect to the Y-axis, the X-axis, and the origin. Similarly, the fact that $(-x, y)$ satisfies the equation $y = \cos x$ whenever (x, y) does tells us that the cosine curve is symmetric with respect to the Y-axis, and the fact that $(-x, -y)$ satisfies the equation $y = \tan x$ whenever (x, y) does tells us that the tangent curve is symmetric with respect to the origin.

PROBLEMS
82

1. Suppose that we construct an $\overline{X}\overline{Y}$-coordinate system from an XY-coordinate system by translating the axes so that the origin is translated to the point $(-2, 3)$. First consider the following pairs of numbers as XY-coordinates of points and find the $\overline{X}\overline{Y}$-coordinates of the points; then consider the pairs as $\overline{X}\overline{Y}$-coordinates of points and find the XY-coordinates of the points.

   (a) $(2, -5)$ (b) $(-2, 5)$ (c) $(-5, 2)$ (d) $(5, -2)$
   (e) $(-\frac{7}{5}, \frac{5}{7})$ (f) $(\frac{3}{2}, -\frac{3}{2})$ (g) $(\log 3, \log 2)$ (h) $(\cos 1, \sin 1)$

2. Find translations of axes such that the origin of the new axis system is the point of intersection of the given pair of lines. What are the equations of the lines in the $\overline{X}\overline{Y}$-system?

   (a) $y = 2x - 3$ (b) $y = 2$
   $\quad\ x = 2$ $\quad x - 2y = 1$
   (c) $x - 2y = 4$ (d) $2x + y = 0$
   $\quad 3x + y = 5$ $\quad x - 3y = 7$

3. Find transformation of coordinates equations under which a sine curve in the XY-system becomes a cosine curve in the $\overline{X}\overline{Y}$-system.

4. Discuss the symmetry of the graphs of the following equations.

   (a) $3x^2 + 4y^2 = 5$ (b) $3x^2 - 4y^2 = 5$ (c) $3|x| + 4|y| = 5$
   (d) $3\cos x + 4\cos y = 5$ (e) $3x^2 + 4y = 5$ (f) $3x^2 + 4\tan y = 5$

5. Compare the graphs of the equations $y = \sin x$, $y = \sin |x|$, $|y| = \sin |x|$, and $|y| = |\sin |x||$.

6. Show that the points (a, b) and (b, a) are symmetric with respect to the line $y = x$. What point is symmetric to (a, b) with respect to the line $y = -x$?

7. (Do these problems in two ways. First, complete the square and make the translation that the resulting equation suggests. Then replace x with $\bar{x} + h$ and y with $\bar{y} + k$, and choose h and k so that the resulting equation has the desired form.) Find a translation of axes that reduces the equation
   (a) $2x^2 + 2y^2 - 8x + 4y + 2 = 0$ to the form $\bar{x}^2 + \bar{y}^2 = r^2$.

   (b) $x^2 + 4y^2 + 2x - 16y + 13 = 0$ to the form $\dfrac{\bar{x}^2}{a^2} + \dfrac{\bar{y}^2}{b^2} = 1$.

   (c) $9x^2 - 4y^2 - 36x + 8y - 4 = 0$ to the form $\dfrac{\bar{x}^2}{a^2} - \dfrac{\bar{y}^2}{b^2} = 1$.

   (d) $y = 3x^2 + 6x + 2$ to the form $\bar{y} = c\bar{x}^2$.

8. Show that the graph of the equation $2(x - h)^2 + 3(y - k)^2 = 5$ is symmetric with respect to the lines $x = h$ and $y = k$ and also with respect to the point (h, k).

9. What does the equation $y = \displaystyle\sum_{r=0}^{5} \binom{5}{r} x^{5-r} 2^r$ become under the translation defined by the equations $\bar{y} = y$ and $\bar{x} = x + 2$. Sketch the situation.

10. Show that the graph of an odd function is symmetric with respect to the origin, and the graph of an even function is symmetric with respect to the Y-axis.

11. Show that if a graph is symmetric with respect to both of the coordinate axes, then it is symmetric with respect to the origin. Is the converse statement true?

12. Find a pair of translation of axes equations that
   (a) reduces the equation $y = \sin x \cos 2 - \cos x \sin 2 + 3$ to $\bar{y} = \sin \bar{x}$.
   (b) reduces the equation $y = 5 \cdot 10^x - 20$ to $\bar{y} = 10^{\bar{x}}$.
   (c) reduces the equation $y = \tan x$ to $\bar{y} = -\cot \bar{x}$.
   (d) reduces the equation $y = \sin x$ to $\bar{y} = -\sin \bar{x}$.
   (e) reduces the equation $y = \text{Cos}^{-1} x$ to $\bar{y} = -\text{Sin}^{-1} \bar{x}$.
   (f) reduces the equation $y = \log \frac{1}{2} x$ to $\bar{y} = \log \bar{x}$.

83
THE ELLIPSE

A circle can be thought of as a special case of a curve known as an **ellipse**. To describe a circle, we start with a given point and a given positive number, and we say that the circle is the set of points whose distance from the given point is the given number. To describe an ellipse, we start with *two* points and a positive number, and we say that *the ellipse is the set of points, the average of whose distances from the two given points is the given number*. In set notation, if F_1 and F_2 are the two given points, and a is the given positive number, then the ellipse is the set of points $\{P \mid \frac{1}{2}(\overline{PF}_1 + \overline{PF}_2) = a\}$. The points F_1 and F_2 are called the **foci** (singular, *focus*) of the ellipse. If the two foci coincide, the ellipse becomes a circle.

To translate our geometric description of an ellipse into analytic terms, we introduce a coordinate system into the plane, and it will simplify our formulas if we set up the coordinate system so that our foci are the points $(-c, 0)$ and $(c, 0)$ as shown in Fig. 83-1. For each point (x, y) of the ellipse,

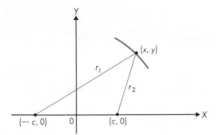

FIGURE 83-1

the average of the distances r_1 and r_2 is the given number a; that is, $\frac{1}{2}(r_1 + r_2) = a$, or

(83-1)
$$r_1 + r_2 = 2a.$$

The foci are $2c$ units apart, so since the sum of the lengths of two sides of a triangle is greater than the length of the third side, $r_1 + r_2 > 2c$. When we compare this inequality with Equation 83-1, we see that $a > c$.

A simple way to construct our ellipse is to stick thumbtacks into each focus and then place a loop of string $2a + 2c$ units long around the tacks. Now put your pencil point inside the loop of string and draw it taut, thus obtaining the triangle shown in Fig. 83-1. As you move your pencil, keeping the string taut, you will trace out the ellipse.

To find the equation of our ellipse, we use the distance formula to express r_1 and r_2:

$$r_1 = \sqrt{(x + c)^2 + y^2} \qquad \text{and} \qquad r_2 = \sqrt{(x - c)^2 + y^2}.$$

When we substitute these numbers in Equation 83-1, we find that a point (x, y) is one of the points of our ellipse if, and only if, its coordinates satisfy the equation

(83-2)
$$\sqrt{(x + c)^2 + y^2} + \sqrt{(x - c)^2 + y^2} = 2a.$$

We could consider Equation 83-2 as the equation of our ellipse, but a little algebra enables us to reduce this equation to a much simpler form. To eliminate radicals, we transpose the term $\sqrt{(x - c)^2 + y^2}$ and square, thus obtaining

$$(x + c)^2 + y^2 = 4a^2 - 4a\sqrt{(x - c)^2 + y^2} + (x - c)^2 + y^2.$$

When we simplify the last equation, we have $a\sqrt{(x-c)^2+y^2} = a^2 - cx$, so we square again to get

$$a^2[(x-c)^2+y^2] = a^4 - 2a^2cx + c^2x^2.$$

Simplification now yields

$$(a^2-c^2)x^2 + a^2y^2 = a^2(a^2-c^2).$$

We noticed that $a > c$, so $a^2 - c^2 > 0$. We can therefore introduce the positive number $\boldsymbol{b} = \sqrt{\boldsymbol{a}^2-\boldsymbol{c}^2}$, and our equation reduces to

$$b^2x^2 + a^2y^2 = a^2b^2.$$

Now we divide both sides of this equation by a^2b^2 to get the final form of the equation of our ellipse:

(83-3)
$$\frac{\boldsymbol{x}^2}{\boldsymbol{a}^2} + \frac{\boldsymbol{y}^2}{\boldsymbol{b}^2} = \boldsymbol{1}.$$

The graph of this equation is shown in Fig. 83-2.

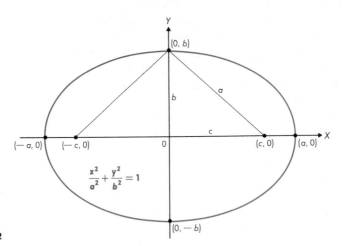

FIGURE 83-2

We see immediately that if the pair (x, y) satisfies Equation 83-3, then the pairs $(-x, y)$, $(x, -y)$, and $(-x, -y)$ do, too, so our ellipse is symmetric with respect to the Y-axis, the X-axis, and the origin. The point of symmetry of an ellipse (here the origin) is called the **center** of the ellipse. An ellipse has two lines of symmetry, and the four points of intersection of the ellipse with its lines of symmetry are the **vertices** (singular, *vertex*) of the ellipse. The vertices of the ellipse whose equation is Equation 83-3 are the points $(a, 0)$, $(-a, 0)$, $(0, b)$, and $(0, -b)$. A line segment that contains the center of an ellipse and whose terminal points are points of the ellipse is called a

378

diameter of the ellipse. The longest diameter of an ellipse is its **major diameter,** and the shortest diameter is its **minor diameter.** Figure 83-2 shows that the major diameter of our ellipse joins the vertices $(a, 0)$ and $(-a, 0)$ and is $2a$ units long, whereas the minor diameter joins the vertices $(0, b)$ and $(0, -b)$ and is $2b$ units long. Notice that the foci are points of the major diameter.

The graph of Equation 83-3 is also an ellipse if $b > a$. In that case, the roles of b and a are interchanged. The foci are the points $(0, -c)$ and $(0, c)$, where $c = \sqrt{b^2 - a^2}$, of the Y-axis. The major diameter is now the segment terminating in the points $(0, -b)$ and $(0, b)$, while the minor diameter terminates in $(-a, 0)$ and $(a, 0)$. The average of the distances of a point of this ellipse from the two foci is b, rather than a.

EXAMPLE 83-1   Find the equation of the ellipse whose foci are the points $(0, 4)$ and $(0, -4)$ and which has $(0, 5)$ and $(0, -5)$ as vertices.

Solution   The desired equation takes the form of Equation 83-3; we are to find the numbers a and b. Because the foci are points of the Y-axis, $b > a$. Here, $b = 5$ and $c = 4$, and so $a = \sqrt{25 - 16} = 3$. The equation of our ellipse is

$$\frac{x^2}{9} + \frac{y^2}{25} = 1,$$

and a sketch is shown in Fig. 83-3.

Equation 83-3 represents an ellipse whose center is the origin. Now let us suppose that we have an ellipse whose center is the point (h, k). We still

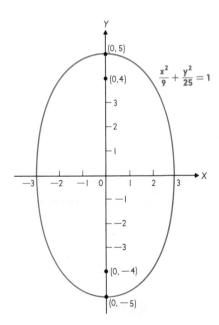

FIGURE 83-3

379

assume that the major and minor diameters of the ellipse are parallel to the coordinate axes. If we choose a translated $\overline{X}\overline{Y}$-coordinate system whose origin is the point (h, k), then the equation of the ellipse relative to this new coordinate system takes the form of Equation 83-3; that is,

$$\frac{\overline{x}^2}{a^2} + \frac{\overline{y}^2}{b^2} = 1.$$

Now we use the translation equations $\overline{x} = x - h$ and $\overline{y} = y - k$ to find the equation of our ellipse in the XY-coordinate system:

(83-4)
$$\frac{(x - h)^2}{a^2} + \frac{(y - k)^2}{b^2} = 1.$$

EXAMPLE 83-2 Write the equation of the ellipse whose vertices are the points $(1, 1)$, $(5, 1)$, $(3, 6)$, and $(3, -4)$, and sketch its graph.

Solution We first plot the given vertices (Fig. 83-4). These four points are the ends

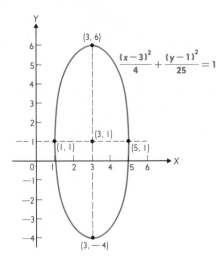

$$\frac{(x-3)^2}{4} + \frac{(y-1)^2}{25} = 1$$

FIGURE 83-4

of the major and minor diameters of the ellipse. The diameter that is parallel to the Y-axis is 10 units long, and the diameter that is parallel to the X-axis is 4 units long. Thus, $a = 2$ and $b = 5$. The center of an ellipse is the point in which its major and minor diameters intersect. In this case, the center is the point $(3, 1)$. Thus, $h = 3$ and $k = 1$, and Equation 83-4 becomes

$$\frac{(x - 3)^2}{4} + \frac{(y - 1)^2}{25} = 1.$$

EXAMPLE 83-3 Reduce the equation $9x^2 + 4y^2 + 18x - 16y - 11 = 0$ to a simpler form by a translation of axes. What can you say about its graph?

Solution   We first rewrite our equation by completing the square:

$$9x^2 + 4y^2 + 18x - 16y - 11 = 0,$$
$$9(x^2 + 2x \quad\;) + 4(y^2 - 4y \quad\;) - 11 = 0,$$
$$9(x^2 + 2x + 1) + 4(y^2 - 4y + 4) - 11 = 9 + 16,$$
$$9(x + 1)^2 + 4(y - 2)^2 = 36.$$

So if we set $\bar{x} = x + 1$ and $\bar{y} = y - 2$, our equation becomes $9\bar{x}^2 + 4\bar{y}^2 = 36$, which reduces to the standard form (relative to the $\overline{XY}$-coordinate system) of the equation of an ellipse:

$$\frac{\bar{x}^2}{4} + \frac{\bar{y}^2}{9} = 1.$$

Here, $a = 2$, $b = 3$, and $c = \sqrt{b^2 - a^2} = \sqrt{5}$. The coordinates of the vertices of our ellipse in the $\overline{XY}$-coordinate system are $(-2, 0)$, $(2, 0)$, $(0, -3)$, and $(0, 3)$, and its foci are the points with $\overline{XY}$-coordinates $(0, -\sqrt{5})$ and $(0, \sqrt{5})$. The XY- and the $\overline{XY}$-coordinates are related by the equations $x = \bar{x} - 1$ and $y = \bar{y} + 2$. Therefore, the XY-coordinates of the vertices are $(-3, 2)$, $(1, 2)$, $(-1, -1)$, and $(-1, 5)$. The foci are the points $(-1, 2 - \sqrt{5})$ and $(-1, 2 + \sqrt{5})$.

PROBLEMS
83

1. Find the equation of, and then sketch, the ellipse whose

    (a) center is $(0, 0)$, with major diameter a part of the X-axis and 8 units long, and minor diameter 6 units long.

    (b) center is $(0, 0)$, with major diameter a part of the Y-axis and 10 units long, and minor diameter 2 units long.

    (c) center is $(2, -1)$, with major diameter parallel to the X-axis and 6 units long, and minor diameter 4 units long.

    (d) center is $(2, -1)$, with major diameter parallel to the Y-axis and 12 units long, and minor diameter 10 units long.

2. Find the equations of the ellipses whose foci are the given points and whose lengths of major diameters are the given numbers.

    (a) $(-1, 0)$, $(0, 1)$, 6   (b) $(0, -2)$, $(0, 2)$, 8
    (c) $(0, 4)$, $(2, 4)$, 4   (d) $(1, 0)$, $(1, 8)$, 16

3. Find the equations of the ellipses whose vertices are the given points.

    (a) $(-5, 0)$, $(5, 0)$, $(0, -3)$, $(0, 3)$ (b) $(2, 1)$, $(2, 5)$, $(-1, 3)$, $(5, 3)$
    (c) $(0, -1)$, $(10, -1)$, $(5, -3)$, $(5, 1)$ (d) $(1, -2)$, $(1, 6)$, $(-1, 2)$, $(3, 2)$

4. Find the foci of each of the following ellipses.

    (a) $9x^2 + 25y^2 = 225$ (b) $5x^2 + y^2 = 5$
    (c) $x^2 + 2y^2 + 6x + 7 = 0$ (d) $3x^2 + 7y^2 + 6x - 28y + 10 = 0$

5. Find the equation of the ellipse whose

    (a) minor diameter is 4 units long, and whose major diameter ends with the points $(-1, 2)$ and $(9, 2)$.

    (b) major diameter ends with the points $(-4, -1)$ and $(2, -1)$, and whose foci are 4 units apart.

(c) foci are the points (1, 4) and (1, 0), and whose minor diameter is one unit long.

(d) major diameter ends with the points (0, 1) and (6, 1), and which contains the point (3, 0).

6. The segment that an ellipse cuts from a line that contains a focus and is perpendicular to the major diameter is called a **latus rectum** of the ellipse. Show that the length of a latus rectum of the ellipse whose equation is Equation 83-3 and whose major diameter lies along the X-axis is $\dfrac{2b^2}{a}$.

7. (a) A point moves so that the line segments joining it to two given points are always perpendicular. Describe the path of the point.
    (b) Now suppose that the point moves so that the product of the slopes of the line segments joining it to the two given points is $-k^2$, where k is a given number. Describe the path of the point.

8. In his barnyard, Farmer Jones put a salt block and a watering trough 50 feet apart. Then he built a fence so that from any point inside it his cow could walk to the salt block, then to the watering trough, and back to her starting point without traveling more than 150 feet. Describe the fence.

9. An ellipse is to be constructed on an $8\frac{1}{2}'' \times 11''$ sheet of paper by the string and thumbtack method so that its major diameter is parallel to the $11''$ edge. Where should we place the tacks and how long a piece of string should we use in order to draw the largest possible ellipse on the paper?

10. Give a geometric description of the set of points

$$\{(x, y) \mid x = 5 \cos t, \, y = 3 \sin t, \, 0 \leq t \leq \pi\}.$$

11. Discuss the graphs of the following equations.
    (a) $\sin \pi \sqrt{x^2 + 4y^2} = 0$
    (b) $4x^2 + 9y^2 - 8|x| - 36|y| + 4 = 0$
    (c) $x(x + |x|) + 4y(y - |y|) = 8$
    (d) $[\![x^2/25 + y^2/16]\!] = 0$

12. Circumscribe a circle of radius a about an ellipse whose major diameter is $2a$ units long and whose minor diameter is $2b$ units long. Now draw a line perpendicular to the major diameter of the ellipse, and suppose that the segment of the line the ellipse cuts off is L units long, and the (longer) segment cut off by the circle is M units long. Show that $L/M = b/a$. Can you use this result to convince yourself that the ratio of the area of the elliptical region to the circular region is also b/a? What, therefore, is the area of the elliptical disk?

84
THE
HYPERBOLA

As with the ellipse, the geometric definition of the hyperbola starts with two given points F_1 and F_2 (again called the **foci**) and a given positive number a. An ellipse is the set of points, the *sum* of whose distances from the foci is $2a$, and a **hyperbola** *is the set of points, the differences of whose distances from the foci is $2a$.* In set notation, if F_1 and F_2 are the foci, and a is a positive number,

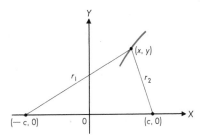

FIGURE 84-1

the hyperbola is the set of points $\{P \mid |\overline{PF}_1 - \overline{PF}_2| = 2a\}$. In Fig. 84-1 we have chosen our coordinate system so that the foci of the hyperbola are the points $(-c, 0)$ and $(c, 0)$. A point (x, y) at distances of r_1 and r_2 from the foci is a point of our hyperbola if, and only if,

(84-1)
$$|r_1 - r_2| = 2a.$$

The sides of the triangle whose vertices are the points $(-c, 0)$, $(c, 0)$, and (x, y) have lengths of $2c$, r_1, and r_2. Since the difference in the lengths of two sides of a triangle is less than the length of the third side, $|r_1 - r_2| > 2c$. When we compare this inequality with Equation 84-1, we see that for a hyperbola, $a < c$.

Just as in the case of the ellipse, we can construct an arc of a hyperbola by using a pencil, some string, and two thumbtacks. The tacks are placed at the foci, and the pencil is tied in the middle of the string. Then the string is looped around the tacks as shown in Fig. 84-2 (the pencil point is at P). We

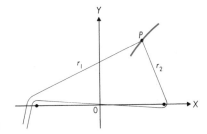

FIGURE 84-2

keep the string taut and move the pencil, paying out both strands of the string together. Thus, the difference $r_1 - r_2$ is the same for any two points of the arc traced out by the pencil, so the arc must be part of a hyperbola.

To find the equation that tells us when a point (x, y) belongs to our hyperbola, we need only make the substitutions

$$r_1 = \sqrt{(x+c)^2 + y^2} \quad \text{and} \quad r_2 = \sqrt{(x-c)^2 + y^2}$$

in Equation 84-1. But the resulting equation is so complicated that it is to our advantage to rearrange Equation 84-1 a little before we make the

383

substitution. We first observe that Equation 84-1 is equivalent to the equation $(r_1 - r_2)^2 = 4a^2$. Because $r_1{}^2$ and $r_2{}^2$ are "nicer" expressions than r_1 and r_2, we write our last equation as

$$2r_1r_2 = 4a^2 - r_1{}^2 - r_2{}^2$$

and square both sides. We leave it to you to show that the resulting equation can be written as

$$8a^2(r_1{}^2 + r_2{}^2) = (r_1{}^2 - r_2{}^2)^2 + 16a^4.$$

Now we substitute for r_1 and r_2:

$$8a^2[(x + c)^2 + y^2 + (x - c)^2 + y^2] = [(x + c)^2 - (x - c)^2]^2 + 16a^4.$$

When we simplify, this equation becomes

(84-2) $$(c^2 - a^2)x^2 - a^2y^2 = a^2(c^2 - a^2).$$

Since $c > a$, the equation $\boldsymbol{b} = \sqrt{\boldsymbol{c^2 - a^2}}$ defines a positive number b, in terms of which we can write Equation 84-2 as $b^2x^2 - a^2y^2 = a^2b^2$; that is,

(84-3) $$\frac{\boldsymbol{x}^2}{\boldsymbol{a}^2} - \frac{\boldsymbol{y}^2}{\boldsymbol{b}^2} = \boldsymbol{1}$$

is the equation of our hyperbola.

Replacing x with $-x$ and y with $-y$, separately or together, in Equation 84-3 does not alter the equation, so we see that our hyperbola is symmetric with respect to the Y-axis, the X-axis, and the origin. The point of symmetry of a hyperbola (in this case the origin) is the **center** of the hyperbola. From Equation 84-3 we see that our hyperbola intersects one of its lines of symmetry, the X-axis, in the points $(-a, 0)$ and $(a, 0)$. These points are the **vertices** of the hyperbola. The line segment joining the two vertices is the **transverse diameter** of the hyperbola. Since there is no real number y that satisfies Equation 84-3 when $x = 0$, we see that our hyperbola does not intersect its other line of symmetry, the Y-axis. Figure 84-3 shows the graph of Equation 84-3.

For the moment, let us consider only points in the first quadrant and solve Equation 84-3 for y to obtain $y = \frac{b}{a}\sqrt{x^2 - a^2}$, or equivalently,

(84-4) $$y = \frac{bx}{a}\sqrt{1 - \frac{a^2}{x^2}}.$$

Since $\sqrt{1 - a^2/x^2} < 1$, we see that $y < bx/a$. Hence the arc of the hyperbola

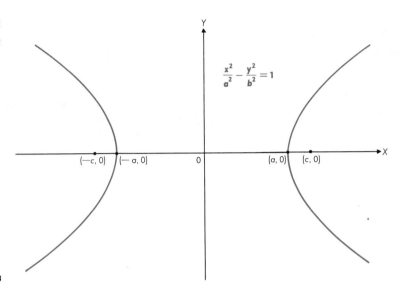

$$\frac{x^2}{a^2} - \frac{y^2}{b^2} = 1$$

$(-c, 0)$ $(-a, 0)$ $(a, 0)$ $(c, 0)$

FIGURE 84-3

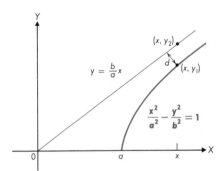

$y = \frac{b}{a}x$

(x, y_2)

d

(x, y_1)

$$\frac{x^2}{a^2} - \frac{y^2}{b^2} = 1$$

FIGURE 84-4

that lies in the first quadrant is below the line $y = (b/a)x$ (Fig. 84-4). But if x is very large, $\sqrt{1 - a^2/x^2}$ is close to 1, and Equation 84-4 suggests that the equation of our hyperbola does not differ a great deal from the linear equation $y = (b/a)x$.

To verify that this statement is indeed correct, let x be a large number and suppose that (x, y_1) is the corresponding point of the hyperbola. We wish to show that the (perpendicular) distance d between this point and our line is small. Now if (x, y_2) is the point of the line that corresponds to x, it is clear (Fig. 84–4) that

(84-5)
$$d < y_2 - y_1.$$

By definition, $y_2 = (b/a)x$, and according to Equation 84-4,

(84-6)
$$y_1 = \frac{b}{a}\sqrt{x^2 - a^2}.$$

385

Hence,

$$y_2 - y_1 = \frac{b}{a}\left(x - \sqrt{x^2 - a^2}\right)$$

$$= \frac{b}{a}\left(x - \sqrt{x^2 - a^2}\right)\left[\frac{x + \sqrt{x^2 - a^2}}{x + \sqrt{x^2 - a^2}}\right]$$

$$= \frac{b}{a}\left[\frac{x^2 - (x^2 - a^2)}{x + \sqrt{x^2 - a^2}}\right]$$

$$= \frac{ab}{x + \sqrt{x^2 - a^2}} < \frac{ab}{x}.$$

Therefore, $d < ab/x$, and we see that d is small when x is large. Thus, for points in the first quadrant far to the right on our graph, the hyperbola practically coincides with the line $y = (b/a)x$. This line is called an **asymptote** of the hyperbola. It is not hard to see that for points in the third quadrant remote from the origin, the hyperbola also practically coincides with the line $y = (b/a)x$, whereas for points in the second and fourth quadrants, the line $y = -(b/a)x$ is an asymptote of the hyperbola.

The rectangle whose sides are parallel to the axes and contain the points $(-a, 0)$, $(a, 0)$, $(0, -b)$, and $(0, b)$ is called the **auxiliary rectangle** of our hyperbola. The asymptotes of the hyperbola are extensions of the diagonals of this rectangle. Our hyperbola, with the asymptotes drawn in, is the curve on the left in Fig. 84-5. If the auxiliary rectangle is a square, then the hyperbola is an **equilateral hyperbola.**

If we choose the coordinate axes so that the foci of a hyperbola are the points $(0, c)$ and $(0, -c)$, then its vertices are points of the Y-axis. We use $2b$ (rather than $2a$) to denote the difference of the focal distances of a point of this hyperbola, and we obtain the following equation:

(84-7)
$$\frac{y^2}{b^2} - \frac{x^2}{a^2} = 1,$$

where again $a^2 + b^2 = c^2$. The points $(0, -b)$ and $(0, b)$ are the vertices of this hyperbola. We also see that there is no real number x for which $y = 0$:

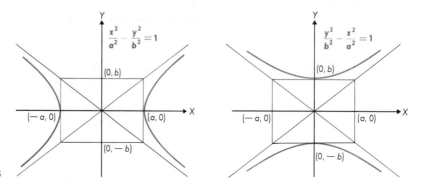

FIGURE 84-5

that is, the curve does not intersect the X-axis. The graph of Equation 84-7 is shown on the right in Fig. 84-5. The asymptotes of both the hyperbolas in Fig. 84-5 are the same lines, and we say that these hyperbolas form a pair of **conjugate hyperbolas.**

EXAMPLE 84-1 Find the equation of the hyperbola whose foci are the points $(5, 0)$ and $(-5, 0)$ and whose vertices are $(4, 0)$ and $(-4, 0)$.

Solution We obtain the equation of this hyperbola by setting $c = 5$, $a = 4$, and $b = \sqrt{c^2 - a^2} = 3$ in Equation 84-3:

$$\frac{x^2}{16} - \frac{y^2}{9} = 1.$$

Its asymptotes are the lines $y = \frac{3}{4}x$ and $y = -\frac{3}{4}x$, and it looks like the hyperbola on the left in Fig. 84-5.

Now suppose we have a hyperbola whose center is the point (h, k) and whose transverse diameter is parallel to the X-axis. If we choose a translated $\overline{X}\overline{Y}$-coordinate system whose origin is the point (h, k), then the equation of the hyperbola relative to this coordinate system is

$$\frac{\overline{x}^2}{a^2} - \frac{\overline{y}^2}{b^2} = 1.$$

Since $\overline{x} = x - h$ and $\overline{y} = y - k$, the equation of the hyperbola in the XY-coordinate system is

(84-8)
$$\frac{(x - h)^2}{a^2} - \frac{(y - k)^2}{b^2} = 1.$$

In a similar manner, the equation of a hyperbola whose center is the point (h, k) and whose transverse diameter is parallel to the Y-axis is

(84-9)
$$\frac{(y - k)^2}{b^2} - \frac{(x - h)^2}{a^2} = 1.$$

EXAMPLE 84-2 The foci of an equilateral hyperbola are the points $(-3, 4)$ and $(-3, -2)$. Find the equation of the hyperbola and sketch its graph.

Solution We are to replace the letters a, b, h, and k in Equation 84-9 with appropriate numbers. The center of the hyperbola is midway between the two foci, so it is the point $(-3, 1)$. Hence, $h = -3$ and $k = 1$ (Fig. 84-6). Since the hyperbola is equilateral, $a = b$. Now c is the distance between the center of the hyperbola and a focus, so in the present case, $c = 3$. Hence the equation $c^2 = a^2 + b^2$ becomes $9 = 2a^2$, and therefore $a^2 = b^2 = \frac{9}{2}$. The equation of our hyperbola is

$$\frac{2(y - 1)^2}{9} - \frac{2(x + 3)^2}{9} = 1;$$

its graph is shown in Fig. 84-6.

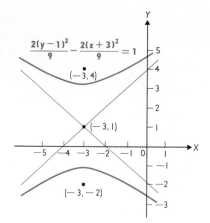

$$\frac{2(y-1)^2}{9} - \frac{2(x+3)^2}{9} = 1$$

$(-3, 4)$

$(-3, 1)$

$(-3, -2)$

FIGURE 84-6

PROBLEMS
84

1. Find the foci, vertices, and asymptotes of the following hyperbolas, and sketch them.

(a) $x^2 - 4y^2 = 4$ (b) $4x^2 - y^2 = 4$

(c) $x^2 - 4y^2 = -4$ (d) $4x^2 - y^2 = -4$

(e) $\dfrac{(x+1)^2}{16} - \dfrac{(y-2)^2}{9} = 1$ (f) $4x^2 - y^2 = 1$

(g) $\dfrac{(y+2)^2}{9} - \dfrac{(x-1)^2}{16} = 1$ (h) $y^2 - 9x^2 = 1$

2. Find the equation of the hyperbola whose foci and vertices are the given points.

(a) foci: $(-2, 0)$, $(2, 0)$, vertices: $(-1, 0)$, $(1, 0)$

(b) foci: $(0, -3)$, $(0, 3)$, vertices: $(0, -2)$, $(0, 2)$

(c) foci: $(0, 1)$, $(6, 1)$, vertices: $(1, 1)$, $(5, 1)$

(d) foci: $(0, -1)$, $(0, 5)$, vertices: $(0, 0)$, $(0, 4)$

3. Find the foci of each of the following hyperbolas.

(a) $9x^2 - 16y^2 + 36x + 32y - 124 = 0$

(b) $16x^2 - 9y^2 + 64x + 18y - 89 = 0$

(c) $9x^2 - 16y^2 + 36x + 32y + 164 = 0$

(d) $16x^2 - 9y^2 + 64x + 18y + 199 = 0$

4. Find the equation of, and then sketch, the hyperbola whose

(a) center is the origin, transverse diameter is part of the X-axis and is 12 units long, and whose foci are 16 units apart.

(b) transverse diameter ends with the points $(0, 0)$ and $(0, 8)$, and whose foci are 10 units apart.

(c) vertices are the points $(1, 0)$ and $(1, 10)$ and that contains the point $(4, 15)$.

(d) foci are the points $(-2, -2)$ and $(-2, 8)$ and that has $(-2, 0)$ as one vertex.

5. Find the equation of the hyperbola whose asymptotes are the given lines and which contains the given point.

(a) $y = 2x$ and $y = -2x$, $(1, 1)$ (b) $y = 2x$ and $y = -2x$, $(0, 1)$

(c) $y = x + 1$ and $y = -x + 3$, $(2, 4)$ (d) $y = x + 1$ and $y = -x + 1$, $(0, 0)$

6. (a) Find the equation of the hyperbola whose foci are the vertices of the ellipse $11x^2 + 7y^2 = 77$ and whose vertices are the foci of this ellipse.
   (b) Find the equation of the ellipse whose vertices are the foci of the hyperbola $11x^2 - 7y^2 = 77$ and whose foci are the vertices of this hyperbola.

7. The segment cut by a hyperbola from a line that contains a focus and is perpendicular to the transverse diameter is called a **latus rectum** of the hyperbola. Show that the length of a latus rectum of the hyperbola whose equation is Equation 84-3 is $\dfrac{2b^2}{a}$.

8. A point moves so that the product of the slopes of the line segments that join it to two given points is k^2. Describe the curve traced out by the point.

9. In Problem 79-10 we said that a *convex* set is one that contains the line segment joining each of its points. Explain why the set
$$\{(x, y) \mid x \geq 0,\ 9x^2 - 16y^2 \geq 144\}$$
is convex.

10. A man standing 2200 feet from the base of a cliff fires a rifle, and another man hears the echo 3 seconds after he hears the sound of the shot itself. Where is the second man standing? (Sound travels at 1100 feet per second.)

11. Describe the graph of the equation $\dfrac{x^4}{a^4} - \dfrac{y^4}{b^4} + \dfrac{2y^2}{b^2} = 1$.

12. Sketch the graphs of the following equations.

   (a) $\left| \dfrac{x^2}{4} - y^2 \right| = 1$ (b) $\sin \pi \sqrt{x^2 - 4y^2} = 0$

   (c) $\dfrac{x\,|x|}{4} + y\,|y| = 1$ (d) $[\![x^2/4 - y^2]\!] = 0$

85 THE PARABOLA

For our geometric descriptions of the ellipse and the hyperbola, we started with two given points and a given positive number. In the case of the **parabola,** we start with a given point and a given line. The point is the **focus** of the parabola, the line is its **directrix,** and the *parabola is the set of points, each of which is equidistant from the focus and the directrix.*

To translate this geometric description of a parabola into an analytic one, let us choose our coordinate axes so that the focus is the point $(c, 0)$ and the directrix is the line $x = -c$. In Fig. 85-1 we have shown the focus and directrix if $c > 0$, but our discussion is valid if c is either positive or negative. The (perpendicular) distance between a point (x, y) and the directrix is $|x + c|$. The distance formula tells us that the distance between the points (x, y) and $(c, 0)$ is $\sqrt{(x - c)^2 + y^2}$. By definition, (x, y) is a point of our parabola if, and only if, these two numbers are equal; that is,

$$\sqrt{(x - c)^2 + y^2} = |x + c|.$$

389

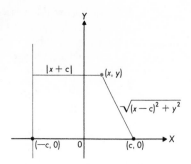

FIGURE 85-1

When we square both sides of this equation and simplify, we get the equation

(85-1)
$$y^2 = 4cx.$$

Since we can replace y with $-y$ without changing Equation 85-1, we see that our parabola is symmetric with respect to the X-axis. The line of symmetry of a parabola (in this case the X-axis) is called the **axis** of the parabola. The axis contains the focus and is perpendicular to the directrix. The point of intersection of a parabola with its axis (in this case the origin) is the **vertex** of the parabola. The **focal length** of the parabola is the distance between the vertex and the focus (in this case $|c|$).

The "orientation" of the parabola of Equation 85-1 depends on the sign of c. For a typical positive value of c, we have the graph shown in Fig. 85-2, while if c is negative, our parabola is oriented as shown in Fig. 85-3.

If we choose the coordinate axes so that the focus of the parabola is the point $(0, c)$ of the Y-axis and the directrix is the line $y = -c$ parallel to the X-axis, then the vertex of the parabola is again the origin. However, the roles of x and y are now interchanged, so the line of symmetry of the parabola is the Y-axis, and its equation is

(85-2)
$$x^2 = 4cy.$$

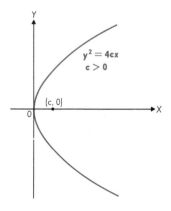

FIGURE 85-2

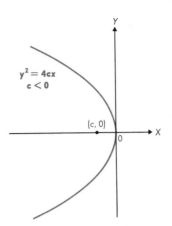

$$y^2 = 4cx$$
$$c < 0$$

$(c, 0)$

FIGURE 85-3

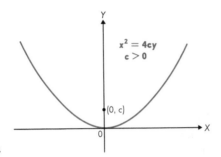

$$x^2 = 4cy$$
$$c > 0$$

$(0, c)$

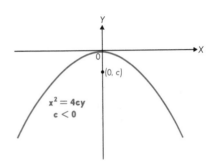

$(0, c)$

$$x^2 = 4cy$$
$$c < 0$$

FIGURE 85-4

If $c > 0$, the parabola opens up as shown on the left in Fig. 85-4. If $c < 0$, the parabola opens down as shown on the right in Fig. 85-4.

EXAMPLE 85-1 The vertex of a parabola is the origin, its focal length is 3, it is symmetric with respect to the X-axis, and it does not intersect the first quadrant. What is its equation?

Solution Since the vertex of the parabola is the origin and it is symmetric with respect to the X-axis, we know that its equation is $y^2 = 4cx$. We must determine the number c. The absolute value of c is the focal length of the parabola, and hence $|c| = 3$. Therefore, $c = 3$ or -3. We select $c = -3$ because (see Fig. 85-3) we are told that our parabola does not intersect the first quadrant. Therefore, its equation is $y^2 = -12x$.

The equation

(85-3) $$(y - k)^2 = 4c(x - h)$$

is also an equation of a parabola. If we use the translation equations $\bar{y} = y - k$ and $\bar{x} = x - h$ to introduce an $\overline{X}\overline{Y}$-coordinate system, our equation becomes $\bar{y}^2 = 4c\bar{x}$, which tells us that the vertex of the parabola is the origin of the $\overline{X}\overline{Y}$-system (that is, the point (h, k) of the XY-system), and the directrix is the line $\bar{x} = -c$ (that is, the line $x = h - c$ in the XY-system).

391

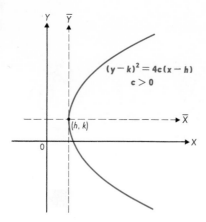

FIGURE 85-5

Figure 85-5 shows what the parabola looks like if $c > 0$. Similarly, the equation

(85-4) $$(x - h)^2 = 4c(y - k)$$

represents a parabola whose vertex is the point (h, k) and whose directrix is the line $y = k - c$ (see Fig. 85-6 for the case $c > 0$).

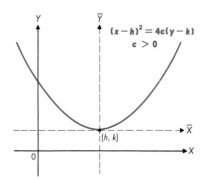

FIGURE 85-6

EXAMPLE 85-2   Write the equation of the parabola whose directrix is the line $y = -5$ and whose focus is the point $(4, -1)$.

Solution   Since the directrix is horizontal, the axis of our parabola is vertical. The vertex is midway between the focus and the point in which the axis intersects the directrix. In this case the vertex is the point $(4, -3)$. Clearly (see Fig. 85-7), the focal length is $c = 2$. Thus we obtain the equation of our parabola from Equation 85-4 by taking $h = 4$, $k = -3$, and $c = 2$; that is,

$$(x - 4)^2 = 8(y + 3).$$

This equation can also be written

$$x^2 = 8x + 8y + 8.$$

392

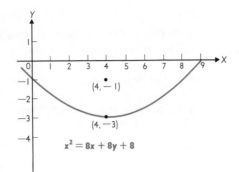

FIGURE 85-7

EXAMPLE 85-3 Show that the graph of the equation $y = ax^2 + bx + c$ is a parabola if $a \neq 0$.

Solution We replace x with $\bar{x} + h$ and y with $\bar{y} + k$, where h and k are numbers to be determined, and we obtain the equation

$$\bar{y} + k = a\,(\bar{x} + h)^2 + b(\bar{x} + h) + c$$
$$= a\bar{x}^2 + (2ah + b)\bar{x} + ah^2 + bh + c.$$

Now if we let $h = -b/2a$ and then $k = ah^2 + bh + c = (4ac - b^2)/4a$, we obtain the standard form of the equation of a parabola,

$$\bar{x}^2 = \frac{1}{a}\,\bar{y}.$$

If $a > 0$, the parabola opens up; if $a < 0$, the parabola opens down. It would be a good exercise for you to obtain these translation equations by completing the square. (Compare this calculation with the derivation of the Quadratic Formulas in Section 49.)

**PROBLEMS
85**

1. Find the focus and directrix of, and sketch, each of the following parabolas.
   (a) $y^2 = 16x$ (b) $x^2 = -8y$
   (c) $(x - 2)^2 = 24(y + 3)$ (d) $y^2 = 4(x - 2)$
   (e) $x^2 + 2x + 4y + 5 = 0$ (f) $y^2 - 8x + 2y + 1 = 0$

2. Find the equation of the parabola whose focus is the given point and whose directrix is the given line.
   (a) $(4, 0)$, $x = -4$ (b) $(0, 6)$, $y = 0$
   (c) $(1, 2)$, $x = -5$ (d) $(-3, 2)$, $y = -6$

3. Find the equation of the parabola whose
   (a) focus is the point $(2, 4)$ and whose vertex is the point $(2, 0)$.
   (b) directrix is the Y-axis and whose vertex is the point $(4, 2)$.
   (c) axis is parallel to the Y-axis, whose vertex is the point $(-1, 3)$, and that contains the point $(1, 1)$.
   (d) focus is the point $(-2, 2)$, that contains the point $(-2, 4)$, and whose axis is parallel to a coordinate axis.

393

4. The segment that a parabola cuts off the line that contains the focus and is perpendicular to the axis is called the **latus rectum** of the parabola.

(a) A parabola has a focal length of c. Show that the length of its latus rectum is $4c$.

(b) Determine the length of the latus rectum of each of the parabolas in Problem 85-2.

(c) Find the equation of the circle that contains the vertex and the two endpoints of the latus rectum of the parabola $y^2 = 4cx$.

5. Find the points of intersection (if any) of the following pairs of curves. Sketch the situation.

(a) $y = x^2$, $y = x + 2$
(b) $y = x^2$, $y = 8 - x^2$
(c) $4x^2 + y^2 = 4$, $y^2 = 4x + 5$
(d) $4x^2 + y^2 = 4$, $y^2 = 16x + 20$

6. Show that the parabola that contains a given set $\{(x_1, y_1), (x_2, y_2), (x_3, y_3)\}$ of three points has the equation

$$y = y_1 \frac{(x - x_2)(x - x_3)}{(x_1 - x_2)(x_1 - x_3)} + y_2 \frac{(x - x_1)(x - x_3)}{(x_2 - x_1)(x_2 - x_3)} + y_3 \frac{(x - x_1)(x - x_2)}{(x_3 - x_1)(x_3 - x_2)}.$$

7. In the course of a yacht race, a boat is required to pass between a buoy and a straight shore 200 yards away. Assuming that the skipper stays as far as possible from both buoy and shore, describe his path as he rounds the buoy.

8. Discuss the graphs of the following equations.

(a) $y^2 = 8|x|$
(b) $y|y| = 8x$
(c) $y^2 = 4(x + |x|)$
(d) $(y + |y|)^2 = 16x$

9. Describe the following sets of pairs of numbers geometrically.

(a) $\{(x, y) \mid y^2 < 8x\}$
(b) $\{(x, y) \mid y > 4x^2 + 24x + 38\}$
(c) $\{(x, y) \mid x = \sin t, y = \sin^2 t, t \in R^1\}$
(d) $\{(x, y) \mid x = \sin t, y = \cos^2 t, t \in R^1\}$

10. From the geometric definition of a parabola, find the directrix of the "tilted" parabola whose vertex is the origin and whose focus is the point $(1, 1)$. Can you find the equation of the parabola?

86
CONICS

We have defined a parabola as the set of points, each of which is equidistant from a given point (the focus) and a given line (the directrix). Another way to word this definition is to say that the ratio of the distance between a point of our parabola and the focus to the distance between the point and the directrix is 1 for each point of the parabola.

Ellipses and hyperbolas have a similar property. Associated with each focus of an ellipse or a hyperbola is a line (called a *directrix*) such that the ratio of the distance between a point of the curve and the focus to the distance between the point and the directrix is a number e that is the same for every point of the curve. (We are not talking about the number $e = 2.71828 \ldots$ that we mentioned in the chapter on logarithms!)

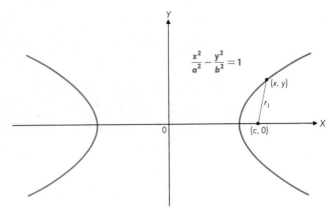

$$\frac{x^2}{a^2} - \frac{y^2}{b^2} = 1$$

FIGURE 86-1

Let us look at a hyperbola to see exactly what we mean. In Fig. 86-1 we have drawn the hyperbola

(86-1)
$$\frac{x^2}{a^2} - \frac{y^2}{b^2} = 1.$$

We have also shown the focus $(c, 0)$, where

(86-2)
$$c = \sqrt{a^2 + b^2}.$$

Now we will compute the number $r_1{}^2$, where r_1 is the distance between a point (x, y) of our hyperbola and the focus $(c, 0)$. [Even though we have shown (x, y) as a point of the "right branch" in our figure, it could be a point of either branch of the hyperbola.] According to the distance formula,

$$r_1{}^2 = (x - c)^2 + y^2 = x^2 - 2cx + c^2 + y^2.$$

In this equation we may replace y^2 with $(b^2 x^2 / a^2) - b^2$ (from Equation 86-1), and get

$$r_1{}^2 = x^2 - 2cx + c^2 + \frac{b^2 x^2}{a^2} - b^2$$

$$= \left(\frac{a^2 + b^2}{a^2}\right) x^2 - 2cx + c^2 - b^2.$$

Equation 86-2 tells us that $a^2 + b^2 = c^2$, and $c^2 - b^2 = a^2$, so

$$r_1{}^2 = \frac{c^2}{a^2} x^2 - 2cx + a^2$$

$$= \frac{c^2}{a^2}\left(x^2 - \frac{2a^2}{c} x + \frac{a^4}{c^2}\right)$$

$$= \frac{c^2}{a^2}\left(x - \frac{a^2}{c}\right)^2.$$

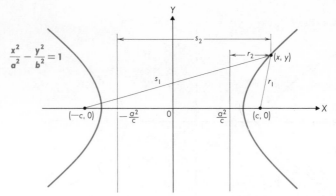

$$\frac{x^2}{a^2} - \frac{y^2}{b^2} = 1$$

FIGURE 86-2

Thus, we see that

(86-3)
$$r_1 = \frac{c}{a}\left| x - \frac{a^2}{c} \right|.$$

The number $|x - a^2/c|$ is the distance between the point (x, y) and the line parallel to the Y-axis and a^2/c units to the right of it (Fig. 86-2). We have labeled this distance as r_2, and so Equation 86-3 can be written as

(86-4)
$$\frac{r_1}{r_2} = \frac{c}{a}.$$

Thus, associated with the focus $(c, 0)$ we have found a line, $x = a^2/c$, such that the ratio of the distance between a point of our hyperbola and the focus to the distance between the point and the line is a number, c/a, that is independent of the choice of the point of the hyperbola. It is easy to show that the line a^2/c units to the left of the Y-axis is associated with the focus $(-c, 0)$ in the same way. That is, if s_1 is the distance between the point (x, y) and the focus $(-c, 0)$, and s_2 is the distance between the point and the associated line, then the ratio s_1/s_2 is again the number c/a. The lines $x = a^2/c$ and $x = -a^2/c$ are called the **directrices** of the hyperbola, and the ratio $e = c/a$ is the **eccentricity** of the hyperbola.

The same reasoning leads to similar results for an ellipse. Let us consider the ellipse.

(86-5)
$$\frac{x^2}{a^2} + \frac{y^2}{b^2} = 1,$$

where $a > b$. This ellipse is shown in Fig. 86-3. Associated with the focus $(c, 0)$ is the directrix parallel to the Y-axis and a^2/c units to the right of it. The directrix a^2/c units to the left of the Y-axis is associated with the focus $(-c, 0)$. In the problems at the end of the section, we ask you to show that if (x, y) is any point of our ellipse, then the ratio r_1/r_2 of the distance between (x, y) and $(c, 0)$ to the distance between (x, y) and the associated directrix

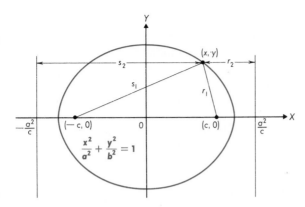

FIGURE 86-3

is the eccentricity $e = c/a$. The ratio s_1/s_2 of the distances to the other focus and directrix is also e.

We have shown that the ellipse, parabola, and hyperbola all share a common "ratio" property. Associated with a focus of one of these curves is a line, called a directrix, such that the ratio of the distance between a point of the curve and the focus to the distance between the point and the directrix is a number e (the eccentricity) that is independent of the choice of the point of the curve. In the case of our ellipse and hyperbola, we saw that the number e was given by the formula $e = c/a$. Since $c < a$ for an ellipse and $c > a$ for a hyperbola, we see that *the eccentricity of an ellipse is a number that is less than 1, the eccentricity of a hyperbola is greater than 1, and the eccentricity of a parabola is equal to* 1.

Ellipses, parabolas, and hyperbolas make up the family of curves known as the *conics*. The members of this family can be geometrically described as follows: *A **conic** is determined by a given point F (a **focus**), a given line d not containing F (the **directrix** associated with F), and a number e (the **eccentricity**). A point P is a point of the conic if, and only if, the ratio of the distance $\overline{FP}$ to the distance between P and the line d is the number e.* The name "conic" comes from another geometric description of these curves. They are the curves that result when an ordinary cone in three-dimensional space is cut by a plane. Since we are not venturing into three dimensions, we will not prove this assertion.

A conic is an ellipse, parabola, or hyperbola depending on whether $e < 1$, $e = 1$, or $e > 1$. In the case of our hyperbola of Equation 86-1 and of our ellipse of Equation 86-5 (in which $a > b$), the directrices are the lines

$$(86\text{-}6) \qquad x = -\frac{a^2}{c} \quad \text{and} \quad x = \frac{a^2}{c},$$

and the eccentricity is the number

$$(86\text{-}7) \qquad e = \frac{c}{a}.$$

397

If a hyperbola or ellipse has a different orientation, these formulas must be modified accordingly. For example, if $b > a$ in Equation 86-5 of the ellipse, then $e = c/b$, and the directrices are the lines $y = -b^2/c$ and $y = b^2/c$.

EXAMPLE 86-1   Find the equation of the conic that has an eccentricity of 2 and has the point $(3, 0)$ as a focus for which the corresponding directrix is the Y-axis.

Solution   In Fig. 86-4 we show the typical point (x, y) of our conic and also the focus

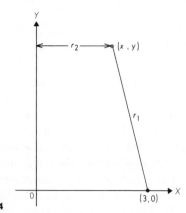

FIGURE 86-4

$(3, 0)$. The geometric description of a conic tells us that $r_1/r_2 = 2$; that is,

$$r_1{}^2 = 4r_2{}^2.$$

Since $r_1{}^2 = (x - 3)^2 + y^2$ and $r_2{}^2 = x^2$, the equation of our conic is

$$(x - 3)^2 + y^2 = 4x^2.$$

This equation can be written as $3x^2 - y^2 + 6x - 9 = 0$, or, after completing the square,

$$\frac{(x+1)^2}{4} - \frac{y^2}{12} = 1.$$

In this form, it is plain that the equation represents a hyperbola, but we knew from the start that our conic was a hyperbola, since its eccentricity is greater than 1.

EXAMPLE 86-2   Find the eccentricity and locate the directrices of the conic $9x^2 + 25y^2 = 1$.

Solution   If we write this equation as

$$\frac{x^2}{\frac{1}{9}} + \frac{y^2}{\frac{1}{25}} = 1,$$

we recognize it as the standard form of the equation of an ellipse for which $a^2 = \frac{1}{9}$ and $b^2 = \frac{1}{25}$. Hence,

$$c = \sqrt{a^2 - b^2} = \sqrt{\frac{1}{9} - \frac{1}{25}} = \sqrt{\frac{16}{225}} = \frac{4}{15}.$$

Equation 86-7 gives the eccentricity of our ellipse as $e = c/a = \frac{4}{15}/\frac{1}{3} = \frac{4}{5}$, and we see from Equations 86-6 that the directrices are the lines $x = -\frac{5}{12}$ and $x = \frac{5}{12}$.

1. Find the eccentricity and directrices of each of the following conics.
   - (a) $3x^2 + 4y^2 = 12$
   - (b) $y^2 = 12x$
   - (c) $9x^2 - 16y^2 = 144$
   - (d) $12x^2 + y = 3$
   - (e) $x^2 - y^2 = 1$
   - (f) $5x^2 + 9y^2 = 45$
   - (g) $16y^2 - 9x^2 = 144$
   - (h) $9x^2 + 5y^2 = 45$

2. Use our geometric description to find the conic with the given focus, corresponding directrix, and eccentricity.
   - (a) focus $(2, 0)$, directrix $x = -2$, $e = 2$
   - (b) focus $(2, 0)$, directrix $x = -2$, $e = 1$
   - (c) focus $(2, 0)$, directrix $x = -2$, $e = \frac{1}{2}$
   - (d) focus $(0, 4)$, directrix $y = 0$, $e = 3$
   - (e) focus $(1, 0)$, directrix $x = 2$, $e = 1$
   - (f) focus $(1, 0)$, directrix $x = 4$, $e = \frac{1}{2}$
   - (g) focus $(1, 2)$, directrix $y = -2$, $e = \frac{1}{3}$
   - (h) focus $(0, 4)$, directrix $y = 1$, $e = 2$

3. Find the eccentricity and directrices of each of the following conics.
   - (a) $3x^2 + 4y^2 - 6x + 16y + 7 = 0$
   - (b) $9x^2 - 16y^2 + 18x - 32y = 151$
   - (c) $x^2 - y^2 = 2x$
   - (d) $5x^2 + 9y^2 - 20x - 36y + 11 = 0$

4. Find the eccentricity and the directrices of the hyperbola $a^2y^2 - b^2x^2 = a^2b^2$.

5. (a) Tell how the shape of an ellipse changes as the eccentricity varies from near 0 to near 1.
   (b) Tell how the shape of a hyperbola changes as the eccentricity varies from near 1 to a very large number.

6. Suppose you are told that the Y-axis is a directrix of an equilateral hyperbola and that the corresponding focus is a point of the X-axis that is 3 units away from this directrix. Can you find the equation of the hyperbola?

7. What is the eccentricty of an ellipse in which the distance between its foci is one-half the distance between its directrices?

8. A vertex of a certain conic is the origin, and the nearest focus is the point $(3, 4)$. What is the corresponding directrix if
   (a) the conic is a parabola?
   (b) the conic is a hyperbola of eccentricity 2?
   (c) the conic is an ellipse of eccentricity $\frac{1}{2}$?

9. Mr. Jones owns a small rectangular woodlot 200 yards wide. One side is bounded by a very noisy highway, and right in the middle of the opposite edge is a particularly foul-smelling foundry. Describe the path Mr. Jones takes on his Sunday morning walks if
   (a) he dislikes the highway and foundry equally.
   (b) the wind is such that the smell of the foundry is twice as annoying as the sound of the highway.

(c) the wind is such that the smell of the foundry is half as annoying as the sound of the highway.

10. Use algebra to show that the distance d between a point (x, y) of the ellipse $x^2/a^2 + y^2/b^2 = 1$ (with $a > b$) and the focus $(c, 0)$ is given by the equation

$$d = e \left| x - \frac{a^2}{c} \right|.$$

87

POLAR COORDINATES

The fundamental idea of plane analytic geometry is the introduction of a coordinate system into the plane so that we have a "reference frame" with respect to which we can locate points. Our cartesian coordinate system consisting of two number scales meeting at right angles is only one of many possible coordinate systems. Now we will look at another one, the system of **polar coordinates.**

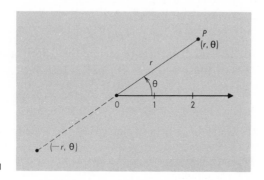

FIGURE 87-1

We start with a given half-line with endpoint O, as shown in Fig. 87-1. We will think of this half-line as half a number scale so that we have a unit of distance available. The point O is the **pole,** and the half-line is the **polar axis** of our polar coordinate system. If P is any point of the plane, we can assign to it the pair (r, θ) of polar coordinates, where r is the distance between O and P, and θ is an angle that is determined by rotating the polar axis into the half-line that contains O and P. This idea is perfectly straightforward, and you will have no difficulty verifying the coordinates of the points we have plotted in Fig. 87-2. To illustrate both possibilities, we have measured some of our angles in radians and some in degrees. In a given problem, of course, it is best to use just one unit of angular measurement. You will also notice that a given point has more than one pair (in fact, it has infinitely many pairs) of polar coordinates. For we can add any integral multiple of 360° to the angular coordinate of a point or use negative angles. Thus, the pairs (2, 45°), (2, 405°), (2, −315°), and so on, all represent the same point. We also occasionally find it convenient to use a negative number for the radial polar coordinate of a point. To plot the point (r, θ) with $r > 0$, we proceed r units from the point O along the terminal side of the polar angle. To plot the point

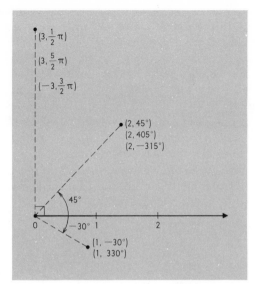

FIGURE 87-2

$(-r, \theta)$, we proceed r units from the pole along the extension of the terminal side through O (Fig. 87-1). Thus, (r, θ) and $(-r, \theta + 180°)$ are polar coordinates of the same point. The coordinates $(0, \theta)$ represent the pole for every θ.

The graph of an equation in x and y is the set of points whose XY-coordinates satisfy the equation. Similarly, the graph of an equation in the polar coordinates r and θ is the set of those points that have polar coordinates that satisfy the equation.

EXAMPLE 87-1   Sketch the graph of the equation $r = 2 \cos 3\theta$.

Solution   We proceed in the usual way. That is, we set up a table of values, plot the corresponding points, and join them to obtain the desired curve. Our table and the resulting graph appear in Fig. 87-3. Naturally, the more experience you have with polar coordinates, the easier it is to draw a correct graph using a relatively small number of plotted points.

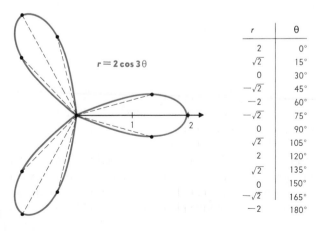

r	θ
2	0°
$\sqrt{2}$	15°
0	30°
$-\sqrt{2}$	45°
-2	60°
$-\sqrt{2}$	75°
0	90°
$\sqrt{2}$	105°
2	120°
$\sqrt{2}$	135°
0	150°
$-\sqrt{2}$	165°
-2	180°

$r = 2 \cos 3\theta$

FIGURE 87-3

401

We won't have time to discuss many of the details of how to draw the graph of an equation in polar coordinates. As with equations in rectangular coordinates, we can often recognize that graphs are symmetric, and so on. One item does deserve mention, however. We have said that a point belongs to the graph of an equation in r and θ if it has a pair of polar coordinates that satisfy the equation. Since each point has infinitely many pairs of polar coordinates, it may be true that some of these pairs satisfy the equation while others don't. For example, the pair of coordinates $(1, 0)$ satisfies the equation $r = 2^\theta$, so the point with polar coordinates $(1, 0)$ belongs to the graph of this equation. But this point also has polar coordinates $(-1, \pi)$ and $(1, 2\pi)$, and neither of these pairs satisfies the equation. Thus, you must keep your eyes open when using polar coordinates and be alert for such situations.

EXAMPLE 87-2   Sketch the graph of the equation $r = 2(1 - \cos \theta)$.

Solution   In this example we can lighten our labor by noting that, since $\cos(-\theta) = \cos \theta$, the point $(r, -\theta)$ belongs to the graph if the point (r, θ) does. Therefore, our graph is symmetric about the line lying along the polar axis. We can sketch the graph by plotting points whose angular coordinates lie between $0°$ and $180°$ and then reflecting this portion about the line of symmetry. The table accompanying Fig. 87-4 lists the coordinates we used to sketch the curve. This heart-shape curve is called a **cardioid.**

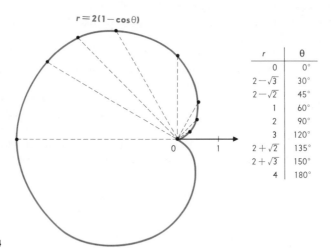

$r = 2(1 - \cos\theta)$

r	θ
0	0°
$2 - \sqrt{3}$	30°
$2 - \sqrt{2}$	45°
1	60°
2	90°
3	120°
$2 + \sqrt{2}$	135°
$2 + \sqrt{3}$	150°
4	180°

FIGURE 87-4

Usually, we introduce a polar coordinate system in conjunction with a rectangular system. Then we can use our knowledge of rectangular coordinates, too. Figure 87-5 shows a rectangular coordinate system, and we will suppose that its origin is the pole and its positive X-axis is the polar axis of a polar coordinate system. Then a given point P has cartesian coordinates (x, y) and polar coordinates (r, θ), as shown. You recall from our study of

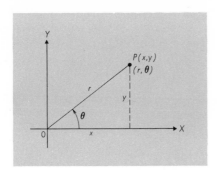

FIGURE 87-5

trigonometry (Equations 40-1) that these coordinates are related by the equations

(87-1) $$x = r \cos \theta \quad \text{and} \quad y = r \sin \theta.$$

It is easy to use these equations to find the rectangular coordinates of a point whose polar coordinates are given and somewhat more difficult to use them to find polar coordinates of a point if we know its rectangular coordinates. A sketch of the situation is very helpful.

We can use Equations 87-1 to change equations in rectangular coordinates to equations in polar coordinates, and conversely. For example, the line whose equation in rectangular coordinates is $\sqrt{3}x - y = 2$ becomes $\sqrt{3}\, r \cos \theta - r \sin \theta = 2$ in polar coordinates. We can write the left-hand side of this equation (see Equation 35-5) as

$$r(\sqrt{3} \cos \theta - \sin \theta) = 2r \cos (\theta + 30°),$$

and so our line has the simple polar equation

$$r \cos (\theta + 30°) = 1.$$

Geometrically (see Fig. 87-6), this equation says that our given line contains the point $(1, -30°)$ and is perpendicular to the line that contains that point and the origin. In the next section we will have more to say about the polar equations of other familiar curves. Frequently, our knowledge of cartesian equations of curves helps us with polar equations.

EXAMPLE 87-3   Discuss the graph of the equation

$$r = 4 \cos \theta - 2 \sin \theta.$$

Solution   Instead of plotting points as we did in Examples 87-1 and 87-2, we will write this equation in terms of cartesian coordinates x and y. First, we multiply both sides of the given equation by r to obtain the equation

$$r^2 = 4r \cos \theta - 2r \sin \theta.$$

403

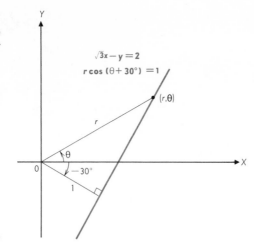

$\sqrt{3}x - y = 2$

$r \cos(\theta + 30°) = 1$

FIGURE 87-6

From Equations 87-1, we see that $r^2 = x^2 + y^2$, and so we have

$$x^2 + y^2 = 4x - 2y.$$

We now complete the square,

$$(x^2 - 4x + 4) + (y^2 + 2y + 1) = 4 + 1,$$
$$(x - 2)^2 + (y + 1)^2 = 5,$$

and we find that our equation represents the circle whose center is the point $(2, -1)$ and whose radius is $\sqrt{5}$.

PROBLEMS
87

1. Find the cartesian coordinates of the points that have the following polar coordinates.

(a) $(8, 30°)$ (b) $(4, -420°)$ (c) $(-10, \frac{3}{4}\pi)$

(d) (π, π) (e) $(60, 60°)$ (f) $(-4, \frac{17}{6}\pi)$

2. Find a set of polar coordinates for each of the points with the following rectangular coordinates.

(a) $(-3, 0)$ (b) $(0, -3)$ (c) $(\sqrt{2}, -\sqrt{2})$

(d) (π, π) (e) $(4, -4\sqrt{3})$ (f) $(-4, 4\sqrt{3})$

3. Transform the following equations to cartesian coordinates and sketch their graphs.

(a) $r = 4 \cos \theta$ (b) $r = 2 \sin \theta$ (c) $r = 3 \sec \theta$

(d) $r = -2 \csc \theta$ (e) $\tan \theta = 2$ (f) $r \cos \theta = \tan \theta$

4. Sketch the graphs of the following equations.

(a) $r = 4 \cos 2\theta$ (b) $r = 4 \cos 4\theta$ (c) $r = 2 - \cos \theta$

(d) $r = 1 - 2 \cos \theta$ (e) $r = 1 - \cos \theta$ (f) $r^2 = 4 \cos \theta$

5. Convince yourself that the graph of an equation in r and θ is

(a) symmetric with respect to the polar axis if $(r, -\theta)$ satisfies the equation whenever (r, θ) does.

(b) symmetric with respect to the line $\theta = \frac{1}{2}\pi$ if $(r, \pi - \theta)$ satisfies the equation whenever (r, θ) does.

(c) symmetric with respect to the pole if $(-r, \theta)$ or $(r, \theta + \pi)$ satisfies the equation whenever (r, θ) does.

Convince yourself that the graphs of

(d) all the equations of Problem 87-4 are symmetric with respect to the polar axis.

(e) the equations of parts (a) and (b) of Problem 87-4 are symmetric with respect to the line $\theta = \frac{1}{2}\pi$.

(f) the equations of parts (a), (b), and (f) of Problem 87-4 are symmetric with respect to the pole.

6. Find a polar equation of the line that contains the points with polar coordinates $(1, \pi)$ and $(2, \frac{1}{2}\pi)$. (*Hint:* You can write the cartesian coordinate equation first and then change to polar coordinates.)

7. Does it follow from Equations 87-1 that $r = \sqrt{x^2 + y^2}$ and $\theta = \text{Tan}^{-1}(y/x)$?

8. Find the points of intersection of the graphs of the following pairs of equations. Make a sketch, because this problem is somewhat tricky.

(a) $r = 4 \cos 2\theta, r = 2$ (b) $r = 4 \cos 2\theta, r = 4 \cos \theta$

(c) $r^2 = \sin^2 2\theta, \tan \theta = 1$ (d) $r = 1 - \cos \theta, r = \sin \frac{1}{2}\theta$

(e) $r = 2(1 - \cos \theta), r = -6 \cos \theta$ (f) $r = \cos \theta, r = \sin \theta$

9. Show that the distance between the points (r_1, θ_1) and (r_2, θ_2) is given by the formula

$$d = \sqrt{r_1^2 + r_2^2 - 2r_1 r_2 \cos(\theta_2 - \theta_1)}.$$

10. Sketch the graph of the equation $r\theta = 1$ (where θ is measured in radians). Use Table II to convince yourself that if θ is close to 0, the Y-coordinate of a point (r, θ) of the graph is close to 1.

11. Sketch the graphs of the following equations.

(a) $r = [\![\theta]\!]$ (b) $[\![r]\!] = \theta$ (c) $[\![r]\!] = [\![\theta]\!]$

12. Show that the equations $r = 2(1 - \cos \theta)$ and $r = -2(1 + \cos \theta)$ have the same graph.

88
LINES, CIRCLES, AND CONICS IN POLAR COORDINATES

In earlier sections we discuss the cartesian equations of certain common curves—lines, circles, and conics. Now we will consider the representation of these particular curves in polar coordinates. It helps tie our work with polar coordinates to our previous work with rectangular coordinates if we suppose that our polar coordinate system is superimposed on a cartesian system, as it was in the preceding section.

If α is a given angle, then it is clear that every point that has polar coordinates (r, θ) such that

$$(88\text{-}1) \qquad\qquad \theta = \alpha$$

belongs to the line that contains the pole and makes an angle of α with the

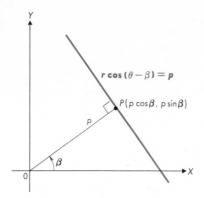

FIGURE 88-1

polar axis, and conversely. Thus the graph of Equation 88-1 is a line that contains the pole.

Now let us look at lines that do not contain the pole O (Fig. 88-1). A line is determined by a given point and a given direction. So we will start with a given point P and find the equation of the line that contains P and is perpendicular to the line segment joining P to the pole. Suppose that P has polar coordinates (p, β), where $p \neq 0$. Then the segment OP has slope $\tan \beta$, and the cartesian equation of any line perpendicular to OP has the form $(\cos \beta)x + (\sin \beta)y = c$ (the slope of this line is $-1/\tan \beta$). Our line contains the point P whose cartesian coordinates are $(p \cos \beta, p \sin \beta)$. Hence, $(\cos \beta)(p \cos \beta) + (\sin \beta)(p \sin \beta) = c$; that is, $p = c$. Therefore,

(88-2) $$x \cos \beta + y \sin \beta = p$$

is the cartesian equation of the line that contains the point P with polar coordinates (p, β) and is perpendicular to OP. By replacing x with $r \cos \theta$ and y with $r \sin \theta$ and simplifying, we obtain

(88-3) $$r \cos (\theta - \beta) = p$$

as the polar equation of our line. (What if $\beta = 0°$, $90°$, $180°$, or $270°$?)

EXAMPLE 88-1   Find the distance between the origin and the line $2x - 3y + 7 = 0$.

Solution   We write

$$-2x + 3y = 7$$

and divide both sides of this equation by

$$\sqrt{(-2)^2 + 3^2} = \sqrt{13}$$

to obtain

$$-\frac{2}{\sqrt{13}} x + \frac{3}{\sqrt{13}} y = \frac{7}{\sqrt{13}}.$$

406

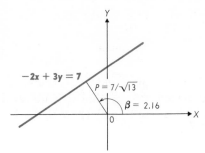

$-2x + 3y = 7$

$P = 7/\sqrt{13}$

$\beta = 2.16$

FIGURE 88-2

This last equation is in the form of Equation 88-2, with $p = 7/\sqrt{13}$, $\cos \beta = -2/\sqrt{13}$, and $\sin \beta = 3/\sqrt{13}$. Thus the distance between the origin and the line is $7/\sqrt{13}$, and one possible choice for β is (approximately) $\beta = 2.16$. This problem is illustrated in Fig. 88-2.

The cartesian equation of a circle with a radius of a and whose center is a point with polar coordinates (c, α) is

$$(x - c \cos \alpha)^2 + (y - c \sin \alpha)^2 = a^2.$$

If we replace x with $r \cos \theta$ and y with $r \sin \theta$ and simplify, we obtain the polar equation of our circle—namely,

(88-4) $$r^2 - 2rc \cos (\theta - \alpha) + c^2 = a^2.$$

If the center of the circle is the pole O, then $c = 0$ and Equation 88-4 reduces to $r^2 = a^2$, which is equivalent to the polar equation

(88-5) $$r = a.$$

If the circle contains the origin (see Figure 88-3), then $c^2 = a^2$ and Equation 88-4 is equivalent to the equation

(88-6) $$r = 2c \cos (\theta - \alpha).$$

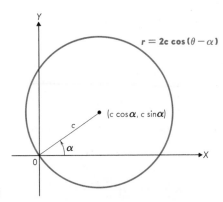

$r = 2c \cos (\theta - \alpha)$

$(c \cos \alpha, c \sin \alpha)$

c

α

FIGURE 88-3

407

In particular, if we set $\alpha = 0$, we obtain the equation

(88-7)
$$r = 2c \cos \theta$$

that represents a circle with a radius of $|c|$ and whose center is the point of the X-axis with cartesian coordinates $(c, 0)$. Similarly, if we set $\alpha = \frac{1}{2}\pi$, we get the equation of a circle with a radius of $|c|$ and whose center is the point of the Y-axis with cartesian coordinates $(0, c)$:

(88-8)
$$r = 2c \sin \theta.$$

Polar coordinates are particularly well suited for representing conics. In Section 86 we found how a conic is determined by a point called a *focus*, a line called a *directrix*, and a positive number called the *eccentricity* of the conic. A point belongs to the conic if, and only if, the ratio of the distance between the point and the focus to the distance between the point and the directrix is the eccentricity. Suppose we know the focus F, the corresponding directrix d, located p units from F, and the eccentricity e of a certain conic. Let us introduce polar coordinates so that the pole is the focus F and so that the directrix is perpendicular to the polar axis at the point with polar coordinates (p, π), where $p > 0$. We have sketched an arc of our conic in Fig. 88-4.

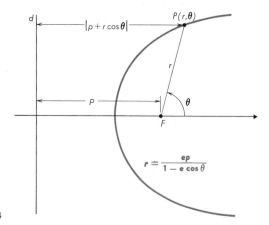

FIGURE 88-4

If our conic is a hyperbola ($e > 1$), it will also have a branch lying to the left of d, but in the case of an ellipse ($e < 1$) or a parabola ($e = 1$), the entire conic lies to the right of the directrix. Suppose that P is a point of our conic, and let (r, θ) be polar coordinates of P. The cartesian equation of the directrix is $x = -p$, so the distance between P and d is $|p + r \cos \theta|$. Thus the definition of a conic tells us that

$$\frac{|r|}{|p + r \cos \theta|} = e,$$

so that either

$$(88\text{-}9) \qquad \frac{r}{p + r \cos\theta} = e \quad \text{or} \quad \frac{r}{p + r \cos\theta} = -e.$$

Now it is easy to show (we ask you to in Problem 88-9) that if the coordinates (r, θ) of a certain point satisfy either one of these equations, then the coordinates $(-r, \theta + \pi)$ *of the same point* satisfy the other equation. It follows that we do not need both equations to describe our conic; either one will do. We will take the equation that is obtained by solving the first of Equations 88-9 for r to be the standard form of the polar equation of our conic:

$$(88\text{-}10) \qquad r = \frac{ep}{1 - e \cos\theta}.$$

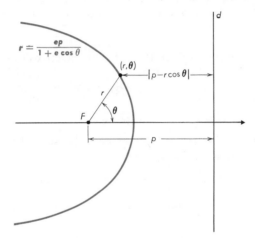

FIGURE 88-5

By reasoning as we did above, you can find that if the focus is the pole, and if the directrix is to the right of the pole (see Fig. 88-5), then we can take the standard form of the equation of our conic to be

$$(88\text{-}11) \qquad r = \frac{ep}{1 + e \cos\theta}.$$

If the directrix is parallel to the polar axis, we take the standard polar equation of the conic to be

$$(88\text{-}12) \qquad r = \frac{ep}{1 + e \sin\theta}$$

if the directrix is above the focus, and

$$(88\text{-}13) \qquad r = \frac{ep}{1 - e \sin\theta}$$

if the directrix is below the focus (see Fig. 88-6).

409

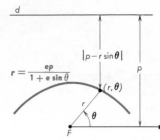

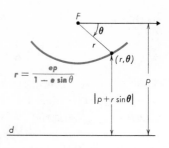

FIGURE 88-6

In each case the focus is the pole of our coordinate system, and the directrix that corresponds to this focus is p units away from it.

EXAMPLE 88-2   Write a polar equation of the parabola whose focus is the pole, whose axis lies along the polar axis, which opens to the right, and which contains the point $(2, \frac{1}{3}\pi)$.

Solution   Since our parabola opens to the right, its directrix lies to the left of the focus (the pole), and hence Equation 88-10 is the equation we want. We are dealing with a parabola, so $e = 1$, and our equation is

$$r = \frac{p}{1 - \cos \theta}.$$

We determine p by setting $r = 2$ and $\theta = \frac{1}{3}\pi$, since the point $(2, \frac{1}{3}\pi)$ belongs to the parabola. Thus,

$$2 = \frac{p}{1 - \cos \frac{1}{3}\pi} = \frac{p}{1 - \frac{1}{2}} = 2p,$$

and hence $p = 1$. Therefore, the equation of our parabola is

$$r = \frac{1}{1 - \cos \theta}.$$

EXAMPLE 88-3   Discuss the graph of the equation

(88-14)
$$r = \frac{16}{5 + 3 \sin \theta}.$$

Solution   If we divide the numerator and the denominator of our fraction by 5, we can write Equation 88-14 in the form of Equation 88-12:

$$r = \frac{\frac{16}{5}}{1 + \frac{3}{5} \sin \theta} = \frac{\frac{3}{5} \cdot \frac{16}{5}}{1 + \frac{3}{5} \sin \theta}.$$

Here, $e = \frac{3}{5}$ and $p = \frac{16}{3}$. Our curve is an ellipse, since $e < 1$, and its major diameter contains the pole and is perpendicular to the polar axis. Thus we find the vertices of the ellipse by setting $\theta = \frac{1}{2}\pi$ and $\theta = \frac{3}{2}\pi$ in Equation 88-14. We obtain

$$r = \frac{16}{5 + 3 \cdot 1} = 2 \quad \text{when} \quad \theta = \frac{1}{2}\pi,$$

and

$$r = \frac{16}{5 + 3 \cdot (-1)} = 8 \quad \text{when} \quad \theta = \frac{3}{2}\pi.$$

410

Therefore, the vertices are the points with polar coordinates $(2, \frac{1}{2}\pi)$ and $(8, \frac{3}{2}\pi)$. The major diameter is $8 + 2 = 10$ units long. In Section 86 we saw that the eccentricity of an ellipse is the ratio of the distance between its foci to the length of its major diameter. The major diameter of our present ellipse is $2b = 10$ units long, and if its foci are $2c$ units apart, we have the equation $e = 2c/10$; that is, $\frac{3}{5} = 2c/10$. Hence, $c = 3$. Finally, if the minor diameter of our ellipse is $2a$ units long, then $a = \sqrt{b^2 - c^2} =$

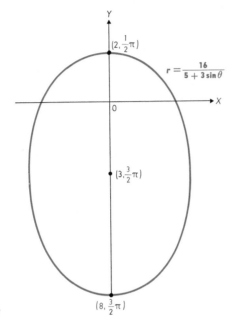

FIGURE 88-7

$\sqrt{25 - 9} = 4$. Our ellipse is shown in Fig. 88-7. You may readily verify that its cartesian equation is

$$\frac{x^2}{16} + \frac{(y + 3)^2}{25} = 1.$$

PROBLEMS
88

1. Find a polar equation of the line of which the given point is nearest the origin, and sketch the situation.
   (a) $(6, 30°)$ (b) $(6, 150°)$ (c) $(8, -\frac{1}{4}\pi)$ (d) $(3, 27°)$

2. Find the distance between
   (a) the line $3x + 4y = 30$ and the origin.
   (b) the line $4x - 3y = 100$ and the origin.
   (c) the lines $3x + 4y = 30$ and $3x + 4y = -25$.
   (d) the line $3x + 4y = 30$ and the point $(2, 1)$.

3. Find a polar equation of the circle whose center has rectangular coordinates $(-1, 1)$ and
   (a) that contains the pole.

411

(b) that has a radius of 3.

(c) that contains the point with rectangular coordinates (2, 3).

(d) that contains the point with polar coordinates $(2\sqrt{2}, -\frac{1}{4}\pi)$.

4. Sketch the graphs of the following equations. In each case, label the vertices (or vertex) and the foci (or focus) of the conic.

(a) $r = \dfrac{16}{3 + 5\sin\theta}$

(b) $r = \dfrac{16}{5 - 3\cos\theta}$

(c) $r = \dfrac{4}{1 + \sin\theta}$

(d) $r = \dfrac{16}{5 + 3\cos\theta}$

(e) $r\sin\theta = 1 - r$

(f) $3r\sin\theta = 5r - 16$

5. A conic has the origin of a cartesian coordinate system as a focus and a corresponding directrix whose equation is $x = -4$. Identify the conic and find a polar equation for it if it contains the following point (cartesian coordinates).

(a) (5, 12)　　　　(b) $(5, 2\sqrt{14})$　　　　(c) $(1, 2\sqrt{2})$　　　　(d) $(0, -2)$

6. Show that the graph of the equation $r\sin(\theta - \alpha) = q\sin(\beta - \alpha)$ is the line that contains the point with polar coordinates (q, β) and has a slope of $m = \tan\alpha$.

7. Describe the conic for which $p = 5{,}000{,}000$ and $e = 1/1{,}000{,}000$.

8. Express Equation 88-10 in cartesian coordinates.

9. Show that if the coordinates (r, θ) of a certain point satisfy one of Equations 88-9, then the coordinates $(-r, \theta + \pi)$ of the same point satisfy the other equation.

10. Show that the standard form of the polar equation of a parabola can be written as

$$r = \frac{p}{2}\csc^2\frac{\theta}{2}.$$

11. A chord that contains a focus of a conic is divided into two segments by the focus. Prove that the sum of the reciprocals of the lengths of these two segments is the same no matter what chord is chosen.

12. Suppose that $e < 1$ in Equation 88-10, and let $2a$ denote the length of the major diameter and $2c$ the distance between the foci of the ellipse represented by that equation. Show that

$$a = \frac{ep}{1 - e^2} \quad \text{and} \quad c = \frac{e^2 p}{1 - e^2}.$$

89
ROTATION
OF
AXES

In Section 82 we found that certain equations could be simplified considerably by translating the axes. Other transformations of coordinates also reduce complicated equations to simple ones, and we will devote this section to a must useful class of transformations, the *rotations*.

　　Figure 89-1 shows two cartesian systems, an XY-system and an $\overline{X}\,\overline{Y}$-system. These systems have the same origin, but the $\overline{X}$-axis makes an angle of

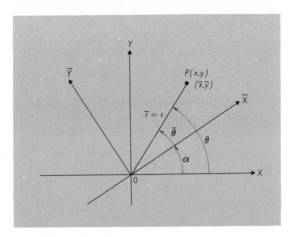

FIGURE 89-1

α with the X-axis. We say that the $\overline{X}\overline{Y}$-system is obtained from the XY-system by a **rotation** through α. Naturally, we should explain why anyone would *want* to rotate axes, but you will see why in the next section. Here, we are going to concentrate on answering the technical question, "If a point P has coordinates (x, y) and $(\overline{x}, \overline{y})$ relative to the two coordinate systems, how are the numbers x, y, and $\overline{x}$, $\overline{y}$ related?"

Our first step seems to complicate matters. We introduce two *more* coordinate systems, this time polar coordinate systems, both of which have 0 as the pole. In one system the positive X-axis is the polar axis, and in the other system the positive $\overline{X}$-axis is the polar axis. Then our point P will have polar coordinates (r, θ) and $(\overline{r}, \overline{\theta})$ in addition to its two pairs of cartesian coordinates. We know the relationship between (x, y) and (r, θ) and between $(\overline{x}, \overline{y})$ and $(\overline{r}, \overline{\theta})$, so we can find the equations connecting (x, y) and $(\overline{x}, \overline{y})$ from the equations connecting (r, θ) and $(\overline{r}, \overline{\theta})$. But these latter equations are obvious from a glance at Fig. 89-1:

$$\overline{r} = r \qquad \text{and} \qquad \overline{\theta} = \theta - \alpha.$$

Therefore, the equations

$$\overline{x} = \overline{r} \cos \overline{\theta} \qquad \text{and} \qquad \overline{y} = \overline{r} \sin \overline{\theta}$$

give us

(89-1) $$\overline{x} = r \cos (\theta - \alpha) \qquad \text{and} \qquad \overline{y} = r \sin (\theta - \alpha).$$

Using the trigonometric identities (Equations 34-2)

$$\cos (\theta - \alpha) = \cos \theta \cos \alpha + \sin \theta \sin \alpha$$

and

$$\sin (\theta - \alpha) = \sin \theta \cos \alpha - \cos \theta \sin \alpha,$$

413

we can write Equations 89-1 as

$$\bar{x} = r \cos \theta \cos \alpha + r \sin \theta \sin \alpha$$
$$\bar{y} = r \sin \theta \cos \alpha - r \cos \theta \sin \alpha.$$

(89-2)

Now we replace $r \cos \theta$ with x and $r \sin \theta$ with y in Equations 89-2, and we obtain our transformation equations

$$\bar{x} = x \cos \alpha + y \sin \alpha$$
$$\bar{y} = -x \sin \alpha + y \cos \alpha.$$

(89-3)

You can solve these equations for x and y (see Problem 59-10) and thus obtain the inverse transformation equations

$$x = \bar{x} \cos \alpha - \bar{y} \sin \alpha,$$
$$y = \bar{x} \sin \alpha + \bar{y} \cos \alpha.$$

(89-4)

EXAMPLE 89-1   Suppose the $\overline{X}\overline{Y}$-axes are obtained by rotating the XY-axes through an angle of 45°. To what does the equation $xy = 1$ transform?

Solution   Here $\alpha = 45°$, so Equations 89-4 become

$$x = \frac{\bar{x}}{\sqrt{2}} - \frac{\bar{y}}{\sqrt{2}},$$

$$y = \frac{\bar{x}}{\sqrt{2}} + \frac{\bar{y}}{\sqrt{2}}.$$

Hence

$$xy = \frac{(\bar{x} - \bar{y})(\bar{x} + \bar{y})}{2} = \frac{(\bar{x}^2 - \bar{y}^2)}{2}.$$

Thus the equation $xy = 1$ is transformed into the equation

$$\frac{\bar{x}^2}{2} - \frac{\bar{y}^2}{2} = 1.$$

This equation is the standard form for an equilateral hyperbola whose asymptotes bisect the quadrants in the $\overline{X}\overline{Y}$-coordinate system. Thus the equation $xy = 1$ represents an equilateral hyperbola whose asymptotes are the X- and Y-axes (see Fig. 89-2).

EXAMPLE 89-2   Find the distance between the point $P(1, 4)$ and the line $2y - x = 2$.

Solution   Figure 89-3 shows the given point and line. The slope of the line is $\tan \alpha = \frac{1}{2}$, where α is the angle of inclination. Now let us introduce an $\overline{X}\overline{Y}$-coordinate system in such a way that the $\overline{X}$-axis is parallel to our given line. We can obtain such an $\overline{X}\overline{Y}$-system by a rotation of axes through α. Since $\tan \alpha = \frac{1}{2}$, we see that $\sin \alpha = 1/\sqrt{5}$ and $\cos \alpha = 2/\sqrt{5}$. Thus the Trans-

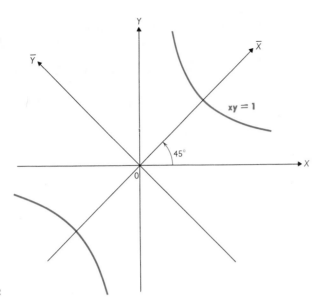

FIGURE 89-2

formation Equations 89-3 and 89-4 become

$$\bar{x} = \frac{2x}{\sqrt{5}} + \frac{y}{\sqrt{5}}, \quad \bar{y} = \frac{-x}{\sqrt{5}} + \frac{2y}{\sqrt{5}}$$

$$x = \frac{2\bar{x}}{\sqrt{5}} - \frac{\bar{y}}{\sqrt{5}}, \quad y = \frac{\bar{x}}{\sqrt{5}} + \frac{2\bar{y}}{\sqrt{5}}.$$

When we substitute these expressions for x and y in the equation of our line, it becomes $\bar{y} = \frac{2}{\sqrt{5}}$. Furthermore, these transformation equations tell us that the $\bar{Y}$-coordinate of the point P is $\frac{7}{\sqrt{5}}$. Since the line is parallel to the $\bar{X}$-axis, we find the distance d between it and P by subtraction:

$$d = \frac{7}{\sqrt{5}} - \frac{2}{\sqrt{5}} = \frac{5}{\sqrt{5}} = \sqrt{5}.$$

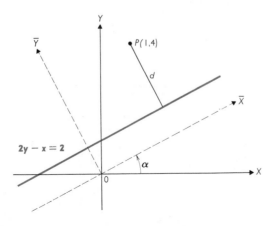

FIGURE 89-3

The **general rigid transformation** equations

(89-5)

$$\bar{x} = x \cos \alpha + y \sin \alpha - h$$
$$\bar{y} = -x \sin \alpha + y \cos \alpha - k$$

are the transformation equations that result from a rotation through α *followed* by a translation.

EXAMPLE 89-3   Sketch the two cartesian coordinate systems that are related by the equations

(89-6)

$$\bar{x} = \tfrac{3}{5}x + \tfrac{4}{5}y - 3$$
$$\bar{y} = -\tfrac{4}{5}x + \tfrac{3}{5}y - 1.$$

Draw the XY-system in the "usual" position.

Solution   The points whose $\bar{X}\bar{Y}$-coordinates are $(0,0)$ and $(1,0)$ determine the $\bar{X}$-axis, and the points whose $\bar{X}\bar{Y}$-coordinates are $(0,0)$ and $(0,1)$ determine the $\bar{Y}$-axis. We will locate these points by finding their XY-coordinates from Equations 89-6 and then plotting them in the XY-coordinate

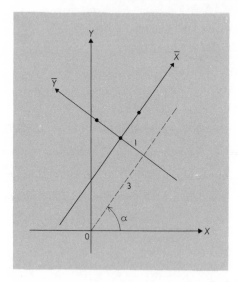

FIGURE 89-4

system of Fig. 89-4. The XY-coordinates of the point whose $\bar{X}\bar{Y}$-coordinates are $(0,0)$ satisfy the system of equations

$$0 = \tfrac{3}{5}x + \tfrac{4}{5}y - 3$$
$$0 = -\tfrac{4}{5}x + \tfrac{3}{5}y - 1.$$

When we solve this system, we obtain $(x, y) = (1, 3)$. Similarly, we find that the point whose $\bar{X}\bar{Y}$-coordinates are $(1, 0)$ has XY-coordinates $(\tfrac{8}{5}, \tfrac{19}{5})$. The point whose $\bar{X}\bar{Y}$-coordinates are $(0, 1)$ has XY-coordinates $(\tfrac{1}{5}, \tfrac{18}{5})$. Now we plot these points and draw in the $\bar{X}\bar{Y}$-axes. We see that the $\bar{X}\bar{Y}$-axes are obtained by first rotating the axes through an angle α, where $\cos \alpha = \tfrac{3}{5}$ and $\sin \alpha = \tfrac{4}{5}$, and then translating the origin 3 units in the "new X-direction" and 1 unit in the "new Y-direction."

1. Find the $\overline{X}\overline{Y}$-coordinates of the point $(-4, 6)$ under a rotation of axes through the given angle.

   (a) $30°$ (b) $135°$ (c) $240°$ (d) $-180°$

   (e) π (f) $-\frac{1}{2}\pi$ (g) 2π (h) $-\frac{1}{4}\pi$

2. Under a rotation of axes through an angle of $60°$, a point has the given $\overline{X}\overline{Y}$-coordinates. What are its XY-coordinates?

   (a) $(2, 0)$ (b) $(0, 2)$ (c) $(-2, 4)$ (d) $(-\sqrt{12}, 6)$

3. (a) Sketch the XY- and $\overline{X}\overline{Y}$-axes if they are related by the transformation equations
$$\bar{x} = 0.8x + 0.6y$$
$$\bar{y} = -0.6x + 0.8y.$$

   (b) Solve for x and y in terms of $\bar{x}$ and $\bar{y}$.
   (c) What angle does the $\overline{X}$-axis make with the X-axis?
   (d) What is the equation of the line $4x + 3y = 10$ in the $\overline{X}\overline{Y}$-system?
   (e) What is the equation of the circle $x^2 + y^2 = 16$ in the $\overline{X}\overline{Y}$-system? (You can answer this question with no computation!)

4. Find the distance between the line $3x - 4y = 10$ and

   (a) the origin.
   (b) the point $(-5, 15)$.
   (c) the line $3x - 4y + 15 = 0$.
   (d) the line $3x - 4y = 15$.
   (e) the center of the circle $x^2 + y^2 = 10x$.
   (f) the focus of the parabola $y^2 + 25 = 10y + 20x$.

5. (a) What are the transformation of axes equations for a rotation through $90°$?
   (b) What are the transformation of axes equations for a rotation through $180°$?
   (c) What are the transformation of axes equations for a rotation through $270°$?
   (d) Do the equations $\bar{x} = y$ and $\bar{y} = x$ represent a rotation of axes?

6. Suppose that the $\overline{X}\overline{Y}$-axes are obtained by a rotation through $30°$.

   (a) Express $\bar{x}$ and $\bar{y}$ in terms of x and y.
   (b) Express x and y in terms of $\bar{x}$ and $\bar{y}$.
   (c) Express the line $4x - 6y = 5$ in $\overline{X}\overline{Y}$-coordinates.
   (d) Express the circle $x^2 + y^2 - 2x = 2$ in $\overline{X}\overline{Y}$-coordinates. (*Hint:* You can find the center and radius very quickly.)

7. (a) Sketch the $\overline{X}\overline{Y}$- and XY-axes if they are related by the equations $\bar{x} = .9611x + .2764y$ and $\bar{y} = -.2764x + .9611y$.
   (b) What is the angle of rotation?
   (c) What is the slope of the $\overline{Y}$-axis in the XY-coordinate system?
   (d) What is the slope of the X-axis in the $\overline{X}\overline{Y}$-coordinate system?
   (e) What is the angle of inclination of the line $x + y = 3$ in the $\overline{X}\overline{Y}$-system?
   (f) What is the angle of inclination of the line $\bar{x} + \bar{y} = 3$ in the XY-system?

8. If z is the complex number $x + iy$, w is the number $u + iv$, and r is the number $\cos \alpha - i \sin \alpha$, show that the equation $w = rz$ is a compact way of writing Equations 89-3. (We didn't want to use the bar notation here for fear of

introducing some confusion with the complex conjugate.) Observe that $1/r = \cos\alpha + i \sin\alpha$, and hence that $z = (1/r)w$ is a compact way of writing Equations 89-4.

9. Explain geometrically why moving the bars from the left- to the right-hand sides of Equations 89-3 and replacing α with $-\alpha$ produces Equations 89-4.

10. Solve Equations 89-5 for x and y.

11. Give a geometric description of the transformation that is determined by the equations $\bar{x} = hx$ and $\bar{y} = ky$, where h and k are given positive numbers.

12. Suppose that points P_1 and P_2 have coordinates (x_1, y_1), $(\bar{x}_1, \bar{y}_1)$ and (x_2, y_2), $(\bar{x}_2, \bar{y}_2)$ relative to two coordinate systems. It is geometrically obvious that if the coordinate transformation is a translation or a rotation, the distance formula should be the same in each coordinate system; that is,

$$\sqrt{(x_1 - x_2)^2 + (y_1 - y_2)^2} = \sqrt{(\bar{x}_1 - \bar{x}_2)^2 + (\bar{y}_1 - \bar{y}_2)^2}.$$

Prove it.

90
THE
GENERAL
QUADRATIC
EQUATION

For certain choices of the numbers $A, B, C, D, E,$ and F we already know that the graph of the quadratic equation

(90-1) $$A x^2 + Bxy + Cy^2 + Dx + Ey + F = 0$$

is a circle, a conic, or a line. For example, if we choose $A = 1$, $B = 0$, $C = -1$, $D = 0$, $E = 0$, and $F = -1$, then Equation 90-1 becomes

$$x^2 - y^2 = 1,$$

which we recognize as the equation of an equilateral hyperbola. But other choices of $A, B, C, D, E,$ and F lead to equations that we have not yet studied —for example, the equation

(90-2) $$6x^2 + 24xy - y^2 - 12x + 26y + 11 = 0.$$

In this section we shall see that the graph of a quadratic equation is always a familiar figure. In particular, it will turn out that the graph of Equation 90-2 is a hyperbola.

To find the graph of a quadratic equation we will first make a transformation of axes that reduces the equation to one of the standard forms we have already studied. Thus in Example 89-1 we saw that a rotation of axes through 45° transforms the equation $xy = 1$ into the standard equation of an equilateral hyperbola

$$\frac{\bar{x}^2}{2} - \frac{\bar{y}^2}{2} = 1.$$

The question that naturally comes to mind is, "How did we decide to rotate through 45°?" The following example shows how we pick our angle of rotation.

EXAMPLE 90-1   By a suitable rotation of axes, reduce the equation

(90-3)
$$8x^2 - 4xy + 5y^2 = 36$$

to a standard form.

Solution   When we make the rotation given by Equations 89-4, our equation becomes

$$8(\bar{x} \cos \alpha - \bar{y} \sin \alpha)^2 - 4(\bar{x} \cos \alpha - \bar{y} \sin \alpha) \cdot (\bar{x} \sin \alpha + \bar{y} \cos \alpha)$$
$$+ 5(\bar{x} \sin \alpha + \bar{y} \cos \alpha)^2 = 36.$$

Now we expand and collect terms:

(90-4)
$$(8 \cos^2 \alpha - 4 \sin \alpha \cos \alpha + 5 \sin^2 \alpha)\bar{x}^2$$
$$+ (4 \sin^2 \alpha - 6 \sin \alpha \cos\alpha - 4 \cos^2 \alpha)\bar{x}\bar{y}$$
$$+ (8 \sin^2 \alpha + 4 \sin \alpha \cos \alpha + 5 \cos^2 \alpha)\bar{y}^2 = 36.$$

This last equation has the form

(90-5)
$$\overline{A}\bar{x}^2 + \overline{B}\bar{x}\bar{y} + \overline{C}\bar{y}^2 = 36,$$

where the numbers $\overline{A}$, $\overline{B}$, and $\overline{C}$ depend on α. If $\overline{B} = 0$, we would recognize the type of curve we are dealing with, so let us choose α so that $\overline{B} = 0$; that is, so that

$$4 \sin^2 \alpha - 6 \sin \alpha \cos \alpha - 4 \cos^2 \alpha = 0.$$

Since $\cos^2 \alpha - \sin^2 \alpha = \cos 2\alpha$ and $\sin \alpha \cos \alpha = \frac{1}{2} \sin 2\alpha$, this equation can be written as

$$-3 \sin 2\alpha - 4 \cos 2\alpha = 0,$$

from which we see that

$$\cot 2\alpha = -\tfrac{3}{4}.$$

We can select an angle 2α in the range $0° < 2\alpha < 180°$ that satisfies this equation, and thus $0° < \alpha < 90°$. To find the coefficients of $\bar{x}^2$ and $\bar{y}^2$ we find the numbers $\cos \alpha$ and $\sin \alpha$ from the trigonometric identities (Equations 33-5 and 33-6)

$$\cos \alpha = \sqrt{\frac{1 + \cos 2\alpha}{2}} \quad \text{and} \quad \sin \alpha = \sqrt{\frac{1 - \cos 2\alpha}{2}}.$$

Since $\cot 2\alpha = -\tfrac{3}{4}$, it follows that the radial line containing the point $(-3, 4)$ makes an angle of 2α with the positive X-axis. Hence, $\cos 2\alpha = -\tfrac{3}{5}$, and we have

$$\cos \alpha = \sqrt{\frac{1 - \tfrac{3}{5}}{2}} = \sqrt{\tfrac{1}{5}}$$

and

$$\sin \alpha = \sqrt{\frac{1 + \tfrac{3}{5}}{2}} = \sqrt{\tfrac{4}{5}}.$$

When we substitute these numbers in Equation 90-4, the equation becomes

$$4\bar{x}^2 + 9\bar{y}^2 = 36.$$

In other words,

$$\frac{\bar{x}^2}{9} + \frac{\bar{y}^2}{4} = 1.$$

So we see that the graph of Equation 90-3 is an ellipse whose major diameter is 6 units long and whose minor diameter is 4 units long. We have shown our ellipse, together with the rotated axes, in Fig. 90-1.

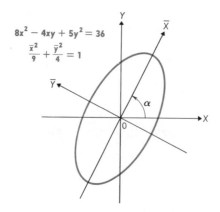

$$8x^2 - 4xy + 5y^2 = 36$$

$$\frac{\bar{x}^2}{9} + \frac{\bar{y}^2}{4} = 1$$

FIGURE 90-1

We can employ the same procedure to reduce any quadratic expression

$$Ax^2 + Bxy + Cy^2$$

in which $B \neq 0$ to the form

$$\bar{A}\bar{x}^2 + \bar{C}\bar{y}^2.$$

We simply replace x and y in terms of $\bar{x}$ and $\bar{y}$ according to the rotation Equations 89-4 and obtain the expression $A(\bar{x} \cos \alpha - \bar{y} \sin \alpha)^2 + B(\bar{x} \cos \alpha - \bar{y} \sin \alpha)(\bar{x} \sin \alpha + \bar{y} \cos \alpha) + C(\bar{x} \sin \alpha + \bar{y} \cos \alpha)^2$. When we multiply out and collect the coefficients of $\bar{x}^2$, $\bar{x}\bar{y}$, and $\bar{y}^2$, we find that

$$Ax^2 + Bxy + Cy^2 = \bar{A}\bar{x}^2 + \bar{B}\bar{x}\bar{y} + \bar{C}\bar{y}^2,$$

where

(90-6)
$$\bar{A} = A \cos^2 \alpha + B \sin \alpha \cos \alpha + C \sin^2 \alpha,$$
$$\bar{B} = (C - A) \sin 2\alpha + B \cos 2\alpha,$$
$$\bar{C} = A \sin^2 \alpha - B \sin \alpha \cos \alpha + C \cos^2 \alpha.$$

420

We see from the second of Equations 90-6 that $\overline{B}$ will be 0 if we choose α so that

$$\cot 2\alpha = \frac{A - C}{B}.$$

Clearly, we can always choose α so that $0° < 2\alpha < 180°$. We sum up our results as a theorem.

Theorem 90-1 *Every quadratic expression $Ax^2 + Bxy + Cy^2$ in which $B \neq 0$ can be reduced to the form $\overline{A}\overline{x}^2 + \overline{C}\overline{y}^2$ by rotating axes through an angle α, where $0° < \alpha < 90°$, and $\cot 2\alpha = (A - C)/B$.*

By rotating axes through the angle α, as described in Theorem 90-1, we can reduce any quadratic equation of the form of Equation 90–1 in which $B \neq 0$ to a quadratic equation of the form

$$\overline{A}\overline{x}^2 + \overline{C}\overline{y}^2 + \overline{D}\overline{x} + \overline{E}\overline{y} + \overline{F} = 0.$$

If we don't recognize the graph of this equation, a suitable translation will bring it to a form that we do recognize. We will illustrate the entire procedure by an example. It will pay you to review our work on translation in Section 82.

EXAMPLE 90-2 Describe the graph of Equation 90-2.

Solution Here $A = 6$, $B = 24$, and $C = -1$. So we first rotate the axes through the acute angle α such that

$$\cot 2\alpha = \tfrac{7}{24}.$$

It follows that $\cos 2\alpha = \tfrac{7}{25}$, and hence

$$\cos \alpha = \sqrt{\frac{1 + \cos 2\alpha}{2}} = \frac{4}{5}$$

and

$$\sin \alpha = \sqrt{\frac{1 - \cos 2\alpha}{2}} = \frac{3}{5}.$$

Therefore, our rotation equations are

$$x = \frac{4\overline{x}}{5} - \frac{3\overline{y}}{5} \quad \text{and} \quad y = \frac{3\overline{x}}{5} + \frac{4\overline{y}}{5}.$$

When we substitute these quantities in Equation 90-2 we obtain the equation

$$15\overline{x}^2 - 10\overline{y}^2 + 6\overline{x} + 28\overline{y} + 11 = 0.$$

Now we will translate the axes by means of the equations

$$\overline{x} = \overline{\overline{x}} + h \quad \text{and} \quad \overline{y} = \overline{\overline{y}} + k,$$

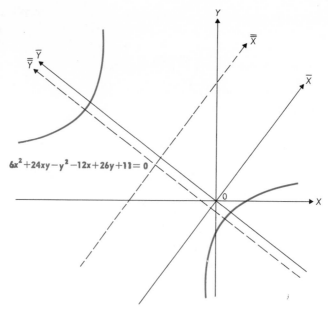

$6x^2 + 24xy - y^2 - 12x + 26y + 11 = 0$

FIGURE 90-2

where h and k are to be determined so as to make our resulting equation "simpler" than Equation 90-8. In terms of $\overline{\overline{x}}$ and $\overline{\overline{y}}$, Equation 90-8 reads

(90-9)
$$15\overline{\overline{x}}^2 - 10\overline{\overline{y}}^2 + (30h + 6)\overline{\overline{x}} + (-20k + 28)\overline{\overline{y}} + 15h^2$$
$$-10k^2 + 6h + 28k + 11 = 0.$$

Now we select h and k so that $30h + 6 = 0$ and $-20k + 28 = 0$; that is, $h = -\frac{1}{5}$ and $k = \frac{7}{5}$. Then Equation 90-9 becomes

$$15\overline{\overline{x}}^2 - 10\overline{\overline{y}}^2 + 30 = 0.$$

Finally, we write this equation in the standard form

$$\frac{\overline{\overline{y}}^2}{3} - \frac{\overline{\overline{x}}^2}{2} = 1,$$

and we see that our graph is a hyperbola. The "specifications"—diameter, length of latus rectum, focal length, and so on—of our hyperbola can be read from this equation. The graph of our equation is shown in Fig. 90-2, together with the three different coordinate axes involved.

In the preceding example we began with a quadratic equation

(90-10)
$$Ax^2 + Bxy + Cy^2 + Dx + Ey + F = 0,$$

and rotated axes to obtain an equation of the form

(90-11)
$$\overline{A}\overline{x}^2 + \overline{C}\overline{y}^2 + \overline{D}\overline{x} + \overline{E}\overline{y} + \overline{F} = 0.$$

It is easy to verify, by using the proper translation, that if $\bar{A}$ and $\bar{C}$ have opposite signs ($\bar{A}\bar{C} < 0$), then the graph of Equation 90-11 is a hyperbola. If $\bar{A}$ and $\bar{C}$ have the same signs ($\bar{A}\bar{C} > 0$), our curve is an ellipse or a circle. And if one of the numbers $\bar{A}$ or $\bar{C}$ is 0 ($\bar{A}\bar{C} = 0$), we are dealing with a parabola. Here we are allowing for *degenerate* conics, such as the "hyperbola" $x^2 - 4y^2 = 0$, or the "ellipse" $\dfrac{x^2}{4} + \dfrac{y^2}{9} + 1 = 0$.

In Problem 90-9 we ask you to verify that

$$\bar{B}^2 - 4\bar{A}\bar{C} = B^2 - 4AC,$$

regardless of the angle of rotation. In particular, if we choose α so that $\bar{B} = 0$, then we find that

$$-4\bar{A}\bar{C} = B^2 - 4AC.$$

Therefore, the sign of $\bar{A}\bar{C}$ is determined by the sign of $B^2 - 4AC$. Thus from our remarks concerning the graph of Equation 90-11, we can conclude that *the graph of Equation 90-10 is*

(i) *a parabola if* $B^2 - 4AC = 0$,

(ii) *an ellipse if* $B^2 - 4AC < 0$,

(iii) *a hyperbola if* $B^2 - 4AC > 0$,

with the understanding that degenerate cases may occur.

The number $B^2 - 4AC$ is called the *discriminant* of the quadratic Equation 90-10. Thus the discriminant of Equation 90-2 is $(24)^2 - 4 \cdot 6 \cdot (-1) = 600$, and we found that the graph of this equation is a hyperbola. The discriminant of Equation 90-3 is $(-4)^2 - 4 \cdot 8 \cdot 5 = -144$, and we found that the graph of this equation is an ellipse.

1. Reduce the following quadratic expressions to the form $\bar{A}\bar{x}^2 + \bar{C}\bar{y}^2$.
   (a) $x^2 + 4xy + y^2$ (b) $9x^2 + 12xy + 4y^2$
   (c) $4x^2 + 4xy + y^2$ (d) $6x^2 + 5xy - 6y^2$
   (e) $19x^2 - 7xy - 5y^2$ (f) $2x^2 - 5xy - 10y^2$

2. Rotate the axes so that the following equations take the form $\bar{A}\bar{x}^2 + \bar{C}\bar{y}^2 = \bar{F}$. Sketch the graphs of these equations. Find the algebraic sign of the discriminant of each equation, and use the result to check the form of your graph.
   (a) $11x^2 + 24xy + 4y^2 = 20$
   (b) $25x^2 + 14xy + 25y^2 = 288$
   (c) $3x^2 + 4xy = 4$
   (d) $9x^2 - 24xy + 16y^2 - 400x - 300y = 0$

3. Use translations and rotations to help you sketch the graphs of the following equations. Find the algebraic sign of the discriminant of each equation, and use the result to check the form of your graph.
   (a) $5x^2 + 6xy + 5y^2 - 32x - 32y + 32 = 0$
   (b) $x^2 - 4xy + 4y^2 + 5y - 9 = 0$
   (c) $9x^2 - 24xy + 16y^2 - 56x - 92y + 688 = 0$
   (d) $x^2 + 2xy + y^2 - 2x - 2y - 3 = 0$

4. Use the discriminant to identify the "conic" whose equation is $2x^2 + xy - y^2 + 6y - 8 = 0$. Write the equation in factored form, $(x + y - 2)(2x - y + 4) = 0$, and describe the graph of this "conic."

5. What are the graphs of the following equations (degeneracies allowed)?
   (a) $\pi x^2 + \pi y^2 + \sqrt{17}x - (\tan 5)y + \log 7 = 0$
   (b) $(\cos 3)x^2 + 2(\sin 3)xy + (\cos 3)y^2 + 5x - 7y + \tan 5 = 0$
   (c) $(\cos 3)x^2 + 2(\cos 3)xy + (\sin 3)y^2 + 5x - 7y + \tan 5 = 0$
   (d) $(\log 2)x^2 + 2(\log 3)xy + (\log 4)y^2 + 5x - 7y + \tan 5 = 0$

6. How must A, B, and C be related algebraically for the graph of the equation $Ax^2 + Bxy + Cy^2 + Dx + Ey + F = 0$ to be a circle (or a degenerate circle)?

7. Give a complete discussion of the graph of the equation $Ax^2 + Bxy + Cy^2 = 0$.

8. Identify the conic that contains the given points.
   (a) $(-3, -2)$, $(2, -2)$, $(2, 1)$, $(0, -5)$, $(3, 0)$
   (b) $(1, 2)$, $(1, 8)$, $(-1, 6)$, $(-1, 0)$, $(-5, 0)$
   (c) $(3, 4)$, $(-3, -2)$, $(-3, 0)$, $(-2, 0)$, $(2, 4)$
   (d) $(0, 0)$, $(4, 0)$, $(1, 1)$, $(3, 1)$, $(0, -2)$

9. Use Equations 90-6 to show that $\bar{A} + \bar{C} = A + C$, and $\bar{B}^2 - 4\bar{A}\bar{C} = B^2 - 4AC$.

10. Show that if we pick α to satisfy Equation 90-7, then
$$\bar{A} = \tfrac{1}{2}[A + C + (B/|B|)\sqrt{(A - C)^2 + B^2}]$$
and
$$\bar{C} = \tfrac{1}{2}[A + C - (B/|B|)\sqrt{(A - C)^2 + B^2}].$$

REVIEW PROBLEMS, CHAPTER ELEVEN

You should be able to answer the following questions without referring back to the text.

1. Find the equation of the tangent line to the circle $x^2 + y^2 - 8x - 12y + 47 = 0$ at the point $(3, 8)$.

2. The endpoints of a diameter of a certain circle have polar coordinates $(4, 60°)$ and $(4, 180°)$. Find the equation of the circle in cartesian coordinates.

3. Find the points of intersection of the following curves. Make a sketch that illustrates the problem.

(a) $y = \sqrt{3}x$ and $r = \cos\theta$

(b) $y^2 = 3x^2$ and $r = \cos\theta$

(c) $y = 2x^2$ and $r = \cos\theta$

(d) $\dfrac{x^2}{169} + \dfrac{y^2}{9} = 1$ and $x^2 + y^2 = 25$

4. If the circle $x^2 + y^2 = 4y$ is rolled to the right exactly one revolution along the X-axis, what is its new center?

5. Show that the graph of the equation $r = 2a\cos\theta + 2b\sin\theta$ is a circle containing the origin, and find its radius. What are polar coordinates of the center? What is a polar equation of the tangent line that contains the origin? What is a polar equation of the line that contains the center and is parallel to this tangent?

6. The slope of the segment that joins a point P to the origin is 4 times the X-coordinate of P. Where is P?

7. An artificial earth satellite moves in an elliptical orbit with the center of the earth as one focus. The minimum distance from the center of the earth to the satellite's path is d_1 miles, and the maximum distance is d_2 miles. Find the formula for the eccentricity of the elliptical path in terms of d_1 and d_2.

8. A farmer has 100 yards of fencing and two fence posts with which he wants to enclose a rectangular pasture along a straight stone wall. He will drive each post x yards from the wall and use the wall as one side of the pasture and the fencing for the other three sides. Make a graph showing how the area of his pasture depends on the distance x of the fence posts from the wall.

9. Mr. Jones rents a car in Phoenix to make a business trip to Flagstaff 150 miles away. If he pays the rental agency for 450 miles of driving, describe the region in which he can do a little free sight-seeing on his way to Flagstaff and back.

10. The Union admiral attacking Magnolia Bay in 1865 knows that each of the two forts (200 yards apart) guarding the harbor is down to one cannon ball, and between them have only enough powder to shoot these cannon balls a total of 600 yards. He doesn't know how the forts have divided up their powder. Where is he in danger of being hit by both forts? Where is he safe from being hit by either? And where is it possible to get hit once but not twice?

MISCELLANEOUS PROBLEMS, CHAPTER ELEVEN

These problems are designed to test your ability to apply your knowledge of analytic geometry to somewhat more difficult problems.

1. Discuss the graphs of the following equations.

(a) $\dfrac{x\,|x|}{a^2} + \dfrac{y\,|y|}{b^2} = 1$

(b) $\tan x^2 = \tan y^2$

(c) $r = \dfrac{ep}{1 - e\,|\cos\theta|}$

(d) $r + |r| = 2\cos\theta$

2. Suppose that x and y are very large, and divide both sides of the equation

$$x^2 + xy - 6y^2 - 26x + 12y - 11 = 0$$

by xy. The resulting equation will be approximately $(x/y) + 1 - 6(y/x) = 0$. Use this approximation to figure out the slopes of the asymptotes of the given hyperbola.

3. Can you convince yourself that for first quadrant points far enough to the right of the Y-axis, the hyperbola $x^2/a^2 - y^2/b^2 = 1$ will be above the parabola $y^2 = 4cx \, (c > 0)$?

4. One focus and one vertex of the hyperbola $b^2x^2 - a^2y^2 = a^2b^2$ are points of the positive X-axis. Find the equation of the parabola with the same vertex and focus. Show that the parabola is "eaten by" the right-hand branch of the hyperbola.

5. Let $a > b$ so that one focus and one vertex of the ellipse $b^2x^2 + a^2y^2 = a^2b^2$ are points of the positive X-axis. Find the equation of the parabola with the same vertex and focus. Show that the parabola "eats" the ellipse.

6. Let (a, b) be a point of the parabola $y^2 = 4cx$. Show that the line $4cx - 2by + b^2 = 0$ also contains the point (a, b) and that the parabola is entirely on one side of the line. Can you therefore convince yourself that the line is tangent to the parabola at the point (a, b)?

7. Show that the graph of an equation of the form $a|z|^2 + bz + \bar{b}\bar{z} + c = 0$, where a and c are real numbers and b is a complex number such that $|b|^2 > ac$, is a circle or a line, and conversely. (Here, of course, $z = x + iy$.)

8. Discuss the graph of the inequality $r \leq \cos \theta$. How does this graph differ from the graph of the inequality $|r| \leq \cos \theta$?

9. Find a polar coordinate equation for the conic with eccentricity e whose focus is the pole and whose directrix is the line $r \cos (\theta - \beta) = p$, where p is a given positive number and β is a given angle.

10. Given a triangle ABC, choose a coordinate plane so that the vertices are $A(-a, 0)$, $B(0, b)$, and $C(c, 0)$. For any point $P_1(x_1, y_1)$ of AB determine the point Q_1 of BC, and then the point R_1 of CA, such that P_1Q_1 is parallel to AC and Q_1R_1 is parallel to AB. Find the coordinates of the point P_2 of AB (in terms of x_1 and y_1) such that R_1P_2 is parallel to BC. If the above procedure is repeated, starting with P_2, to obtain a point P_3 of AB, what are the coordinates of P_3?

tables

TABLE
1

Common Logarithms of Numbers

n	0	1	2	3	4	5	6	7	8	9
1.0	.0000	.0043	.0086	.0128	.0170	.0212	.0253	.0294	.0334	.0374
1.1	.0414	.0453	.0492	.0531	.0569	.0607	.0645	.0682	.0719	.0755
1.2	.0792	.0828	.0864	.0899	.0934	.0969	.1004	.1038	.1072	.1106
1.3	.1139	.1173	.1206	.1239	.1271	.1303	.1335	.1367	.1399	.1430
1.4	.1461	.1492	.1523	.1553	.1584	.1614	.1644	.1673	.1703	.1732
1.5	.1761	.1790	.1818	.1847	.1875	.1903	.1931	.1959	.1987	.2014
1.6	.2041	.2068	.2095	.2122	.2148	.2175	.2201	.2227	.2253	.2279
1.7	.2304	.2330	.2355	.2380	.2405	.2430	.2455	.2480	.2504	.2529
1.8	.2553	.2577	.2601	.2625	.2648	.2672	.2695	.2718	.2742	.2765
1.9	.2788	.2810	.2833	.2856	.2878	.2900	.2923	.2945	.2967	.2989
2.0	.3010	.3032	.3054	.3075	.3096	.3118	.3139	.3160	.3181	.3201
2.1	.3222	.3243	.3263	.3284	.3304	.3324	.3345	.3365	.3385	.3404
2.2	.3424	.3444	.3464	.3483	.3502	.3522	.3541	.3560	.3579	.3598
2.3	.3617	.3636	.3655	.3674	.3692	.3711	.3729	.3747	.3766	.3784
2.4	.3802	.3820	.3838	.3856	.3874	.3892	.3909	.3927	.3945	.3962
2.5	.3979	.3997	.4014	.4031	.4048	.4065	.4082	.4099	.4116	.4133
2.6	.4150	.4166	.4183	.4200	.4216	.4232	.4249	.4265	.4281	.4298
2.7	.4314	.4330	.4346	.4362	.4378	.4393	.4409	.4425	.4440	.4456
2.8	.4472	.4487	.4502	.4518	.4533	.4548	.4564	.4579	.4594	.4609
2.9	.4624	.4639	.4654	.4669	.4683	.4698	.4713	.4728	.4742	.4757
3.0	.4771	.4786	.4800	.4814	.4829	.4843	.4857	.4871	.4886	.4900
3.1	.4914	.4928	.4942	.4955	.4969	.4983	.4997	.5011	.5024	.5038
3.2	.5051	.5065	.5079	.5092	.5105	.5119	.5132	.5145	.5159	.5172
3.3	.5185	.5198	.5211	.5224	.5237	.5250	.5263	.5276	.5289	.5302
3.4	.5315	.5328	.5340	.5353	.5366	.5378	.5391	.5403	.5416	.5428
3.5	.5441	.5453	.5465	.5478	.5490	.5502	.5514	.5527	.5539	.5551
3.6	.5563	.5575	.5587	.5599	.5611	.5623	.5635	.5647	.5658	.5670
3.7	.5682	.5694	.5705	.5717	.5729	.5740	.5752	.5763	.5775	.5786
3.8	.5798	.5809	.5821	.5832	.5843	.5855	.5866	.5877	.5888	.5899
3.9	.5911	.5922	.5933	.5944	.5955	.5966	.5977	.5988	.5999	.6010
4.0	.6021	.6031	.6042	.6053	.6064	.6075	.6085	.6096	.6107	.6117
4.1	.6128	.6138	.6149	.6160	.6170	.6180	.6191	.6201	.6212	.6222
4.2	.6232	.6243	.6253	.6263	.6274	.6284	.6294	.6304	.6314	.6325
4.3	.6335	.6345	.6355	.6365	.6375	.6385	.6395	.6405	.6415	.6425
4.4	.6435	.6444	.6454	.6464	.6474	.6484	.6493	.6503	.6513	.6522
4.5	.6532	.6542	.6551	.6561	.6571	.6580	.6590	.6599	.6609	.6618
4.6	.6628	.6637	.6646	.6656	.6665	.6675	.6684	.6693	.6702	.6712
4.7	.6721	.6730	.6739	.6749	.6758	.6767	.6776	.6785	.6794	.6803
4.8	.6812	.6821	.6830	.6839	.6848	.6857	.6866	.6875	.6884	.6893
4.9	.6902	.6911	.6920	.6928	.6937	.6946	.6955	.6964	.6972	.6981
5.0	.6990	.6998	.7007	.7016	.7024	.7033	.7042	.7050	.7059	.7067
5.1	.7076	.7084	.7093	.7101	.7110	.7118	.7126	.7135	.7143	.7152
5.2	.7160	.7168	.7177	.7185	.7193	.7202	.7210	.7218	.7226	.7235
5.3	.7243	.7251	.7259	.7267	.7275	.7284	.7292	.7300	.7308	.7316
5.4	.7324	.7332	.7340	.7348	.7356	.7364	.7372	.7380	.7388	.7396

TABLE
1

Common Logarithms of Numbers (Cont.)

n	0	1	2	3	4	5	6	7	8	9
5.5	.7404	.7412	.7419	.7427	.7435	.7443	.7451	.7459	.7466	.7474
5.6	.7482	.7490	.7497	.7505	.7513	.7520	.7528	.7536	.7543	.7551
5.7	.7559	.7566	.7574	.7582	.7589	.7597	.7604	.7612	.7619	.7627
5.8	.7634	.7642	.7649	.7657	.7664	.7672	.7679	.7686	.7694	.7701
5.9	.7709	.7716	.7723	.7731	.7738	.7745	.7752	.7760	.7767	.7774
6.0	.7782	.7789	.7796	.7803	.7810	.7818	.7825	.7832	.7839	.7846
6.1	.7853	.7860	.7868	.7875	.7882	.7889	.7896	.7903	.7910	.7917
6.2	.7924	.7931	.7938	.7945	.7952	.7959	.7966	.7973	.7980	.7987
6.3	.7993	.8000	.8007	.8014	.8021	.8028	.8035	.8041	.8048	.8055
6.4	.8062	.8069	.8075	.8082	.8089	.8096	.8102	.8109	.8116	.8122
6.5	.8129	.8136	.8142	.8149	.8156	.8162	.8169	.8176	.8182	.8189
6.6	.8195	.8202	.8209	.8215	.8222	.8228	.8235	.8241	.8248	.8254
6.7	.8261	.8267	.8274	.8280	.8287	.8293	.8299	.8306	.8312	.8319
6.8	.8325	.8331	.8338	.8344	.8351	.8357	.8363	.8370	.8376	.8382
6.9	.8388	.8395	.8401	.8407	.8414	.8420	.8426	.8432	.8439	.8445
7.0	.8451	.8457	.8463	.8470	.8476	.8482	.8488	.8494	.8500	.8506
7.1	.8513	.8519	.8525	.8531	.8537	.8543	.8549	.8555	.8561	.8567
7.2	.8573	.8579	.8585	.8591	.8597	.8603	.8609	.8615	.8621	.8627
7.3	.8633	.8639	.8645	.8651	.8657	.8663	.8669	.8675	.8681	.8686
7.4	.8692	.8698	.8704	.8710	.8716	.8722	.8727	.8733	.8739	.8745
7.5	.8751	.8756	.8762	.8768	.8774	.8779	.8785	.8791	.8797	.8802
7.6	.8808	.8814	.8820	.8825	.8831	.8837	.8842	.8848	.8854	.8859
7.7	.8865	.8871	.8876	.8882	.8887	.8893	.8899	.8904	.8910	.8915
7.8	.8921	.8927	.8932	.8938	.8943	.8949	.8954	.8960	.8965	.8971
7.9	.8976	.8982	.8987	.8993	.8998	.9004	.9009	.9015	.9020	.9025
8.0	.9031	.9036	.9042	.9047	.9053	.9058	.9063	.9069	.9074	.9079
8.1	.9085	.9090	.9096	.9101	.9106	.9112	.9117	.9122	.9128	.9133
8.2	.9138	.9143	.9149	.9154	.9159	.9165	.9170	.9175	.9180	.9186
8.3	.9191	.9196	.9201	.9206	.9212	.9217	.9222	.9227	.9232	.9238
8.4	.9243	.9248	.9253	.9258	.9263	.9269	.9274	.9279	.9284	.9289
8.5	.9294	.9299	.9304	.9309	.9315	.9320	.9325	.9330	.9335	.9340
8.6	.9345	.9350	.9355	.9360	.9365	.9370	.9375	.9380	.9385	.9390
8.7	.9395	.9400	.9405	.9410	.9415	.9420	.9425	.9430	.9435	.9440
8.8	.9445	.9450	.9455	.9460	.9465	.9469	.9474	.9479	.9484	.9489
8.9	.9494	.9499	.9504	.9509	.9513	.9518	.9523	.9528	.9533	.9538
9.0	.9542	.9547	.9552	.9557	.9562	.9566	.9571	.9576	.9581	.9586
9.1	.9590	.9595	.9600	.9605	.9609	.9614	.9619	.9624	.9628	.9633
9.2	.9638	.9643	.9647	.9652	.9657	.9661	.9666	.9671	.9675	.9680
9.3	.9685	.9689	.9694	.9699	.9703	.9708	.9713	.9717	.9722	.9727
9.4	.9731	.9736	.9741	.9745	.9750	.9754	.9759	.9763	.9768	.9773
9.5	.9777	.9782	.9786	.9791	.9795	.9800	.9805	.9809	.9814	.9818
9.6	.9823	.9827	.9832	.9836	.9841	.9845	.9850	.9854	.9859	.9863
9.7	.9868	.9872	.9877	.9881	.9886	.9890	.9894	.9899	.9903	.9908
9.8	.9912	.9917	.9921	.9926	.9930	.9934	.9939	.9943	.9948	.9952
9.9	.9956	.9961	.9965	.9969	.9974	.9978	.9983	.9987	.9991	.9996

TABLE
2
The Circular Functions

t	sin t	cos t	tan t	cot t	sec t	csc t
.00	.0000	1.0000	.0000	. . . . .	1.000	. . . . .
.01	.0100	1.0000	.0100	99.997	1.000	100.00
.02	.0200	.9998	.0200	49.993	1.000	50.00
.03	.0300	.9996	.0300	33.323	1.000	33.34
.04	.0400	.9992	.0400	24.987	1.001	25.01
.05	.0500	.9988	.0500	19.983	1.001	20.01
.06	.0600	.9982	.0601	16.647	1.002	16.68
.07	.0699	.9976	.0701	14.262	1.002	14.30
.08	.0799	.9968	.0802	12.473	1.003	12.51
.09	.0899	.9960	.0902	11.081	1.004	11.13
.10	.0998	.9950	.1003	9.967	1.005	10.02
.11	.1098	.9940	.1104	9.054	1.006	9.109
.12	.1197	.9928	.1206	8.293	1.007	8.353
.13	.1296	.9916	.1307	7.649	1.009	7.714
.14	.1395	.9902	.1409	7.096	1.010	7.166
.15	.1494	.9888	.1511	6.617	1.011	6.692
.16	.1593	.9872	.1614	6.197	1.013	6.277
.17	.1692	.9856	.1717	5.826	1.015	5.911
.18	.1790	.9838	.1820	5.495	1.016	5.586
.19	.1889	.9820	.1923	5.200	1.018	5.295
.20	.1987	.9801	.2027	4.933	1.020	5.033
.21	.2085	.9780	.2131	4.692	1.022	4.797
.22	.2182	.9759	.2236	4.472	1.025	4.582
.23	.2280	.9737	.2341	4.271	1.027	4.386
.24	.2377	.9713	.2447	4.086	1.030	4.207
.25	.2474	.9689	.2553	3.916	1.032	4.042
.26	.2571	.9664	.2660	3.759	1.035	3.890
.27	.2667	.9638	.2768	3.613	1.038	3.749
.28	.2764	.9611	.2876	3.478	1.041	3.619
.29	.2860	.9582	.2984	3.351	1.044	3.497
.30	.2955	.9553	.3093	3.233	1.047	3.384
.31	.3051	.9523	.3203	3.122	1.050	3.278
.32	.3146	.9492	.3314	3.018	1.053	3.179
.33	.3240	.9460	.3425	2.920	1.057	3.086
.34	.3335	.9428	.3537	2.827	1.061	2.999
.35	.3429	.9394	.3650	2.740	1.065	2.916
.36	.3523	.9359	.3764	2.657	1.068	2.839
.37	.3616	.9323	.3879	2.578	1.073	2.765
.38	.3709	.9287	.3994	2.504	1.077	2.696
.39	.3802	.9249	.4111	2.433	1.081	2.630
.40	.3894	.9211	.4228	2.365	1.086	2.568
.41	.3986	.9171	.4346	2.301	1.090	2.509
.42	.4078	.9131	.4466	2.239	1.095	2.452
.43	.4169	.9090	.4586	2.180	1.100	2.399
.44	.4259	.9048	.4708	2.124	1.105	2.348

TABLE
2

The Circular Functions (Cont.)

t	$\sin t$	$\cos t$	$\tan t$	$\cot t$	$\sec t$	$\csc t$
.45	.4350	.9004	.4831	2.070	1.111	2.299
.46	.4439	.8961	.4954	2.018	1.116	2.253
.47	.4529	.8916	.5080	1.969	1.122	2.208
.48	.4618	.8870	.5206	1.921	1.127	2.166
.49	.4706	.8823	.5334	1.875	1.133	2.125
.50	.4794	.8776	.5463	1.830	1.139	2.086
.51	.4882	.8727	.5594	1.788	1.146	2.048
.52	.4969	.8678	.5726	1.747	1.152	2.013
.53	.5055	.8628	.5859	1.707	1.159	1.978
.54	.5141	.8577	.5994	1.668	1.166	1.945
.55	.5227	.8525	.6131	1.631	1.173	1.913
.56	.5312	.8473	.6269	1.595	1.180	1.883
.57	.5396	.8419	.6310	1.560	1.188	1.853
.58	.5480	.8365	.6552	1.526	1.196	1.825
.59	.5564	.8309	.6696	1.494	1.203	1.797
.60	.5646	.8253	.6841	1.462	1.212	1.771
.61	.5729	.8196	.6989	1.431	1.220	1.746
.62	.5810	.8139	.7139	1.401	1.229	1.721
.63	.5891	.8080	.7291	1.372	1.238	1.697
.64	.5972	.8021	.7445	1.343	1.247	1.674
.65	.6052	.7961	.7602	1.315	1.256	1.652
.66	.6131	.7900	.7761	1.288	1.266	1.631
.67	.6210	.7838	.7923	1.262	1.276	1.610
.68	.6288	.7776	.8087	1.237	1.286	1.590
.69	.6365	.7712	.8253	1.212	1.297	1.571
.70	.6442	.7648	.8423	1.187	1.307	1.552
.71	.6518	.7584	.8595	1.163	1.319	1.534
.72	.6594	.7518	.8771	1.140	1.330	1.517
.73	.6669	.7452	.8949	1.117	1.342	1.500
.74	.6743	.7358	.9131	1.095	1.354	1.483
.75	.6816	.7317	.9316	1.073	1.367	1.467
.76	.6889	.7248	.9505	1.052	1.380	1.452
.77	.6961	.7179	.9697	1.031	1.393	1.437
.78	.7033	.7109	.9893	1.011	1.407	1.422
.79	.7104	.7038	1.009	.9908	1.421	1.408
.80	.7174	.6967	1.030	.9712	1.435	1.394
.81	.7243	.6895	1.050	.9520	1.450	1.381
.82	.7311	.6822	1.072	.9331	1.466	1.368
.83	.7379	.6749	1.093	.9146	1.482	1.355
.84	.7446	.6675	1.116	.8964	1.498	1.343
.85	.7513	.6600	1.138	.8785	1.515	1.331
.86	.7578	.6524	1.162	.8609	1.533	1.320
.87	.7643	.6448	1.185	.8437	1.551	1.308
.88	.7707	.6372	1.210	.8267	1.569	1.297
.89	.7771	.6294	1.235	.8100	1.589	1.287

TABLE
2   *The Circular Functions (Cont.)*

t	$\sin t$	$\cos t$	$\tan t$	$\cot t$	$\sec t$	$\csc t$
.90	.7833	.6216	1.260	.7936	1.609	1.277
.91	.7895	.6137	1.286	.7774	1.629	1.267
.92	.7956	.6058	1.313	.7615	1.651	1.257
.93	.8016	.5978	1.341	.7458	1.673	1.247
.94	.8076	.5898	1.369	.7303	1.696	1.238
.95	.8134	.5817	1.398	.7151	1.719	1.229
.96	.8192	.5735	1.428	.7001	1.744	1.221
.97	.8249	.5653	1.459	.6853	1.769	1.212
.98	.8305	.5570	1.491	.6707	1.795	1.204
.99	.8360	.5487	1.524	.6563	1.823	1.196
1.00	.8415	.5403	1.557	.6421	1.851	1.188
1.01	.8468	.5319	1.592	.6281	1.880	1.181
1.02	.8521	.5234	1.628	.6142	1.911	1.174
1.03	.8573	.5148	1.665	.6005	1.942	1.166
1.04	.8624	.5062	1.704	.5870	1.975	1.160
1.05	.8674	.4976	1.743	.5736	2.010	1.153
1.06	.8724	.4889	1.784	.5604	2.046	1.146
1.07	.8772	.4801	1.827	.5473	2.083	1.140
1.08	.8820	.4713	1.871	.5344	2.122	1.134
1.09	.8866	.4625	1.917	.5216	2.162	1.128
1.10	.8912	.4536	1.965	.5090	2.205	1.122
1.11	.8957	.4447	2.014	.4964	2.249	1.116
1.12	.9001	.4357	2.066	.4840	2.295	1.111
1.13	.9044	.4267	2.120	.4718	2.344	1.106
1.14	.9086	.4176	2.176	.4596	2.395	1.101
1.15	.9128	.4085	2.234	.4475	2.448	1.096
1.16	.9168	.3993	2.296	.4356	2.504	1.091
1.17	.9208	.3902	2.360	.4237	2.563	1.086
1.18	.9246	.3809	2.427	.4120	2.625	1.082
1.19	.9284	.3717	2.498	.4003	2.691	1.077
1.20	.9320	.3624	2.572	.3888	2.760	1.073
1.21	.9356	.3530	2.650	.3773	2.833	1.069
1.22	.9391	.3436	2.733	.3659	2.910	1.065
1.23	.9425	.3342	2.820	.3546	2.992	1.061
1.24	.9458	.3248	2.912	.3434	3.079	1.057
1.25	.9490	.3153	3.010	.3323	3.171	1.054
1.26	.9521	.3058	3.113	.3212	3.270	1.050
1.27	.9551	.2963	3.224	.3102	3.375	1.047
1.28	.9580	.2867	3.341	.2993	3.488	1.044
1.29	.9608	.2771	3.467	.2884	3.609	1.041
1.30	.9636	.2675	3.602	.2776	3.738	1.038
1.31	.9662	.2579	3.747	.2669	3.878	1.035
1.32	.9687	.2482	3.903	.2562	4.029	1.032
1.33	.9711	.2385	4.072	.2456	4.193	1.030
1.34	.9735	.2288	4.256	.2350	4.372	1.027

TABLE
2   *The Circular Functions (Cont.)*

t	$\sin t$	$\cos t$	$\tan t$	$\cot t$	$\sec t$	$\csc t$
1.35	.9757	.2190	4.455	.2245	4.566	1.025
1.36	.9779	.2092	4.673	.2140	4.779	1.023
1.37	.9799	.1994	4.913	.2035	5.014	1.021
1.38	.9819	.1896	5.177	.1931	5.273	1.018
1.39	.9837	.1798	5.471	.1828	5.561	1.017
1.40	.9854	.1700	5.798	.1725	5.883	1.015
1.41	.9871	.1601	6.165	.1622	6.246	1.013
1.42	.9887	.1502	6.581	.1519	6.657	1.011
1.43	.9901	.1403	7.055	.1417	7.126	1.010
1.44	.9915	.1304	7.602	.1315	7.667	1.009
1.45	.9927	.1205	8.238	.1214	8.299	1.007
1.46	.9939	.1106	8.989	.1113	9.044	1.006
1.47	.9949	.1006	9.887	.1011	9.938	1.005
1.48	.9959	.0907	10.938	.0910	11.029	1.004
1.49	.9967	.0807	12.350	.0810	12.390	1.003
1.50	.9975	.0707	14.101	.0709	14.137	1.003
1.51	.9982	.0609	16.428	.0609	16.458	1.002
1.52	.9987	.0508	19.670	.0508	19.965	1.001
1.53	.9992	.0408	24.498	.0408	24.519	1.001
1.54	.9995	.0308	32.461	.0308	32.476	1.000
1.55	.9998	.0208	48.078	.0208	48.089	1.000
1.56	.9999	.0108	92.620	.0108	92.626	1.000
1.57	1.0000	.0008	1255.8	.0008	1255.8	1.000
1.58	1.0000	−.0092	−108.65	−.0092	−108.65	1.000
1.59	.9998	−.0192	−52.067	−.0192	−52.08	1.000
1.60	.9996	−.0292	−34.233	−.0292	−34.25	1.000

TABLE

3   *Trigonometric Functions*

θ	$\sin \theta$	$\tan \theta$	$\cot \theta$	$\cos \theta$	
0°	.0000	.0000	. . . . .	1.0000	90°
1°	.0175	.0175	57.290	.9998	89°
2°	.0349	.0349	28.636	.9994	88°
3°	.0523	.0524	19.081	.9986	87°
4°	.0698	.0699	14.301	.9976	86°
5°	.0872	.0875	11.430	.9962	85°
6°	.1045	.1051	9.5144	.9945	84°
7°	.1219	.1228	8.1443	.9925	83°
8°	.1392	.1405	7.1154	.9903	82°
9°	.1564	.1584	6.3138	.9877	81°
10°	.1736	.1763	5.6713	.9848	80°
11°	.1908	.1944	5.1446	.9816	79°
12°	.2079	.2126	4.7046	.9781	78°
13°	.2250	2309	4.3315	.9744	77°
14°	.2419	2493	4.0108	.9703	76°
15°	.2588	.2679	3.7321	.9659	75°
16°	.2756	.2867	3.4874	.9613	74°
17°	.2924	.3057	3.2709	.9563	73°
18°	.3090	.3249	3.0777	.9511	72°
19°	.3256	.3443	2.9042	.9455	71°
20°	.3420	.3640	2.7475	.9397	70°
21°	.3584	.3839	2.6051	.9336	69°
22°	.3746	.4040	2.4751	.9272	68°
23°	.3907	.4245	2.3559	.9205	67°
24°	.4067	.4452	2.2460	.9135	66°
25°	.4226	.4663	2.1445	.9063	65°
26°	.4384	.4877	2.0503	.8988	64°
27°	.4540	.5095	1.9626	.8910	63°
28°	.4695	.5317	1.8807	.8829	62°
29°	.4848	.5543	1.8040	.8746	61°
30°	.5000	.5774	1.7321	.8660	60°
31°	.5150	.6009	1.6643	.8572	59°
32°	.5299	.6249	1.6003	.8480	58°
33°	.5446	.6494	1.5399	.8387	57°
34°	.5592	.6745	1.4826	.8290	56°
35°	.5736	.7002	1.4281	.8192	55°
36°	.5878	.7265	1.3764	.8090	54°
37°	.6018	.7536	1.3270	.7986	53°
38°	.6157	.7813	1.2799	.7880	52°
39°	.6293	.8098	1.2349	.7771	51°
40°	.6428	.8391	1.1918	.7660	50°
41°	.6561	.8693	1.1504	.7547	49°
42°	.6691	.9004	1.1106	.7431	48°
43°	.6820	.9325	1.0724	.7314	47°
44°	.6947	.9657	1.0355	.7193	46°
45°	.7071	1.0000	1.0000	.7071	45°
	$\cos \theta$	$\cot \theta$	$\tan \theta$	$\sin \theta$	θ

TABLE

4   *Logarithms of Factorials*

n	log n!	n	log n!	n	log n!
1	0.0000	35	40.0142	69	98.2333
2	0.3010	36	41.5705		
3	0.7782	37	43.1387	70	100.0784
4	1.3802	38	44.7185	71	101.9297
		39	46.3096	72	103.7870
5	2.0792			73	105.6503
6	2.8573	40	47.9116	74	107.5196
7	3.7024	41	49.5244		
8	4.6055	42	51.1477	75	109.3946
9	5.5598	43	52.7812	76	111.2754
		44	54.4246	77	113.1619
10	6.5598			78	115.0540
11	7.6012	45	56.0778	79	116.9516
12	8.6803	46	57.7406		
13	9.7943	47	59.4127	80	118.8547
14	10.9404	48	61.0939	81	120.7632
		49	62.7841	82	122.6770
15	12.1165			83	124.5961
16	13.3206	50	64.4831	84	126.5204
17	14.5511	51	66.1906		
18	15.8063	52	67.9066	85	128.4498
19	17.0851	53	69.6309	86	130.3843
		54	71.3633	87	132.3238
20	18.3861			88	134.2683
21	19.7083	55	73.1037	89	136.2177
22	21.0508	56	74.8519		
23	22.4125	57	76.6077	90	138.1719
24	23.7927	58	78.3712	91	140.1310
		59	80.1420	92	142.0948
25	25.1906			93	144.0632
26	26.6056	60	81.9202	94	146.0364
27	28.0370	61	83.7055		
28	29.4841	62	85.4979	95	148.0141
29	30.9465	63	87.2972	96	149.9964
		64	89.1034	97	151.9831
30	32.4237			98	153.9744
31	33.9150	65	90.9163	99	155.9700
32	35.4202	66	92.7359	100	157.9700
33	36.9387	67	94.5620		
34	38.4702	68	96.3945		

selected answers

1. (a) -1; (c) 31. **2.** (a) -4; (c) x^3; (e) $y - x$. **3.** The associative law of addition. **4.** (a) 0; (c) $x - 2$; (e) $20x$. **5.** (a) $x^3 - y^3$; (c) $8x^3 - 12x^2y + 6xy^2 - y^3$. **6.** (a) $(x - y)(x^2 + xy + y^2)$; (c) $(3y + 2/x)(9y^2 - 6y/x + 4/x^2)$; (e) $(3x - .1) \times (3x + .1)$; (g) $2(2 - x)(2 + x)(4 + 2x + x^2)(4 - 2x + x^2)$; (i) $(3z - 2x + y) \times (3z + 2x - y)$; (k) $2z(x + 3y)(3x - z)$. **7.** (a) An *even* integer if x is 3 times an odd integer, a *negative* integer if x is one of the numbers -9, -12, -15, and so on; (c) An *even* integer if x is odd, a *negative* integer only if $x = 0$. **9.** Equations 1-1, 1-2, 1-5, and 1-6 hold.

1. (a) $\frac{17}{1}$; (c) $\frac{14}{3}$; (e) $\frac{1}{3}$; (g) $\frac{1776}{100}$. **2.** $\frac{21}{11}$. **4.** (a) u/v; (c) $1/a^2$; (e) $1/(x - y + 1)$; (g) $1/(x + 1)$. **5.** (a) $1/(x + 1)$; (c) $(x + 1)/x$; (e) x. **6.** (a) x; (c) $2/(x + 3) \times (x + 2)$; (e) $4/x(1 - x)$; (g) $3/x$. **7.** Not true for $x = 1$. **8.** $0/a = 0$. **9.** a. **10.** (d) is an identity. **11.** (a) x can be any integer; (b) $x = -1$; (c) $x = y/(1 - 2y)$, where y can be any integer; (d) $x \neq -\frac{1}{2}$.

1. (a) $\frac{13}{3}$; (b) $-\frac{1}{3}$; (c) $\frac{1}{3}(6+x)$; (d) $\frac{1}{3}x+\frac{2}{3}c$. **2.** (a) $\frac{10}{7}$; (b) $-\frac{4}{7}$. **4.** .090909 ...
5. $b = \frac{142}{999}$, $c = 3$. **6.** Yes, no, irrational. **7.** Yes, no, irrational unless the rational factor is 0. **8.** Yes, no. **10.** All irrational.

1. (a), (c), (d), (e), (f), (h) are true. **2.** (a) Yes; (b) yes; (c) not necessarily; (d) not necessarily. **3.** Yes. **4.** $x = \frac{999}{1000}$, $y = \frac{1001}{1000}$, for example. **5.** Notice that $(\sqrt{x} - 1/\sqrt{x})^2 \geq 0$. **6.** $21.743 \leq$ perimeter ≤ 21.765. **8.** $\frac{50}{9} \leq R_2 \leq \frac{25}{2}$.
10. 4.

1. (a) $\{3 < x < 5\}$; (c) $\{-1 < x \leq 4\}$; (d) $\{2 < x \leq 3\}$. **2.** (a) $\{8, 10\}$;
(b) A; (c) $5+4 = 7+2$. **3.** (a) $\{x < -\frac{1}{2}\}$; (c) $\{x < 6\}$; (e) $\varnothing$; (g) $\{\frac{1}{2} < x < 3\}$.
4. (b) $2^{10} = 1024$. **5.** (a) $\varnothing$; (c) $\{x \neq 3\}$; (e) $\{1\}$; (g) $\{0 < x < 1\}$. **6.** (a) a,
b, c; (b) a, c; (c) c; (d) a, c; (e) maybe none; (f) a, c. **7.** All except (f) are true.
8. $\{x < m\} \subseteq \{x < a\} \subseteq \{x < M\}$ and $\{x < m\} \subseteq \{x < b\} \subseteq \{x < M\}$. **9.** (a) A_6;
(c) A_6; (e) A_6. **10.** Yes, and also the distributive law $A \cup (B \cap C) = (A \cup B) \cap$
$(A \cup C)$.

1. (a) $\{x \leq -3\}$; (c) $\{-1 < x < 1\}$; (e) $\{x \mid x \leq 2, x \neq 0\}$. **2.** (a) $\{-1 < x <$
$2\}$; (c) $\varnothing$; (e) $\{x > -1\} \cup \{x \leq -3\}$. **3.** (a) $\varnothing$; (c) $\{x \leq 3\}$; (d) $\{x \leq -1\} \cup$
$\{x \geq 5\}$; (f) $\{-1, 1\}$. **4.** (a) and (d) are true. **5.** (a) $\{x < -2\} \cup \{0 < x < 1\}$;
(c) $\{-3 \leq x \leq -2\} \cup \{-4 \leq x < 5\}$. **7.** (a) $A \cap cB$; (b) cA; (c) $cA \cup B$;
(d) $\varnothing$. **8.** True if, and only if, $B \subseteq A$.

1. (a) $\{-\frac{3}{2}, \frac{3}{2}\}$; (b) $\varnothing$; (c) $\{-2\}$; (d) $\{-1, 1\}$; (e) $\{-3, 3\}$; (f) $\{-3, 3\}$; (g) $\{0, 2\}$;
(h) $\{-3, 1\}$; (i) $\{2, 4\}$; (j) $\{-\frac{2}{3}, 2\}$. **2.** (a) $\{x \geq 0\}$; (c) $\{x \geq 3\}$; (e) $\{x \geq 0\}$;
(g) $\{x \leq -1\} \cup \{x \geq 1\}$. **3.** 1 unit or 5 units. **4.** (a) 3 or 5; (c) 1 or 9.
6. (a) All x; (b) $x \leq 0$; (c) all x; (d) $x \leq 0$; (e) $a \geq b$; (f) $a \leq b$; (g) $ab \geq 0$; (h)
$ab \geq b^2$. **7.** (a) $-\frac{2}{3} < x < \frac{2}{3}$; (c) $-2 < x < 4$; (e) $1 < x < 3$; (g) $-\frac{1}{6} \leq x \leq \frac{7}{6}$;
(i) $-1 \leq x \leq 1$. **8.** (a) $\{x < -\frac{3}{2}\} \cup \{x > \frac{3}{2}\}$; (c) $\{x < -7\} \cup \{x > 9\}$; (e)
$\{x \leq -1\} \cup \{x \geq 2\}$. **11.** (a) $\{x < -5\} \cup \{x > 5\}$; (c) $\{-3 < x < -1\} \cup$
$\{1 < x < 3\}$. **12.** (a) $\{x > \frac{1}{4}\}$; (b) $\{x < -1\} \cup \{x > \frac{1}{5}\}$; (c) $\{\frac{1}{3} < x < 1\}$.

1. (a) $|x - 2| < 1$; (c) $|x - \frac{5}{4}| < \frac{5}{4}$; (e) $|8x + 1| < 7$. **2.** $|x - 2| < 2$. **3.** $|x^2 +$
$x - 6| < 24$. **7.** (a) R^1; (b) $\varnothing$; (c) $\varnothing$; (d) $\{x > 0\}$. **8.** (a) True; (b) let $x = 0$;
(c) let $x = 0$; (d) true; (e) true; (f) true. **9.** (a) False; (b) false. **10.** $\frac{1}{50}$, for
example.

1. (a) $-\frac{1}{8}$; (c) $\frac{11}{18}$; (e) 1; (g) x; (i) $\frac{2}{3}x^2y^{-5}$; (k) x^6. **2.** (a) 1; (c) $1/a^7$; (e) $x+y$; (g) $\frac{13}{72}$; (i) $(x+1)/x^3$; (k) $1/(c-b)$. **3.** $(1/x^{-2})-(1/y^{-2})$; (c) $(1/x^{-2})+y^{-3}$; (e) $(x/3)^{-2}$; (g) $x^{-2}y^{-2}/(x^{-2}+y^{-2})$. **4.** (e) and (f). **5.** (a) $\{\frac{2}{3}<x<\frac{4}{3}\}$; (c) $\{x<-2\}\cup\{x>2\}$; (e) $\{x\geq 3\}$. **7.** Rules 9-2 and 9-3 would be preserved. **9.** $(-a)^n=-a^n$ if n is odd.

1. (a) 9; (c) 16; (e) $\frac{1}{32}$; (g) 5; (i) $-\frac{1}{1024}$; (k) $\frac{1}{8}$. **3.** (a) $\sqrt[4]{x^3}/x$; (c) $\sqrt[3]{a^2}\,x/a^2$; (e) $(x-\sqrt{y})/(x^2-y)$. **4.** $\frac{13}{60}$ and $\frac{17}{60}$. **5.** (a) $x^{1/3}y^{-2/3}$; (c) $x^{11/12}y^{-1/3}$; (e) $\sqrt{3}+\sqrt{2}$; (g) $(x-y)(\sqrt{x}+\sqrt{y})$. **6.** (a) $(5+2\sqrt{6})$; (c) $-\frac{1}{7}(11+6\sqrt{2})$. **7.** 512. **11.** If $\sqrt[n]{a}<\sqrt[n]{b}$, then $a<b$ for any positive integer n.

REVIEW PROBLEMS, CHAPTER ONE

1. (a) $\{-1<x\leq 2\}$; (b) $\{0<x<1\}$; (c) $\{x\geq 0\}$. **2.** (a) $3(x-2)(x+2)$; (b) $2(x-2)(x-3)$; (c) $2(x-2)(x+2)(x^2+2x+4)(x^2-2x+4)$; (d) $(x+1)\times(x^2-2)$; (e) $(a^3+2b^2)(a^6-2a^3b^2+4b^4)$. **3.** (a) $4x^{1/4}y^{1/2}$; (b) $2/xy^{1/4}z^{1/3}$. **4.** $(2n+1)/2n(n+1)$. **5.** Either x is to the right of y or to the left of $-y$. **6.** (a) $\{0<x<\frac{4}{3}\}$; (b) $\{x<0\}\cup\{x>2\}$. **7.** $m=7$, $p=4$. **8.** (a) Any pair of non-negative numbers; (b) x or y is negative and the other is at most 0. **10.** If $a>b$, then a/b is larger.

MISCELLANEOUS PROBLEMS, CHAPTER ONE

1. (a) is true, (b) is false, others may be either true or false. **3.** (2, 2). **4.** $\{-1<x<0\}$. **6.** $(x-1)(x-2)(x+3)$. **7.** Copper is cheaper on a cost-per-year basis. If cost per year is the same, then the iron pipe costs less than $100. **10.** Toward the train. **11.** Each inequality can be written in the form $(a-b)^2p(a,b)\geq 0$, where $p(a,b)\geq 0$.

1. (a) $f(x)=-x$; (c) $h(x)=-x+1/x$; (d) $k(x)=2$. **2.** (a) $\{x\neq 0\}$; (b) $\{x\neq 0\}\cap\{x\neq 1\}$; (c) $\{x\neq -\frac{5}{2}\}$; (d) R^1; (e) $\{x\leq -2\}\cup\{x\geq 1\}$; (f) $\{0<x\leq 5\}$. **3.** (a) $\{3\}$; (c) $\{\frac{1}{2},1,2\}$; (e) $\{-1,2,-3\}$. **4.** (a) $2\sqrt{y}-3$; (c) $2/z-3$; (e) $4x-9$. **5.** (a) 4; (b) 5; (c) 39; (d) 84; (e) 6; (f) $2\sqrt{3}$; (g) 19; (h) $\frac{523}{169}$. **7.** No. **8.** They are the same. **9.** (b) and (c). **10.** $f(h)=h\sqrt{4h-h^2}$.

1. (a) 0; (b) -4; (c) 0; (d) -3; (e) -3; (f) 2; (g) -2; (h) 1. **2.** (a), (c), and (d) are true. **3.** All are true. **4.** (b) is true. **5.** $y=2x+6000/x$, $\{230, 220, 219, 220\}$. **7.** $C=5x^2+160/x$, $\{157, 207\}$. **9.** (a) $V=\frac{4}{3}\pi(24-\frac{1}{60}t)^3$; (c) $V=\frac{4}{3}\pi(576-\frac{1}{12}t)^{3/2}$. **10.** (a) Domain is R^1, range is $\{-1, 0\}$. **12.** $200+50t-50|t-4|$.

1. (a) 5; (c) 13; (e) $\sqrt{97}$; (f) $\pi + 1/\pi$. **2.** (a) $2\sqrt{a^2 + b^2}$; (c) $2\sqrt{a+b}$; (e) $|t|$ $(t^2 + 1)$. **3.** (a) $(5, 0)$; (b) $(0, 5)$. **5.** $(6, 1)$ and $(6, 5)$, $(-2, 1)$ and $(-2, 5)$, $(0, 3)$ and $(4, 3)$. **6.** $(a + b - c, b)$ and $(a + b - c, c)$, $(a + c - b, b)$ and $(a + c - b, c)$, $(a + \frac{1}{2}c - \frac{1}{2}b, \frac{1}{2}b + \frac{1}{2}c)$ and $(a - \frac{1}{2}c + \frac{1}{2}b, \frac{1}{2}b + \frac{1}{2}c)$. **7.** 8. **8.** $x = 5$ or $x = -1$. **9.** $(4, 0)$. **11.** (b) and (c) collinear.

3. $(-3, 2)$. **4.** (b, a), $(-a, -b)$, $(-b, -a)$. **5.** (a) 1.2; (c) 1.2. **6.** $\sqrt{37}$. **8.** (a) $(1, 0)$ and $(4, 4)$; (b) $(4, 3)$.

1. (a) $f(x) = \frac{9}{4}x$; (c) $f(x) = \frac{9}{16}x^2$. **2.** (b) No; (c) no; (d) no; (e) no. **3.** [Let us assume that $f(1) \neq 0$. What does this assumption amount to?] (a) $\{x \neq 0\}$; (b) $\{y \neq 0\}$; (d) not unless $f(x) = 1/x$; (e) no; (f) no. **4.** $34\frac{2}{7}$ mph. **6.** Not unless $f(x) = 0$ for every x. **7.** $4\pi r^2$. **8.** (a) $g(-2)$; (c) $f(\frac{1}{2})$. **9.** $F = 980$ m. **10.** 8 amperes, $R > 240$ ohms. **11.** (a) Directly proportional to x^2; (c) directly proportional to x. **13.** $365(\frac{142}{93})^{3/2}$. **14.** 819 feet.

1. (a) $f(x) = 2x + 3$; (c) $f(x) = (1/k)x + k$. **2.** (a) Slope is -2, Y-intercept is 3, zero of f is $\frac{3}{2}$; (c) slope is 2, Y-intercept is 3, zero of f is $-\frac{3}{2}$. **3.** (a) $f(x) = mx$; (b) $f(x) = x + b$; (c) $f(x) = x + b$; (d) $f(x) = mx + 2 - 3m$; (e) $f(x) = \frac{1}{4}x + \frac{5}{4}$; (f) $f(x) = -\frac{8}{3}x + \frac{4}{3}$; (g) $f(x) = b$, where $b \geq 0$, or $f(x) = mx$, where $m \geq 0$; (h) $f(x) = x + k$, where k is an integer. **4.** $\{(3, 3)\}$. **5.** It is the graph of a function, but not of a linear function. **6.** $2x - y - 3 = 0$. **7.** (a) 9; (b) 1; (c) 27; (d) $-\sqrt[3]{3}$. **10.** $v = -9.8t + 100$, $\frac{500}{49}$ seconds after firing. **11.** 10°. **12.** $A = 2.75t + 50$, $18\frac{2}{11}$ years. **13.** 80 calories.

1. (a) $\frac{1}{2}(x - 3)$; (b) $\frac{1}{2}(3 - x)$; (c) no inverse; (d) $\frac{1}{2}\sqrt[5]{x}$; (e) no inverse; (f) $3/x$; (g) no inverse; (h) no inverse; (i) $|x|^{3/2}/x$. **2.** Domain is $\{x \neq -2\}$, range is $\{x \neq -1\}$, $f^{-1}(x) = (3 - 2x)/(1 + x)$. **6.** $f^{-1}(y) = 2 + \frac{1}{2}\sqrt{y - 1}$ if $y \geq 1$, $f^{-1}(y) = 2 - \frac{1}{2}\sqrt{1 - y}$ if $y < 1$. **8.** $p^{-1}(x) = f^{-1}(g^{-1}(x))$. **10.** $f(x) = x + 1$ if $-1 \leq x \leq 0$ and $f(x) = -x$ if $0 < x \leq 1$.

REVIEW PROBLEMS, CHAPTER TWO

1. $g(2) = 0$, domain of g is the set $\{x \neq -1\}$. **2.** (All answers here are approximations, of course.) (a) $\{0 < x < 6\}$; (b) $\{-\frac{3}{2} < y \leq 3\}$; (c) 2.1; (d) $\frac{4}{3}$ and $\frac{7}{3}$; (e) .2, 2.8, or 5; (f) 5.2, 3.8, or 4.2. **3.** Two strips along the coordinate axes, each 2 units wide. **4.** $f(x) = 4x + 1$. **5.** $A = 2(120 - h)(\pi h + 4h + 120\pi)/(\pi + 2)^2$. **6.** $F^{-1}(x) = \sqrt{1 + x}$. **7.** The line segment between the points $(-1, -2)$ and $(\frac{1}{3}, 2)$. **8.** uv is independent of x. **9.** Yes, yes. **10.** $\{3, -1\}$. **12.** $|x|\sqrt{1 + x^2}$. **13.** $f(i) = \frac{1}{12}i$. **14.** $(0, -4)$, $(-2, 2)$, $(4, 6)$.

2. $R = 10x/(3x + 10)$. **3.** $L = L_0(mt + 1)$. **4.** Domain of F^{-1} is the set $\cup_k\{2k \le y < 2k + 1\}$, $F^{-1}(y) = y - \frac{1}{2}[\![y]\!]$. **6.** $f(x) = |x - 2[\![(x + 1)/2]\!]|$. **8.** (All answers are approximations, of course.) (a) $g(4) = f(2)$, $f(3.2) = f(4.3) = g(2)$; (b) domain is $\{2 \le x \le 5\}$, range is $\{1.2 \le y \le 3.7\}$. **9.** $\varnothing$. **11.** All the equations are valid.

PROBLEMS
18

1. (a) 8; (c) 1. **3.** (a) π; (c) $2\sqrt{2}$; (e) any positive number; (g) .8. **4.** (a) 2.7; (c) 2.3. **6.** (a) One; (c) none. **7.** (d) Domain is R^1, range is $\{-1 < y < 1\}$. **8.** $3000(.6)^t$, 233 cubic feet. **9.** $N = N_0(.8)^t$, 3.1 minutes. **10.** One half as large.

PROBLEMS
19

1. (a) $\log_3 81 = 4$; (c) $\log_M 5 = k$; (e) $\log_5 k = M$; (g) $\log_k 5 = M$. **2.** (a) 125; (c) 2; (e) 2; (g) -3; (i) $\{-5, 5\}$. **3.** (a) 1; (c) the set of positive numbers; (e) 5; (g) 2. **4.** (a) 1.6; (c) $-.4$. **6.** -2. **7.** -2. **9.** (a) 5; (c) 5; (e) 7; (g) $\log_2 b^2$. **10.** (a) 3; (c) 5.

PROBLEMS
20

2. (a) $-.41$; (c) 1.38; (e) 2.20; (g) 3.17; (i) 4.50; (k) -1.10; (m) .08. **3.** (a) $\frac{5}{2}\log_b x$; (c) $\log_b 3x^{5/4}$; (e) $\log_b |x^2 - xy + y^2|$. **4.** (a) $\frac{100}{69}$; (c) $\frac{69}{122}$. **5.** (a) .301; (c) -1.602. **6.** (a) 14; (c) a^2; (e) $\{\frac{1}{25}, 25\}$. **9.** $\log_{10} 4 = .6020$, $\log_{10} 5 = .6990$, $\log_{10} 6 = .7781$, $\log_{10} 8 = .9030$, $\log_{10} 9 = .9542$. **12.** The functions have different domains.

PROBLEMS
21

1. (a) 1.58; (c) $-.58$; (e) -2.3; (g) 1.15. **2.** (a) $\{x > 1\}$. **3.** (a) 2; (c) $\frac{2}{3}$; (e) $\frac{1}{6}$; (g) 3. **4.** 10. **6.** (a) $\log_2 2x = 1 + \log_2 x$; (b) $\log_2 \sqrt{x} = \frac{1}{2}\log_2 x$. **7.** (a) $\{x > 0\}$; (b) $\{x > 0\} \cup \{x < -1\}$. **9.** (a) 0; (b) 1; (c) -1. **10.** 1.65.

PROBLEMS
22

1. (a) 4.8627; (c) $.9600 - 5$; (e) -2.2. **2.** (a) .9499; (c) $.2330 - 2$; (e) 3.6031. **3.** (a) 155,000; (c) 2.52; (e) .000693. **4.** (a) 61.1; (c) 6.31. **5.** (a) .00455; (c) .0753; (e) 6.84; (g) 4130. **7.** (a) $.9031 - 1$; (c) $.5740 - 1$; (e) 1.4843. **8.** (a) 1.43; (c) .356; (e) 3.89. **9.** (a) .3633; (c) $-.6367$; (e) -1.2676.

PROBLEMS
23

1. (a) .9550; (c) $.3938 + 3$; (e) $.5712 - 6$; (g) $.1416 + 3$; (i) $.971 + 4$. **2.** (a) 3.565; (c) .5226; (e) 2.154. **3.** (a) $-.4005$; (c) 1.3644; (e) $-.4771$. **4.** (a) 2.483; (c) 1.585; (e) .001913. **5.** 3.25, too small. **7.** $\frac{38}{9}$. **9.** $\frac{5}{8} \approx \frac{4}{9}$, $\frac{41}{800} \approx \frac{4}{81}$. **10.** $1.5 \approx 1.414$.

PROBLEMS
24

1. (a) 3424; (c) .544; (e) .00148; (g) 2.001; (i) .626; (k) $335 + .0009594 - 66870$. **2.** (a) $.5834 - 1$; (c) .4971; (e) .946. **3.** (a) $25^{7/6}$; (c) 3^{21}. **4.** .707 .5, .354 micrograms. **5.** $A = p(1.06)^n$, about \$10,000. **6.** 3.532×10^{-9}. **7.** .99 meters. **9.** 2, 2.594, 2.691. **11.** (a) 6365; (b) 5.1×10^{10}.

PROBLEMS   page 103
25

1. (a) .63; (c) 1.76; (e) −1.07; (g) .129; (i) $\{x \mid 1.346 \leq x < 1.369\}$. **2.** (a) .37; (c) .836; (e) −.7; (g) 2. **3.** (a) .46; (c) .00476; (e) $\{\frac{1}{10}\sqrt{101}, -\frac{1}{10}\sqrt{101}, \frac{3}{10}\sqrt{11},$ $-\frac{3}{10}\sqrt{11}\}$. **5.** $\{-2, -\frac{2}{7}\}$. **6.** 6 hours and 20 minutes. **7.** $N = 7$, $k = .301$. **8.** $N = 7$, $k = 2$. **9.** A_0, 1730 years. **10.** 6% (approximately). **11.** .14. **12.** $\{x \mid 10^k \leq x < 10^k + 1, k = 0, 1, 2, \ldots\}$.

REVIEW PROBLEMS, CHAPTER THREE page 104

1. Domain of f is R^1, domain of g is $\{x > 0\}$, range of f is $\{y > 0\}$, range of g is R^1; (a), (d), (e), (f), (g), (h) are true. **2.** (a) $\{1, 100\}$; (c) $\{x \geq 1\}$. **3.** e^π. **4.** (a) $\{x < -2\} \cup \{x > 1\}$; (b) $\{x < -4\} \cup \{x > 3\}$. **5.** Divide each number in the body of the table by 2. **6.** 5.87×10^{12}. **7.** (a) $\frac{1}{9}$; (b) −.9542. **8.** 10. **9.** Equations can also be written as $y = -x$ and $y = 1/x$. **10.** $r < \frac{1}{10}$.

MISCELLANEOUS PROBLEMS, CHAPTER THREE page 104

2. $\varnothing$, $\{\frac{1}{10}, 10\}$. **3.** $b = f(1)$. **5.** $\{(2, 2)\}$. **7.** $\{\frac{1}{5} < x < 1\} \cup \{x > 5\}$. **8.** $r > .5$. **10.** (d) $x = 10^k$, where k is a positive integer.

PROBLEMS   page 112
26

1. (a) $(-1, 0)$; (c) $(0, -1)$; (e) $(-1, 0)$. **2.** (a) $(-\frac{1}{2}\sqrt{2}, -\frac{1}{2}\sqrt{2})$; (c) $(\frac{1}{2}\sqrt{2}, \frac{1}{2}\sqrt{2})$; (e) $(-\frac{1}{2}\sqrt{2}, -\frac{1}{2}\sqrt{2})$. **3.** (a) $(\frac{1}{2}, -\frac{1}{2}\sqrt{3})$; (c) $(\frac{1}{2}, -\frac{1}{2}\sqrt{3})$; (e) $(-\frac{1}{2}, -\frac{1}{2}\sqrt{3})$. **4.** (a) $(-\frac{1}{2}\sqrt{3}, \frac{1}{2})$; (c) $(\frac{1}{2}\sqrt{3}, -\frac{1}{2})$; (e) $(\frac{1}{2}\sqrt{3}, \frac{1}{2})$. **5.** (.54, .84). **6.** (a) Negative; (c) positive; (e) negative; (g) positive. **8.** (a) $\{t \mid t = 3 + 2k\pi, k$ an integer$\}$; (c) $\{\frac{1}{2}\pi - 2 + 2k\pi\} \cup \{2 - \frac{1}{2}\pi + 2k\pi\}$. **9.** (a) $\sqrt{2}$; (c) $\sqrt{2}$. **11.** (a) $y = -x + 1$; (c) $y = \frac{1}{2}\sqrt{2}$. **12.** 1 square unit.

PROBLEMS   page 116
27

1. (a) $\sin t = -\frac{4}{5}$, $\tan t = -\frac{4}{3}$; (c) $\sec t = -3$; (e) $\tan t = -\frac{4}{3}$. **3.** (a) .14; (c) .01; (e) −2.2. **4.** (a), (c), (f) are positive. **5.** $\frac{1}{2}[\sqrt{2} + 3(4 - \pi)(\sqrt{3} - \sqrt{2})/\pi] < \sin 1$. **6.** (a) $\cos(\sin 2) > \sin(\cos 2)$; (c) $\cos(\sin 0) > \sin(\cos 0)$. **7.** $\tan 2 \cdot \frac{1}{3}\pi = -\sqrt{3}$ and $2\tan\frac{1}{3}\pi = 2\sqrt{3}$, for example. **10.** (a) $\{k\pi \mid k$ an integer$\}$; (c) $\{t \mid (2k - 1)\pi < i < 2k\pi\}$.

PROBLEMS   page 119
28

1. (a) .2345; (c) 1.120; (e) .3662. **2.** (a) −36.293; (c) .777; (e) −.152. **3.** (a) .384; (c) 1.21; (e) 1. **5.** (a) .5312; (c) −.0298. **6.** cos, cot, csc. **9.** (a) 1.25, 1.26, 1.27; (c) 1.11, 1.115, 1.12.

PROBLEMS   page 124
29

1. (a) $\frac{1}{2}\sqrt{2}$; (c) $2/\sqrt{3}$; (e) $\frac{1}{2}\sqrt{3}$; (g) $\sqrt{3}$. **2.** (a) −.9131; (c) .8437; (e) 1.381; (g) −.3342. **3.** (a) $(-.6960, .7180)$; (c) $(-.4635, -.8861)$; (e) $(.4076, .9132)$;

(g) $(-.9511, .3089)$. **4.** (a) $-.4314$; (c) $.9964$. **5.** (a) $.7446$; (c) $.4745$.
6. (a) $\frac{13}{6}\pi$; (c) $\frac{5}{4}\pi$; (e) $\frac{13}{6}\pi$. **8.** (a) $d = k + \sin 2\pi T$; (b) $k - \frac{1}{2}\sqrt{2}$, $k + \frac{1}{2}\sqrt{3}$,
$k + .7630$. **9.** (b) $\frac{1}{2}(\sqrt{33} - 1)$, $\frac{1}{2}(\sqrt{33} - 1)$, 3.16.

PROBLEMS
30

3. (a) $-.96$; (c) -2.2; (e) $-.48$. **4.** (a) 2; (c) 2; (e) 2. **5.** .65. **7.** It becomes
the cosine curve. It remains unchanged. **8.** One, one, infinitely many. **11.**
(a) $f(-x) = f(x)$; (b) $f(-x) = -f(x)$.

PROBLEMS
31

page 132

2. $\cos^2 u + \sin^2 u = 1$. **5.** (a) $\cot(-t) = -\cot t$; (b) $\sec(-t) = \sec t$; (c) $\csc$
$(-t) = -\csc t$. **6.** $(\frac{1}{4}\sqrt{2}(1 - \sqrt{3}), \frac{1}{4}\sqrt{2}(1 + \sqrt{3}))$. **9.** (a) $\frac{1}{6}\pi$; (b) $\frac{1}{2}\pi - \frac{2}{3}$;
(c) $\frac{1}{2}\pi$; (d) $\frac{3}{4}\pi$. **10.** (a) 2; (b) -10; (c) 3; (d) $\frac{1}{2}\pi$. **11.** (b), (c), (d), (e), (f), (h)
are even. **12.** (a), (c), (d), (f) are odd.

PROBLEMS
32

page 135

1. $(\cot u - \tan v)/(1 + \cot u \tan v)$. **2.** $(\tan u - \tan v)/(1 + \tan u \tan v)$.
4. (a) $\frac{1}{4}\sqrt{2}(\sqrt{3} - 1)$; (c) $-\frac{1}{4}\sqrt{2}(\sqrt{3} + 1)$. **5.** (a) $.9975$; (c) $.8415$. **6.** (a) 1;
(c) 0. **7.** (a) $\cos u$; (c) $\cos y$. **8.** $\sin x \cos y \cos z + \cos x \sin y \cos z + \cos x$
$\cos y \sin z - \sin x \sin y \sin z$. **12.** Domain is the set $\{x > 0\}$, range is the set
$\{-1 \le y \le 1\}$.

PROBLEMS
33

page 137

1. (a) $2 \tan t/(1 - \tan^2 t)$; (b) $(\cot^2 t - t)/2 \cot t$. **2.** $(1 - \cos 2t)/(1 + \cos 2t)$.
3. $\frac{1}{3}\pi$. **4.** (a) 1; (c) $-\cos 4t$. **5.** (a) $.4208$; (c) $.6967$. **6.** (a) $\frac{24}{25}$; (c) $\frac{117}{125}$.
7. (a) $\frac{1}{2}\sqrt{2 + \sqrt{2}}$; (c) $\sqrt{2} - 1$; (e) $-\sqrt{2} - 1$; (g) $2/\sqrt{2 - \sqrt{2}}$. **8.** $\frac{1}{2}\sqrt{2 + \sqrt{3}}$.
9. $3 \sin t \cos^2 t - \sin^3 t$. **10.** (a) $\frac{1}{2}\sqrt{2 - \sqrt{2 + \sqrt{2}}}$; (c) $(2 + \sqrt{2 + \sqrt{2}})/$
$\sqrt{2 - \sqrt{2}}$.

PROBLEMS
34

page 141

1. (a) $\sin t$; (c) $\cos u$; (e) $2 \csc 2t$. **2.** (a) $1/\sqrt{1 - \sin^2 t}$; (c) $\sqrt{1 - \sin^2 t}/\sin t$.
3. (a) $-\cot 2t$; (c) 1; (e) $\cot^2 x \cos^2 x$; (g) $\cos^4 t$. **5.** $\frac{1}{4}\pi$. **6.** (a) $1 + \frac{1}{2}\sin 2t$;
(c) $\sec 2t$. **10.** (a) $\{0, \frac{1}{4}\pi, \frac{1}{2}\pi\}$; (c) $\{\frac{1}{8}\pi, \frac{3}{8}\pi, \frac{1}{2}\pi\}$. **11.** Maximum, $2 \cos 2$; minimum, $-2 \cos 2$.

PROBLEMS
35

page 146

2. (a) Frequency is 1, phase shift is $-\frac{1}{2}$; (c) frequency is 1, phase shift is $\frac{1}{2}$.
3. (a) Frequency is $3/2\pi$, phase shift is $\frac{1}{6}\pi$, amplitude is 4; (c) frequency is 1,
phase shift is 2, amplitude is 3. **4.** Frequency is 1, phase shift is $\frac{1}{2}$. **5.** (a)
$2 \sin(2\pi x + 1 + \pi)$; (c) $\sin(4x - 3)$. **7.** (a) $\sqrt{2} \sin(x + \frac{1}{4}\pi)$; (c) $5 \sin(\pi x + .93)$

PROBLEMS
36

page 150

1. (a) $\frac{1}{3}\pi$; (c) 1.03; (e) $\frac{1}{2}\pi$; (g) $\frac{1}{2}\pi$. **2.** (a) $.6$; (c) $.96$. **3.** (a) $\frac{1}{2}\sqrt{3}$; (c) 0. **7.** (a) 1;
(b) $\frac{16}{65}$. **9.** (a) $\mathrm{Sin}^{-1} v$; (c) $\mathrm{Sin}^{-1}(v - u)$ (if $v - u \le 1$).

1. (a) $\frac{1}{3}\pi$; (c) $-\frac{1}{4}\pi$; (e) .25; (g) .05. **2.** (a) $\frac{4}{5}$; (c) $\frac{24}{25}$. **3.** (a) x; (c) $\sqrt{1-x^2}$; (e) $1/\sqrt{1+x^2}$; (g) $x/\sqrt{1+x^2}$. **4.** (a) $2x/(1-x^2)$; (c) $2x/(1+x^2)$. **6.** Arccot $(1/x) = \pi + $ Arctan x, Arctan $x + $ Arccot $x = \frac{1}{2}\pi$. **7.** (a) $\frac{1}{2}\sqrt{2}$; (b) $-\frac{1}{2}\sqrt{2}$. **8.** $\frac{1}{2}(\sqrt{5}-1)$.

1. (a) $\varnothing$; (c) $\{\frac{5}{6}\tau + k\pi\}$; (e) $\{2k\pi \le x < \frac{1}{2}\pi + 2k\pi\} \cup \{\frac{1}{2}\pi + 2k\pi < x \le \pi + 2k\pi\}$; (g) $\{\frac{1}{6}\pi + k\pi\} \cup \{-\frac{1}{6}\pi + k\pi\}$; (i) $\{.46 + k\pi\} \cup \{2.03 + k\pi\}$. **2.** (a) $\{k\pi\} \cup \{\frac{1}{6}\pi + 2k\pi\} \cup \{\frac{5}{6}\pi + 2k\pi\}$; (c) $\{k\pi\}$; (e) $\{k\pi\} \cup \{\frac{2}{3}\pi + 2k\pi\} \cup \{-\frac{2}{3}\pi + 2k\pi\}$; (g) $\{\frac{1}{6}\pi + \frac{1}{3}k\pi\}$; (i) $\{\frac{1}{6}\pi + 2k\pi\} \cup \{\frac{3}{2}\pi + 2k\pi\}$; (k) $\varnothing$. **3.** (a) $\{(k\pi, 2k\pi)\}$; (c) $\{(\frac{1}{3}\pi + 2k\pi, 8)\} \cup \{(\frac{4}{3}\pi + 2k\pi, -8)\}$. **4.** (a) $\frac{1}{2}$; (c) $\{1 + \frac{1}{2}\sqrt{2}, 1 - \frac{1}{2}\sqrt{2}\}$; (e) the right-hand side of the equation is meaningless; (f) 0. **6.** $\{\sqrt{6}, .62, \frac{1}{4}\pi\}$. **7.** (a) $\{\ldots, -\pi, -\frac{1}{4}\pi, 0, \frac{1}{4}\pi, \pi, \ldots\}$; (c) $(\ldots, -.78, -.5, .5, .78, \ldots)$; (e) $\{-.72, 0, .72\}$; (g) $\{-.53, .51, .95\}$; (i) $\{-1, -.9, -.24, .24, .9, 1\}$; (k) $\{.74\}$; (m) $\{-\frac{1}{2}, 0, \frac{1}{2}\}$. **8.** $\{-.93, .93\}$.

REVIEW PROBLEMS, CHAPTER FOUR

1. (a) $\frac{3}{5}$ or $-\frac{3}{5}$; (b) $\frac{3}{5}$; (c) $\frac{24}{25}$; (d) $-\frac{7}{25}$. **2.** (a) 0; (b) $(-1)^n$; (c) $(-1)^n$; (d) 0. **6.** (a) $\{x > 0\}$; (b) $\{2k\pi < x < (2k+1)\pi\}$. **7.** (a) cos (sin 1); (b) cos (log 1); (c) log (tan 1); (d) cos $[\![\sqrt{5}]\!]$. **8.** (a) 0; (b) no solution; (c) 0; (d) .5403. **9.** (a) $\{\frac{1}{4}(2k+1)\pi\}$; (b) $\{-\frac{3}{4}\pi < x < \frac{1}{4}\pi\}$ and so on; (c) $\{0 < x < \frac{1}{2}\pi\} \cup \{\pi < x < \frac{3}{2}\pi\}$ and so on.

MISCELLANEOUS PROBLEMS, CHAPTER FOUR

1. $p = \frac{1}{2}(a+b)$, $q = \frac{1}{2}(a-b)$. **4.** $\{k\} \cup \{k+.74\}$. **6.** $N(m) = 2[\![\frac{1}{2}(m+1)]\!]$. **7.** $(0, 0)$, $(4.07, 1.33)$. **8.** $d = .4$.

1. $\{(\frac{1}{6}\pi, 30°), (\frac{1}{4}\pi, 45°), (\frac{1}{3}\pi, 60°), (\frac{1}{2}\pi, 90°), (\frac{2}{3}\pi, 120°), (\frac{3}{4}\pi, 135°), (\frac{5}{6}\pi, 150°), (\pi, 180°)\}$. **2.** (a) 15°; (c) 172°. **3.** (a) 27° 12'; (c) 57° 17' 45''. **4.** (a) .017; (c) .0000048; (e) 2.6. **5.** 51° 25' 43''. **6.** $\frac{5}{2}$, 143.24°. **7.** 3456 miles. **9.** $\frac{62}{225}\pi \times 10^6$ miles. **10.** $\frac{42}{5}\pi$ inches. **11.** 2.86 feet. **12.** $x = r \cos \omega T + \sqrt{L^2 - r^2 \sin^2 \omega T}$. **13.** $\frac{9}{11}\pi$ square inches.

1. (a) .6508; (c) -11.430; (e) .0584; (g) 1.07; (i) 3.86. **2.** (a) cos 3°; (c) sin 7. **6.** $\frac{1}{4}\sqrt{2}(\sqrt{3}+1)$. **7.** (a) (4.096, 2.868); (c) $(-.4890, -1.040)$. **8.** (a) 53°; (c) 16°. **9.** (a) $b = 4.2, r = 6.5$; (c) $a = 4.1, r = 9.17$. **10.** 89 feet. **11.** (a) $22\frac{1}{2}$°; (c) 36°. **12.** (a) 63° 26'; (c) 71° 35'.

1. (a) $\alpha = 75°, c = 5, b = 2.6$; (c) $\gamma = 95° 42', a = 6.9, b = 9.7$. **2.** (a) $\beta = 38° 41', \gamma = 111° 19', c = 7.4$ or $\beta = 141° 19', \gamma = 8° 41', c = 1.2$; (c) no triangle.

3. (a) $c = \sqrt{7}$, $\alpha = 79°$, $\beta = 41°$; (c) $a = \sqrt{65}$, $\beta = 38°$, $\gamma = 7°$; (e) $c = \sqrt{101}$, $\beta = 23° 25'$, $\alpha = 63° 20'$. **4.** (a) $60°$, $38° 13'$, $81° 47'$; (c) $90°$, $58° 7'$, $31° 53'$. **5.** 270 miles. **6.** 1.24 miles. **7.** 6.4 inches and 5 inches. **8.** Either 226 feet or 63.7 feet. **9.** 4.3 miles. **11.** 308 feet. **12.** Approximately 5500 feet. **13.** $22°$, 103 miles.

PROBLEMS 42

page 190

1. (a) $\frac{35}{2}$; (c) 17. **2.** (a) 3.3; (c) 30. **3.** (a) $A = \frac{3}{4}\sqrt{15}$, $r = \frac{1}{6}\sqrt{15}$; (c) $A = \frac{1}{4}\sqrt{\log 30 \log \frac{15}{2} \log \frac{10}{3} \log \frac{6}{5}}$. **5.** No. **7.** $4\sqrt{66}$ rods.

REVIEW PROBLEMS, CHAPTER FIVE

page 191

2. About 776. **3.** $(\cos 2\varphi, \sin 2\varphi)$. **4.** (a) $90°$, $53°$, $37°$; (b) $15°$, $15°$, $150°$; (c) $30°$, $60°$, $90°$; (d) $45°$, $45°$, $90°$. **5.** 1165 miles. **6.** About 2:20. **7.** $A = \frac{1}{2}(R^2 - r^2)t = \frac{1}{2}(R - r)(S + s)$.

MISCELLANEOUS PROBLEMS, CHAPTER FIVE

page 192

3. $b = 2a \cos \alpha$. **6.** (a) $\frac{1}{6}r^2(4\pi - 3\sqrt{3})$; (b) 4.91. **9.** 84.6 inches, 91.8 inches. **11.** $5r^2 \tan 18°$.

PROBLEMS 43

page 198

1. (a) $6 - 3i$; (c) $-8 - 5i$. **2.** (a) $8 - 4i$; (c) $10i$. **3.** $\{-1, i, -i\}$. **4.** (a) $-1 - 12i$; (c) 7; (e) $2i$; (g) 8. **6.** $\{-2 + 2i, 2 - 2i\}$. **7.** (a) $\{4 + 6i, 4 + i\}$. **8.** (a) 4; (b) b; (c) yes; (d) the set of all complex numbers; (e) R^1. **9.** $(a, b)(c, d) = (ac - bd, bc + ad)$. **10.** $u = 2i$, $v = 1 - i$. **13.** $a = 100$, $b = \log 5/\log 2 \approx 2.3$.

PROBLEMS 44

page 201

2. (a) $\frac{22}{13} - \frac{6}{13}i$; (c) $-1 - 2i$. **3.** (a) $\frac{7}{2} + \frac{5}{2}i$; (c) $-\frac{8}{25} + \frac{6}{25}i$. **6.** $\cos t - i \sin t$. **8.** (a) $\{-\frac{3}{2}i\}$; (c) $\{\frac{2}{5} + \frac{1}{5}i\}$; (e) $\{\frac{1}{3}i, \frac{3}{2} - i\}$. **9.** (a) The set of all complex numbers; (b) the set of all complex numbers; (c) yes; (d) yes; (e) we can't have $v = 0$, of course; (f) z. **10.** (a) $2 - i$. **11.** They are equal.

PROBLEMS 45

page 205

1. (a) 2, $-30°$; (c) 1, $180°$; (e) 2, $120°$. **2.** (a) $\sqrt{3} + i$; (c) $\sqrt{2}(1 - i)$; (e) $-6.58 + 2.39i$. **9.** (a) $-12 - 12i$; (c) 2. **11.** (a) Circle of radius 2 and center the origin; (c) disk consisting of all points less than 1 unit from the origin; (e) $\{(x, y) \mid x \geq 0, y = x\}$.

PROBLEMS 46

page 210

1. (a) 1; (c) 2^{12}; (e) $\cos 5 + i \sin 5$. **2.** (a) $\{2, -1 + \sqrt{3}i, -1 - \sqrt{3}i\}$; (c) $\{-2, 1 - \sqrt{3}i, 1 + \sqrt{3}i\}$. **3.** (a) $8i$; (c) $\sqrt{\sqrt{2} + 1} + \sqrt{\sqrt{2} - 1}i$; (e) $(1 - i)/\sqrt{2}$. **8.** (a) $\{1 + i, -1 - i\}$; (c) $\{-i, \frac{1}{2}(-\sqrt{3} + i), \frac{1}{2}(\sqrt{3} + i)\}$. **9.** $\cos 3\theta = \cos^3 \theta - 3 \cos \theta \sin^2 \theta$, $\sin 3\theta = 3 \cos^2 \theta \sin \theta - \sin^3 \theta$. **12.** $p + q = 8$.

3. (a) The set of all complex numbers; (b) $\{ir \mid r$ a real number$\}$; (c) yes; (d) yes; (e) no. **4.** The line segment Oz is rotated $45°$ in the counter-clockwise direction. **5.** All but (b). **6.** (a) The line $3y = 4x$; (b) the line $3y = -4x$; (c) $\{0\}$. **8.** $\{4k + 2 \mid k$ an integer$\}$. **10.** All except (d).

1. (a) The circle of radius 3 and center $(2, 0)$; (b) the lines $y = 2$ and $y = -2$; (c) the part of the line $4x + 3y = 0$ that lies in the fourth quadrant; (d) the region between the circles of radius 1 and 4 and with center $(3, 0)$. **2.** (a) $\{z \mid |z| = 1\}$. **3.** $\{1 + 2i, -1 - 2i\}$. **6.** (a) $\{z \mid |z| \le 2\}$; (b) $\{z \mid |z| \le 1\}$. **9.** $\{2 + i, -2 + 2i, -3 - 2i, 1 - 3i\}$.

PROBLEMS 47

1. (a) $x^4 + x^2 + x - 15$; (c) $x^6 + 3x^5 - 2x^4 - 5x^3 + 3x^2 - 2x - 6$. **2.** (a) $2x^5 - 3x^4 + 3x^3 - 9x^2 + 10x - 3$; (c) $7x^3 - 3x^2 - 4x - 35$. **3.** (a) 3268; (b) 1776; (c) 4476. **4.** (b) $R(x) = -\frac{1}{32}x^2 + \frac{1}{16}x - \frac{1}{4}$, $S(x) = \frac{1}{32}x$. **5.** (a) $q = 9$, $r = 2$; (c) $q = 13$, $r = 8$. **7.** (a) $Q(x) = x^2 - x + 1$, $R(x) = x$; (c) $Q(x) = 1$, $R(x) = -2x^2 - 2$; (e) $Q(x) = -ix^2 - 3x - 8i$, $R(x) = -3 + 16i$. **8.** (a) $Q(x) = x^2 - x + 2$, $R = 0$; (c) $Q(x) = x^6 - x^5 + x^4 - x^3 + x^2 - x + 1$, $R = 0$; (e) $Q(x) = 3x^2 - 4ix + 6$, $R = 3i$.

PROBLEMS 48

1. (a) 2; (c) $-\frac{7}{25} - \frac{1}{25}i$; (e) $a - b$, $b \ne -a$; (g) $S(1 - r)/(1 - r^n)$, $r^n \ne 1$. **2.** $5/2\pi$ inches. **3.** 420, 14 hours. **4.** $\frac{35}{12}$ minutes. **5.** (a) $\frac{700}{3} \approx 18$ gallons; (b) 70. **6.** $\frac{480}{11}$ minutes after 8. **7.** (a) $\{1000\}$; (c) $\{.3310\}$; (e) $\{3 \le x < 4\}$; (g) $\{\frac{1}{4}\pi + k\pi\}$.

PROBLEMS 49

1. (a) $\{0, -\frac{2}{3}\}$; (c) $\{3, -5\}$. **2.** (a) $\{-2 + 2\sqrt{2}, -2 - 2\sqrt{2}\}$; (c) $\{4 + 2i, 4 - 2i\}$. **3.** (a) $\{\frac{5}{2} + \frac{3}{2}\sqrt{5}, \frac{5}{2} - \frac{3}{2}\sqrt{5}\}$; (c) $\{\frac{1}{2} + \frac{1}{2}\sqrt{21}, \frac{1}{2} - \frac{1}{2}\sqrt{21}\}$; (e) $\{-\sqrt{3} + \sqrt{2}, -\sqrt{3} - \sqrt{2}\}$. **4.** (a) $\{(-b + \sqrt{b^2 - ac})/a, (-b - \sqrt{b^2 - ac})/a\}$; (c) $\{(v_0 + \sqrt{v_0^2 - 2gs})/g, (v_0 - \sqrt{v_0^2 - 2gs})/g\}$. **5.** (a) Positive; (c) negative; (e) negative. **6.** (a) $\{\frac{1}{4}(-i + \sqrt{7}), \frac{1}{4}(-i - \sqrt{7})\}$; (c) $\{-1, 1 + 2i\}$. **7.** (a) Two; (c) two. **8.** (a) $\{(-1 + \sqrt{2})y, (-1 - \sqrt{2})y\}$; (c) $\{y, z\}$. **10.** (a) $\{0 < x < 6\}$ (c) $\{x < 2 - \sqrt{7}\} \cup \{x > 2 + \sqrt{7}\}$; (e) $\{3 - 2\sqrt{2} < x < 3 + 2\sqrt{2}\}$. **11.** (a) $\{k < -\sqrt{5}\} \cup \{k > \sqrt{5}\}$; (c) $\{k \le \frac{4}{7}\}$. **13.** 420 miles per hour **14.** $\frac{1}{4}(5 - \sqrt{13})$. **16.** $q - p^2$.

PROBLEMS 50

1. (a) $\{2, -2, i, -i\}$; (c) $\{\sqrt{2}, -\sqrt{2}\}$; (e) $\{\frac{1}{2}, -\frac{1}{2}, \frac{1}{3}i, -\frac{1}{3}i\}$. **2.** (a) $\{1, -32\}$; (c) $\{1, 9\}$; (e) $\{1\}$. **3.** (a) $\{0, 3\}$; (c) $\{\frac{1}{100}, 10\}$; (e) $\{\frac{1}{6}\pi + 2k\pi\} \cup \{\frac{5}{6}\pi + 2k\pi\}$. **4.** $\{3\}$; (c) $\{0\}$. **7.** $\{2, -1 + i\sqrt{3}, -1 - i\sqrt{3}, -1, \frac{1}{2}(1 - i\sqrt{3}), \frac{1}{2}(1 + i\sqrt{3})\}$. **8.** $6(\sqrt{2} - 1)$ inches. **9.** $\{\frac{1}{2}(1 + i\sqrt{3}), \frac{1}{2}(1 - i\sqrt{3})\}$. **10.** $1 + \sqrt{2}$ inches.

11. (a) True for $=, \subseteq, \supseteq$; (c) true for $\supseteq$; (e) not necessarily true for any of the symbols. **12.** (a) $\{\frac{1}{2}\}$; (c) $\{-3 \le x \le 2\}$.

1. (a) $b = 1$, $c = -6$; (b) $\{2, -3\}$; (c) $(x - 2)(x + 3)$. **2.** (a) 1; (c) 8 or -27. **3.** (a) $x^2 - 2x - 3$; (c) $x^2 - 2x + 5$; (e) $x^2 - 1 + 2\sqrt{2}i$. **4.** (a) $(2x - i\sqrt{3})(2x + i\sqrt{3})$; (c) $(x - 3 + i)(x + 3 + i)$. **5.** (a) $(x - 1 - \sin t)(x - 1 + \sin t)$; (c) $(x \sec t - \tan t - i)(x - \sin t + i \cos t)$; (e) $(x \sin t - i \cos t - i)(x - i \cot t + i \csc t)$. **6.** (a) $(x - (2 - \sqrt{3})y)(x - (2 + \sqrt{3})y)$; (c) $(x - k) \times (kx + 1)$; (e) $2(x \sin t - \cos t)(x \cos t - \sin t)$. **7.** $5x^2 - 18x + 9$. **9.** $ax^2 + (b - 2a)x + c - b + a$. **10.** $b^2 - 4ac$ is the square of an integer.

1. (a) $Q(x) = x^4 + 2x^3 + 4x^2 + 8x + 15$, $R = 31$; (c) $Q(x) = x^2 + \sqrt{3}x + 3$, $R = 3\sqrt{3} + 3$; (e) $Q(x) = x^2 + ix + i - 1$, $R = -1 - 3i$. **2.** (a) 15; (c) 1. **3.** (a) 1,000,005; (c) 1. **4.** (a) 2; (c) -2. **7.** $b - mi$. **9.** (b) $P(x) = Q(x)(x^2 - r^2) + P(r)$; (c) $P(x) = Q(x)(x^2 - r^2) + (P(r)/r)x$.

1. (a) $x^3 - x^2 + x - 1$; (c) $x^3 - 3ix^2 - 4x + 2i$. **2.** $P(0) = 4$. **3.** (a) $(x + 2) \times (x - 2)(x + 2i)(x - 2i)$; (c) $(3x - 2)(2x - 1)^2$. **4.** (a) Not necessarily; (b) no. **6.** No. **7.** $\{-2, -1, 1\}$. **8.** (b) $P(x) = -\frac{1}{2}x^2 - \frac{5}{2}x + 1$; (c) $P(x) = 3x^2 - 2x$.

1. (a) $x^3 - 3x^2 + 4x - 12$; (c) $x^4 + 5x^2 + 4$; (e) $x^6 - 2x^5 + 3x^4 - 4x^3 + 3x^2 - 2x + 1$. **2.** (a) $x^2 - 2$; (c) $x^2 + 2$. **3.** (a) $\{-\frac{1}{2}, i, -i\}$. (c) $\{-1, 2, 1 - \sqrt{5}i, 1 + \sqrt{5}i\}$; (e) $\{-1, 2, i\}$. **4.** (a) $(x + 2)(x^2 - 2x + 4)$; (c) $(x - 1)(x^2 + 2x + 2)$. **8.** $\sqrt{3}x^{17} + 1492x^9 - 3x^5 - \pi x^2 + 2^{-1/3}$. **9.** $P(r) = 0$ if, and only if, $\overline{P}(\bar{r}) = 0$.

1. (a) $\{-1, \frac{1}{2}, 1\}$; (c) $\{-\frac{1}{2}, -\frac{1}{3}, 0, \frac{1}{2}\}$; (e) none. **2.** (a) $\{-1, 2\}$; (c) $\{-1, 7, -2i, 2i\}$; (e) $\{-\frac{1}{2}, \frac{1}{3}, \frac{1}{2}, \frac{3}{4}\}$. **3.** (a) $\{\frac{3}{4}\}$; (c) $\{-\frac{5}{3}, \frac{1}{2}, 1\}$. **4.** (a) $\{2, 3\}$; (c) $\{2, 3, 5, 7\}$. **6.** $\{\frac{1}{6}\pi + 2k\pi\} \cup \{\frac{5}{6}\pi + 2k\pi\}$. **7.** No. **8.** No. **9.** $\{-1, 1, -2, 2, -3, 3, -6, 6, -\frac{1}{3}, \frac{1}{3}, -\frac{2}{3}, \frac{2}{3}\}$. **10.** $P(x) = x^3 - 2$.

1. (a) Between -1 and 0, 0 and 1, 2 and 3; (c) 1 is a zero, between 1 and 2; (e) between -9 and -8, 3 and 4. **2.** (a) 1.2; (c) 1.5. **3.** (a) .80; (c) .61. **7.** Yes, near 0. **8.** 2.69. **9.** About 6 inches. **10.** .42.

REVIEW PROBLEMS, CHAPTER SEVEN

1. $A(x) = a_0$ and $B(x) = 1/a_0$ for some non-zero number a_0. **2.** (a) k is odd; (b) $k \in \{2 + 4m \mid m$ an integer$\}$. **3.** $c = -1$. **4.** (a) $(1 - c^2 r)/(3c^2 r + 1)$; (b) $\{1 + 1/k, 1 - 1/k\}$ $(k \ne 0)$; (c) no solution; (d) $\{2, 38\}$. **5.** 3150 miles.

6. $\{a, -(a^2+1)/2a\}$. **7.** 5. **8.** $3x^5 - \frac{21}{2}x^2 + \frac{9}{2}x + 6$. **9.** $x^2 + 2x + 4 + a = 0$.
10. $\{a, -2a, -\frac{1}{2}a\}$. **11.** $b = -2, c = -1$. **12.** $\{1+i, 1-i, -\frac{5}{2}\}$. **14.** $\{x < -\frac{1}{3}\} \cup \{x > 1\}$. **15.** (a) $-b/c$; (b) $(b^2 - 2ac)/a^2$; (c) $(3abc - b^3)/a^3$; (d) $\sqrt{b^2 - 4ac}/|a|$.

MISCELLANEOUS PROBLEMS, CHAPTER SEVEN page 253

1. $\{0\}$. **4.** $\{\frac{3}{5}, -\frac{12}{5}\}$. **5.** $\{x < -1\} \cup \{0 < x < 7\}$. **6.** $\{k \le -7\} \cup \{k \ge 3\}$.
7. 12.2 **9.** (b) $\sqrt{10} \approx \frac{721}{228} = 3.16228\ldots$, $\sqrt{10} = 3.162278\ldots$ **10.** (b) 25 feet from the wall and 50 feet apart. **11.** (a) $-\sqrt{19}/\pi$; (b) a_{n-2}/a_n. **12.** $ab > 0$ and $ac > 0$. **13.** (a) $a_0 x^n + a_1 x^{n-1} + \cdots + a_n$; (b) if k is odd, replace a_k with $-a_k$; (c) $P(x - r)$; (d) $P(x/r)$.

PROBLEMS 57 page 258

1. (a) Yes; (c) yes; (e) yes. **2.** (a) $(x, y) = (1, -2)$; (c) $\{(100, 2), (\frac{1}{10}, -1)\}$; (e) $\{(-1, -2), (2, 1)\}$. **3.** All except (e). **4.** (a) $x^2 = 1, y^2 = 1$; (c) $x + y = 2$, $x^2 - xy + y^2 = 3$. **5.** 3 feet and 5 feet. **6.** $a = 7, b = -6$. **7.** The system has no solution. **8.** Yes. **9.** (a) $\{(0, 0)\}$; (c) $\{(0, 0), (1, 1), (\omega, \omega^2), (\omega^2, \omega)\}$, where $\omega = \frac{1}{2}(-1 + i\sqrt{3})$.

PROBLEMS 58 page 262

1. (a) $(-2, 1)$; (c) $(1, 1)$; (e) $(3, -4)$. **2.** (a) $(i, 2)$; (c) $(2+i, i)$. **3.** (a) $(x, y, z) = (2, -1, 3)$; (c) $(x, y, z, w) = (1, 1, 1, 1)$. **4.** (a) $(\frac{1}{2}, -\frac{1}{3})$; (c) $(100, \frac{1}{10})$. **5.** 60% cheap and 40% expensive. **6.** 58. **9.** (a) $\{(x, y) \mid 2 \le x < 3, -1 \le y < 0\}$; (b) $\varnothing$; (c) $\{(100, \frac{1}{10})\}$; (d) $\varnothing$.

PROBLEMS 59 page 267

1. (a) $\begin{bmatrix} 1 & -2 & 0 & | & 2 \\ 0 & 1 & -3 & | & 3 \\ 1 & 0 & -4 & | & -3 \end{bmatrix}$; (c) $\begin{bmatrix} 1 & 0 & -1 & 0 & | & 2 \\ 0 & 1 & 0 & -1 & | & -1 \\ -1 & 0 & 0 & 1 & | & 3 \\ 0 & -1 & 1 & 0 & | & 0 \end{bmatrix}$.

2. (a) $(-2, 1)$; (c) $(3, -2)$; (e) $(1, -1)$. **3.** (a) $(1, -1, 2)$; (c) $(0, 0, 0)$; (e) $(x, y, z, w) = (1, -1, 1, -1)$. **4.** (a) $(0, 0, 0)$. **5.** $(6u - v - 10w, -u + 2w, -5u + v + 9w)$. **6.** 100°, 50°, 30°. **7.** $(a, b, c) = (-1, 2, 3)$. **8.** (a) $\frac{120}{13}$ hours; (b) 24 hours. **9.** No. **10.** $(x, y) = (u \cos\alpha - v \sin\alpha, u \sin\alpha + v \cos\alpha)$.

PROBLEMS 60 page 271

1. (a) $(2, -2, -5)$; (b) $(3, -5, -10)$; (c) $(1 + i, 1 - 3i, -5i)$; (d) $(1 - i, 1 + 3i, 5i)$. **2.** Yes. **3.** Dependent; (c) determinative; (e) determinative. **4.** (a) Inconsistent. **5.** (a) $(1, 1)$; (c) do not intersect. **6.** (a) $\{(x, y, z) = (t, \frac{1}{5} - \frac{3}{5}t, \frac{9}{5} - \frac{7}{5}t)\}$; (c) $\varnothing$. **7.** $k = 2$. **8.** 572, no. No.

PROBLEMS 61 page 274

1. (a) $\{(1, -1), (-13, -8)\}$; (c) $\{(1, 2), (9, -6)\}$; (e) $\{(t, t) \mid t \text{ any number}\}$. **2.** (a) $\{(1, 2), (1, -2), (-1, 2), (-1, -2)\}$; (c) $\{(1 + i, 2 + i), (-1 - i, -2 - i), (-1 - i, 2 + i), (1 + i, -2 - i)\}$; (e) $\{(0, 0)\}$. **3.** (a) $\{(2i, 0), (-2i, 0), (4, 2), (-4, -2)\}$; (c) $\{(-1, -2), (1, 2), (-2, 2), (2, -2)\}$. **4.** 16 inches and 30 inches.

5. 9 inches by 12 inches. **6.** $\{(2, 1),\ (-1, -2)\}$. **7.** $\sqrt{abc}$. **8.** $\{(-2, -1),$ $(-1, 1),\ (1, 2),\ (3, 0)\}$.

page 280

PROBLEMS
62

1. (a) 11; (c) -11; (e) 1. **2.** (a) $\{8\}$; (c) $\{-1, 3\}$. **3.** (a) $\{x > 8\}$; (b) $\{x < -1\} \cup \{x > 3\}$. **4.** $\{(0, 0),\ (4, 1),\ (-4, -1)\ldots\}$. **9.** No, yes. **10.** -25. **11.** $r = c/d$.

page 284

PROBLEMS
63

1. (a) 4; (c) $-\sec^2 x$. **2.** (a) -6; (c) -1. **3.** (a) 0; (c) -1. **4.** (a) -32; (c) 0. **6.** (a) $(1, 3, -1)$. **7.** $\{0, -1, 1\}$. **8.** $k \neq -1$. **9.** $\{-2, 0, 1\}$.

REVIEW PROBLEMS, CHAPTER EIGHT page 286

1. $\{\frac{1}{10}, 100\}$. **2.** $a = 2,\ b = -1,\ c = 3$. **3.** $s = 2,\ t = -1$. **4.** $\{(t, 7t, -5t) \mid t$ any number$\}$. **6.** $\{-1, 3\}$. **9.** 120 6's and 230 8's. **10.** Bet \$40 on A, \$48 on B, \$60 on C, and \$80 on D. **11.** The half-line lying wholly in the first quadrant that terminates in the point $(2, 5)$ and has slope 3. **12.** (a) $\{(x, y, z) \mid (t, 7t - 2, 11t - 4)\}$; (b) $(x, y) = (1, 1)$; (c) $\varnothing$; (d) $\{(x_1, x_2, x_3, x_4, x_5, x_6, x_7, x_8) = (a, b, c, d, e, f, 1 - a - c - e, 1 - b - d - f)\}$.

MISCELLANEOUS PROBLEMS, CHAPTER EIGHT page 287

1. (a) $\{(2, 2),\ (i, -4i),\ (-i, 4i)\}$; (b) $\{(1, \frac{1}{5}),\ (-5, -1)\}$. **2.** $(3, 7, 2, 6, 1, 5, 0, 4, -1)$. **3.** (a) $\{-1, 3 + 2\sqrt{3},\ 3 - 2\sqrt{3}\}$; (b) $\{3, -2i, 2i\}$. **4.** 4 nickels, 3 dimes, and 3 quarters, or 1 nickel, 7 dimes, and 2 quarters. **5.** No, not always. **8.** (a) $ad \neq bc$. **10.** Yes.

page 293

PROBLEMS
64

1. (a) 990; (c) $n/(n + 1)$; (e) $n + 1 + 1/n$. **2.** (a) 2.05×10^{36}; (c) $1.033 \times 10^{40} - 5.04 \times 10^3 \approx 1.033 \times 10^{40}$. **3.** 4. **4.** (a) 9000. **5.** 8. **7.** 15,675. **8.** 60!/35!, (10!/5!)(50!/30!). **9.** 18. **10.** (a) $26 \cdot 25 \cdot 25 \cdot 25 = 406{,}250$. **11.** (a) 7.

page 297

PROBLEMS
65

1. (a) 9.812×10^9; (c) 4.959×10^{14}; (e) 1.
3. $\binom{52}{13, 13, 13, 13}$. **4.** (a) $\binom{100}{39, 28, 33}$.
5. $30 \cdot 29 \cdot 28 \cdot 27$ (in real life, no one volunteers). **6.** (a) 1260. **7.** (a) 630; (b) 1260. **8.** 9!, 9!/4!3!2!. **9.** (a) 90,090; (c) 23,760. **10.** (a) 5!.

page 301

PROBLEMS
66

1. (a) 20; (c) n; (e) 1. **2.** (a) 10; (c) 6. **5.** 120 possible outcomes. **6.** 90.
7. 6.35×10^{11}. **8.** $\binom{30}{6}\binom{23}{4} < \binom{53}{10}$. **9.** (a) $\binom{7}{2}$. **10.** (a) $\binom{10}{3}\binom{10}{7}$.
11. $\binom{23}{5}\binom{18}{3} = \binom{23}{5, 3, 18}$.

1. (a) $70x^4y^4$; (c) $252x^{10}y^4$. **2.** (a) $-160(x/y)^3$; (c) $35(ab)^{3/2}(ia^{1/2}+b^{1/2})$. **3.** (a) $-8i$; (c) $a^4-6a^2b^2+b^4+i(4a^3b-4ab^3)$. **4.** (a) $-672u^3$; (c) -280 $\tan x$; (e) $-20(216)^x$. **5.** (a) 1.219; (c) 1.104×10^{10}. **6.** (a) $15a^2x^2$; (c) $2160x^2$ **7.** $(a-b)^n=[a+(-b)]^n$. **11.** $\sin 4\theta=4\cos^3\theta\sin\theta-4\cos\theta\sin^3\theta$. **12.** (a) $a^3+b^3+c^3+3a^2b+3a^2c+3ab^2+3ac^2+3b^2c+6abc$; (e) $\sin 2t+2i(\cos t+\sin t)$.

1. (a) $495a^8b^4$; (c) $70x^4$. **2.** (a) $x^{100}+100x^{99}+4950x^{98}+161{,}700x^{97}$; (c) $x^{19}+19x^{17}+17x^{15}+969x^{13}$. **3.** $x^6-6x^4+15x^2-20+15x^{-2}-6x^{-4}+x^{-6}$. **4.** (a) $\cos 1+\cos 2+\cos 3+\cos 4$; (c) $x^5-x^4+x^3-x^2+x-1$; (e) x^3+3x^2+3x+1. **7.** $nx^{n-1}+n(n-1)x^{n-2}h/2!+\ldots+h^{n-1}$. **9.** 49. **10.** (a) -5; (c) $\{-3,i\sqrt{3},-i\sqrt{3}\}$.

1. $\frac{1}{18}$. **2.** $\frac{1}{26}$. **3.** (a) $\frac{5}{16}$; (c) $\frac{1}{12}$. **4.** $\binom{20}{13}\Big/\binom{52}{13}$. **5.** (a) $\frac{2}{13}$; (b) $\frac{11}{13}$. **6.** $\frac{308}{1615}$. **7.** (a) $\frac{1}{8}$. **8.** (a) $\frac{5}{12}$. **9.** (a) $\frac{1}{6}$; (c) $\frac{4}{9}$. **10.** (a) .18; (c) 16. **11.** (a) $\frac{1}{6}$; (b) 1.17. **12.** 1.90. **13.** (a) $4\Big/\binom{52}{5}$; (c) $4\left[\binom{13}{5}-10\right]\Big/\binom{52}{5}$.

1. (a) $\frac{9}{25}$; (c) 0. **2.** $\frac{15}{36}$. **3.** (a) $\frac{1}{4}$; (c) $\frac{1}{2}$. **4.** (a) $(\frac{5}{36})^2$; (c) $(\frac{5}{36})^2+(\frac{3}{36})^2$. **5.** (a) $\frac{1}{19}$; (c) $\frac{27}{38}$. **6.** $(\frac{2}{9})^5$. **7.** (a) $\frac{2}{5}$; (b) $\frac{93}{275}$. **8.** $\frac{611}{1352},\frac{611}{2704},\frac{949}{2704}$. **9.** (a) 6.38; (c) 5.25. **10.** $\frac{37}{75}$. **11.** 25. **12.** (a) $\frac{3}{8}$; (c) 0.

1. (a) $\frac{24}{169}\approx.14$; (c) $\frac{32}{221}\approx.145$. **2.** .474. **3.** (a) $\frac{386}{1024}\approx.376$; (b) $\frac{176}{1024}\approx.172$. **4.** (a) $\frac{63}{256}$; (c) 5. **5.** 2, .296. **6.** (a) 6; (c) 6. **7.** No, the probability of his winning is .6. **9.** (a) 4. **10.** (a) .0000; (c) .0415; (e) .7522. **11.** $(\frac{35}{18})(\frac{5}{6})^4$. **12.** $x=7$ or $x=8$, $p(7)=p(8)=.19$, .69.

1. 1, $\frac{1}{2}$, $\frac{1}{1024}$. **2.** (a) $\frac{1}{128}$; (b) $\binom{7}{4}(\frac{1}{2})^7=\frac{35}{128}$; (c) $\frac{1}{2}$. **3.** (a) $(\frac{4}{5})^{10}$; (c) .0063. **4.** .22. **5.** $\sum_{r=0}^{7}\binom{30}{r}(\frac{2}{5})^{30-r}(\frac{3}{5})^r$. **6.** $(\frac{3}{4})^{30}$, 7. **7.** (a) $\frac{36}{125}$; (b) 1. **8.** 17, practically zero. **9.** No, $p_{10}=.126$ is "close to" $p_8=.167$. Yes, $p_3=.011$ is "not close to" p_8. **10.** 7, 8, 9; no; pick 7, 8, 9, 10.

REVIEW PROBLEMS, CHAPTER NINE

1. (a) 10; (b) 8. **2.** (a) 1; (b) -4. **3.** 5, $n(n-3)/2$. **4.** $26!/19!$, $7\cdot25!/19!$. **5.** Chances of failing are $\frac{1}{3}$. **7.** $(\frac{7}{8})^5$. **8.** 1. **9.** $(\frac{7}{9})^3\approx.47$. **10.** $\frac{1}{50{,}400}$, $\frac{1}{140}$. **11.** 5, .017. **12.** $\frac{1}{2}$.

1. $\{(k, n) \mid (6, 2), (5, 3)\}$. **3.** 0. **4.** No. **5.** $\frac{657}{3025}$. **6.** 1. **7.** $\frac{7}{12}, \frac{5}{12}$. **10.** 7.30, $5.00. **11.** 11¢.

1. (a) 0, 2, 0, 2, 0; (c) 1, 1/3, $1/3 \cdot 5$, $1/3 \cdot 5 \cdot 7$, $1/3 \cdot 5 \cdot 7 \cdot 9$; (e) 0, -1, 0, 1, 0.
2. (a) $3, 3 \cdot 2, 3 \cdot 2^2, 3 \cdot 2^3, 3 \cdot 2^4$; (c) $2, \frac{1}{2}, 2, \frac{1}{2}, 2$; (e) 1, 1, -1, 1, -1; (g) 0, 2, 1, 1, 1.
3. (a) $\{5 \cdot 2^{-n}\}$; (b) $\{3n + 1\}$; (c) $\{n^2 + 1\}$; (d) $\{\sin \frac{1}{2}n\pi\}$; (e) $\{\exp_5 (2^n - 1)\}$;
(f) $\{(-1)^{[\![(n-1)/2]\!]}\}$. **4.** (a) $S_n = n$; (c) $S_n = ((-1)^n - 1)/2$; (e) $S_n = (n + 1)!$.
5. (a) $a_1 = 0$, $a_n = -2/n(n + 1)$ for $n > 1$; (c) $a_n = 1$; (e) $a_n = n$.
6. (a) $\frac{1}{2}(1 + (-1)^n)$; (c) 2^n. **7.** 21. **8.** 0, 7, 6, 9, 2, 3, 0, 7; $a_{1984} = 9$. **10.** $\frac{1}{2}n$
$\sin (2\pi/n); \pi$.

2. All except (b). **3.** (a) $(\frac{1}{2})^{n-2}$; (c) $2(-1)^{n-1}/n$; (e) $2 \cdot 3^{n-1}n!$. **9.** $a_n = (3n^2 - 2n + 1)a_{n-1}/(3n^2 - 8n + 6)$, $a_n = a_{n-1} + 6n - 5$.

1. (a) $a_7 = -16$, $S_7 = -49$; (c) $a_7 = -22$, $d = -4$; (e) $a_1 = 6$, $a_{12} = 28$.
2. 23, 3588. **3.** 16, 5. **4.** (a) 55; (c) $25 + 15i$. **5.** (a) 110; (c) -60. **6.** 15 feet.
7. (b) $n(n + 1)$. **8.** 1600 feet, $s = 16t^2$. **9.** $600, $3900, $13,000. **11.** $a_{2k-1} = 5k - 3$, $a_{2k} = 5k - 1$, $b_{2k-1} = 5k - 4$, $b_{2k} = 5k - 1$.

1. (a) $a_5 = 1875$, $S_5 = 2343$; (c) $n = 8$; (d) $r \in \{2, -2, 2i, -2i\}$, $S_5 \in \{155, 55, 65 - 30i, 65 + 30i\}$; (e) at most 12. **2.** $S_n = na_1$. **3.** (a) 93; (c) $\frac{255}{256}$; (e) 92.
4. 7.7, 1994. **5.** 2046. **6.** $a = b = c$. **8.** 1092. **11.** $4716. **12.** 0.

1. (a) 8, 10; (c) 7, $5 + 5i$; (e) $2 \cdot 3^{x-1} + 3^{y-1}$, $3^{x-1} + 2 \cdot 3^{y-1}$. **2.** (a) 24, 36, 54;
(c) 4.49, 6.71, 10.03. **3.** (a) $\frac{3}{4}$; (c) $\frac{100}{7}$; (e) 1.469. **4.** (a) $\frac{2}{9}$; (c) $\frac{10}{99}$. **5.** $a = 5$, $b = 8$, $c = 12$. **6.** If x is not an integral multiple of π, the sum is $\csc^2 x$.
7. (a) $(\frac{1}{2})^9$; (c) $2 - (\frac{1}{2})^9$.

4. $N = 6$. **6.** Yes, $\{i, j\} = \{1, 2\}$.

2. (a) Yes, no, no. **3.** (a) 32.4237; (c) 76.3604. **4.** 1. **5.** (a) -1; (c) $\frac{1}{2}\sqrt{2}$.
7. $N = 6$. **8.** No. **9.** $2^{64} - 1 \approx 1.837 \times 10^{19}$. **10.** 35 years old, 152,570
cigarettes.

1. $n \geq 25$. **2.** $a_1{}^n r^{n(n-1)/2}$. **3.** 35. **4.** (a) $1, \frac{1}{2}, \frac{1}{3}, \frac{1}{4}$; (b) $2, \frac{3}{2}, \frac{6}{5}, 1$; (c) $2ab/(a+b)$. **7.** Solve the equation $1 + (1+r) + \cdots + (1+r)^{11} = 11(1+r)^{12}$ for r. **8.** We don't cover all points, for example, not $0, \frac{1}{3}, \frac{1}{9}$, and many, many more. Even so, the eventual colored set is 1 unit "long." **10.** $a_{1776} = -2 \times 6^{887}$, $b_{1493} = -6^{746}$.

PROBLEMS 79

1. (a) The upper half-plane; (c) the strip between the lines $y = -2$ and $y = 1$. **2.** (a) The coordinate axes; (c) the line $y = -x$. **3.** (a) $x > 0$; (c) $x^2 + y^2 = 9$; (e) $(0, 0)$. **6.** (a) $(x-2)^2 + (y+3)^2 = 1$; (c) $3x^2 + 3y^2 - 14x + 4y + 15 = 0$; (e) $y = -3$. **7.** (a) $\{(0, 0), (1, 1)\}$; (c) $\{(k\pi, 0)\}$; (e) $\{(1, 1), (-\frac{1}{2}, \frac{1}{4})\}$. **9.** (b) $|y| = 2|x|$; (d) $|y| > 2|x|$. **10.** (a) and (c).

PROBLEMS 80

1. (a) Slope is 2, Y-intercept is -3; (c) slope is $-\frac{2}{3}$, Y-intercept is $\frac{4}{3}$. **2.** (a) $x - 2y + 4 = 0$; (c) $y = -3x + 10$; (e) $y = -\sqrt{3}x - 1 + \sqrt{3}$. **3.** (a) $2x + 3y = -6$; (c) $3x - 4y = 25$; (e) $3x + 2y = 1$. **4.** (a) Negative; (c) positive. **5.** (a) 5 units. **8.** $\frac{1}{11}x + \frac{1}{7}y = 1$. **9.** $(3, 2)$ and $(27, 42)$.

PROBLEMS 81

1. (a) $(x-2)^2 + (y+3)^2 = 25$; (c) $x^2 + y^2 = 2x \cos 3 + 2y \sin 3$. **2.** (a) Center $(-2, 1)$, radius 3; (c) center $(\frac{1}{2}, 1)$, radius $\frac{1}{2}\sqrt{5}$. **3.** (a) $(x-2)^2 + (y+3)^2 = 17$; (c) $(x+1)^2 + (y-2)^2 = 10$; (e) $(x-8)^2 + (y-14)^2 = 100$; (g) $x^2 + (y-5)^2 = 25$ or $x^2 + (y+1)^2 = 25$. **4.** (a) The exterior of the disk of radius 3 whose center is $(0, 0)$; (c) the ring between the circles of radii 2 and 5 whose center is $(0, 0)$. **5.** Circles tangent to the Y-axis. **6.** Yes. **7.** The circle of radius 3 whose center is $(0, 0)$. **9.** $a^2 + b^2 > 4c$. **10.** Replace x with $x - h$ and y with $y - k$.

PROBLEMS 82

1. (a) $(4, -8), (0, -2)$; (c) $(-3, -1), (-7, 5)$; (e) $(\frac{3}{5}, -\frac{16}{7}), (-\frac{17}{5}, \frac{26}{7})$; (g) $(\log 300, \log .002), (\log .03, \log 200)$. **2.** (a) $(2, 1), \bar{y} = 2\bar{x}, \bar{x} = 0$; (c) $(2, 1), \bar{x} = 2\bar{y}, 3\bar{x} = -\bar{y}$. **3.** $\bar{y} = y, \bar{x} = x - \frac{1}{2}\pi$. **4.** (a) Symmetric about both axes and the origin; (c) symmetric about both axes and the origin; (e) symmetric about Y-axis. **7.** (a) $\bar{x} = x - 2, \bar{y} = y + 1$; (c) $\bar{x} = x - 2, \bar{y} = y - 1$. **9.** $\bar{y} = \bar{x}^5$. **11.** No. **12.** (a) $\bar{x} = x - 2, \bar{y} = y - 3$; (c) $\bar{x} = x - \frac{1}{2}\pi, \bar{y} = y$; (e) $\bar{x} = x, \bar{y} = y - \frac{1}{2}\pi$.

PROBLEMS 83

1. (a) $x^2/16 + y^2/9 = 1$; (c) $(x-2)^2/9 + (y+1)^2/4 = 1$. **2.** (a) $x^2/9 + y^2/8 = 1$; (c) $(x-1)^2/4 + (y-4)^2/3 = 1$. **3.** (a) $x^2/25 + y^2/9 = 1$; (c) $(x-5)^2/25 + (y+1)^2/4 = 1$. **4.** (a) $(-4, 0), (4, 0)$; (c) $(-4, 0), (-2, 0)$. **5.** (a) $(x-4)^2/25 + (y-2)^2/4 = 1$; (c) $(x-1)^2 + (y-2)^2/5 = 1$. **7.** (a) The curve is a circle, the given points are endpoints of a diameter. **8.** An ellipse with major diameter of 100 feet and minor diameter of $50\sqrt{3}$ feet. **9.** Tacks should be placed in a line through the center of the paper and parallel to the 11-inch edge and at points $\frac{1}{4}\sqrt{195} \approx \frac{7}{2}$ inches from the center. The string should be about 18 inches long. **10.** The upper half of the ellipse $x^2/25 + y^2/9 = 1$. **12.** Area $= \pi ab$.

1. (a) Vertices: $(-2, 0)$, $(2, 0)$, foci: $(-\sqrt{5}, 0)$, $(\sqrt{5}, 0)$, asymptotes: $y = \frac{1}{2}x$, $y = -\frac{1}{2}x$; (c) vertices: $(0, -1)$, $(0, 1)$, foci: $(0, -\sqrt{5})$, $(0, \sqrt{5})$, asymptotes: $y = \frac{1}{2}x$, $y = -\frac{1}{2}x$; (e) vertices: $(-5, 2)$, $(3, 2)$, foci: $(-6, 2)$, $(4, 2)$, asymptotes: $3x + 4y = 5$, $3x - 4y = -11$; (g) vertices: $(1, -5)$, $(1, 1)$, foci: $(1, -7)$, $(1, 3)$, asymptotes: $3x - 4y = 11$, $3x + 4y = -5$. **2.** (a) $3x^2 - y^2 = 3$; (c) $5(x - 3)^2 - 4(y - 1)^2 = 20$. **3.** (a) $(-7, 1)$, $(3, 1)$; (c) $(-2, -4)$, $(-2, 6)$. **4.** (a) $x^2/36 - y^2/28 = 1$; (c) $(y - 5)^2/25 - (x - 1)^2/3 = 1$. **5.** (a) $4x^2 - y^2 = 3$; (c) $(y - 2)^2 - (x - 1)^2 = 3$. **6.** (a) $7y^2 - 4x^2 = 28$. **8.** A hyperbola. **10.** The second man is standing on a point of the hyperbola $3x^2 - y^2 = 550^2$; the first man is at the focus. **11.** The union of an ellipse and a hyperbola.

1. (a) focus: $(4, 0)$, directrix: $x = -4$; (c) focus: $(2, 3)$, directrix: $y = -9$; (e) focus: $(-1, -2)$, directrix: $y = 0$. **2.** (a) $y^2 = 16x$; (c) $(y - 2)^2 = 12(x + 2)$. **3.** (a) $(x - 2)^2 = 16y$; (c) $(y - 3)^2 = 2(x + 1)$. **4.** (b) 16, 12, 12, 16; (c) $x^2 + y^2 = 5cx$. **5.** (a) $(-1, 1)$, $(2, 4)$; (c) $(-\frac{1}{2}, -\sqrt{3})$, $(-\frac{1}{2}, \sqrt{3})$. **7.** A parabola with the buoy as focus and shoreline as directrix. **10.** Directrix: $x + y = -2$, parabola: $(x - y)^2 = 8(x + y)$.

1. (a) $e = \frac{1}{2}$, directrices: $x = 4$, $x = -4$; (c) $e = \frac{5}{4}$, directrices: $x = \frac{16}{5}$, $x = -\frac{16}{5}$; (e) $e = \sqrt{2}$, directrices: $x = \frac{1}{2}\sqrt{2}$, $x = -\frac{1}{2}\sqrt{2}$; (g) $e = \frac{5}{3}$, directrices: $y = \frac{9}{5}$, $y = -\frac{9}{5}$. **2.** (a) $3x^2 - y^2 + 20x + 12 = 0$; (c) $3x^2 + 4y^2 - 20x + 12 = 0$; (e) $y^2 + 2x = 3$; (g) $9x^2 + 8y^2 - 18x - 40y + 41 = 0$. **3.** (a) $e = \frac{1}{2}$, directrices: $x = 5$, $x = -3$; (c) $e = \sqrt{2}$, directrices: $x = 1 + \frac{1}{2}\sqrt{2}$, $x = 1 - \frac{1}{2}\sqrt{2}$. **4.** $e = c/b$, directrices: $y = b^2/c$, $y = -b^2/c$. **5.** (a) An ellipse with eccentricity near 0 is relatively circular, an ellipse with eccentricity near 1 is relatively flat or narrow; (b) a hyperbola with eccentricity near 1 is relatively narrow, a hyperbola with a large eccentricity is relatively open or wide. **6.** $y^2 - x^2 + 6x + 9 = 0$ or $y^2 - x^2 - 6x + 9 = 0$. **7.** $2^{-1/2}$. **8.** (a) $3x + 4y = -25$; (c) $3x + 4y = -50$. **9.** (a) An arc of a parabola; (b) an arc of a hyperbola; (c) an arc of an ellipse.

1. (a) $(4\sqrt{3}, 4)$; (c) $(5\sqrt{2}, -5\sqrt{2})$; (e) $(30, 30\sqrt{3})$. **2.** (a) $(3, \pi)$; (c) $(2, -\frac{1}{4}\pi)$; (e) $(8, -60°)$. **3.** (a) $x^2 + y^2 = 4x$; (c) $x = 3$; (e) $y = 2x$. **6.** $r(\sin \theta - 2 \cos \theta) = 2$. **7.** No. **8.** (a) $(2, 30°)$ and 7 other points; (c) $\{(0, 0), (1, \frac{1}{4}\pi), (1, \frac{5}{4}\pi)\}$; (e) $\{(0, 0), (3, \frac{2}{3}\pi), (3, \frac{4}{3}\pi)\}$.

1. (a) $r \cos (\theta - 30°) = 6$; (c) $r \cos (\theta + \frac{1}{4}\pi) = 8$. **2.** (a) 6; (c) 11. **3.** (a) $r = 2\sqrt{2} \cos (\theta - 135°)$; (c) $r^2 - 2\sqrt{2}r \cos (\theta - 135°) = 11$. **4.** (a) Hyperbola, vertices: $(2, \frac{1}{2}\pi)$, $(-8, \frac{3}{2}\pi)$, foci: $(0, 0)$, $(10, \frac{1}{2}\pi)$; (c) parabola, vertex: $(2, \frac{1}{2}\pi)$, focus: $(0, 0)$; (e) parabola, vertex: $(\frac{1}{2}, \frac{1}{2}\pi)$, focus $(0, 0)$. **5.** (a) $r = 52/(9 - 13 \cos \theta)$; (c) $r = 12/(5 - 3 \cos \theta)$. **8.** $x^2(1 - e^2) + y^2 = 2pe^2x + e^2p^2$.

453

1. (a) $(3 - 2\sqrt{3}, 2 + 3\sqrt{3})$; (c) $(2 - 3\sqrt{3}, -3 - 2\sqrt{3})$; (e) $(4, -6)$; (g) $(-4, 6)$.
2. (a) $(1, \sqrt{3})$; (c) $(-1 - 2\sqrt{3}, 2 - \sqrt{3})$. **3.** (b) $x = .8\bar{x} - .6\bar{y}$, $y = .6\bar{x} + .8\bar{y}$;
(c) about $37°$; (d) $\bar{x} = 2$; (e) $\bar{x}^2 + \bar{y}^2 = 16$. **4.** (a) 2; (b) 17; (c) 5; (d) 1; (e) 1;
(f) 1. **5.** (a) $\bar{x} = y$, $\bar{y} = -x$; (c) $\bar{x} = -y$, $\bar{y} = x$. **6.** (a) $\bar{x} = \frac{1}{2}\sqrt{3}x + \frac{1}{2}y$,
$\bar{y} = -\frac{1}{2}x + \frac{1}{2}\sqrt{3}y$; (c) $(2\sqrt{3} - 3)\bar{x} - (2 + 3\sqrt{3})\bar{y} = 5$. **7.** (b) $.28$ radian;
(c) -3.478; (d) $-.2876$; (e) 2.08 radians; (f) 2.64 radians. **10.** $x = \bar{x}\cos\alpha - \bar{y}\sin\alpha + h\cos\alpha - k\sin\alpha$, $y = \bar{x}\sin\alpha + \bar{y}\cos\alpha + h\sin\alpha + k\cos\alpha$.

1. (a) $3\bar{x}^2 - \bar{y}^2$; (c) $5\bar{x}^2$; (e) $-\frac{11}{2}\bar{x}^2 + \frac{39}{2}\bar{y}^2$. **2.** (a) $4\bar{x}^2 - \bar{y}^2 = 4$; (c) $4\bar{x}^2 - \bar{y}^2 = 4$. **3.** (a) $4\bar{\bar{x}}^2 + \bar{\bar{y}}^2 = 16$; (c) $\bar{\bar{y}}^2 = 4\bar{\bar{x}}$. **4.** Two intersecting lines.
5. (a) Circle; (c) ellipse. **6.** $A = C$, $B = 0$. **7.** The point $(0, 0)$, one line, two
lines, or the whole plane. **8.** (a) Parabola; (c) hyperbola.

REVIEW PROBLEMS, CHAPTER ELEVEN

1. $x + 2y = 19$. **2.** $(x + 1)^2 + (y - \sqrt{3})^2 = 12$. **3.** (a) $(0, 0)$, $(\frac{1}{4}, \frac{1}{4}\sqrt{3})$;
(b) $(0, 0)$, $(\frac{1}{4}, \frac{1}{4}\sqrt{3})$, $(\frac{1}{4}, -\frac{1}{4}\sqrt{3})$; (c) $(0, 0)$, $(\frac{1}{2}, \frac{1}{2})$; (d) $(\frac{13}{3}, \frac{2}{3}\sqrt{14})$, $(-\frac{13}{3}, \frac{2}{3}\sqrt{14})$,
$(\frac{13}{3}, -\frac{2}{3}\sqrt{14})$, $(-\frac{13}{3}, -\frac{2}{3}\sqrt{14})$. **4.** $(4\pi, 2)$. **5.** Radius is $\sqrt{a^2 + b^2}$, center is
$(\sqrt{a^2 + b^2}, \text{Tan}^{-1} b/a)$, tangent line is $\theta = -\text{Tan}^{-1} a/b$, parallel line is $r\cos$
$(\theta - \text{Tan}^{-1} b/a) = \sqrt{a^2 + b^2}$. **6.** A point of the parabola $y = 4x^2$. **7.** $(d_2 - d_1)/(d_2 + d_1)$. **9.** An elliptical disk, foci Phoenix and Flagstaff, major diameter
300 miles and minor diameter $150\sqrt{3}$ miles. **10.** There is an elliptical disk in which
the attacker could get hit twice.

MISCELLANEOUS PROBLEMS, CHAPTER ELEVEN

2. $\frac{1}{2}$ and $-\frac{1}{3}$. **4.** $y^2 = 4(c - a)(x - a)$. **5.** $y^2 = 4(c - a)(x - a)$. **8.** The first
graph is the plane, the second is two circular disks. **9.** $r = ep/[1 + e\cos(\theta - \beta)]$.
10. P_3 is (x_1, y_1).

index

a